Palaeoseismology: Historical and Prehistorical Records of Earthquake Ground Effects for Seismic Hazard Assessment

Geological Society books refereeing procedures

The Society makes every effort to ensure that the scientific and production quality of its books matches that of its journals. Since 1997, all book proposals have been refereed by specialist reviewers as well as by the Society's Books Editorial Committee. If the referees identify weaknesses in the proposal, these must be addressed before the proposal is accepted.

Once the book is accepted, the Society Book Editors ensure that the volume editors follow strict guidelines on refereeing and quality control. We insist that individual papers can only be accepted after satisfactory review by two independent referees. The questions on the review forms are similar to those for *Journal of the Geological Society*. The referees' forms and comments must be available to the Society's Book Editors on request.

Although many of the books result from meetings, the editors are expected to commission papers that were not presented at the meeting to ensure that the book provides a balanced coverage of the subject. Being accepted for presentation at the meeting does not guarantee inclusion in the book.

More information about submitting a proposal and producing a book for the Society can be found on its web site: www.geolsoc.org.uk.

It is recommended that reference to all or part of this book should be made in one of the following ways:

Reichherter, K., Michetti, A. M. & Silva, P. G. (eds) 2009. *Palaeoseismology: Historical and Prehistorical Records of Earthquake Ground Effects for Seismic Hazard Assessment.* Geological Society, London, Special Publications, **316**.

White, S., Stollhofen, H., Stanistreet, I. G. & Lorenz, V. 2009. Pleistocene to Recent rejuvenation of the Hebron Fault, SW Namibia. *In*: Reicherter, K., Michetti, A. M. & Silva, P. G. (eds) *Palaeoseismology: Historical and Prehistorical Records of Earthquake Ground Effects for Seismic Hazard Assessment.* Geological Society, London, Special Publications, **316**, 293–317.

GEOLOGICAL SOCIETY SPECIAL PUBLICATION NO. 316

Palaeoseismology: Historical and Prehistorical Records of Earthquake Ground Effects for Seismic Hazard Assessment

EDITED BY

K. REICHERTER
RWTH Aachen University, Germany

A. M. MICHETTI
Università dell'Insubria, Italy

AND

P. G. SILVA
Universidad de Salamanca, Spain

2009
Published by
The Geological Society
London

THE GEOLOGICAL SOCIETY

The Geological Society of London (GSL) was founded in 1807. It is the oldest national geological society in the world and the largest in Europe. It was incorporated under Royal Charter in 1825 and is Registered Charity 210161.

The Society is the UK national learned and professional society for geology with a worldwide Fellowship (FGS) of over 9000. The Society has the power to confer Chartered status on suitably qualified Fellows, and about 2000 of the Fellowship carry the title (CGeol). Chartered Geologists may also obtain the equivalent European title, European Geologist (EurGeol). One fifth of the Society's fellowship resides outside the UK. To find out more about the Society, log on to www.geolsoc.org.uk.

The Geological Society Publishing House (Bath, UK) produces the Society's international journals and books, and acts as European distributor for selected publications of the American Association of Petroleum Geologists (AAPG), the Indonesian Petroleum Association (IPA), the Geological Society of America (GSA), the Society for Sedimentary Geology (SEPM) and the Geologists' Association (GA). Joint marketing agreements ensure that GSL Fellows may purchase these societies' publications at a discount. The Society's online bookshop (accessible from www.geolsoc.org.uk) offers secure book purchasing with your credit or debit card.

To find out about joining the Society and benefiting from substantial discounts on publications of GSL and other societies worldwide, consult www.geolsoc.org.uk, or contact the Fellowship Department at: The Geological Society, Burlington House, Piccadilly, London W1J 0BG: Tel. +44 (0)20 7434 9944; Fax +44 (0)20 7439 8975; E-mail: enquiries@geolsoc.org.uk.

For information about the Society's meetings, consult *Events* on www.geolsoc.org.uk. To find out more about the Society's Corporate Affiliates Scheme, write to enquiries@geolsoc.org.uk.

Published by The Geological Society from:
The Geological Society Publishing House, Unit 7, Brassmill Enterprise Centre, Brassmill Lane, Bath BA1 3JN, UK

(*Orders*: Tel. +44 (0)1225 445046, Fax +44 (0)1225 442836)
Online bookshop: www.geolsoc.org.uk/bookshop

British Library Cataloguing in Publication Data

A catalogue record for this book is available from the British Library.
ISBN 978-1-86239-276-2

Typeset by Techset Composition Ltd., Salisbury, UK
Printed by MPG Books Ltd., Bodmin, UK

Distributors

North America
For trade and institutional orders:
The Geological Society, c/o AIDC, 82 Winter Sport Lane, Williston, VT 05495, USA
Orders: Tel. +1 800-972-9892
Fax +1 802-864-7626
E-mail: gsl.orders@aidcvt.com

For individual and corporate orders:
AAPG Bookstore, PO Box 979, Tulsa, OK 74101-0979, USA
Orders: Tel. +1 918-584-2555
Fax +1 918-560-2652
E-mail: bookstore@aapg.org
Website: http://bookstore.aapg.org

India
Affiliated East-West Press Private Ltd, Marketing Division, G-1/16 Ansari Road, Darya Ganj, New Delhi 110 002, India
Orders: Tel. +91 11 2327-9113/2326-4180
Fax +91 11 2326-0538
E-mail: affiliat@vsnl.com

Contents

Foreword

Earthquakes are one of the greatest natural hazards humans face. During the twentieth century alone, over two million people died during strong ground shaking, attendant fires, tsunamis and landslides. Most recently, in May 2008, about 80 000 people died in an earthquake in Sichuan Province in China and, earlier, on 26 December 2004, more than 200 000 people lost their lives to the tsunami resulting from the great earthquake off the west coast of Sumatra in Indonesia. In December 2003, the ancient city of Bam in Iran was destroyed by an earthquake, with the loss of over 30 000 lives. The worst disaster in modern times occurred in China in July 1976, when an entire city was destroyed and over 240 000 people killed in less than six minutes. Earlier, in 1556, an earthquake in north-central China killed an estimated 800 000 people, one of the worst natural disasters in recorded history.

Given the tremendous toll in human lives and attendant economic losses, it is appropriate that scientists are working hard to better understand earthquakes, with the ultimate aim of forecasting and, ultimately, predicting them. Their research is broadly of three types. First are instrumental studies of earthquakes by seismologists. Most countries have seismic networks, and seismologists use these records to characterize earthquakes in time and space. Their research allows them to continually improve understanding of seismic risk. Second are geodetic studies of contemporary surface deformation. Modern, GPS-based measurements of changes in the position of fixed points on Earth's surface are providing important insights into crustal stress that can be linked to the instrumental earthquake record. Third are studies of active tectonics and of the geological evidence left by historic and prehistoric earthquakes. These studies provide valuable context for interpreting contemporary seismicity and crustal strain accumulation. They also are the only means of extending the instrumental earthquake record into prehistory, which is particularly important in areas such as western North America where written accounts of earthquakes are limited to the past 150 years.

The papers in this Special Publication fall into the third group mentioned above, a field of research termed 'palaeoseismology'. Palaeoearthquake research is a broad endeavour, with roots in geology, seismology, tectonics, structural geomorphology, geomorphology, stratigraphy and sedimentology. Its practitioners are interdisciplinary scientists who emphasize field research and typically have strong interests in risk, a topic that lies outside physical science.

This Special Publication is the 'brain child' of its three editors, Klaus Reicherter, Alessandro Michetti and Pablo Silva Barroso. It stems from presentations and lively discussions at two recent events sponsored by the INQUA Subcommission on Palaeoseismology: a session at the 2006 EGU in Vienna titled '3000 years of earthquake ground effects in Europe'; and an ICTP/IAEA workshop held in 2006 at Trieste on seismic hazard analyses for critical facilities. A key contribution of these meetings was to show how the systematic study of earthquake surface rupture, liquefaction, tsunami deposits, and other ground effects can be integrated with traditional seismological and tectonic information to provide a better understanding of seismic hazards and risk.

Some words about INQUA seem appropriate here, because this organization enabled the scientific meetings and discussions that led to this Special Publication. INQUA (the International Union for Quaternary Research) is a member union of the International Council of Science. Its primary objectives are to encourage the interdisciplinary study of all aspects of the Quaternary Period (the last two million years), and to facilitate and coordinate international cooperation for this study. The Quaternary is a unique period in Earth history – humans appeared at the beginning of the Quaternary, and their evolution was driven by frequent large changes in global climate with environmental conditions very different from those of today. These climatic fluctuations led to major global reorganization of terrestrial geography, ocean circulation and structure, and biotic communities. An important part of INQUA's remit is fostering research on hazardous Earth processes, including earthquakes, tsunamis, landslides, floods, and severe storms. The research is carried out under the aegis of INQUA's Commission on Terrestrial Deposits and Processes, of which the Subcommission on Palaeoseismology is part.

In the introduction the editors comment on each of the papers in the volume. Their summary makes it clear that the papers, although diverse in subject and scope, group around several themes. Many of the papers are concerned with the effects of earthquakes on the natural environment, and in particular with the application of the recently introduced 'INQUA ESI scale' to large historic earthquakes in different tectonic settings. The ESI scale is based on primary and secondary ground effects of earthquakes rather than traditional effects on people and infrastructure, and was developed by the Subcommission on Palaeoseismology over the past five years. Another group of papers examines regional earthquake histories in relation to the tectonic environments in which they occur. A third group is concerned with earthquakes and archaeology.

By nature of its subject, this volume can provide only a sample of modern palaeoseismological research. It cannot possibly cover the entire breadth of palaeoseismology, for such a volume would be an encyclopedia not a single volume. Nevertheless, this 'sampler' will whet the appetite of readers interested in learning what palaeoearthquake research can bring to the table of earthquake research. To those readers, I say 'bon appetite'.

JOHN J. CLAGUE
Past-President, INQUA
Director, Centre for Natural Hazard Research,
Simon Fraser University

Palaeoseismology: historical and prehistorical records of earthquake ground effects for seismic hazard assessment

K. REICHERTER[1]*, A. M. MICHETTI[2] & P. G. SILVA BARROSO[3]

[1]*Lehr- und Forschungsgebiet Neotektonik und Georisiken, Geowissenschaften, RWTH Aachen, Lochnerstr. 4-20, D-52064 Aachen, Germany*

[2]*Dipartimento di Scienze Chimiche e Ambientali, Università dellÍnsubria, Via Valleggio 11, 22100 Como, Italy*

[3]*Departamento de Geología, Universidad de Salamanca, Escuela Politécnica Superior de Ávila, Avda. Hornos Caleros, 50. 05003-Ávila, Spain*

**Corresponding author (e-mail: k.reicherter@nug.rwth-aachen.de)*

This volume grew particularly out of two meetings held in 2006 (European Geosciences Union General Assembly 2006, Session TS4.4, '3000 years of earthquake ground effects in Europe: geological analysis of active faults and benefits for hazard assessment', Vienna, Austria, April 2006; and the ICTP/IAEA workshop on 'The conduct of seismic hazard analyses for critical facilities', Trieste, Italy, May 2006) that brought together geoscientists who have explored and studied palaeoseismicity and its environmental effects in several parts of the world. This publication contains 18 papers based on a selection of presentations, and addresses a wide range of topics related to both a) palaeoseismological studies, and b) the assessment of a new macroseismic intensity scale based only on the natural phenomena associated with an earthquake, that is the ESI 2007 scale.

In 1999, during the 15th INQUA (International Union for Quaternary Research) Congress in Durban, the Subcommission on Palaeoseismicity promoted the compilation of a new scale of macroseismic intensity based only on environmental effects. A working group including geologists, seismologists and engineers compiled a first version of the scale that was presented at the 16th INQUA Congress in Reno in 2003, and updated one year later at the 32nd International Geological Congress in Florence (Michetti *et al.* 2004). To this end, the INQUA TERPRO (Commission on Terrestrial Processes) approved a specific project (INQUA Scale Project 2007). The revised version was ratified during the 17th INQUA Congress in Cairns in 2007. This revised version of the scale, which is formally named the Environmental Seismic Intensity scale–ESI 2007, is composed of two parts.

(1) The definition of intensity degrees on the basis of coseismic ground effects (see Appendix). ESI 2007 is a 12-degree macroseismic scale (Michetti *et al.* 2007), which follows the same basic structure of the original Mercalli–Cancani–Sieberg scale (MCS scale; Sieberg 1912), and of the subsequent, widely used, Modified Mercalli macroseismic scales MM-31; (Wood & Neumann 1931) and MM-56 (Richter 1958), MSK-64 (Medvedev–Sponheuer–Karnik scale; Medvedev *et al.* 1964), and EMS-98 (European Macroseismic Scale; Grünthal 1998).

(2) The guidelines, which aim at better clarifying: (i) the background of the scale and the scientific concepts that support the introduction of such a new macroseismic scale; (ii) the procedure to use the scale alone or integrated with damage-based, traditional scales; (iii) how the scale is organized; (iv) the descriptions of diagnostic features required for intensity assessment, and the meaning of idioms, colours and fonts.

The main advantage of the ESI 2007 scale is the classification, quantification and measurement of several known geological, hydrological, botanical and geomorphic features for different intensity degrees, differentiating two main categories of earthquake effects on the environment: (a) primary (fault surface ruptures and tectonic uplift/subsidence); and (b) secondary (including ground cracks, slope movements, liquefaction processes, anomalous waves and tsunamis, hydrogeological anomalies, and tree shaking). Primary effects triggered by surface faulting are almost absent for intensity degrees below VIII, are characteristic, but moderate for intensities between VIII and X, and diagnostic for the stronger top intensities of XI and XII (Fig. 1). This differentiation subdivides the earthquakes into three main categories (A, B, C), in which the absence (A), occurrence (B) and dimensions (B, C) of fault surface offsets allow

From: REICHERTER, K., MICHETTI, A. M. & SILVA, P. G. (eds) *Palaeoseismology: Historical and Prehistorical Records of Earthquake Ground Effects for Seismic Hazard Assessment*. The Geological Society, London, Special Publications, **316**, 1–10. DOI: 10.1144/SP316.1 0305-8719/09/$15.00

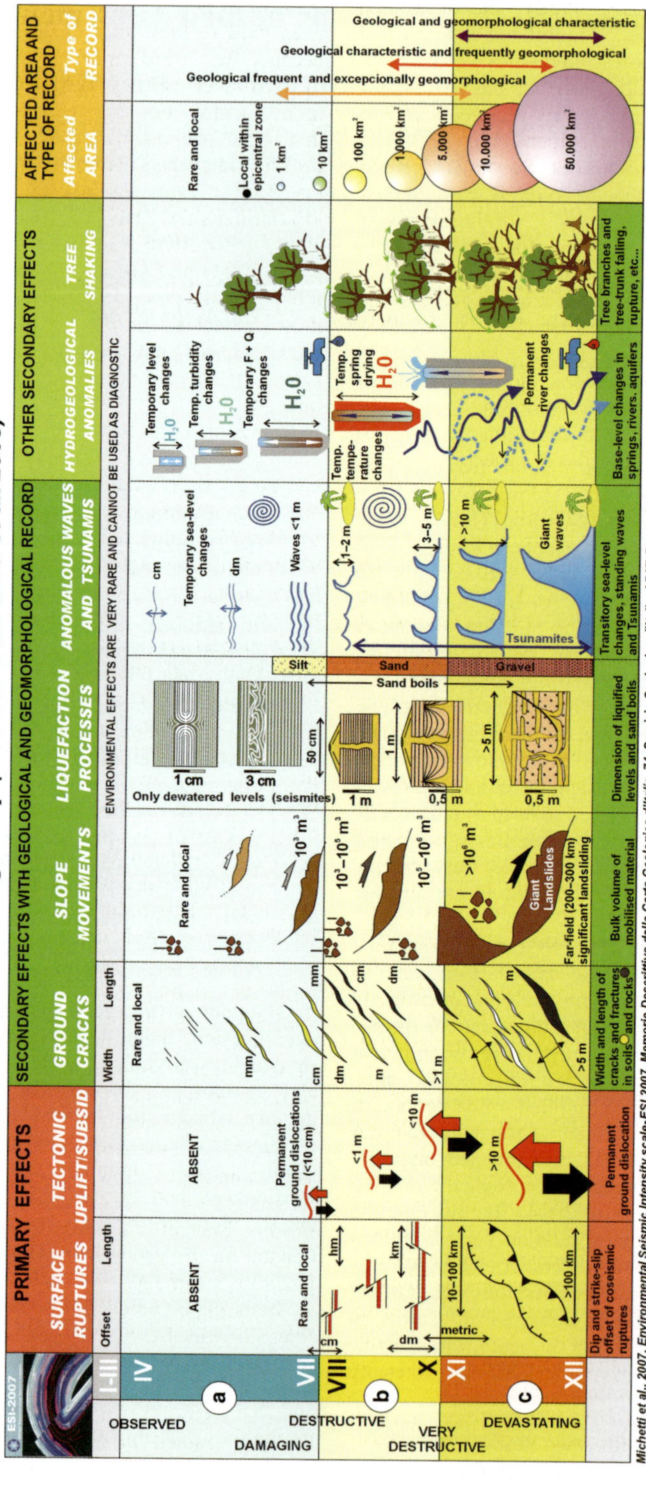

Fig. 1. The ESI 2007 chart summarizes the main features and dimensions of the more relevant Earthquake Environmental Effects (modified after Silva *et al.* 2008).

the assignment of intensity to present and past seismic events. Complementarily, the dimension (width, length, volume of mobilized material) of secondary effects allows intensities to be constrained for type A and B earthquakes; while the extension of the area affected by secondary effects allows assessment of the epicentral intensity for type A, B and C earthquakes. Secondary effects are typically diagnostic for type B earthquakes, but frequently saturate for type C. In the same way primary effects are diagnostic for type C earthquakes, when structural damage to human constructions and engineering facilities saturate. In principle, both the total area affected by secondary effects and the dimensions (surface rupture length, displacement, amount of coseismic uplift or subsidence) of primary effects do not saturate for the large earthquakes. The combination of the ESI 2007 scale with other classic intensity scales (MSK, EMS, MM, MCS) helps to compare recorded structural damage with the dimension of observed or reported (past earthquakes) environmental effects, and consequently exports the obtained seismic records to past prehistoric events. Figure 1 summarizes the ESI 2007 chart (Silva *et al.* 2008), which illustrates the different categories of earthquakes as well as the main characteristic features for the different types of effects. Also, this chart gives a qualitative approach for the affected areas, type of geological and geomorphologic record, and their respective degree of preservation through time.

There is one very important aspect in introducing a new intensity scale into the practice. A great deal of work in seismic hazard assessment is accomplished in the world, and intensity is a basic parameter in this. Any 'new word' in this research field must not result in dramatic changes. Intensity VIII, for instance, has to mean more or less the same 'strength' of the earthquake, regardless of which macroseismic phenomena (anthropic or geological) it is assessed from. Obviously the ESI 2007 scale is not intended to replace the existing scales. We are simply affording a means to factor in the modifications induced by the earthquake on the physical environment, and then to compare them with the effects taken into account by other scales. There, indeed, the combined observations of widely varied effects is most likely to yield a more representative estimate of intensity, which in turn, using modern events as test cases, can then be collated with such instrumental measurements as magnitude and seismic moment. A more detailed description of the relationships between the ESI 2007 scales and the other scales is beyond the scope of this introduction; for a complete analysis of this point see Michetti *et al.* (2004, 2007).

The authors of this introduction do not ignore, however, that the use of macroseismic effects on natural surroundings is controversial. Over the past 40 years at least, proper attention has not been paid to these effects in estimating intensity, because they were reputed to be too variable, and likewise because they were not properly weighted in the scales. For example, recent data indicate that some phenomena occur, or start to occur, at degrees other than the ones they are assigned to in the scales: liquefaction, for instance, starts at lower intensities (VI–VII, or even V; e.g. Keefer 1984; Galli 2000; Porfido *et al.* 2002; Rodriguez *et al.* 2002) and not at VII or IX as indicated in most scales. We argue that the existence of similar inconsistencies in the available macroseismic scales should not lead to the conclusion that ground effects are useless for assessing earthquake intensity.

These uncertainties lead to an increasing lack of confidence in using ground effects as diagnostics, and progressively the effects on human perception and the anthropic environment (mainly buildings) became the only sensors analysed for intensity assessment. Exemplifying this logic, in the latest proposal by the European Seismological Commission to revise the MSK scale (Grünthal 1998), these effects are not reported in the scale per se, only in a brief appendix. We believe, however, that if this orientation is pursued, intensity will come to reflect mainly the economic development of the area that experienced the earthquake instead of its 'strength' (Serva 1994). It is also our belief that by ignoring ground effects, it will not be possible to assess intensity accurately in sparsely populated areas and/or areas inhabited by people with different modes of existence, such as nomads. This point has been very clearly made by Dengler & McPherson (1993). The ESI scale is the logical extension of their approach. Furthermore the main problems arise for the highest degrees, XI and XII, where ground effects are the only ones that permit a reliable measurement of the severity of earthquake. All the scales, in fact, show that in this range of intensity ground effects predominate.

We believe the new ESI 2007 scale needs wider dissemination to allow a full scientific debate about its application to take place. One of the purposes of this Special Publication is thus to open the debate on a 'ground effects' scale for seismic hazard assessment.

It should be noted that the motivation for a new intensity scale based only on one class of macroseismic information, the effects on nature, rests exactly on the dramatic progress of our knowledge about the coseismic ground effects, and notably about surface faulting, gained in the last 30 years thanks to the growth of palaeoseismological studies. In the monograph *Active Tectonics: Impacts on Society*

(Wallace 1986), the first book that can be regarded as an overview of palaeoseismology, several papers made absolutely clear the quantitative relations that link the physical phenomena induced by earthquakes in the natural environment and the earthquake size. It has become a global, standard practice for palaeoseismologists since the late 1970s to survey in the field immediately after an earthquake the distribution of landslides, liquefactions, hydrological changes, coastal uplift and subsidence, and especially the characters and dimensions of tectonic ground ruptures. This is particularly true for the environmental effects generated by large earthquakes that break the ground surface (e.g. Allen 1986). Today, for instance, we have about 40 large earthquakes for which the geometry of surface faulting and the slip distribution along the fault strike have been mapped in detail (Wesnousky 2008). In this way, ground effects can be estimated from observations and regression analyses of historical earthquakes and a) fault displacement (Slemmons & dePolo 1986), b) liquefaction (Galli 2000), c) landslides (Keefer 1984), and several other features. This knowledge was not available at the time of early macroseismic scales, which very wisely included environmental effects in the different intensity degrees, but obviously without a detailed quantitative description due to the poor available dataset. There now exists an entirely new catalogue of information that allows us to update the macroseismic intensity observations by incorporating a wealth of palaeoseismological data. Vice-versa, the new macroseismic intensity scale based on environmental effects becomes a valuable tool and a guide for the palaeoseismologist. The lessons learned from intensity observations are educational for palaeoseismic analyses and interpretations, because they encourage the specialist to cross-check the results obtained using one particular evidence of palaeoseismicity. Once an ESI 2007 intensity degree has been assessed from a particular palaeoseismic feature, consistency with the whole spectrum of ground effects included in the same intensity degree should be ensured.

In our opinion, this illustrates quite well the scope of the present Special Publication and the basic idea behind all the presented contributions. The volume is divided into two sections. The first section focuses on the analysis of the coseismic ground effects from contemporary and historical earthquakes, and the implementation and refinement of the ESI 2007 scale. The second section is devoted to the analysis of individual case histories illustrating the different geological, geomorphological, geophysical techniques and field-survey methods used to identify causative and capable faults, and seismic hazard, from seismological and palaeoseismological approaches.

Papanikolaou *et al.* revise the macroseismic information for several earthquakes in Greece in order to calibrate the ESI 2007 scale against the traditional, damage-based scales. Their results show how the ESI 2007 scale, following the same criteria for all earthquakes, can compare not only events from different settings, but also contemporary and future earthquakes with historical events. This is of particular value for seismic hazard assessment in countries with a long record of seismicity such as Greece.

Two papers take advantage of a very large number of fault trench exposures to draw inference on earthquake hazard and fault behaviour along major strike-slip structures. **Rockwell *et al.*** illustrate extensive fault trenching across the trace of the coseismic ground ruptures associated with the large earthquakes of 9 August 1912 and 17 August 1999 along the North Anatolian Fault, west and east of the Marmara Sea, respectively. This allows better resolution of the history of surface ruptures for the past 400 years around Istanbul. A better quantitative assessment of coseismic environmental effects such as fault displacement is critical for the mitigation of earthquake risk in one of the largest metropolitan areas of the Earth.

Mouslopoulou *et al.* use fault data from 20 trenches to explore whether changes in late Quaternary fault kinematics principally arise due to earthquake rupture arrest and/or variations in slip vector pitch during individual earthquakes that span the kinematic transition zone occurring along the North Island Fault System, New Zealand, near the intersection with the active Taupo Rift.

Ground effects from four large earthquakes in Japan and Taiwan have been compiled by **Ota *et al.*** in order to assess the ESI 2007 scale. The new resulting maps show more detailed intensity patterns than those previously available for the four areas. Calibration exercise also reveals, however, that the ESI 2007 intensity scale needs some methodological improvement. This is somewhat expected and is needed for the better implementation of this new intensity scale in the future.

A similar exercise is proposed by **Tatevossian *et al.***, who used examples from the Altai (27 September 2003) and the Neftegorsk (27 May 1995) earthquakes. One of the main points made by these authors is the relevance of the environmental effects for intensity assessment in the near field of strong earthquakes. We argue that this is the very fundamental concept which provides reliable relations between palaeoseismology, macroseismic intensity and seismic hazard assessment. The results of Tatevossian *et al.* should be compared with those presented by Ota *et al.*, Mosquera-Machado *et al.* and Zahid *et al.* The epicentral intensity (I_0) based on the ESI 2007 scale can be two to four times higher than I_0 assessed without taking into account the ground effects. This indicates that by excluding the environmental effects, especially primary effects, we not only miss a valuable piece of information, sometimes the only one available in sparsely populated areas, but we are also missing the low frequency (static) part of an earthquake impact. In the epicentral area of strong seismic events, where the static offset reaches the order of several metres, intensity assessments ignoring this component are useless.

The integrated identification and analysis of archeoseismic and palaeoseismic evidence at the Roman site of Baelo Claudia, Gibraltar Strait (south Spain), is the purpose of the work by **Silva *et al.*** These authors combine observations on damage and secondary environmental effects in order to assess the local seismic hazard in terms of expected recurrence of intensity values within a specific time window.

A similar potential archeoseismic case history in a region with moderate seismicity is presented by **Hinzen & Weiner**, who apply geotechnical modelling to test the coseismic hypthesis for the damage to a Neolithic wooden well recently excavated near Erkelenz, in the Lower Rhine Embayment (NW Germany).

Two papers revise earthquake ground effects and active faulting in sparsely populated regions. **Mosquera-Machado *et al.*** studied the Mw 7.3 Murindo earthquake (18 October 1992) in NW Colombia, which provides relevant data for the application of the ESI 2007 scale. The resulting new isoseismal map is relevant for the assessment of future seismic risk in this part of Colombia where intensity assessment based on traditional damage-based scales cannot give a detailed picture of the earthquake severity. The Mw 7.8 Kunlun earthquake (14 November 2001) occurred in northern Tibet, in a remote, high-mountain region. **Lin & Guo** documented for the first time the palaeoseismic history of this region based on evidence of liquefaction within the trace of the 450-km-long surface rupture zone generated by this large event.

The analysis of the coseismic effects on the natural environment along the 110-km-long zone of surface thrust faulting associated with the M 7.6 Muzaffarabad, Pakistan, earthquake of 8 October 2005, is the topic covered by **Ali *et al.*,** also discussed from the seismotectonic point of view by **MonaLisa**. The macroseismic intensity distribution for this event shows a remarkable correlation with the trace of the surface rupture. Near Muzaffarabad, intensity XI in the MM, EMS-98 and ESI 2007 scales has been consistently assessed at sites where maximum values of fault displacement (in the order of 4 m) were observed.

Both **Gregersen & Voss** and **Mörner** provide a comprehensive seismological and palaeoseismological framework for the understanding and interpretation, in terms of seismic hazard, of the remarkable evidence of post-glacial palaeoseismicity available in Scandinavia.

A particular category of ground effects, that is found in the endokarstic terrains, is explored by **Pérez-López *et al.***, starting from the observation of the collapse that occurred within the Benis Cave (−213 m; Murcia, SE Spain), during the Mula earthquake (mb = 4.8, MSK VII, 2 February 1999).

Also in SE Spain (Almería Region), the stratigraphic and sedimentological evidence of past tsunamis in the western Mediterranean is discussed by **Reicherter & Becker-Heidmann**. The authors used shallow drilling in the lagoon of Cabo de Gata for identifying possible tsunamites associated with the 1522 Almería earthquake.

Trenching along the Vilariça segment of the Manteigas-Bragança Fault in NE Portugal, allows **Rockwell *et al.*** to identify evidence of a cluster of surface faulting earthquakes in the latest Pleistocene to early Holocene. This holds relevant implications for the seismic hazard of this region, characterized by moderate historical seismicity. Likewise, **White *et al.*** discuss the evidence for recent activity and related seismic hazard along the Hebron Fault in SW Namibia, within a stable continental area.

In summary, the set of papers included in this volume is basically devoted to the analysis of environmental earthquake effects linked to recent, past and prehistoric strong seismic events. The understanding of the type and dimensions of earthquake ground effects linked to different levels of seismic shaking and earthquake magnitude is the only prudent and consistent way to incorporate past strong events, only witnessed in the geological and geomorphological record, into the classic seismic catalogues, which are the basis of most of the seismic hazard studies and assessments. The efforts of the palaeoseismological community are directed to expanding back in time, and refining in terms of completeness, the seismic history of individual faults and/or seismic regions, in order to achieve a better understanding of the pulse (regularity and/or clustering) of seismic cycles in different tectonic settings, and its further implementation in hazard studies. Although the ESI 2007 scale is properly devoted to its application to past earthquakes, its application to recent events is critical, since it will allow refining the scale, and therefore improving maximum intensities recorded during past events. This volume offers to the scientific community a new tool to assign intensities, and a wide variety of geological methods to identify and measure earthquake environmental effects.

Many thanks are due to the armada of reviewers, who help to shape and focus the scope of this volume (in alphabetical order): P. Alfaro, F. Audemard, J. Cabral, R. Caputo, J. Dolan, F. Dramis, M. Ferry, A. Gorshkov, L. Guerrieri, K. Hinzen, R. Jibson, E. Kagan, J. Lario, T. Little, B. Lund, S. Marco, E. Masana, B. Mohammadioun, K. Okumura, C. Pascal, S. Pavlides, L. Piccardi, S. Porfido, Y. Quinif, G. Roberts, M. Rodriguez Pascua, L. Serva, M. Sintubin, A. Smedile, B. Shyu, I. Stewart, V. Trifonov and J. van der Woerd.

Appendix: ESI 2007 scale definition of intensity degrees

Text in italic indicates effects that can be used directly to define an intensity degree.

From I to III

There are no environmental effects that can be used as diagnostic.

IV Largely observed/First unequivocal effects in the environment

Primary effects are absent.

Secondary effects

(a) Rare small variations of the water level in wells and/or of the flow rate of springs are locally recorded, as well as extremely rare small variations of chemical–physical properties of water and turbidity in springs and wells, especially within large karstic spring systems, which appear to be most prone to this phenomenon.
(b) In closed basins (lakes, even seas) seiches with height not exceeding a few centimetres may develop, commonly observed only by tidal gauges, exceptionally even by naked eye, typically in the far field of strong earthquakes. Anomalous waves are perceived by all people on small boats, few people on larger boats, most people on the coast. Water in swimming pools swings and may sometimes overflows.
(c) Hair-thin cracks (millimetre wide) might be occasionally seen where lithology (e.g. loose alluvial deposits, saturated soils) and/or morphology (slopes or ridge crests) are most prone to this phenomenon.
(d) Exceptionally, rocks may fall and small landslides may be (re)activated, along slopes where the equilibrium is already near the limit state, e.g. steep slopes and cuts, with loose and generally saturated soil.
(e) Tree limbs shake feebly.

V Strong/Marginal effects in the environment

Primary effects are absent.

Secondary effects

(a) Rare variations of the water level in wells and/or of the flow rate of springs are locally recorded, as well as small variations of chemical–physical properties of water and turbidity in lakes, springs and wells.
(b) In closed basins (lakes, even seas) seiches with height of decimetres may develop, sometimes noted also by naked eye, typically in the far field of strong earthquakes. Anomalous waves up to several tens of centimetres high are perceived by all people on boats and on the coast. Water in swimming pools overflows.
(c) Thin cracks (millimetre wide and several centimetres up to 1 metre long) are locally seen where lithology (e.g. loose alluvial deposits, saturated soils) and/or morphology (slopes or ridge crests) are most prone to this phenomenon.
(d) Rare small rockfalls, rotational landslides and slump earth flows may take place, along often but not necessarily steep slopes where equilibrium is near the limit state, mainly loose deposits and saturated soil. Underwater landslides may be triggered, which can induce small anomalous waves in coastal areas of sea and lakes.
(e) Tree limbs and bushes shake slightly, very rare cases of fallen dead limbs and ripe fruit.
(f) Extremely rare cases are reported of liquefaction (sand boil), small in size and in areas most prone to this phenomenon (highly susceptible, recent, alluvial and coastal deposits, near-surface water table).

VI Slightly damaging/Modest effects in the environment

Primary effects are absent.

Secondary effects

(a) Significant variations of the water level in wells and/or of the flow rate of springs are locally recorded, as well as small variations of chemical–physical properties of water and turbidity in lakes, springs and wells.
(b) Anomalous waves up to many tens of centimetres high flood very limited areas nearshore. Water in swimming pools and small ponds and basins overflows.
(c) *Occasionally, millimetre to centimetre-wide fractures and up to several metres long are observed in loose alluvial deposits and/or saturated soils; along steep slopes or riverbanks they can be 1–2 cm wide. A few minor cracks develop in paved (either asphalt or stone) roads.*
(d) Rockfalls and landslides with volume reaching *c.* 10^3 m^3 can take place, especially where equilibrium is near the limit state, e.g. steep slopes and cuts, with loose saturated soil, or highly weathered/fractured rocks. Underwater landslides can be triggered, occasionally provoking small anomalous waves in coastal areas of sea and lakes, commonly seen by intrumental records.
(e) *Trees and bushes shake moderately to strongly; a very few tree tops and unstable dead limbs may break and fall, also depending on species, fruit load and state of health.*

(f) *Rare cases are reported of liquefaction (sand boil), small in size and in areas most prone to this phenomenon (highly susceptible, recent, alluvial and coastal deposits, near surface water table).*

VII Damaging/Appreciable effects in the environment

Primary effects observed very rarely, and almost exclusively in volcanic areas. Limited surface fault ruptures, tens to hundreds of metres long and with centimetric offset, may occur, essentially associated with very shallow earthquakes.

Secondary effects. The total affected area is in the order of 10 km^2.

(a) Significant temporary variations of the water level in wells and/or of the flow rate of springs are locally recorded. Seldom, small springs may temporarily run dry or appear. Weak variations of chemical–physical properties of water and turbidity in lakes, springs and wells are locally observed.

(b) Anomalous waves even higher than a metre may flood limited nearshore areas and damage or wash away objects of variable size. Water overflows from small basins and watercourses.

(c) *Fractures up to 5–10 cm wide and up to a hundred metres long are observed, commonly in loose alluvial deposits and/or saturated soils; rarely in dry sand, sand–clay, and clay soil fractures, up to 1 cm wide. Centimetre-wide cracks are common in paved (asphalt or stone) roads.*

(d) Scattered landslides occur in prone areas, where equilibrium is unstable (steep slopes of loose/saturated soils), while modest rock falls are common on steep gorges, cliffs). Their size is sometimes significant (10^3–10^5 m^3); in dry sand, sand–clay and clay soil, the volumes are usually up to 100 m^3. Ruptures, slides and falls may affect riverbanks and artificial embankments and excavations (e.g. road cuts, quarries) in loose sediment or weathered/fractured rock. Significant underwater landslides can be triggered, provoking anomalous waves in coastal areas of sea and lakes, directly felt by people on boats and ports.

(e) Trees and bushes shake vigorously; especially in densely forested areas, many limbs and tops break and fall.

(f) *Rare cases are reported of liquefaction, with sand boils up to 50 cm in diameter, in areas most prone to this phenomenon (highly susceptible, recent, alluvial and coastal deposits, near surface water table).*

VIII Heavily damaging/Extensive effects in the environment

Primary effects are observed rarely.

Ground ruptures (surface faulting) may develop, up to several hundred metres long, with offsets not exceeding a few centimetres, particularly for very shallow focus earthquakes such as those common in volcanic areas. Tectonic subsidence or uplift of the ground surface with maximum values on the order of a few centimetres may occur.

Secondary effects. The total affected area is in the order of 100 km^2.

(a) Springs may change, generally temporarily, their flow rate and/or elevation of outcrop. Some small springs may even run dry. Variations in water level are observed in wells. Weak variations of chemical–physical properties of water, most commonly temperature, may be observed in springs and/or wells. Water turbidity may appear in closed basins, rivers, wells and springs. Gas emissions, often sulphurous, are locally observed.

(b) Anomalous waves up to 1–2 metres high flood nearshore areas and may damage or wash away objects of variable size. Erosion and dumping of waste is observed along the beaches, where some bushes and even small weak-rooted trees can be uprooted and drift away. Water violently overflows from small basins and watercourses.

(c) *Fractures up to 50 cm wide are and up to hundreds of metres long commonly observed in loose alluvial deposits and/or saturated soils; in rare cases fractures up to 1 cm can be observed in competent dry rocks. Decimetric cracks common in paved (asphalt or stone) roads, as well as small pressure undulations.*

(d) Small to moderate (10^3–10^5 m^3) landslides are widespread in prone areas; rarely they can occur also on gentle slopes. Where equilibrium is unstable (steep slopes of loose/saturated soils; rock falls on steep gorges, coastal cliffs) their size is sometimes large (10^5–10^6 m^3). Landslides can occasionally dam narrow valleys causing temporary or even permanent lakes. Ruptures, slides and falls affect riverbanks and artificial embankments and excavations (e.g. road cuts, quarries) in loose sediment or weathered/fractured rock. Frequent occurrence of landslides below sea level in coastal areas.

(e) *Trees shake vigorously; branches may break and fall, trees even uprooted, especially along steep slopes.*

(f) *Liquefaction may be frequent in the epicentral area, depending on local conditions; sand boils up to c. 1 m in diameter; apparent water fountains in still waters; localized lateral spreading and settlements (subsidence up to c. 30 cm), with fissuring parallel to waterfront areas (river banks, lakes, canals, seashores).*

(g) *In dry areas, dust clouds may rise from the ground in the epicentral area.*

(h) Stones and even small boulders and tree trunks may be thrown in the air, leaving typical imprints in soft soil.

IX Destructive/Effects in the environment are a widespread source of considerable hazard and become important for intensity assessment

Primary effects are observed commonly.

Ground ruptures (surface faulting) develop, up to a few kilometres long, with offsets generally in the order of several centimetres. Tectonic subsidence or uplift of the ground surface with maximum values in the order of a few decimetres may occur.

Secondary effects. The total affected area is in the order of 1000 km^2.

(a) *Springs can change, generally temporarily, their flow rate and/or location to a considerable extent. Some modest springs may even run dry. Temporary variations of water level are commonly observed in wells. Water temperature often changes in springs and/or wells. Variations of chemical–physical properties of water, most commonly temperature, are observed in springs and/or wells. Water turbidity is common in closed basins, rivers, wells and springs. Gas emissions, often sulphurous, are observed, and bushes and grass near emission zones may burn.*

(b) *Waves several metres high develop in still and running waters. In flood plains water streams may even change their course, also because of land subsidence. Small basins may appear or be emptied. Depending on shape of sea bottom and coastline, dangerous tsunamis may reach the shores with runups of up to several metres flooding wide areas. Widespread erosion and dumping of waste is observed along the beaches, where bushes and trees can be uprooted and drift away.*

(c) *Fractures up to 100 cm wide and up to hundreds of metres long are commonly observed in loose alluvial deposits and/or saturated soils; in competent rocks they can reach up to 10 cm. Significant cracks are common in paved (asphalt or stone) roads, as well as small pressure undulations.*

(d) *Landsliding widespread in prone areas, also on gentle slopes; where equilibrium is unstable (steep slopes of loose/saturated soils; rock falls on steep gorges, coastal cliffs) their size is frequently large (10^5 m^3), sometimes very large (10^6 m^3). Landslides can dam narrow valleys causing temporary or even permanent lakes. Riverbanks, artificial embankments and excavations (e.g. road cuts, quarries) frequently collapse. Frequent large landslides below sea level in coastal areas.*

(e) Trees shake vigorously; branches and thin tree trunks frequently break and fall. Some trees might be uprooted and fall, especially along steep slopes.

(f) *Liquefaction and water upsurge are frequent; sand boils up to 3 m in diameter; apparent water fountains in still waters; frequent lateral spreading and settlements (subsidence of more than c. 30 cm), with fissuring parallel to waterfront areas (river banks, lakes, canals, seashores).*

(g) In dry areas, dust clouds commonly rise from the ground.

(h) *Small boulders and tree trunks may be thrown in the air and move away from their site for metres, also depending on slope angle and roundness, leaving typical imprints in soft soil.*

X Very destructive/Effects on the environment become a leading source of hazard and are critical for intensity assessment

Primary effects become leading.

Surface faulting can extend for a few tens of kilometres, with offsets from tens of centimetres up to a few metres. Gravity grabens and elongated depressions develop; for very shallow focus earthquakes in volcanic areas rupture lengths might be much lower. Tectonic subsidence or uplift of the ground surface with maximum values in the order of a few metres may occur.

Secondary effects. The total affected area is in the order of 5000 km^2.

(a) Many springs significantly change their flow rate and/or elevation of outcrop. Some springs may run temporarily or even permanently dry. Temporary variations of water level are commonly observed in wells. Even strong variations of chemical–physical properties of water, most commonly temperature, are observed in springs and/or wells. Often water becomes very muddy in even large basins, rivers, wells and springs. Gas emissions, often sulphurous, are observed, and bushes and grass near emission zones may burn.

(b) *Waves several metres high develop in even big lakes and rivers, which overflow from their beds. In flood plains rivers may change their course, temporarily or even permanently, also because of widespread land subsidence. Basins may appear or be emptied. Depending on shape of sea bottom and coastline, tsunamis may reach the shores with runups exceeding 5 m flooding flat areas for thousands of metres inland. Small boulders can be dragged for many metres. Widespread deep erosion is observed along the shores, with noteworthy changes of the coastline profile. Trees nearshore are uprooted and drift away.*

(c) *Open ground cracks up to more than 1 m wide and up to hundreds of metres long are frequent, mainly in loose alluvial deposits and/or saturated soils; in competent rocks opening is reach several decimetres. Wide cracks develop in paved (asphalt or stone) roads, as well as pressure undulations.*

(d) *Large landslides and rock-falls (>10^5–10^6 m^3) are frequent, almost regardless of equilibrium state of the slopes, causing temporary or permanent barrier lakes. River banks, artificial embankments, and* *sides of excavations typically collapse. Levees and earth dams may even incur serious damage. Frequent large landslides below sea level in coastal areas.*

(e) *Trees shake vigorously; many branches and tree trunks break and fall. Some trees might be uprooted and fall.*

(f) *Liquefaction, with water upsurge and soil compaction, may change the aspect of wide zones; sand volcanoes even more than 6 m in diameter; vertical subsidence even >1 m; large and long fissures due to lateral spreading are common.*

(g) In dry areas, dust clouds may rise from the ground.

(h) *Boulders (diameter in excess of 2–3 m) can be thrown in the air and move away from their site for hundreds of metres down even gentle slopes, leaving typical imprints in soil.*

XI Devastating/Effects on the environment become decisive for intensity assessment, due to saturation of structural damage

Primary effects are dominant.

Surface faulting extends from several tens of kilometres up to more than one hundred kilometres, accompanied by offsets reaching several metres. Gravity graben, elongated depressions and pressure ridges develop. Drainage lines can be seriously offset. Tectonic subsidence or uplift of the ground surface with maximum values in the order of numerous metres may occur.

Secondary effects. The total affected area is in the order of 10 000 km^2.

(a) Many springs significantly change their flow rate and/or elevation of outcrop. Many springs may run temporarily or even permanently dry. Temporary or permanent variations of water level are generally observed in wells. Even strong variations of chemical–physical properties of water, most commonly temperature, are observed in springs and/or wells. Often water becomes very muddy in even large basins, rivers, wells and springs. Gas emissions, often sulphurous, are observed, and bushes and grass near emission zones may burn.

(b) *Large waves develop in big lakes and rivers, which overflow from their beds. In flood plains rivers can change their course, temporarily or even permanently, also because of widespread land subsidence and landsliding. Basins may appear or be emptied. Depending on shape of sea bottom and coastline, tsunamis may reach the shores with runups reaching 15 m and more devastating flat areas for kilometres inland. Even metre-sized boulders can be dragged for long distances. Widespread deep erosion is observed along the shores, with noteworthy changes of the coastal morphology. Trees nearshore are uprooted and drift away.*

(c) Open ground cracks up to several metres wide are very frequent, mainly in loose alluvial deposits and/or saturated soils. In competent rocks they can reach 1 m. Very wide cracks develop in paved (asphalt or stone) roads, as well as large pressure undulations.

(d) *Large landslides and rock-falls (>10^5–10^6 m^3) are frequent, practically regardless to equilibrium state of the slopes, causing many temporary or permanent barrier lakes. River banks, artificial embankments, and sides of excavations typically collapse. Levees and earth dams incur serious damage. Significant landslides can occur at 200–300 km distance from the epicentre. Frequent large landslides below sea level in coastal areas.*

(e) *Trees shake vigorously; many branches and tree trunks break and fall. Many trees are uprooted and fall.*

(f) *Liquefaction changes the aspect of extensive zones of lowland, determining vertical subsidence possibly exceeding several metres, numerous large sand volcanoes, and severe lateral spreading features.*

(g) In dry areas dust clouds rise from the ground.

(h) *Big boulders (diameter of several metres) can be thrown in the air and move away from their site for long distances down even gentle slopes, leaving typical imprints in soil.*

XII Completely devastating/Effects in the environment are the only tool for intensity assessment

Primary effects are dominant.

Surface faulting is at least a few hundreds of kilometres long, accompanied by offsets reaching several tens of metres. Gravity graben, elongated depressions and pressure ridges develop. Drainage lines can be seriously offset. Landscape and geomorphological changes induced by primary effects can attain extraordinary extent and size (typical examples are the uplift or subsidence of coastlines by several metres, appearance or disappearance from sight of significant landscape elements, rivers changing course, origination of waterfalls, formation or disappearance of lakes).

Secondary effects. The total affected area is in the order of 50 000 km^2 and more.

(a) Many springs significantly change their flow-rate and/or elevation of outcrop. Temporary or permanent variations of water level are generally observed in wells. Many springs and wells may run temporarily or even permanently dry. Strong variations of chemical–physical properties of water, most commonly temperature, are observed in springs and/or wells. Water becomes very muddy in even large basins, rivers, wells and springs. Gas emissions, often sulphurous, are observed, and bushes and grass near emission zones may burn.

(b) *Giant waves develop in lakes and rivers, which overflow from their beds. In flood plains rivers change their course and even their flow direction, temporarily or even permanently, also because of widespread land subsidence and landsliding. Large basins may appear or be emptied. Depending on shape of sea bottom and coastline, tsunamis may reach the shores with runups of several tens of metres devastating flat areas for many kilometres inland. Big boulders can be dragged for long distances. Widespread deep erosion is observed along the shores, with outstanding changes of the coastal morphology. Many trees are uprooted and drift away. All boats are torn from their moorings and swept away or carried onshore even for long distances. All people outdoors are swept away.*

(c) Ground open cracks are very frequent, up to one metre or more wide in the bedrock, up to more than 10 m wide in loose alluvial deposits and/or saturated soils. These may extend up to several kilometres in length.

(d) *Large landslides and rock-falls (>10^5–10^6 m^3) are frequent, practically regardless of equilibrium state of the slopes, causing many temporary or permanent barrier lakes. River banks, artificial embankments, and sides of excavations typically collapse. Levees and earth dams incur serious damage. Significant landslides can occur at more than 200–300 km distance from the epicentre. Frequent very large landslides below sea level in coastal areas.*

(e) Trees shake vigorously; many branches and tree trunks break and fall. Many trees are uprooted and fall.

(f) *Liquefaction occurs over large areas and changes the morphology of extensive flat zones, determining vertical subsidence exceeding several metres, widespread large sand volcanoes, and extensive severe lateral spreading features.*

(g) In dry areas dust clouds rise from the ground.

(h) Very big boulders can be thrown in the air and move for long distances even down very gentle slopes, leaving typical imprints in soil.

References

ALLEN, C. R. 1986. Seismological and paleoseismological techniques of research in active tectonics. *In*: WALLACE, R. E. (ed.) *Active Tectonics: Impacts on Society*. National Academy Press, Washington, D.C., Studies in Geophysics, 148–154.

DENGLER, L. & MCPHERSON, R. 1993. The 17 August 1991 Honeydew earthquake, North Coast California: a case for revising the Modified Mercalli scale in sparsely populated areas. *Bulletin of the Seismological Society of America*, **83**, 1081–1094.

GALLI, P. 2000. New empirical relationships between magnitude and distance for liquefaction. *Tectonophysics*, **324**, 169–187.

GRÜNTHAL, G. (ed.) 1998. *European Macroseismic Scale 1998 (EMS-98)*. European Seismological Commission, Subcommission on Engineering Seismology, Working Group Macroseismic Scales. Conseil de l'Europe, Luxembourg, Cahiers du Centre Européen de Géodynamique et de Séismologie, **15**.

INQUA Scale Project, 2007. Available online at http://www.apat.gov.it/site/en-GB/Projects/INQUA_ Scale

KEEFER, D. K. 1984. Landslides caused by earthquakes. *Geological Society of America Bulletin*, **95**, 406–421.

MEDVEDEV, S., SPONHEUER, W. & KARNÍK, V. 1964. Neue seismische Skala. 7. Tagung der Europäischen Seismologischen Kommission vom 24.9. bis 30.9. 1962 in Jena, Veröff. *Institut für Bodendynamik und Erdbebenforschung in Jena, Deutsche Akademie der Wissenschaften zu Berlin*, **77**, 69–76.

MICHETTI, A. M., ESPOSITO, E. *ET AL.* 2004. The INQUA Scale. An innovative approach for assessing earthquake intensities based on seismically-induced ground effects in natural environment. *In*: VITTORI, E. & COMERCI, V. (eds) *Special Paper, APAT, Memorie Descrittive della Carta geologica d'Italia*, **LXVII**. SystemCart Srl, Roma, Italy.

MICHETTI, A. M., ESPOSITO, E. *ET AL.* 2007. Environmental Seismic Intensity Scale 2007 – ESI 2007. *In*: VITTORI, E. & GUERRIERI, L. (eds) *Memorie Descrittive della Carta Geologica d'Italia*, **LXXIV**. Servizio Geologico d'Italia, Dipartimento Difesa del Suolo, APAT, SystemCart Srl, Roma, Italy, 7–54.

PORFIDO, S., ESPOSITO, E. *ET AL.* 2002. Areal distribution of ground effects induced by strong earthquakes in the southern Apennines (Italy). *Surveys in Geophysics*, **23**, 529–562.

RICHTER, C. F. 1958. *Elementary Seismology*. W. H. Freeman, San Francisco.

RODRIGUEZ, L. M., AUDEMARD, F. A. & RODRIGUEZ, J. A. 2002. Casos històricos y contemporaneos de licuefacion de sedimentos inducidos por sismos en Venezuela desde 1530. *III Jornadas Venezolanas de Sismologìa Historica, Serie Tecnica*, **1**, 4–10.

SERVA, L. 1994. Ground effects in the intensity scales. *Terra Nova*, **6**, 414–416.

SIEBERG, A. 1912. Über die makroseismische Bestimmung der Erdbebenstärke. *Gerlands Beitrage Geophysik*, **11**, 227–239.

SILVA, P. G., RODRÍGUEZ PASCUA, M. A. *ET AL.* 2008. Catalogación de los efectos geológicos y ambientales de los terremotos en España en la Escala ESI-2007 y su aplicación a los estudios paleosismológicos. *Geotemas*, **6**, 1063–1066.

SLEMMONS, D. B. & DEPOLO, C. M. 1986. Evaluation of active faulting and associated hazards. *In*: WALLACE, R. E. (ed.) *Active Tectonics: Impacts on Society*. National Academy Press, Washington, D.C., Studies in Geophysics, 45–62.

WALLACE, R. E. 1986. *Active Tectonics: Impacts on Society*, National Academy Press, Washington, D.C., Studies in Geophysics.

WOOD, H. O. & NEUMANN, F. 1931. Modified Mercalli Intensity Scale of 1931. *Bulletin of the Seismological Society of America*, **21**(4), 277–283.

WESNOUSKY, S. G. 2008. Displacement and geometrical characteristics of earthquake surface ruptures: issues and implications for seismic-hazard analysis and the process of earthquake rupture. *Bulletin of the Seismological Society of America*, **98**(4), 1609–1632.

Advances and limitations of the Environmental Seismic Intensity scale (ESI 2007) regarding near-field and far-field effects from recent earthquakes in Greece: implications for the seismic hazard assessment

I. D. PAPANIKOLAOU[1,2]*, D. I. PAPANIKOLAOU[2] & E. L. LEKKAS[2]

[1]*Benfield-UCL Hazard Research Centre, Department of Earth Sciences, University College London, Gower Street, WC1E 6BT London, UK*

[2]*Natural Hazards Laboratory, Department of Dynamic, Tectonic and Applied Geology, Faculty of Geology and Geoenvironment, National and Kapodistrian University of Athens, Panepistimioupolis Zografou, 157-84 Athens, Greece*

**Corresponding author (e-mail: i.papanikolaou@ucl.ac.uk)*

Abstract: The new Environmental Seismic Intensity scale (ESI 2007), introduced by INQUA, incorporates the advances and achievements of palaeoseismology and earthquake geology and evaluates earthquake size and epicentre solely from the earthquake environmental effects (EEE). This scale is tested and compared with traditional existing scales for the 1981 Alkyonides earthquake sequence in the Corinth Gulf (Ms = 6.7, Ms = 6.4, Ms = 6.3), the 1993 Pyrgos event (Ms = 5.5) and the 2006 Kythira event (Mw = 6.7). These earthquakes were of different magnitudes, focal mechanisms and focal depths and produced well-documented environmental effects. The ESI 2007 intensity values and the isoseismal pattern for the 1993 Pyrgos and the 2006 Kythira events are similar to those resulting from the traditional scales, demonstrating that for moderate intensity levels (VII and VIII) the ESI 2007 and the traditional scales comply well. In contrast, the 1981 Alkyonides earthquake sequence shows that there is an inconsistency between the ESI 2007 and the traditional scales both in the epicentral area, where higher ESI 2007 intensity values have been assigned, and for the far-field effects. The ESI 2007 scale offers higher objectivity in the process of assessing macroseismic intensities, particularly in the epicentral area, than traditional intensity scales that are influenced by human parameters. The ESI 2007 scale follows the same criteria–environmental effects for all events and can compare not only events from different settings, but also contemporary and future earthquakes with historical events. A reappraisal of historical earthquakes so as to constrain the ESI 2007 scale may prove beneficial for seismic hazard assessment by reducing the uncertainty implied in the attenuation laws, which constitute one of the most important seismic hazard parameters.

A macroseismic intensity value represents the macroseismic information obtained by the quantification of the effects and damage produced by an earthquake. The macroseismic intensity is not solely used for the description of earthquake effects, but is a major seismic hazard parameter as well. The use of macroseismic intensity as a seismic hazard parameter predominates internationally, and more than 60% of countries have hazard assessment exclusively expressed in terms of intensity (McGuire 1993). The reason is that the historical record and the attenuation laws for large earthquakes are usually expressed in intensity values (e.g. Grandori *et al.* 1991), whereas in case of seismic risk management and earthquake loss estimation seismic intensity is preferred due to its direct representation of earthquake damage (Coburn & Spence 2002).

However, when using the effects on man and the manmade environment to assess the macroseismic intensity, then intensity will tend to reflect mainly the economic development and the cultural setting of the area that experienced the earthquake, instead of its 'strength' (Serva 1994). This led to the development and implementation of the Environmental Seismic Intensity 2007 scale. The newly introduced ESI 2007 scale (Michetti *et al.* 2007) is developed within the INQUA Subcommission on Palaeoseismicity, is the result of the revisions of previous versions, provisionally named as INQUA EEE scale (e.g. Michetti *et al.* 2004) and aims at evaluating earthquake size and epicentre solely from the earthquake environmental effects (EEE). The EEE are not influenced by human parameters such as effects on people and the manmade environment as the traditional intensity

From: REICHERTER, K., MICHETTI, A. M. & SILVA, P. G. (eds) *Palaeoseismology: Historical and Prehistorical Records of Earthquake Ground Effects for Seismic Hazard Assessment*. The Geological Society, London, Special Publications, **316**, 11–30. DOI: 10.1144/SP316.2 0305-8719/09/$15.00

scales (MCS, MM, EMS 1992, etc.) predominantly imply. It is a common notion that two earthquakes that produce similar environmental effects, thus having the same ESI 2007 intensity degree, but occur on sites that are different in terms of cultural and economic development, usually record significantly different intensity values as far as the traditional scales are concerned. In particular, if one of these earthquakes occurs in a developing country it tends to record a higher macroseismic intensity value compared to a developed seismic-prone country.

Over the last few decades palaeoseismology and earthquake geology have contributed significantly to our understanding concerning the EEE and more importantly they have provided a quantitative analysis and description of these effects. These effects have been incorporated into the ESI 2007 scale (Table 1). Among other advantages this scale: (i) allows the accurate assessment of intensity in sparsely populated areas, (ii) provides a reliable estimation of earthquake size with increasing accuracy towards the highest levels of the scale, where traditional scales saturate and ground effects are the only ones that permit a reliable estimation of earthquake size, and (iii) allows comparison among future, recent and historical earthquakes (Michetti *et al.* 2004). Overall, this scale is intended to integrate existing scales, not to replace them, and can encourage greater objectivity in the process of seismic intensity assessment, through independence from the variable nature of man and his infrastructure (Michetti *et al.* 2004). The use of the ESI 2007 scale alone is recommended only when effects on humans and manmade structures: (i) are absent or too scarce (i.e. desert or sparsely populated areas), and (ii) saturate (i.e. for intensity X to XII) losing their diagnostic value (Michetti *et al.* 2007).

Although the traditional intensity scales consider environmental effects for the evaluation of seismic intensity, these effects are not properly weighted and are systematically neglected. For example, the traditional scales do not differentiate between primary and secondary effects and do not use a quantitative approach for the effects on nature (Michetti *et al.* 2007). This is nicely illustrated with the EMS 1992 (European Macroseismic Scale) that forms an updated version of the traditional intensity scales (predominantly the MSK). The EMS 1992 was developed to be more easily implemented in urban areas giving even more emphasis to manmade structures. In particular, it includes new building types and modern construction materials, offers an easier recognition of the structure vulnerability class and a more precise evaluation of the grade of damage (Grunthal 1993).

In this paper we: (a) assess intensities in the ESI 2007 scale for several sites regarding three relatively recent events that occurred in Greece, (b) test and compare the ESI 2007 scale with existing traditional macroseismic intensity scales, and (c) discuss possible implications for seismic hazard assessment. The success of the newly introduced intensity scale also depends on its impact on seismic hazard assessment.

Selected earthquakes

An earthquake sequence and two events have been chosen for our current study (Fig. 1). These include the 1981 Alkyonides earthquake sequence in the Corinth Gulf (Ms = 6.7, Ms = 6.4, Ms = 6.3), the 1993 Pyrgos event (Ms = 5.5) and the 2006 Kythira event (Mw = 6.7). These earthquakes have been carefully selected so as to include events of different magnitudes, focal mechanisms and focal depths. Moreover, they all produced well documented environmental effects that allow us both to test the newly introduced ESI 2007 scale and compare it with the existing scales. In particular, the 1981 Alkyonides earthquake sequence produced significant primary surface ruptures, the 1993 Pyrgos event was on the threshold of primary faulting, producing only secondary but widespread effects, whereas the 2006 Kythira earthquake was a deep event that generated only minor secondary effects.

The 1993 Pyrgos earthquake

The 26 March 1993 Pyrgos (Ms = 5.5) earthquake in the Western Peloponnese produced a maximum intensity VIII on the EMS 1992 scale (Lekkas 1996), affecting the town of Pyrgos, where about 50% of the buildings suffered some form of damage (Figs 2 and 3a). The main shock of Ms = 5.5 occurred about 3 km south of the town of Pyrgos (Stavrakakis 1996). Two foreshocks of Ms = 5.0 (in the offshore area with thrust faulting) and Ms = 5.1 (normal faulting NE–SW plane dipping southwards) occurred 13 and 2 minutes before the main shock (Stavrakakis 1996). Papanastassiou *et al.* (1994) determined that the fault plane solution is characterized by thrust oblique faulting on a fault plane striking NNE–SSW and dipping SE (strike 14° dip 70° rake 158°), which according to Stavrakakis (1996) best fits the observed macroseismic field. A similar solution has been proposed by Dziewonski *et al.* (1994) (NP1 strike 122°, dip 60°, rake 5°, and NP2 strike 30°, dip 86°, rake 150°) and Melis *et al.* (1994). However, Koukouvelas *et al.* (1996) proposed that the Pyrgos earthquake was caused by oblique-normal slip on a north-dipping WNW-trending fault. Indeed, most outcropping active faults in the area are normal east–west trending faults (e.g. Lekkas *et al.* 2000;

Table 1. *Summary of the Environmental Seismic Intensity scale (ESI 2007) for intensities VII–X (Michetti et al. 2007)*

EEE type/degree	VII Damaging – Appreciable effects on the environment	VIII Heavily damaging – Extensive effects on the environment	IX Destructive – Environmental effects are a widespread source of considerable hazard and become important for intensity assessment	X Very destructive – Effects in the environment become a leading source of hazards and are critical for intensity assessment
Surface faulting	Primary effects observed very rarely.	Primary effects observed rarely. Ground ruptures (surface faulting) may develop, up to several hundred meters long, with offsets not exceeding a few cm, particularly for very shallow focus earthquakes. Tectonic subsidence or uplift with maximum values on the order of a few centimetres may occur.	Primary effects observed commonly. Ground ruptures (surface faulting) develop, up to a few km long, with offsets generally in the order of several cm. Tectonic subsidence or uplift of the ground surface with maximum values in the order of a few decimetres may occur.	Primary ruptures become leading. Surface faulting can extend for few tens of km, with offsets from tens of cm up to a few metres. Gravity grabens and elongated depressions develop; Tectonic subsidence or uplift with maximum values in the order of few meters may occur.
Slope movements	Scattered landslides occur in prone areas; (steep slopes of loose/ saturated soils; rock falls on steep gorges, coastal cliffs) their size is sometimes significant (10^3–10^5 m^3. The affected area is in the order of 10 km^2.	Small to moderate (10^3–10^5 m^3) landslides widespread in prone areas; their size is sometimes large (10^5–10^6 m^3). Ruptures, slides and falls affect riverbanks and artificial embankments in loose sediment or weathered/ fractured rock. The affected area is in the order of 100 km^2.	Landsliding widespread in prone areas, also on gentle slopes; where equilibrium is unstable (steep slopes of loose/ saturated soils; rock falls on steep gorges, coastal cliffs) their size is frequently large (10^5 m^3), sometimes very large (10^6 m^3). Riverbanks, artificial embankments and excavations (e.g. road cuts, quarries) frequently collapse. The affected area is in the order of 1000 km^2.	Large landslides and rock-falls ($>10^5$–10^6 m^3) are frequent, practically regardless of equilibrium state of the slopes, causing temporary or permanent barrier lakes. River banks, artificial embankments, and sides of excavations typically collapse. Levees and earth dams may even incur serious damage. The affected area is in the order of 5000 km^2.

(*Continued*)

Table 1. *Continued*

EEE type/degree	VII Damaging – Appreciable effects on the environment	VIII Heavily damaging – Extensive effects on the environment	IX Destructive – Environmental effects are a widespread source of considerable hazard and become important for intensity assessment	X Very destructive – Effects in the environment become a leading source of hazards and are critical for intensity assessment
Ground cracks	Fractures up to 5–10 cm wide and up to hundred metres long commonly in loose alluvial deposits and/or saturated soils; Centimetre-wide cracks common in paved (asphalt or stone) roads.	Fractures up to 50 cm wide and up to hundred metres long are commonly observed in loose alluvial deposits and/or saturated soils; in rare cases fractures up to 1 cm can be observed in competent dry rocks. Decimetric cracks common in paved roads, as well as small pressure undulations.	Fractures up to 100 cm wide and up to hundred metres long are commonly observed in loose alluvial deposits and/or saturated soils; in competent rocks they can reach up to 10 cm. Significant cracks common in paved (asphalt or stone) roads, as well as small pressure undulations.	Open ground cracks up to more than 1 m wide and up to hundred metres long are frequent, mainly in loose alluvial deposits and/or saturated soils; in competent rocks opening reaches several decimetres. Wide cracks develop in paved (asphalt or stone) roads, as well as pressure undulations.
Ground settlements–collapse/tsunami/other effects	Rare cases of liquefaction, with sand boils up to 50 cm in diameter, in areas most prone to this phenomenon (highly susceptible, recent, alluvial and coastal deposits, shallow water table).	Liquefaction may be frequent in the epicentral area; sand boils up to *c.* 1 m in diameter; localized lateral spreading and settlements (subsidence up to *c.* 30 cm), with fissuring parallel to waterfront areas (river banks, lakes, canals, seashores). Waves up to 1–2 m high develop in nearshore areas and may damage of wash away objects of variable size.	Liquefaction and water upsurge are frequent; sand boils up to 3 m in diameter; frequent lateral spreading and settlements (subsidence of more than *c.* 30 cm), with fissuring parallel to waterfront areas (river banks, lakes, canals, seashores). Metre-high waves develop in still and running waters. Tsunamis may reach the coastal areas with runups of up to several metres flooding wide areas. Small boulders and tree trunks may be thrown in the air.	Liquefaction, with water upsurge and soil compaction, may change the aspect of wide zones; sand volcanoes even more than 6 m in diameter; vertical subsidence even >1 m; large and long fissures due to lateral spreading are common. Metre-high waves develop in still and running waters. Tsunamis may reach the shores with runups exceeding 5 m flooding flat areas for thousands of metres in land. Boulders (diameter in excess of 2–3 m) can be thrown in the air.

This table includes only the definitions of intensities VII–X that are used in the paper.

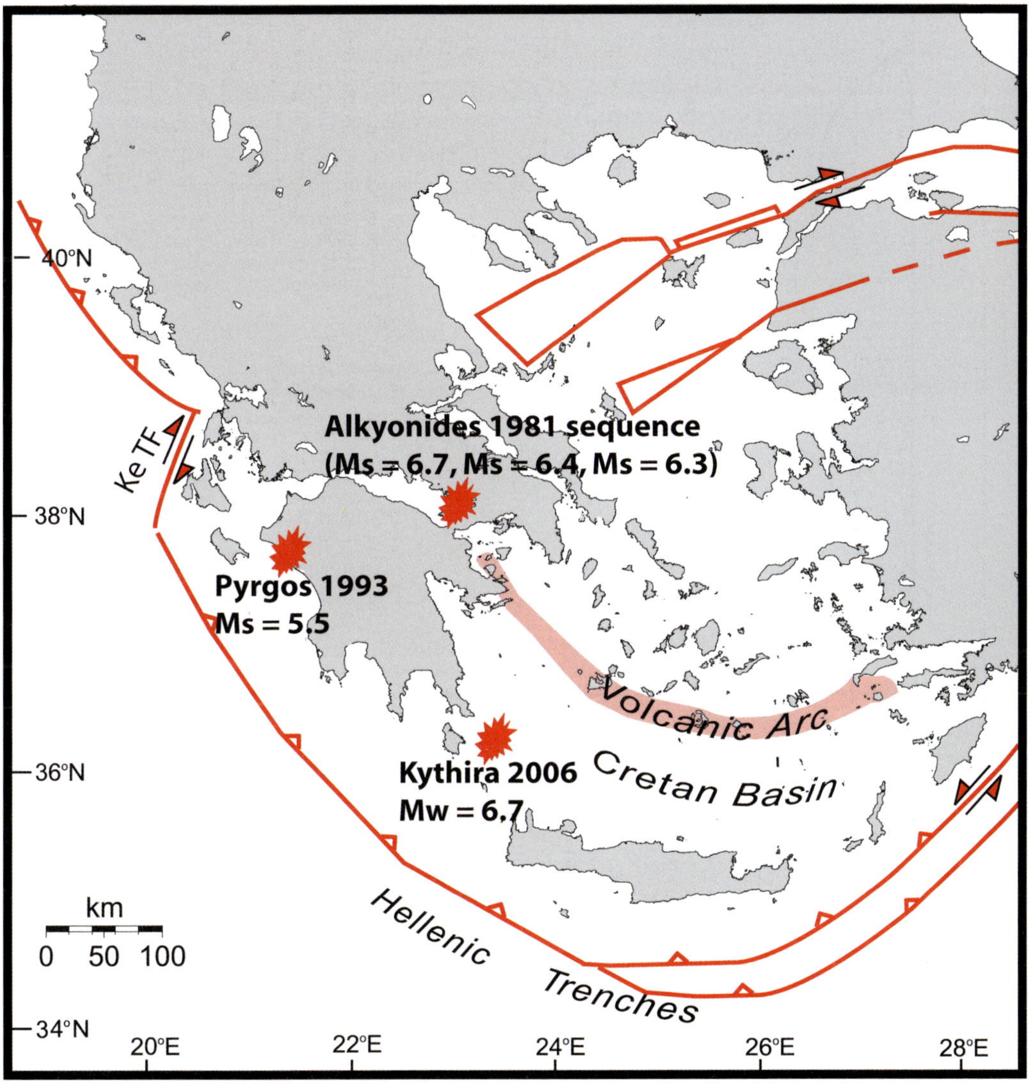

Fig. 1. Regional map showing the locations of the studied earthquakes within the Hellenic Arc. The 1981 Alkyonides earthquake sequence and the 1995 Pyrgos earthquake were shallow events, whereas the 2006 Kythira event occurred on the subduction zone with a focal depth of about 70 km.

Papanikolaou *et al.* 2007), therefore the generation of a Ms = 5.5 thrust event remains a question. Nevertheless, its focal depth (15 km as proposed by Stavrakakis (1996), but not accurately known) implies that it could also be related to the subduction processes of the Hellenic Trench that is situated approximately 70 km to the west. The subduction zone as imaged by Laigle *et al.* (2002) dips at a very low angle up to the Greek mainland where at about 15 km depth the dip of the interplate reflector becomes steeper, forming a deep ramp. The predominant stress field in the offshore area around Zakynthos and Western Peloponnese has been regarded as compressional (e.g. Papazachos 1990) although more recent studies show a predominant dextral strike slip faulting making the situation even more complicated (e.g. Kiratzi & Louvari 2003; Roumelioti *et al.* 2004). Therefore, it is possible that near the coast there is an extensional regime over the upper 10–12 km, whereas at deeper depths there is either a prevailing compressional regime due to the subducting plate (e.g. Laigle *et al.* 2002) or a dextral shear zone that may be linked to the major neighbouring Kefalonia strike-slip

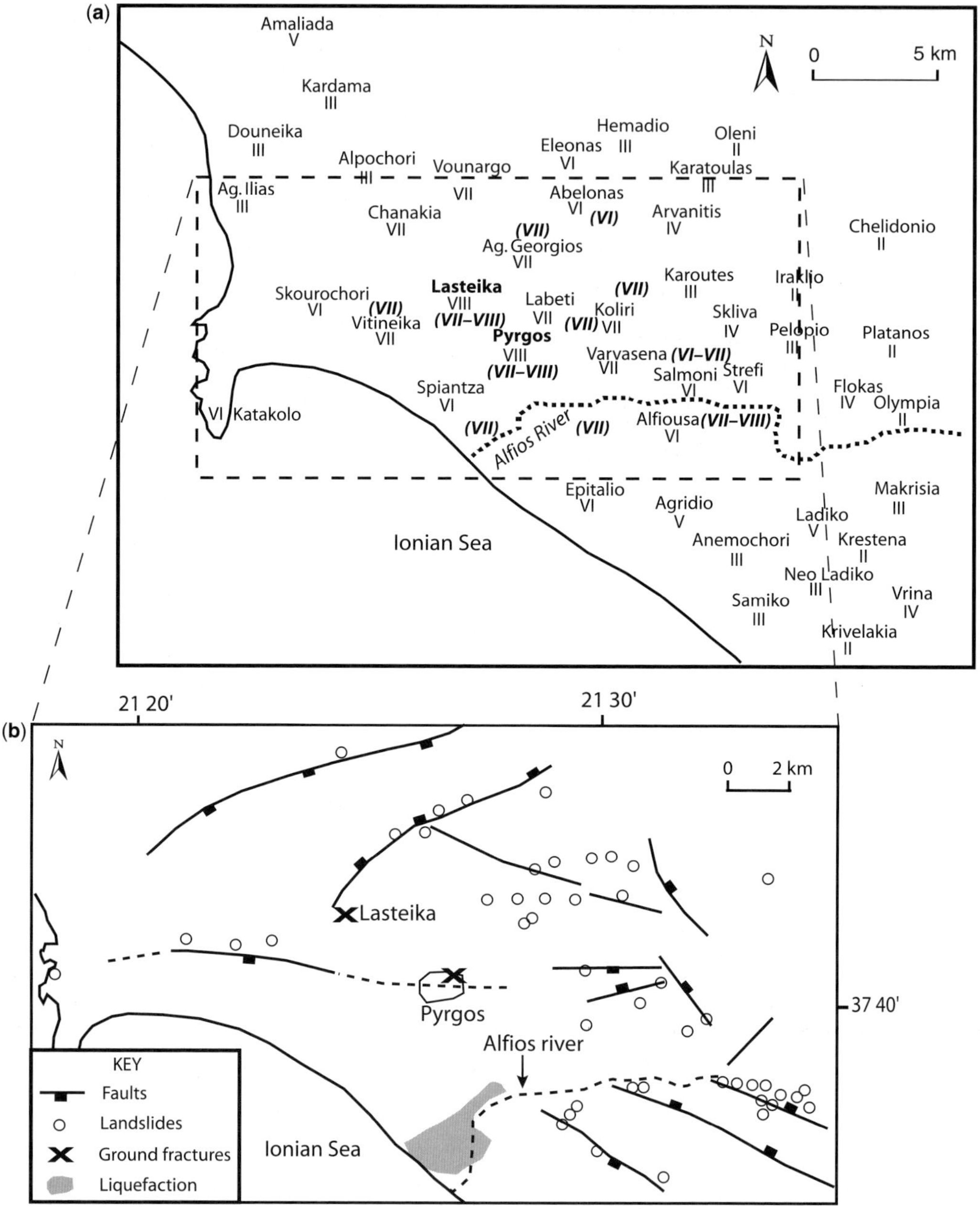

Fig. 2. (**a**) The EMS 1992 (Lekkas *et al.* 2000) and the ESI 2007 scale (this study) distribution of the 1993 Pyrgos earthquake. The ESI 2007 values are in bold italics in parentheses. (**b**) Map of the environmental effects. Ground fractures and liquefaction from Lekkas (1994) and Lekkas *et al.* (2000); landslide localities from Koukouvelas *et al.* (1996).

fault (e.g. Kiratzi & Louvari 2003). The 1993 Pyrgos foreshock activity took place on the SW coast, the main shock under the town of Pyrgos, whereas most aftershocks migrated NE and occurred on planes with a predominant NE–SW direction (Stavrakakis 1996). This probably explains the large spatial distribution of reported secondary effects and particularly landslides.

Although the town of Pyrgos extends over a relatively limited area of about 4 km^2, the distribution

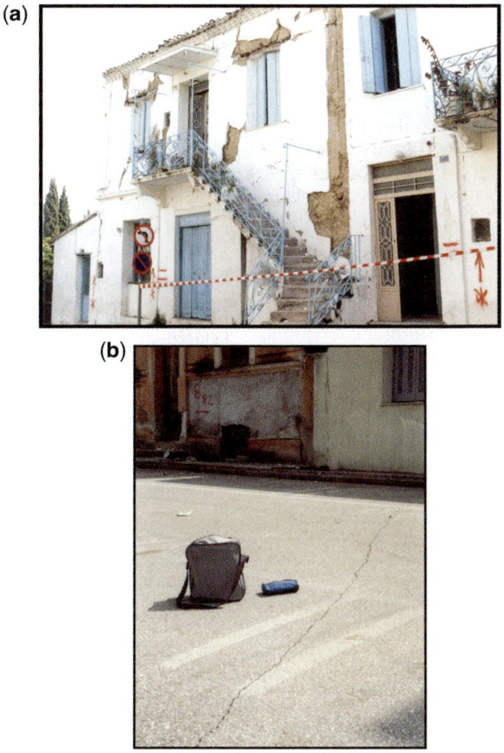

Fig. 3. (**a**) View of the damage inflicted in a traditional two-storey, stone masonry building. Several similar buildings suffered significant damage in the Pyrgos event. (**b**) Ground ruptures on paved road in the city of Pyrgos.

of damage was not uniform (Bouckovalas *et al.* 1996). Approximately 25% of the buildings experienced severe damage, yet no foundation failures were identified (Karantoni & Bouckovalas 1997). The Pyrgos event produced practically no damage to modern reinforced concrete buildings (only 22 buildings experienced light damage), but induced significant damage (Fig. 3a) to traditional buildings of adobe, stone or brick masonry (Karantoni & Bouckovalas 1997). The intensity of seismic motion was affected not only by the local soil conditions, but also by the construction material, the age and the storey number of buildings (Bouckovalas *et al.* 1996).

Several environmental effects were reported (Fig. 2b). No primary surface ruptures were recorded, but ground fractures were observed at the northeastern part of Pyrgos town (Fig. 3b) and Lasteika village (situated 2 km NW of Pyrgos), cutting both paved roads and cultivated land (Lekkas *et al.* 2000). Fractures were arranged partly en echelon. Towards the NE part of the town fractures were trending ENE–WSW, were about 30 m long and 2–3 cm wide, whereas at Lasteika village fractures were trending east–west and NNW–SSE and were up to 60 m long (Lekkas *et al.* 2000). The 1993 event triggered several landslides, the vast majority of which occurred along fault scarps and steep slopes and mostly in the Alfios southern river bank (Koukouvelas *et al.* 1996). Koukouvelas *et al.* (1996) measured landslides at 47 locations within an area of 145 km^2. Significant liquefaction and subsidence phenomena were observed 5 km SW of the town of Pyrgos in the coastal zone in recent coastal and fluvial deposits covering an area of about 5 km^2 (Lekkas 1996). Soil fractures up to 30 m long and sand boils up to 50 cm in diameter were observed (Lekkas 1994). Finally, subsidence was also observed in alluvial unconsolidated deposits within Pyrgos.

All these environmental effects are depicted in Figure 2b and have been assessed in relation to the ESI 2007 scale (in Fig. 2a, values in bold italics in parentheses). According to these effects the maximum ESI 2007 values (VII–VIII) are observed in the town of Pyrgos and Lasteika village, where ground fractures a few of tens of metres long have been recorded. Even though landslides are scattered along a large area ($>$100 km^2), possibly indicating a maximum intensity of VIII, most of them were small, occurred along unstable slopes and fault scarps, and clustered accordingly in certain localities. Therefore, we assign intensity values in these sites, ranging from VI to VII–VIII, depending on the density of the landslides. Liquefaction phenomena were also reported near the coast and the Alfios delta (soil fractures a

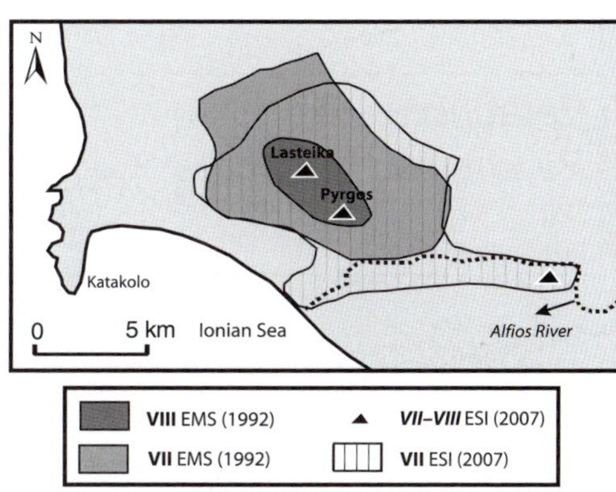

Fig. 4. Comparison of the isoseismal pattern between the EMS 1992 and the ESI 2007 intensity scales. The black triangle symbols show localities where an ESI 2007 intensity VII–VIII degree has been assessed.

few tens of metres long and sand boils up to 50 cm in diameter), a locality highly favourable for liquefaction. These characteristics imply a VII ESI 2007 value.

Maximum EMS 1992 intensity VIII was recorded in the town of Pyrgos and Lasteika village and correlated to the areas where ground ruptures were observed and an ESI 2007 VII–VIII has been assigned (Fig. 2a and Fig. 4). Figure 4 shows the isoseismal pattern of the epicentral region based on the ESI 2007 and the EMS 1992 scales, respectively. The main difference is traced along the Alfios river where liquefaction and sliding phenomena occurred towards the west and the east respectively, extending the ESI 1997 VII isoseismal to the south. However, this difference is mainly due to the lack of EMS 1992 data, since no villages are found in these localities to record an EMS 1992 value. Overall, for the 1993 Pyrgos event, the EMS 1992 and the ESI 2007 scales seem to comply well not only regarding the maximum recorded epicentral intensity, but also with the entire isoseismal pattern.

The 1981 Alkyonides earthquake sequence in the Corinth Gulf

On 24, 25 February and 4 March 1981 three (Ms = 6.7, Ms = 6.4, Ms = 6.3) successive destructive events (20 fatalities and 500 injured) occurred at the eastern end of the Corinth Gulf (Figs 1 and 5) (Jackson *et al.* 1982; Papazachos *et al.* 1982; Taymaz *et al.* 1991). Hubert *et al.* (1996) showed that the last two events of the sequence lie in areas where a positive Coulomb stress increase has been calculated, implying that this earthquake sequence was the result of stress transfer that triggered the second and third events. All three events correspond to normal faulting, accommodating north–south extension. The focal mechanisms that described the coseismic slip at depth (*c.* 10 km), exhibit similar fault plane orientations and kinematics to those measured on the faults at the surface (Morewood & Roberts 2001).

Damages occurred in three different provinces (Beotia, Attica and Corinth) where in total 7701 buildings collapsed or had damage beyond repair,

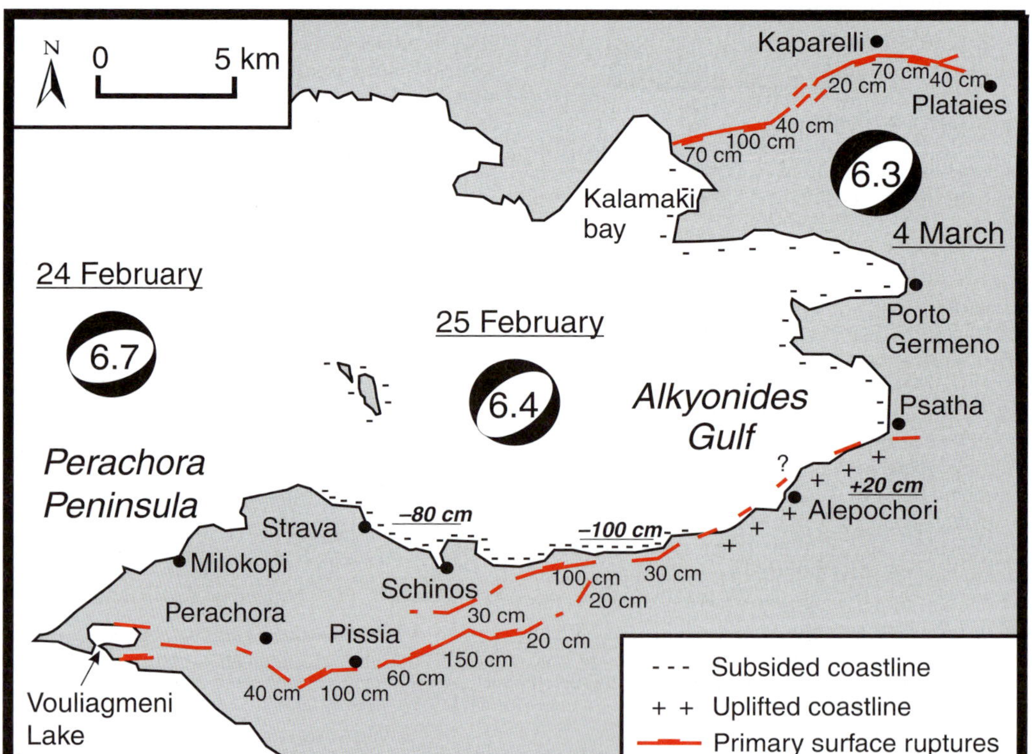

Fig. 5. The 1981 Alkyonides earthquake sequence, eastern Corinth Gulf. View of the epicentral region with emphasis on the primary surface ruptures and coastal uplift/subsidence. Sketch modified from Jackson *et al.* (1982), Mariolakos *et al.* (1982) and Hubert *et al.* (1996).

(a)

(b)

Fig. 6. (**a**) West of Alepochori up to the western part of the bay of Strava, 50–60 cm of subsidence was observed, flooding up to 50 m of the former shore (Mariolakos *et al.* 1982). (**b**) View of the surface ruptures on the Plataies–Kaparelli fault zone during the 4 March event, producing 50–60 cm of throw (70 cm of displacement).

and 20 954 buildings were severely damaged (Antonaki *et al.* 1988). These events produced numerous earthquake environmental effects (EEE) such as rockfalls, landslides (both onshore and offshore), liquefaction, a weak tsunami wave, significant coastal subsidence and uplift, but most importantly extensive primary surface faulting (Figs 5 and 6b). In particular, in the Pissia Fault, surface ruptures were longer than 10 km and displacements were in the range of 50–70 cm with a maximum recorded value of 150 cm, whereas in the Alepochori Fault displacement was about 100 cm high (Jackson *et al.* 1982; Fig. 5). The 4 March event ruptured the Plataie-Kaparelli Fault Zone (*c.* 10 km of surface ruptures) producing an average 50–60 cm of throw (Jackson *et al.* 1982; Mariolakos *et al.* 1982; Figs 5 and 6b) and a maximum heave and throw of 60 and 120 cm respectively between Kaparelli and Plataies (Papazachos *et al.* 1982). West of Alepochori up to the western part of the Bay of Strava, 60 cm of subsidence was observed, flooding up to 50 m of the former shore (Mariolakos *et al.* 1982; Fig. 6a); however, east of Alepochori coastal uplift was observed (Jackson *et al.* 1982; Vita-Finzi & King 1985). In Schinos and Strava coastline, there was disagreement on the amount of subsidence recorded, ranging from 50–80 cm (Andronopoulos *et al.* 1982; Mariolakos *et al.* 1982; Hubert *et al.* 1996) up to 120 and 150 cm (Khoury *et al.* 1983; Vita-Finzi & King 1985). We use a value of 80 cm based on Hubert *et al.*'s (1996) arguments and modelling which show that values higher than 100 cm probably overestimate the coseismic effect.

Subsidence of a few centimetres was also reported away from the epicentral areas in both Loutraki and Kiato coastal area (Andronopoulos *et al.* 1982; see Figs 5 and 7 for localities). Extensive liquefaction occurred at the Kalamaki Bay coastal area (Andronopoulos *et al.* 1982) as well as in Porto-Germeno and in Kineta (Papazachos *et al.* 1982). Ground fissures were reported in Loutraki beach, Vouliagmeni Lake, Porto Germeno, Kiato and Corinth (Papazachos *et al.* 1982).

Submarine slumping in the Alkyonides deep basin and several mass-movement phenomena in the shelf area have also been detected (Perissoratis *et al.* 1984). In particular, a large-scale slump has been documented about 10 km long, 1.5–2 km wide, extending 16 km^2 over a depth of 360 m (Perissoratis *et al.* 1984). Jackson *et al.* (1982) quoted that local people reported a 1 m high tsunami during the main shock in the Alkyonides Gulf. Therefore, it is possible that the tsunami generation can be attributed to the large-scale slumping detected by Perissoratis *et al.* (1984).

All these environmental effects are depicted in Figure 5 and have been assessed according to the ESI 2007 scale (Fig. 8). It should be noted that subsidence values reported by Vita-Finzi & King (1985) around Milokopi and southwards up to the town of Loutraki have been debated (Pirazzoli *et al.* 1994; Hubert *et al.* 1996) and have not been considered in this study. Moreover, there is some controversy as to whether the ruptures near Pissia and Schinos should be ascribed to the first or the second event (Jackson *et al.* 1982; King *et al.* 1985; Taymaz *et al.* 1991; Abercrombie *et al.*

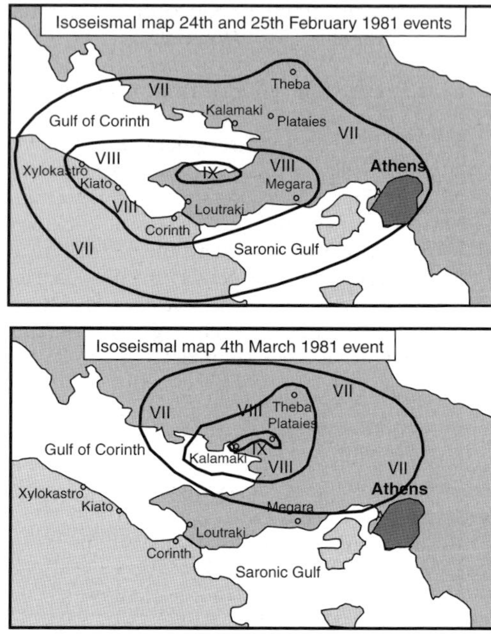

Fig. 7. MS (Mercalli-Sieberg, a version similar to the MCS) intensity distribution of the 1981 Alkyonides earthquake sequence (*Bulletin of the Geodynamic Institute of Athens* 1981; Antonaki et al. 1988). This earthquake sequence had a core of high intensities around the epicentral area and a second maximum of high intensities at 70 km distance, affecting several districts of Athens. On average, Athens experienced intensities VII and VIII; however, in some boroughs and building blocks intensities up to IX were also recorded.

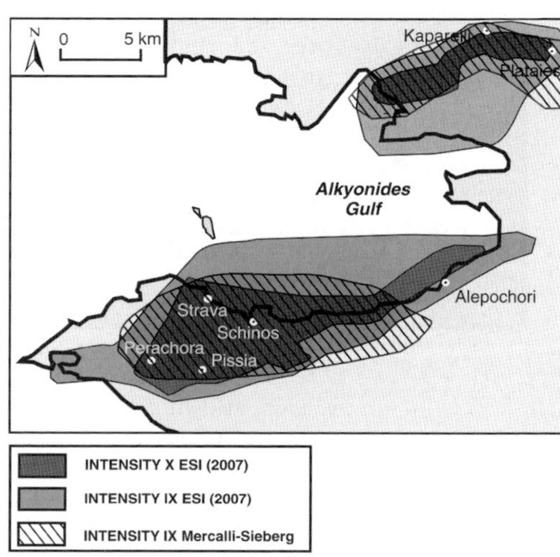

Fig. 8. Comparison of the isoseismal pattern between the MS and the ESI 2007 intensity scales.

1995; Hubert *et al.* 1996), as both occurred at night and only a few hours apart. However, for our study this makes no difference since these two events cannot be separated in terms of their macroseismic effects.

Following the above descriptions a maximum epicentral ESI 2007 value of X is determined in several sites (Fig. 8) and particularly along strike the primary surface ruptures in Pissia, Schinos and in Kaparelli–Plataies where they exceeded a few tens of centimetres in height. Intensity X has also been allocated along the coastal zone from Strava up to Alepochori, where significant subsidence ranging from a few decimetres up to 100 cm has been recorded. Intensity IX is mainly assigned in areas where the surface ruptures were a few tens of centimetres high. Finally, intensity VIII was widely assessed affecting a large area where ground ruptures, extensive landslides, rockfalls and liquefaction phenomena have been observed.

Maximum MS (Mercalli–Sieberg) intensity values (a version similar to the MCS) were also recorded in all villages that were in close proximity to the activated faults (Fig. 8; Perachora IX–X, Plataies IX–X, Schinos IX, Pissia IX, Kaparelli IX). However, no intensity X has been assigned and most of the epicentral villages recorded an epicentral intensity IX (Fig. 8). Figure 8 shows the different isoseismic patterns of the epicentral region based on the ESI 2007 and the Mercalli–Sieberg scales, respectively. Differences are noteworthy, but not substantial. It should be noted that surface geology played a decisive role in the damage distribution and had a significant effect on the intensity observed at a site. On average, under similar circumstances sites located on soil foundations experienced about one intensity degree more shaking than sites located on rock foundations, whereas sites on Neogene sediments experienced about half a degree greater intensity than sites located on rock foundations (Tilford *et al.* 1985).

These shallow normal faulting earthquakes affected not only the Perachora Peninsula (maximum intensity IX–X), Plataies (IX–X) or Kaparelli (IX), but also the city of Athens, located 70 km to the east, where tens of buildings collapsed in certain town districts (Fig. 7). As a result, this earthquake sequence had an anomalous intensity distribution with a core of high intensities around the epicentral area and a second maximum of high intensities at 70 km distance, affecting several districts of Athens. On average, Athens experienced intensities of VII and VIII; however, in some boroughs and building blocks, intensities up to IX were also recorded. In particular, in Athens 1175 buildings collapsed or had to be demolished after the earthquake, whereas 7824 buildings experienced severe damage (Antonaki *et al.* 1988). The degree of damage changed abruptly over short distances due to surface geology. In Athens, the area of damage was highly localized in the boroughs of Chalandri, Anthoupoli, Moschato, Aigaleo, Nea Ionia and Nikaia (suburbs of Athens) mainly due to poor local site conditions (mostly fluvial and alluvial deposits). In particular, in the Chalandri district buildings were of good quality. The highest percentage of damage to single-and two-storey buildings occurred where the depth to bedrock (less than 40 m) and thickness of recent sediments (less than 10 m) were minimum, whereas for the multistorey buildings the higher damage occurred where the depth to bedrock was maximum (greater than 40 m) and the thickness of recent deposits from 5 to 15 m (Christoulas *et al.* 1985). Therefore, damage to multistorey buildings occurred because the dominant period of the soil was approximately equal to the dominant period of these buildings (Christoulas *et al.* 1985).

The 2006 Kythira earthquake

On 8 January 2006 a thrust faulting event $M_w = 6.7$ with considerable strike-slip motion occurred in southwestern Greece (European Mediterranean Seismological Centre, unpubl. data; Konstantinou *et al.* 2006). This event is related to the Hellenic subduction zone (Fig. 1) and the epicentre was located a few tens of kilometres east of the island of Kythira with focal depth estimated at 70 km (United States Geological Survey, unpubl. data). The event was felt throughout Greece and the eastern Mediterranean in general (from Southern Italy and Dalmatian coasts, to Bulgaria, Turkey, Jordan, Israel and Egypt).

No casualties were reported and damage was restricted to the village of Mitata on the island of Kythira (Fig. 9a). Several old stone masonry buildings experienced significant damage (including a few collapses; Fig. 10b); however, modern reinforced concrete buildings did not suffer any damage. The metropolitan church located in the village square sustained severe damage (Fig. 10a) and several stone fences collapsed. A number of rockfalls, landslides and fractures disturbed the local road network (Fig. 11b, c and d), affecting an area of about 15 km^2. However, they were of limited size (Fig. 11d). Fractures a few meters long were observed on paved roads within the village (Fig. 11a). The biggest landslide affected Mitata village square that was partly detached (Fig. 11b), involving a collapsing volume of about 5000 m^3 (Fig. 12; Lekkas & Danamos 2006). Several masses of rock (*c.* 500 m^3 each) were detached for about 100 m and were accumulated at the base of the slope on the Mitata–Viaradika road. A few more rockfalls were observed along the

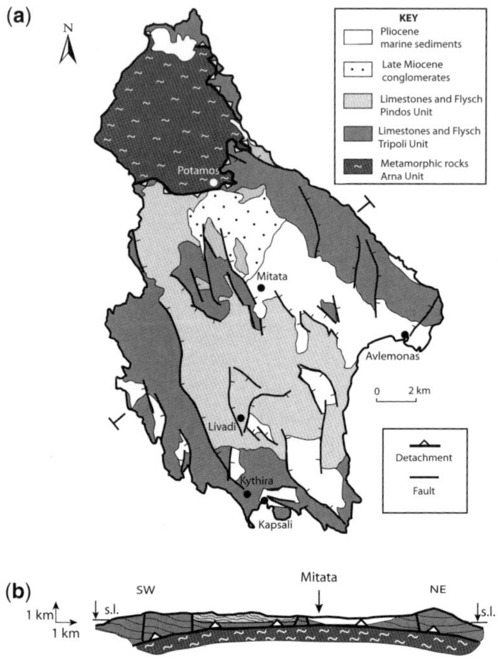

Fig. 9. (**a**) Geological map of Kythira island (modified from Papanikolaou & Danamos 1991). (**b**) Cross-section across Mitata village. Mitata village is situated on the immediate hanging wall of an active fault. In addition, it is founded on Pliocene marine sediments that rest on a large NNW–SSE trending detachment fault that lies a few hundred metres below the village separating the non-metamorphic rocks from the underlying metamorphic rocks of the Arna Unit.

remaining road network of the island, but no liquefaction phenomena were recorded. Therefore, remarkable EEE were observed only in the village of Mitata in compliance with the damage distribution. It is interesting to note that even though several villages of the island are equidistant from the epicentre, only Mitata village was damaged and experienced some noteworthy EEE. In particular, in Potamos village, situated only 35 km from the epicentre (Fig. 9a), the reported MM intensity was V+, whereas in Mitata village situated *c.* 40 km the intensity was VII+ (Konstantinou *et al.* 2006). This intensity distribution was likely due to: (a) the poor foundation conditions of the village, (b) its proximity to an active neotectonic fault that bounds a Late Miocene–Pliocene basin (Papanikolaou & Danamos 1991), and (c) the presence of a large detachment fault at a few hundred metres below the village (Papanikolaou & Danamos 1991). On the other hand, Potamos village is founded on the metamorphic rocks of the Arna Unit and consequently is located under the footwall block of the detachment (Fig. 9).

The NE–SW trending cross-section across Mitata village (Fig. 9b) shows the geological structure of the area. The village of Mitata is founded on Pliocene marine sediments that rest on a large NNW–SSE trending detachment fault. This detachment, situated a few hundred metres below the village, separates the non-metamorphic rocks from the underlying metamorphic rocks of the Arna Unit, belonging to the East Peloponnesus detachment system (Papanikolaou & Royden 2007). Finally, the village of Mitata is situated in the immediate hanging wall of an active NW–SE normal fault. The village of Mitata was devastated (intensity XI) by a similar deep-sourced (*c.* 80 km), but significantly stronger M = 7.9 event that occurred in 1903 (Papazachos & Papazachou 1997). Then, the newly constructed church and the school building collapsed and several ground fissures were reported, of which one was 200 m long and 1 cm wide (Papazachos & Papazachou 1997). This past event also confirms that Mitata village is founded in unfavourable geological conditions.

Based on the environmental effects, an ESI 2007 maximum epicentral intensity of VII–VIII has been assigned, which is similar to the MM reported intensity value.

Discussion: advances and limitations of the ESI 2007 scale

The ESI 2007 scale has been tested in earthquakes that: (i) had different source characteristics (magnitude, focal mechanism and focal depth) and (ii) produced a variety of environmental effects (primary surface faulting, minor and major secondary effects), which help us obtain a spherical view of its performance.

The ESI 2007 scale has been easily applied, leaving no question marks or ‘grey’ areas in all three examples. Above the intensity VII degree when environmental effects become prominent, the ESI 2007 scale can define the intensity degree with a high level of accuracy as also shown in several recent and historic earthquakes worldwide (e.g. Serva *et al.* 2007; Tatevosian *et al.* 2007). Overall, the 1993 Pyrgos and the 2006 Kythira events demonstrate that for moderate intensity levels (VII and VIII) the ESI 2007 and the traditional scales compare fairly well. On the other hand, the 1981 Alkyonides earthquake sequence demonstrates that there is inconsistency between ESI 2007 and traditional scales for the high intensity values (IX, X). This seems to agree with similar examples worldwide, emphasizing the importance and the increasing accuracy of the ESI 2007

Fig. 10. (**a**) Photos of the damage inflicted on the metropolitan church of Mitata village. The church, located in the village square, is constructed of porous limestone blocks cemented with lime wash without concrete columns and experienced significant damage. Severe damage was also inflicted at both bell towers. (**b**) A few collapses of old plain stone masonry buildings were recorded at Mitata.

scale towards the highest levels of the scale in the epicentral area (e.g. Michetti *et al.* 2007; Serva *et al.* 2007).

In particular, for the 1993 Pyrgos event, the EMS 1992 and the ESI 2007 scales seem to comply well regarding not only the maximum recorded epicentral intensity, but also with the overall isoseismal pattern. Nevertheless, it should be mentioned that no villages were founded near the landslide or the liquefaction-prone areas, otherwise it is possible that the traditional intensities would have recorded a higher epicentral intensity value.

In addition, even for the 2006 Kythira deep-sourced event, the ESI 2007 and the traditional macroseismic scales correlate well, suggesting a maximum VII–VIII intensity. A question arises as to whether a more destructive deep-sourced event such as the M = 7.9 earthquake that occurred in Kythira in 1903 ($I_0 = XI$; Papazachos & Papazachou 1997) would have been correctly interpreted in the ESI 2007 scale, since no primary surface ruptures would have been observed to imply a higher intensity value and reported ground fissures were of limited length. In such deep not 'linear morphogenic events' (e.g. Caputo 2005), where ruptures cannot reach the surface, the ESI 2007 intensity level should be correlated with the area where severe environmental effects have been recorded. However, in such cases uncertainties are expected to be higher, implying that the assessment

Fig. 11. (**a**) Ground ruptures on paved road at the village of Mitata. (**b**) View of the landslide that affected Mitata village square, which was partly detached, involving a collapsed volume of about 5000 m^3. (**c**) View of rockfalls disturbing the road. (**d**) View of a minor landslide blocking the road.

of the ESI 2007 scale should probably be considered less precise for deep events.

In the 1981 Alkyonides example, the ESI 2007 intensity scale provides not only a slightly higher maximum epicentral intensity (X), but also a different spatial distribution of the isoseismals, compared to the traditional scales. This implies that current traditional scales possibly underestimate the 'strength' of this kind of earthquake sequence. This occurs partly because the epicentral area, where significant EEE were recorded, was relatively sparsely populated. Indeed, only the small villages of Schinos, Pissia Kaparelli and Plataies were situated a few hundred metres up to a few kilometres

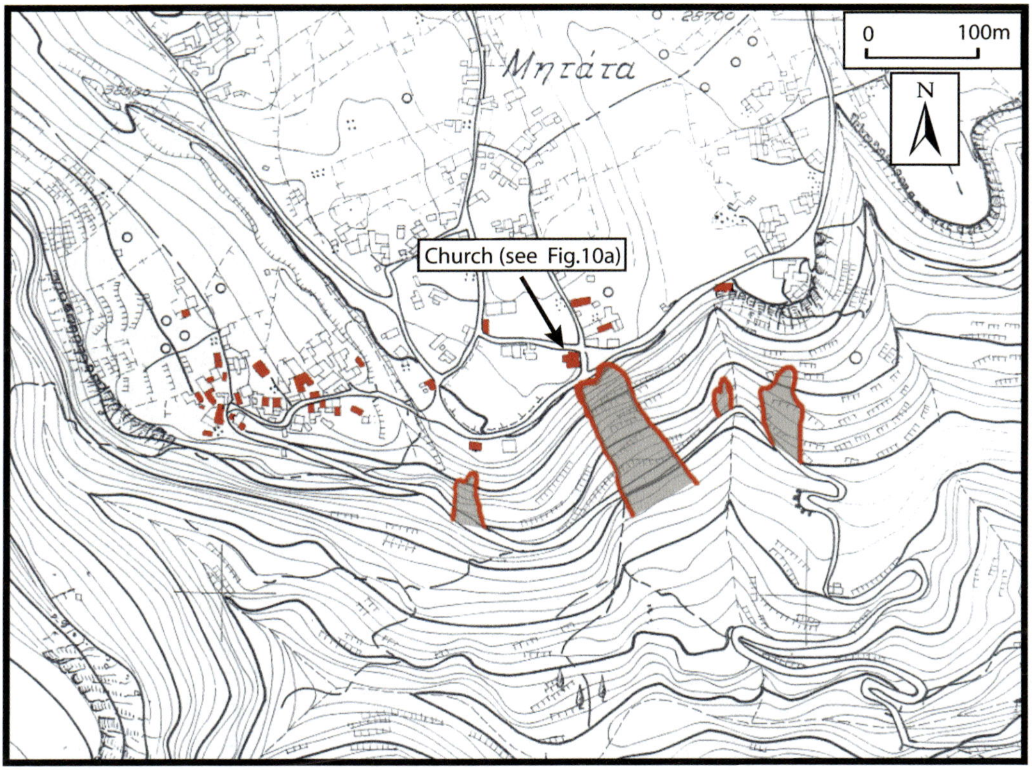

Fig. 12. Landslide (grey areas) and damage (bold blocks) distribution around Mitata village mapped at a 1:5000 scale (Lekkas & Danamos 2006). Thick contours represent intervals of 20 m and thin contours intervals of 4 m.

away from the localities where significant offsets (>20 cm) have been observed. Although Kaparelli was the most proximal village to the surface ruptures of the third event and is founded on a rather unfavourable geological site (Pleistocene fluvial, breccia and slope deposits), it experienced intensity IX. This occurred because it is situated in the immediate footwall of the fault and thus experienced less shaking. Moreover, in Schinos village several houses were founded on the ophiolitic bedrock and experienced minor to moderate damages (Andronopoulos *et al.* 1982), thus lowering the epicentral maximum intensity. The same is probably true for Pissia since part of the village is founded on Holocene scree and Pleistocene breccia deposits, but the remaining part is on the Alpine basement. In addition, Plataies village is also founded on Mesozoic limestones. Tilford *et al.* (1985), following a survey in the area, suggested that surface geology significantly influenced the damage distribution and calculated that on average, buildings located on soil foundations experienced about one degree of intensity more than those located on rock foundations. Previous remarks may explain why an epicentral intensity of X has not been fully implemented. From this perspective, it is argued that for the Alkyonides example the ESI 2007 scale is probably more appropriate for drawing isoseismals of intensity IX and X in the epicentral area.

As far as the far field effects are concerned, there is an inconsistency for the town of Athens between the EEE, which were negligible, and the significant damage that occurred in some town districts. This inconsistency for the far field of the 1981 earthquake sequence between the ESI 2007 and the Mercalli intensity scale can be attributed to several reasons. It could be that intensive earthquake environmental related effects have not been expressed or even recorded in Athens, due to the strictly localized area of damage and its limited geographical coverage. However, it is argued that this inconsistency is probably due to the structural response of multistorey buildings, the bedrock geology and the local site effects (e.g. Tilford *et al.* 1985) in accordance with the long distance from the epicentre. In this case there was a long period resonance because the dominant period of the soil was approximately equal to the dominant period of certain buildings, causing severe damage (e.g. Christoulas *et al.* 1985). In addition, considering that Athens experienced three strong successive events over a few days, the weakness and the vulnerability of the buildings would have been substantially increased. This comparison for the far field effects may be inappropriate since no EEE were recorded in the far field, even though structural damage did occur. However, it is also important to establish that the ESI 2007 scale should be used predominantly in the epicentral area and thus may not accurately describe damage in the far field. Nevertheless, the Kythira example shows that the far field effects are in compliance with the EEE.

A few other events have been studied in Greece and reassessed in terms of the ESI 2007 scale. Papathanasiou & Pavlides (2007) found that in the 2003 Lefkada earthquake (Ms = 6.4), the traditional intensities tend to underestimate the ground shaking because the strengthening of buildings, due to the strict seismic code in the area, resulted in better performance under ground shaking. However, two similar events that occurred in the past before the provision of the seismic code had similar intensity values as the ESI 2007 scale. The 1988 Elia earthquake (Ms = 5.9) in NW Peloponnese shows that the traditional intensities were similar to the ESI 2007 scale, whereas in the 1999 Athens event (Ms = 5.9), the ESI 2007 intensities underestimated the impact of the earthquake (Fokaefs & Papadopoulos 2007). The 2003 Lefkada (Papathanasiou & Pavlides 2007), the 1999 Athens (Fokaefs & Papadopoulos 2007) and the 1981 earthquake sequence (this study) indicate that when the ESI 2007 and the traditional intensity scales disagree, the intensity has to coincide with the highest value between these two independent estimates (see also Serva *et al.* 2007; Michetti *et al.* 2007).

Another important issue that has arisen from the 1981 Alkyonides earthquake sequence is the different isoseismal distributions recorded by several research groups. In several epicentral villages, reported intensity values differ from half (e.g. Perachora and Pissia) up to one degree (Schinos). For example, in Schinos village intensity recordings varied from VIII–IX (Papazachos *et al.* 1982), IX (*Bulletin of the Geodynamic Institute of Athens* 1981) up to IX–X (Andronopoulos *et al.* 1982), providing a rather confusing pattern. This difference can be attributed to several causes. It may be: (i) due to the subjective interpretation of damages or (ii) due to the subjectivity in allocating the predominant damage in a site, or (iii) because the assigned intensities correspond to the maximum observed intensity rather than the mean. From this perspective, the ESI 2007 scale is probably easier to implement and more precise in quantifying macroseismic effects, offering a higher objectivity in the process of assigning intensity values.

Finally, the use of many different intensity scales worldwide (e.g. MM, MCS, MSK, JMA), which are also constantly updated (e.g. EMS 1992) indirectly demonstrates the inefficiency of current earthquake intensity scales in describing the macroseismic earthquake effects.

The ESI 2007 scale and implications for seismic hazard assessment

Intensity is an important seismic hazard parameter. The isoseismal maps are used to derive empirical relations for the decrease of intensity with distance, which then are incorporated into the attenuation laws and used to assess the seismic hazard. One fundamental question that will be posed is why introduce another intensity scale without clear outcomes to the seismic hazard assessment? In this section we show how the ESI 2007 scale could prove beneficial for seismic hazard assessment by reducing the uncertainty in the empirically based attenuation laws and demonstrate: (i) how large the uncertainty is and (ii) how significant is this uncertainty for the seismic hazard maps.

Intensity attenuates with increasing distance from the earthquake source, at first rapidly and then more slowly. Therefore, in order to define the seismic hazard at a given site, it is necessary to know the expected attenuation of intensity with epicentral distance. Thus, attenuation curves are simple empirical relationships that give the largest amplitudes as a function of epicentral intensity and/or earthquake magnitude with distance and are compiled based on the statistical elaboration of historical and instrumental data. However, there is a large variation in the data, which adds uncertainty in the seismic hazard assessment. This variation is nicely portrayed in an empirical magnitude–intensity database compiled by D'Amico *et al.* (1999), which is presented in Table 2. This database is extracted from instrumental catalogues ranging from 1880 to 1980, covering the whole Mediterranean region. For example, Table 2 shows that epicentral intensity X has been produced by significantly different magnitude events, ranging from M = 6.0 to M = 7.0. It is possible that a portion of this variation in the data stems from the way macroseismic effects have been assessed. Therefore, wherever feasible, by reconstructing the macroseismic field of historical earthquakes, through the use of the ESI 2007 scale, the uncertainties may be significantly reduced.

Evidently, apart from the attenuation/amplification relationships, uncertainty in seismic hazard assessment stems from several other factors such as the fault geometry, slip rates, the earthquake occurrence model used etc. However, it seems that

Table 2. *Distribution of the magnitude values for each intensity class in the Mediterranean region*

Intensity	Number of events	Lower magnitude	Upper magnitude	Mean magnitude (Ms)
VIII	161	5.0	5.9	5.4
VIII–IX	20	5.7	6.3	6.0
IX	53	5.8	6.7	6.2
IX–X	**5**	**6.3**	**6.9**	**6.5**
X	**18**	**6.0**	**7.0**	**6.6**

Modified from D'Amico *et al.* 1999

seismic hazard maps are also highly sensitive to the attenuation relationship used. In particular, in some cases uncertainty is large enough so that it overshadows the other factors. For example, following Table 2 on average a M = 6.5 earthquake is expected to produce an epicentral intensity IX or X (or IX–X). However, the exact value of the epicentral intensity (whether IX or X) is crucial for the determination and modelling of the area affected around the epicentre and the attenuation relationships. Indeed, historical data of macroseismic intensity versus epicentral distance published by Grandori *et al.* (1991), covering the whole of the Apennines in Italy, show that earthquakes with epicentral intensity IX ($I_0 = 9$) have a mean radius of 6–7 km for the intensity IX isoseismal, whereas earthquakes with epicentral intensity X ($I_0 = 10$) have a mean radius of 20–21 km for the isoseismal IX (Fig. 13a). Similar values have been reported for other regions such as the Sino-Korean craton (Lee & Kim 2002; Fig. 13b). The significance of this uncertainty can be portrayed and easily assessed in quantitative fault-specific seismic hazard maps from geological fault slip-rate data (Papanikolaou 2003; Roberts *et al.* 2004). Roberts *et al.* (2004) used in Lazio-Abruzzo, central Italy, an average value of 12.5 km radius for the modelled isoseimal IX in the central Appenines, whereas Papanikolaou (2003) used the average value of 12.5 km and an extreme upper value of 20 km, running a sensitivity analysis for the southern Apennines. Sensitivity analysis showed that the error introduced by the implied uncertainty in the dimensions of modelled isoseismals is significantly larger than the fault throw-rate error (of ±20%). This is remarkable because it shows that input parameters such as the isoseismal dimensions, which themselves are derived from attenuation relationships, influence the results more significantly than the uncertainty implied from the fault throw-rate data which govern the earthquake recurrence. Therefore, further evaluation that could constrain the isoseismal dimensions and the attenuation relationships will prove beneficial in limiting the uncertainties implied in the seismic hazard analysis for the regions of central and southern Apennines.

It should be noted that the complication and uncertainty in earthquake ground motion and consequently in the attenuation/amplification relationships is not only related to the way intensity values have been assigned and isoseismal lines

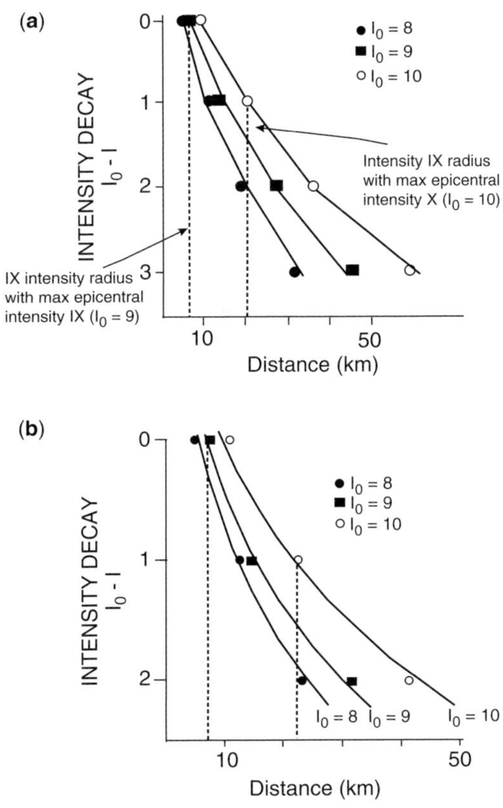

Fig. 13. Attenuation laws for: (**a**) the Apennines in Italy (modified from Grandori *et al.* 1991) and (**b**) the Sino-Korean peninsula (modified from Lee & Kim 2002). Attenuation laws are derived from the statistical elaboration of historical and instrumental data. The Apennines example shows that earthquakes with epicentral intensity IX ($I_0 = 9$) have a mean radius of 6–7 km for the intensity IX isoseismal, whereas earthquakes with epicentral intensity X ($I_0 = 10$) have a mean radius of 20–21 km for the isoseismal IX.

have been drawn, but emerges from several factors, which are critical but not accurately known. These factors vary from source to source, path to path and from site to site introducing a large scatter in numerical values of ground motion (e.g. Hu *et al.* 1996). Moreover, some of these factors are 'highly interdependent' making it difficult to separate near-surface effects from deeper basin effects (Field *et al.* 2000). Studies on the prediction of ground motion and the site effects suggest that site response has a large intrinsic variability with respect to source location (Hartzell *et al.* 1997). The intrinsic variability is caused by basin-edge induced surface waves, focusing and defocusing effects and scattering in general that cannot be reduced in any model (Field *et al.* 2000). Therefore, there is little hope that all uncertainties implied by the attenuation/amplification relationship can be fully reduced. However, it is also probable that part of this uncertainty stems from the intensity evaluation. The ESI 2007 scale offers higher objectivity in the process of assessing macroseismic intensities than traditional intensity scales that are influenced by human parameters. Thus, the ESI 2007 scale can compare not only events from different settings, but also contemporary and future earthquakes with historical and even pre-historical events. This occurs because the ESI 2007 scale follows the same criteria/environmental effects for all events that are independent of the local economy and cultural setting through time. Therefore, a reappraisal of historical earthquakes so as to constrain the INQUA intensity scale may prove beneficial for the seismic hazard assessment by reducing the uncertainty implied in the attenuation laws.

Conclusions

The ESI 2007 scale incorporates the advances and achievements of palaeoseismology and earthquake geology, forming an objective and easily applied tool for measuring the earthquake strength, particularly in the epicentral area.

The ESI 2007 intensity values and the spatial distribution of the isoseismals for the 1993 Pyrgos and the 2006 Kythira events are similar to those resulting from the traditional scales, demonstrating that for moderate intensity levels (VII and VIII) the ESI 2007 and the traditional scales compare fairly well. On the other hand, the 1981 Alkyonides earthquake sequence shows that there is an inconsistency between the ESI 2007 and the traditional scales both in the epicentral area, where higher ESI 2007 values have been assigned, and for the far field effects. In the 1981 Alkyonides sequence, the ESI 2007 scale provides not only a slightly higher maximum epicentral intensity, but also a different isoseismal pattern compared to the traditional scales, implying that current traditional scales underestimate the 'strength' of this earthquake sequence. This occurs partly because the epicentral area, where significant EEE were recorded, was sparsely populated. In addition, several villages located in the epicentral region were founded on bedrock sites and others on the footwall block, experiencing less shaking. The 1981 earthquake sequence emphasizes the importance and the increasing accuracy of the ESI 2007 scale towards the highest levels of the scale in the epicentral area. In contrast, the ESI 2007 scale may not accurately describe the damage in the far field. In Athens, situated about 70 km from the epicentre, the EEE were negligible, but several town districts suffered significant damage, because the dominant period of the soil was approximately equal to the dominant period of certain buildings. This example demonstrates once again that when the ESI 2007 and the traditional intensity scales disagree, the intensity has to coincide with the highest value between these two independent estimates.

Earthquake environmental effects provide higher objectivity in the process of assigning intensity values, so that the ESI 2007 scale is the best tool to compare recent, historic and pre-historic earthquakes as well as earthquakes from different tectonic settings. A reappraisal of historical earthquakes, so as to constrain the ESI 2007 scale, may prove beneficial for the seismic hazard assessment by reducing the uncertainty implied in the attenuation laws.

We thank Alessandro Michetti for discussions concerning the application of the ESI 2007 scale. Reviews from Spyros Pavlides and Riccardo Caputo improved the paper.

References

ABERCROMBIE, R., MAIN, I. G., DOUGLAS, A. & BURTON, P. W. 1995. The nucleation and rupture process of the 1981 Gulf of Corinth earthquakes from deconvolved broad band data. *Geophysical Journal International*, **120**, 393–405.

ANDRONOPOULOS, B., ELEFTHERIOU, A., KOUKIS, G., ROZOS, D. & ANGELIDIS, C. 1982. Macroseismic, geological and tectonic observations in the area affected by the earthquakes of the Corinthian Gulf (February–March 1981). *International Symposium on the Hellenic Arc and Trench (H.E.A.T.)*, **1**, 19–34.

ANTONAKI, R., VLACHOS, I. ET AL. 1988. *Earthquake Economic Impact – Insurance*. Report of Earthquake Planning and Protection Organization, Athens [in Greek].

BOUCKOVALAS, G., ANAGNOSTOPOULOS, A., KAPENIS, A. & KARANTONI, T. 1996. Analysis of soil effects and distribution of damage from the Pyrgos 1993 (Greece) earthquake. *Geotechnical and Geological Engineering*, **14**, 111–128.

CAPUTO, R. 2005. Ground effects of large morphogenic earthquakes. *Journal of Geodynamics*, **40**, 113–118.

CHRISTOULAS, S. G., TSIAMBAOS, G. K. & SABATAKAKIS, N. S. 1985. Engineering geological conditions and the effects of the 1981 earthquake in Athens, Greece. *Engineering Geology*, **22**, 141–155.

COBURN, A. & SPENCE, R. 2002. *Earthquake Protection*. John Wiley, Chichester.

D'AMICO, V., ALBARELLO, D. & MANTOVANI, E. 1999. A distribution-free analysis of magnitude-intensity relationships: an application to the Mediterranean region. *Physics and Chemistry of Earth*, **24**, 517–521.

DZIEWONSKI, A. M., EKSTROM, G. & SALGANIK, M. P. 1994. Centroid-moment tensor solutions for January–March 1993. *Physics of the Earth and Planetary Interiors*, **82**, 9–17.

FIELD, E. H. & SCEC Phase III Working group 2000. Accounting for site effects in probabilistic seismic hazard analyses of southern California: Overview of the SCEC Phase III report. *Bulletin of the Seismological Society of America*, **90**, S1–S31.

FOKAEFS, A. & PAPADOPOULOS, G. 2007. Testing the new INQUA intensity scale in Greek earthquakes. *Quaternary International*, **173–174**, 15–22.

GRANDORI, G., DREI, A., PEROTTI, F. & TAGLIANI, A. 1991. Macroseismic intensity versus epicentral distance: the case of Central Italy. *Tectonophysics*, **193**, 165–171.

GRUNTHAL, G. (ed.) 1993. *European Macroseismic Scale 1992 (updated MSK-scale)*. Conseil de l' Europe, Luxemburg, Cahiers du centre Europeen de Geodynamique et de Sismologie **7**.

HARTZELL, S., HARMSEN, S., FRANKEL, A., CARVER, D. & MEREMONTE, M. 1997. Variability of site response in the Los Angeles urban area. *Bulletin of the Seismological Society of America*, **86**, S186–S192.

HU, Y.-X., LIU, S.-C. & DONG, W. 1996. *Earthquake Engineering*. E & FN Spon, London.

HUBERT, A., KING, G., ARMIJO, R., MEYER, B. & PAPANASTASIOU, D. 1996. Fault-reactivation, stress interaction and rupture propagation of the 1981 Corinth earthquake sequence. *Earth and Planetary Science Letters*, **142**, 573–585.

JACKSON, J. A., GAGNEPAIN, J., HOUSEMAN, G., KING, G. C. P., PAPANIMITRIOU, P., SOUFLERIS, C. & VIRIEUX, J. 1982. Seismicity, normal faulting and the geomorphological development of the Gulf of Corinth (Greece): the Corinth earthquakes of February and March 1981. *Earth and Planetary Science Letters*, **57**, 377–397.

KARANTONI, F. V. & BOUCKOVALAS, G. 1997. Description and analysis of building damage due to Pyrgos, Greece earthquake. *Soil Dynamics and Earthquake Engineering*, **16**, 141–150.

KHOURY, S. G., TILFORD, N. R., CHANDRA, U. & AMICK, D. 1983. The effect of multiple events on isoseismal maps of the 1981 earthquake at the Gulf of Corinth. *Bulletin of the Seismological Society of America*, **73**, 655–660.

KING, G. C. P., OUYANG, Z. X. ET AL. 1985. The evolution of the Gulf of Corinth (Greece): an aftershock study of the 1981 earthquakes. *Geophysical Journal of the Royal Astronomical Society*, **80**, 677–683.

KIRATZI, A. & LOUVARI, E. 2003. Focal mechanisms of shallow earthquakes in the Aegean Sea and the surrounding lands determined by waveform modeling: a new database. *Journal of Geodynamics*, **36**, 251–274.

KONSTANTINOU, K. I., KALOGERAS, I. S., MELIS, N. S., KOUROUZIDIS, M. C. & STAVRAKAKIS, G. N. 2006. The 8 January 2006 Earthquake (Mw 6.7) offshore Kythira island, Southern Greece: Seismological, strong motion and macroseismic observations of an intermediate-depth event. *Seismological Research Letters*, **77**, 544–553.

KOUKOUVELAS, I., MPRESIAKAS, A., SOKOS, E. & DOUTSOS, T. 1996. The tectonic setting and earthquake ground hazards of the 1993 Pyrgos earthquake, Peloponnese, Greece. *Journal of the Geological Society London*, **153**, 39–49.

LAIGLE, M., HIRN, A., SACHPAZI, M. & CLEMENT, C. 2002. Seismic coupling and structure of the Hellenic subduction zone in the Ionian islands region. *Earth and Planetary Science Letters*, **200**, 243–253.

LEE, K. & KIM, J.-K. 2002. Intensity attenuation in the Sino-Korean Craton. *Bulletin of the Seismological Society of America*, **92**, 783–793.

LEKKAS, E. 1994. Liquefaction – Risk zonation and urban development in Western Peloponnese (Greece). *In*: *Proceedings of 7th International Congress, International Association of Engineering Geology*. Balkema, Rotterdam. 2095–2102.

LEKKAS, E. 1996. Pyrgos earthquake damages (based on E.M.S.-1992) in relation with geological and geotechnical conditions. *Soil Dynamics and Earthquake Engineering*, **15**, 61–68.

LEKKAS, E. & DANAMOS, G. 2006. Preliminary observations of the January 8, 2006 Kythira island (SW Greece) earthquake (Mw = 6.9). *Newsletter E.E.R.I.*, 1–20.

LEKKAS, E., FOUNTOULIS, I. & PAPANIKOLAOU, D. 2000. Intensity distribution and Neotectonic macrostructure Pyrgos Earthquake data (26 March 1993, Greece). *Natural Hazards*, **21**, 19–33.

MCGUIRE, R. K. 1993. Computations of seismic hazard. *Annali di Geofisica*, **36**, 181–200.

MARIOLAKOS, I., PAPANIKOLAOU, D., SYMEONIDIS, N., LEKKAS, S., KAROTSIERIS, Z. & SIDERIS, C. 1982. The deformation of the area around the eastern Korinthian gulf, affected by the earthquakes of February–March 1981. *Proceedings of International Symposium on the Hellenic Arc and Trench (H.E.A.T.)*, **1**. National Technical University, Athens, 400–420.

MELIS, N., TSELENTIS, G. & SOKOS, E. 1994. The Pyrgos (March 26, 1993; Ms = 5.2) earthquake sequence as it was recorded by the Patras seismic network. *Bulletin of the Geological Society of Greece*, **30**, 175–180.

MICHETTI, A. M., ESPOSITO, E. ET AL. 2004. The INQUA scale: An innovative approach for assessing earthquake intensities based on seismically-induced ground effects in natural environment. Special paper *Memorie Descrittive della Carta Geologica D'Italia*, **LXVII**.

MICHETTI, A. M., ESPOSITO, E. ET AL. 2007. Intensità scale ESI 2007. *In*: GUERRIERI, L. & VITTORI, E. (eds) *Memorie Descrittive della.* Carta Geologica d'Italia, **74**, Servizio Geologico d'Italia, Dipartimento Difesa del Suolo, APAT, Rome.

MOREWOOD, N. C. & ROBERTS, G. P. 2001. Comparison of surface slip and focal mechanism slip data along normal faults: an example from the eastern Gulf of Corinth, Greece. *Journal of Structural Geology*, **23**, 473–487.

PAPANASTASIOU, D., STAVRAKAKIS, G. N. & DRAKOPOULOS, J. 1994. A study of the March 26, 1993, Pyrgos earthquake sequence and its tectonic Implications. *Proceedings of the XXIV General Assembly of the European Seismological Commision 1994*. Athens.

PAPANIKOLAOU, D. & DANAMOS, G. 1991. The role of the geotectonic location of Kythira and Cyclades in the geodynamic evolution of the Hellenic Arc. *Bulletin of the Geological Society of Greece*, **25**, 65–79.

PAPANIKOLAOU, D. & ROYDEN, L. 2007. Disruption of the Hellenic Arc: Late Miocene extensional detachment faults and steep Pliocene-Quaternary normal faults – or – What happened to Corinth?. *Tectonics*, **26**, TC5003. DOI: 10.1029/2006TC002007.

PAPANIKOLAOU, D., FOUNTOULIS, I. & METAXAS, C. 2007. Active faults, deformation rates and Quaternary paleogeography at Kyparissiakos Gulf (SW Greece) deduced from on-shore and off-shore data. *Quaternary International*, **171–172**, 14–30.

PAPANIKOLAOU, I. D. 2003. *Generation of high-resolution seismic hazard maps in extensional tectonic settings through integration of earthquake geology, fault mechanics theory and GIS techniques*. PhD thesis, University of London.

PAPATHANASSIOU, G. & PAVLIDES, S. 2007. Using the INQUA scale for the assessment of intensity. Case study on the 2003 Lefkada (Ionian Islands), Greece earthquake. *Quaternary International*, **173–174**, 4–14.

PAPAZACHOS, B. C. 1990. Seismicity of the Aegean and surrounding area. *Tectonophysics*, **178**, 287–308.

PAPAZACHOS, B. P. & PAPAZACHOU, C. B. 1997. *The Earthquakes of Greece*. Ziti Editions, Thessaloniki.

PAPAZACHOS, B. C., COMNINAKIS, P. E., MOUNDRAKIS, D. M. & PAVLIDES, S. B. 1982. Preliminary results of an investigation of the February–March 1981 Alkionides Gulf (Greece) earthquakes. *Proceedings of International Symposium on the Hellenic Arc and Trench (H.E.A.T.)*, **2**, National Technical University, Athens, 74–87.

PERISSORATIS, C., MITROPOULOS, D. & ANGELOPOULOS, I. 1984. The role of earthquakes in inducing sediments mass movements in the eastern Korinthiakos Gulf. An example from the February 24-March 4, 1981 activity. *Marine Geology*, **55**, 35–45.

PIRAZZOLI, P. A., STIROS, S. C., ARNOLD, M., LABOREL, J., LABOREL-DEGUEN, F. & PAPAGEORGIOU, S. 1994. Episodic uplift deduced from the Holocene shoreline in the Perachora peninsula. *Tectonophysics*, **229**, 201–209.

ROBERTS, G. P., COWIE, P., PAPANIKOLAOU, I. & MICHETTI, A. M. 2004. Fault scaling relationships, deformation rates and seismic hazards: An example from the Lazio-Abruzzo Apennines, central Italy. *Journal of Structural Geology*, **26**, 377–398.

ROUMELIOTI, Z., BENETATOS, Ch., KIRATZI, A., STAVRAKAKIS, G. & MELIS, N. 2004. A study of the 2 December 2002 (M5.5) Vartholomio (western Peloponnese, Greece) earthquake and of its largest aftershocks. *Tectonophysics*, **387**, 65–79.

SERVA, L. 1994. Ground effects in the intensity scales. *Terra Nova*, **6**, 414–416.

SERVA, L., ESPOSITO, E., GUERRIERI, L., PORFIDO, S., VITTORI, E. & COMMERCI, V. 2007. Environmental effects from five historical earthquakes in southern Apennines (Italy) and macroseismic intensity assessment: Contribution to INQUA EEE Scale project. *Quaternary International*, **173–174**, 30–44.

STAVRAKAKIS, G. N. 1996. Strong motion records and synthetic isoseismals of the Pyrgos, Peloponnisos, southern Greece, earthquake sequence of March 26, 1993. *Pure and Applied Geophysics*, **146**, 147–161.

TATEVOSIAN, R. E. 2007. The Verny, 1887, earthquake in Central Asia: Application of the INQUA scale, based on coseismic environmental effects. *Quaternary International*, **173–174**, 23–29.

TAYMAZ, T., JACKSON, J. & MCKENZIE, D. 1991. Active tectonics of the north and central Aegean Sea. *Geophysical Journal International*, **106**, 433–490.

TILFORD, N. R., CHANDRA, U., AMICK, D. C., MORAN, R. & SNIDER, F. 1985. Attenuation of the intensities and effect of local site conditions on observed intensities during the Corinth, Greece, Earthquake of 24 and 25 February and 4 March 1981. *Bulletin of the Seismological Society of America*, **75**, 923–937.

VITA-FINZI, C. & KING, G. C. P. 1985. The seismicity, geomorphology and structural evolution of the Corinth area of Greece. *Philosophical Transactions of the Royal Society, London*, **A314**, 379–407.

Palaeoseismology of the North Anatolian Fault near the Marmara Sea: implications for fault segmentation and seismic hazard

THOMAS ROCKWELL[1]*, DANIEL RAGONA[1], GORDON SEITZ[1], ROB LANGRIDGE[2], M. ERSEN AKSOY[3], GÜLSEN UCARKUS[3], MATTHIEU FERRY[4,7], ARON J. MELTZNER[5], YANN KLINGER[6], MUSTAPHA MEGHRAOUI[7], DILEK SATIR[3], AYKUT BARKA[3†] & BURCAK AKBALIK[3]

[1]*Geological Sciences, San Diego State University, San Diego, CA 92182, USA*

[2]*Institute of Geological and Nuclear Sciences, PO Box 30-368, Lower Hutt, New Zealand*

[3]*Eurasia Institute of Earth Sciences, Istanbul Technical University, Maslak, Istanbul, Turkey*

[4]*Universidade de Évora, Centro de Geofisica de Évora, Rua Romão Ramalho 59, 7002-554 Évora, Portugal*

[5]*Tectonics Observatory, California Institute of Technology, Pasadena, CA 91125, USA*

[6]*Laboratoire de Tectonique, Institut de Physique du Globe, 4 place Jussieu, 75005, Paris, France*

[7]*Institut de Physique de Globe, 5 rue René Descartes, F-67084 Strasbourg Cedex, France*

[†]*Deceased*

**Corresponding author (e-mail: Trockwel@geology.sdsu.edu)*

Abstract: We conducted palaeoseismic studies along the North Anatolian fault both east and west of the Marmara Sea to evaluate its recent surface rupture history in relation to the well-documented historical record of earthquakes in the region, and to assess the hazard of this major fault to the city of Istanbul, one of the largest cities in the Middle East. Across the 1912 rupture of the Ganos strand of the North Anatolian fault west of the Marmara Sea, we excavated 26 trenches to resolve slip and constrain the earthquake history on a channel–fan complex that crosses the fault at a high angle. A distinctive, well-sorted fine sand channel that served as a marker unit was exposed in 21 trenches totaling over 300 m in length. Isopach mapping shows that the sand is channelized north of the fault, and flowed as an overflow fan complex across a broad fault scarp to the south. Realignment of the feeder channel thalweg to the fan apex required about 9 ± 1 m of reconstruction. Study of the rupture history in several exposures demonstrates that this displacement occurred as two large events. Analysis of radiocarbon dates places the age of the sand channel as post AD 1655, so we attribute the two surface ruptures to the large regional earthquakes of 1766 and 1912. If each was similar in size, then about 4–5 m of slip can be attributed to each event, consistent with that reported for 1912 farther east. We also found evidence for two additional surface ruptures after about AD 900, which probably correspond to the large regional earthquakes of 1063 and 1344 (or 1354). These observations suggest fairly periodic occurrence of large earthquakes (RI = *c.* 283 ± 113 years) for the past millennium, and a rate of *c.* 16 mm/a if all events experienced similar slip.

We excavated six trenches at two sites along the 1999 Izmit rupture to study the past earthquake history along that segment of the North Anatolian fault. One site, located in the township of Köseköy east of Izmit, revealed evidence for three surface ruptures (including 1999) during the past 400 years. The other trench was sited in an Ottoman canal that was excavated (but never completed) in 1591. There is evidence for three large surface rupturing events in the upper 2 m of alluvial fill within the canal at that site, located only a few kilometres from the Köseköy site. One of the past events is almost certainly the large earthquake of 1719, for which historical descriptions of damage are nearly identical to that of 1999. Other earthquakes that could plausibly be attributed to the other recognized rupture of the Izmit segment are the 1754, 1878 or 1894 events, all of which produced damage in the region and for which the source faults are poorly known. Our palaeoseismic observations suggest that the Izmit segment of the North Anatolia fault ruptures every one and a half centuries or so, consistent with the historical record for the region, although the time between ruptures may be as short as 35 years if 1754 broke the Izmit segment.

Release of about 4 m of seismic slip both west and east of the Marmara Sea this past century (1912, 1999) support the contention that Istanbul is at high risk from a pending large earthquake.

From: REICHERTER, K., MICHETTI, A. M. & SILVA, P. G. (eds) *Palaeoseismology: Historical and Prehistorical Records of Earthquake Ground Effects for Seismic Hazard Assessment*. The Geological Society, London, Special Publications, **316**, 31–54. DOI: 10.1144/SP316.3 0305-8719/09/$15.00

In that historical records suggest that the last large central Marmara Sea event occurred in 1766, there may be a similar 4 m of accumulated strain across the Marmara basin segment of the North Anatolian fault.

Information on the size and timing of past earthquakes is important in understanding fault behaviour, a key element in forecasting future seismic activity (Sieh 1996). The North Anatolian fault (NAF) in Turkey (Barka 1992) (Fig. 1) is an ideal candidate for understanding fault behaviour over multiple earthquake cycles because there is a long and excellent historical record of large earthquakes going back over 2000 years (Ambraseys & Finkel 1987*a*, *b*, 1991, 1995; Ambraseys 2002*a*, *b*). Furthermore, like the San Andreas fault of southern California, it is a fast-moving fault (*c.* 24 mm/a; Straub & Kahle 1995; Straub 1996; Reilinger *et al.* 1997; McClusky *et al.* 2000) resulting in many earthquakes during this long historical period. It is also segmented, and nearly all of the segments have ruptured this past century. Thus, there is the opportunity to collect information on the patterns of large earthquake ruptures and their sizes over several earthquake cycles. In this paper, we present new results that further quantify the earthquake history of the NAF both east and west of the Marmara Sea, and discuss their bearing on the seismic hazard to Istanbul.

To the west, near the Gulf of Saros, we extended our earlier palaeoseismic investigations along the Ganos strand and resolved slip for the past two large earthquakes that struck the Gallipoli region in 1766 and 1912. The Saros study area lies within the township of Kavakköy, close to where we completed earlier studies (Rockwell *et al.* 2001). We had determined that four surface ruptures occurred near where the fault passes offshore into the Gulf of Saros during the past 1100 years or so. However, the dating was insufficiently precise to confidently resolve which of the historical events these were. In this new study, we focused on a young (post AD 1655) stream channel that crosses the fault at a high angle and is offset by the fault across a narrow zone. Twenty-six trenches were excavated at this new site to resolve total slip on the channel and to further constrain the timing of ruptures that produced the slip.

East of the Marmara Sea, we continued efforts between Izmit and Lake Sapanca that we had begun prior to the August 1999 Izmit earthquake. Specifically, we excavated new trenches within and adjacent to an Ottoman canal, dated at *c.* 1591 from historical data (Finkel & Barka 1997; C. Finkel, pers. comm.) to resolve the history of surface ruptures for the past 400 years east of Izmit Bay.

Each of these studies bears on the impending seismic hazard to Istanbul, which lies close to

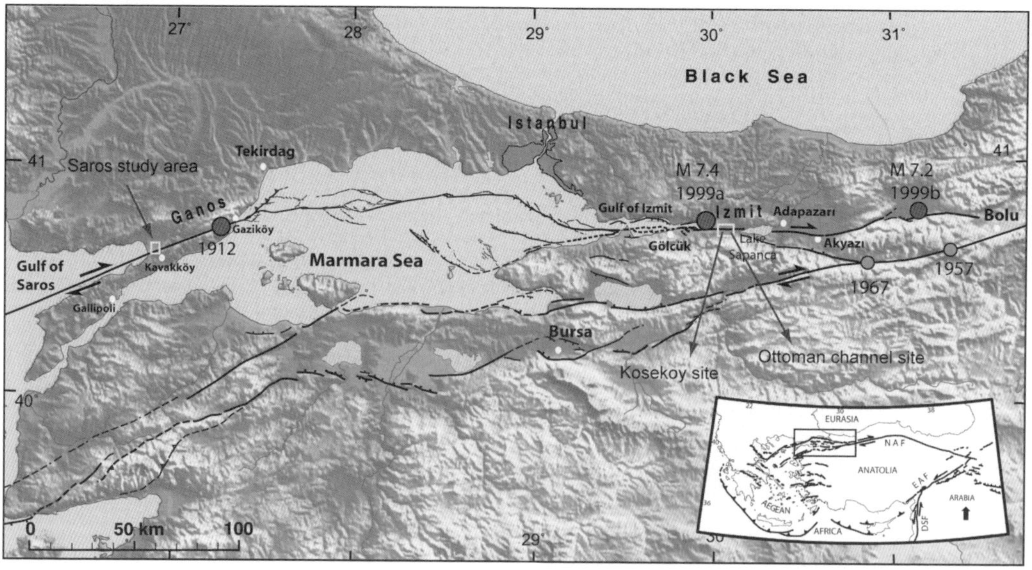

Fig. 1. Generalized map of active faults in NW Turkey, with recent large earthquakes. Note the three areas for this study relative to the Marmara Sea and Istanbul.

the NAF. Our observations support the contention that the NAF near Istanbul should be close to failure based on the distribution and size of earthquakes during the past 400 years, and that the recent earthquakes to the east (Izmit and Düzce events in 1999) have further loaded the fault segments beneath the Marmara Sea (Parsons *et al.* 2000).

The Saros site

Earlier palaeoseismic results at Kavakköy in the Gallipoli Peninsular region near the Gulf of Saros indicate that four earthquakes have ruptured the surface in that area during the past 1000–1200 years, and that two of these post-date a sand channel dated to younger than the fourteenth century AD (Rockwell *et al.* 2001). One of these is almost certainly the surface rupture of the 1912 M7.3 earthquake (Ambraseys 2002*a*), which was photographed east of our site towards Gaziköy (Ambraseys & Finkel 1987*a*) and has been studied in detail by Altunel *et al.* (2004).

We analysed low-altitude stereo photography of the fault along the westernmost few kilometres before it passes offshore into the Gulf of Saros (Fig. 2), to search for the best sites to resolve slip. In the area of T-1 from the Rockwell *et al.* (2001) study, we recognized the presence of an abandoned channel to the Kavak River that crosses the fault at a high angle, and is the likely source for the sand exposed in our original T-1 trench. At the fault, sediments in the abandoned channel appear in the aerial photography to splay out southward across the fault, suggesting that a low scarp is present where the fault crosses the channel. We hoped to expose buried elements of the channel system and resolve slip on buried piercing lines, to better constrain the age of these channel deposits, and to further resolve the

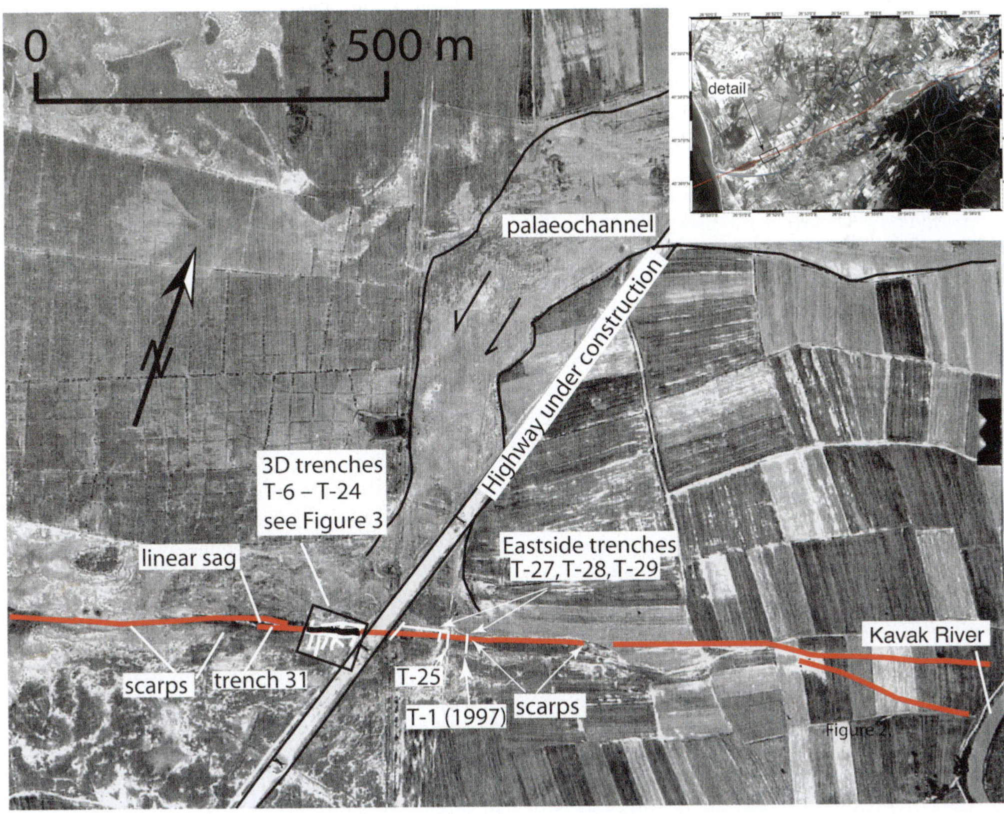

Fig. 2. Annotated aerial photograph of the Saros trench site. Note the Kavak River in the lower right corner of the photo. We trenched in an area that we interpret as a palaeochannel to the Kavak River where it crosses the scarp associated with the NAF. Most trenches are located west of the highway (now completed), with several trenches to the east. The highway was constructed near the thalweg of the main channel and was not open to trenching. The NAF enters the Gulf of Saros 2.6 km west of the highway.

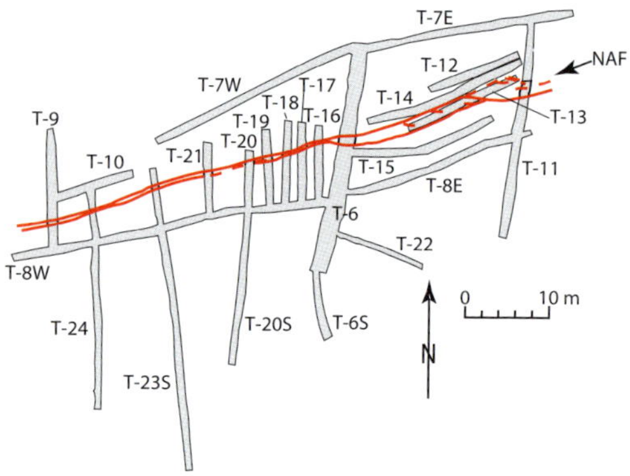

Fig. 3. Map of the 3D trenches on the west side of the highway that were used to resolve lateral slip. The locations of all trenches and fault strands were surveyed with a total station. See Figure 2 for location. The GPS coordinate of the intersection of trenches T-6 and T-8 is 40.6104°N, 26.8643°E.

number of earthquakes that affect the buried sand channel sediments. Towards this aim, we focused our efforts on the margins of the channel and excavated a total of 26 trenches across and parallel to the fault (Figs 2 and 3). The central part of the channel is now occupied by an elevated highway and berm and was no longer available for study.

Several of the trenches that contained important information on past surface ruptures were logged in detail on photographs and later entered into the computer in rectified form. Many of the trenches, however, were excavated for the sole purpose of tracing out the distribution of the distinctive well-sorted sand body (unit 200); data were collected on the thickness of the sand in these trenches but they were not logged in detail. All trench locations, including sand thickness control points, were surveyed with a Wild TC-2000 total station, and all have the same reference elevations established by surveyed horizontal string lines. Further, all critical contacts and relevant stratigraphic pinch-outs were surveyed to precisely locate them in 3D space.

Site stratigraphy

All of the trenches exposed a similar succession of young sediments, with or without the distinctively clean, well-sorted channelized sand. Figure 4 is a detailed log of the east face of trench T-6, and shows the typical stratigraphy of the site, with unit 200 being the distinctive sand that is only exposed on the south side of the fault in this trench. This sand is the same as Unit 3 described in trench T-1 by Rockwell *et al.* (2001), and was also exposed in new trenches near the original T-1 on the east side of the highway (Figs 2 and 5). We use the lateral distribution of this distinctive sand to constrain cumulative slip on the most recent surface ruptures. In this section, we briefly describe each of the primary units and the associated radiometric control on their ages.

We identified several primary units, along with dozens of secondary contacts, within the section exposed in the trenches. Units are given numeric designations ranging from 10 (topsoil; youngest) to 350 (oldest). Within a given trench, correlation of units both laterally along the trench and from wall to wall is fairly certain, and is based on the character of several distinctive strata contained within the section. Conversely, unit designations in the upper section of T-6 may be generally similar to those in the trenches east of the highway, but their correlation is by inference because we did not connect trenches between these sites (because of the highway). Thus, unit 100 at T-6 may not be exactly the same stratum as unit 100 in the East Saros trenches, although it is similar in character and falls in the same part of the section. The only unit which we feel confident to be the same in all exposures is the distinctive well-sorted sand of unit 200. Even this unit, however, may have some variance in age across the overall study area (few decades or less?) as the sand in the eastern group of trenches was associated with the main palaeomeander channel deposits, whereas the sand at T-6 was deposited by a secondary tributary overflow channel, also associated with the palaeomeander shown in Figure 2, as discussed later.

The deepest (oldest) logged stratum, unit 350, was exposed at a depth of about 1.5 m on the north side of the fault in T-6, just above the water table. Units 210 to 350 are generally fine-grained silt and clay strata interpreted as a succession of overbank deposits related to the main Kavak River. The only age control for this part of the section is from two dates on a split sample (T6-6) from near the base of this section that yielded consistent calibrated ages of about one to two centuries BC (Table 1). As this sample could have been reworked or have been resident in the system for some period of time, this represents a maximum age for this part of the section.

We use Unit 200 to constrain lateral slip. The sand is channelized and its distribution is locally restricted or absent. In our analysis of the aerial photography (Fig. 2), we interpret an abandoned or tributary channel to the Kavak River as the primary source for this sand. Within this palaeochannel system, we exposed the sand in most of the trenches and can make some general observations about its extent.

East of the highway, the sand fills a major, broad channel, and the sand is locally over 1 m thick. We excavated a trench parallel to the fault between T-25 and the highway, and the sand locally extended to

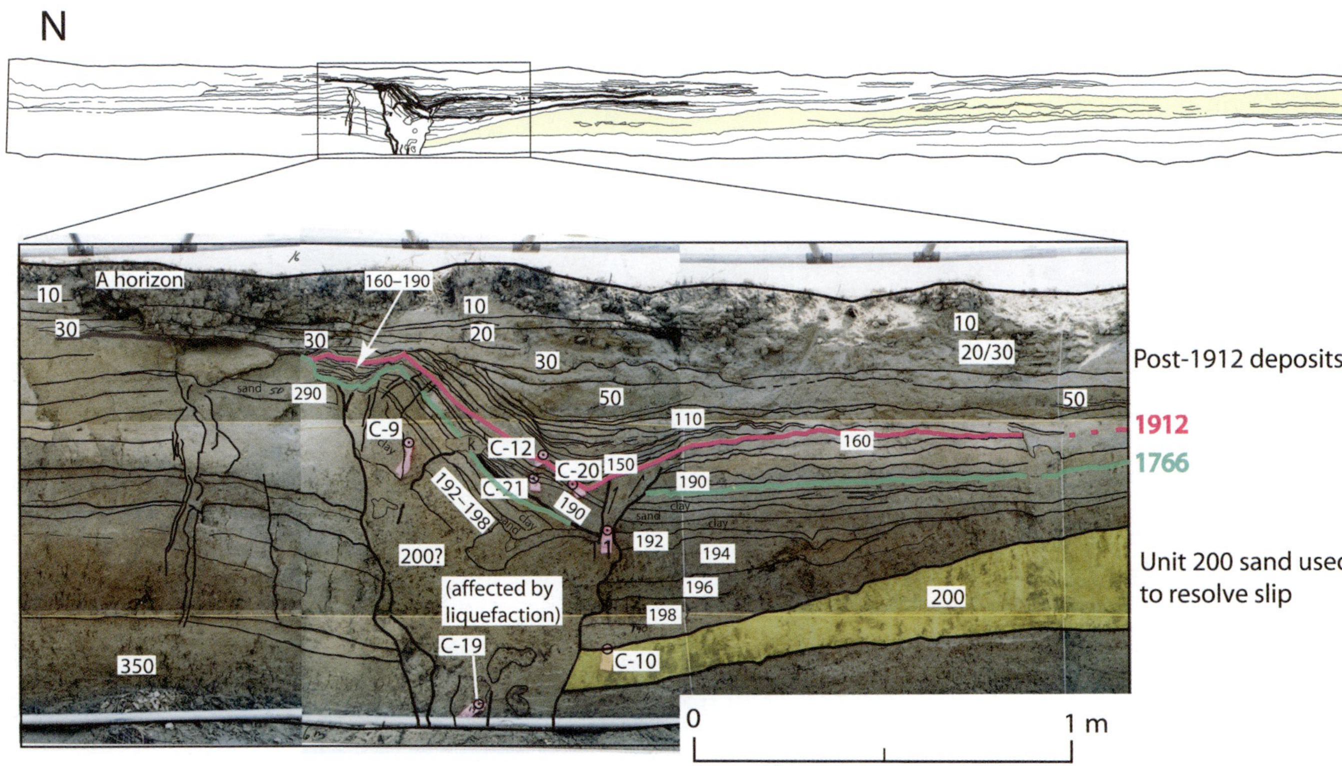

Fig. 4. Log of the east face of trench T-6. Units are described in the text. The labels 1912 and 1766 correspond to the thicker black contacts and represent the 1912 and 1766 event horizons, respectively. The unit 200 sand is the stippled grey unit in the top diagram.

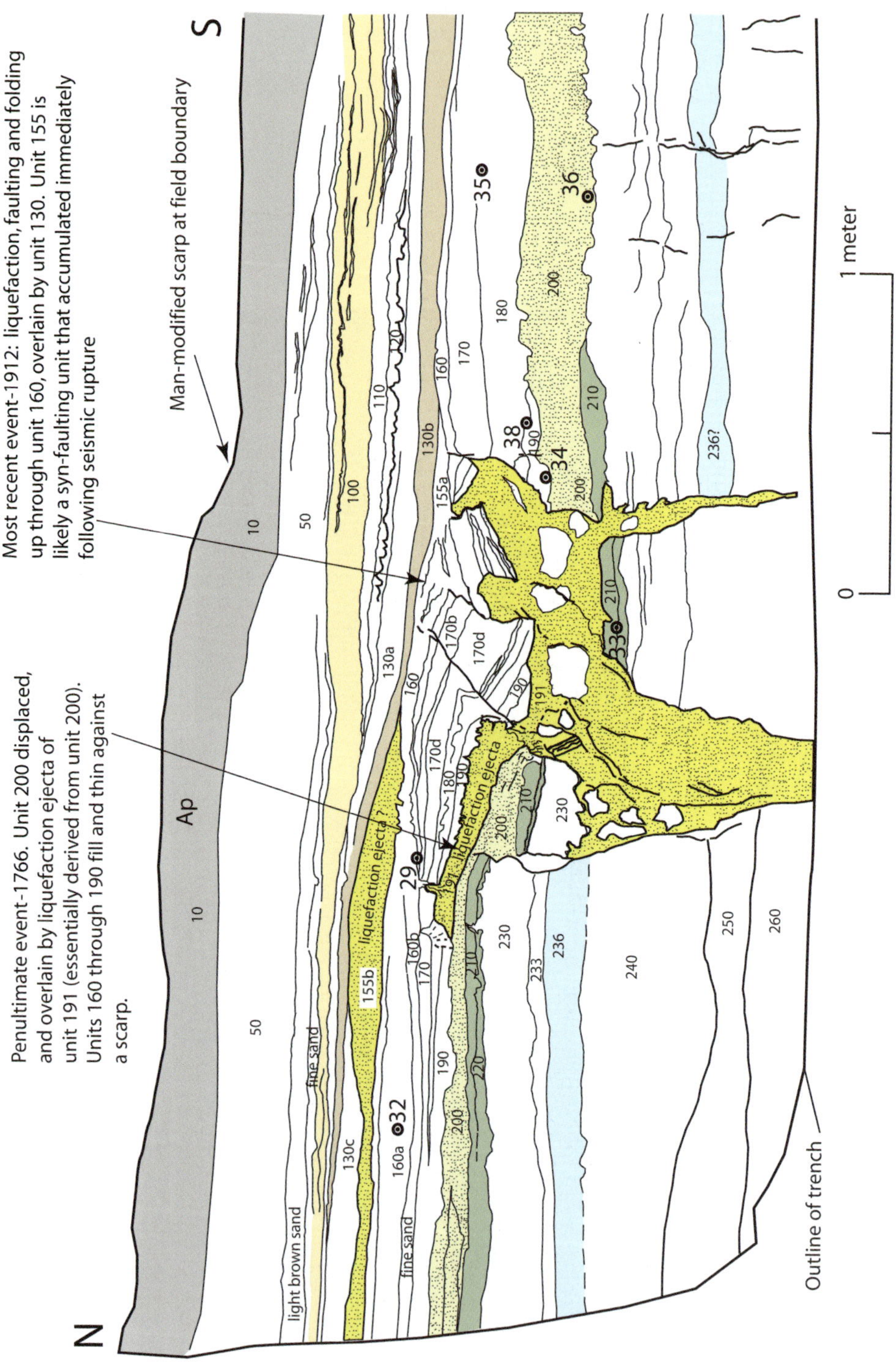

Fig. 5. Log of the east face of trench T-25, west side of the highway. Units are described in the text. Charcoal collection points are indicated as black dots.

Table 1. *Radiocarbon dates from trenches at Saros, the Ottoman Canal, and Köseköy*

CAMS #	Sample name	Trench exposure	Notes	Unit	$\delta^{13}C$	Fraction modern	±	$D^{14}C$	±	^{14}C age	±	Calibrated age range (2σ)
Kavakköy and Saros			**Trenches**									
67291	K-T14-48	Proxy to T6		92	−25	0.9556	0.0048	−44.4	4.8	360	50	AD 1451–1642
67290	K-T6-46	T6	0.15 mgC	95	−25	0.8972	0.0053	−102.8	5.3	870	50	AD 1041–1267
67285	K-T6-12	T6E		*c.* 99	−25	0.8401	0.0042	−159.9	4.2	1400	40	AD 592–703
67289	K-T14-45	Proxy to T6		102	−25	0.9675	0.0043	−32.5	4.3	270	40	AD 1489–1955
67288	K-T14-44	Proxy to T6		105	−24.3	0.9038	0.0043	−96.2	4.3	810	40	AD 1170–1285
67696	K-T13-32	East Saros, East wall		150	−25	0.9698	0.0048	−30.2	4.8	250	50	AD 1489–1955
67694	K-T13-29*	East Saros, East wall	0.12 mgC	base 160	−25	0.9232	0.0065	−76.8	6.5	640	60	AD 1282–1414
67698	K-T13-35	East Saros, East wall		top 180	−25	0.8855	0.0033	−114.5	3.3	980	40	AD 999–1164
68212	K-T13-41	East Saros, East wall	shell	180	0	0.8711	0.0043	−128.9	4.3	1110	40	AD 874–1016
67697	K-T13-34	East Saros, East wall		190	−25	0.9467	0.0048	−53.3	4.8	440	50	AD 1408–1631
67286	K-T6-43	T6W @ M9		200	−25	0.9777	0.0053	−22.3	5.3	180	50	AD 1655–1955
67284	K-T6-3	T6E		200	−25	0.9682	0.0042	−31.8	4.2	260	40	AD 1513–1955
67292	K-T6-43	T6W @ M9	0.04 mgC	200	−25	0.9931	0.016	−6.9	16	60	130	AD 1529–1955
67695	K-T13-30	East Saros, West wall		210	−25	0.9662	0.0048	−33.8	4.8	280	40	AD 1488–1955
67293	K-T6-6	T6E		310	−27.6	0.7638	0.0034	−236.2	3.4	2160	40	360–60 BC
67287	K-T6-6	T6E		310	−25	0.771	0.0034	−229	3.4	2090	40	190–2 BC
Ottoman Canal Berm		(dated at the Arizona AMS facity)										
AA33511	Berm 14C-3	Tepetarla Berm Trench			−25.6	0.9193	0.0055			675	50	AD 1274–1402
AA33512	Berm 14C-7	Tepetarla Berm Trench			−25.2	0.7815	0.0053			1980	55	112 BC–AD 191
AA33513	Berm 14C-8	Tepetarla Berm Trench			−24.7	0.7976	0.005			1815	50	AD 85–371
AA33640	Berm 14C-5	Tepetarla Berm Trench			−25.7	0.7985	0.0047			1805	45	AD 120–373
Köseköy Trench 1												
68683	T1-19	Above E2		3b	−25	0.9809	0.0043	−19.1	4.3	150	40	AD 1670–1955
68684	T1-6	Below E2		4	−25	0.9526	0.0043	−47.4	4.3	390	40	AD 1441–1634
68686	T1-27			5c	−25	0.976	0.0047	−24	4.7	200	40	AD 1647–1955
68685	T1-25			7	−25	0.9922	0.0048	−7.8	4.8	60	40	AD 1688–1927
68687	T1-23			10	−25	0.8183	0.004	−181.7	4	1610	40	AD 381–555
68688	T1-16			11	−25	0.8281	0.004	−171.9	4	1520	40	AD 444–630

below the depth of the 2-m-deep trench. This was an exploratory trench to determine the character of the sand and was not logged in detail because of safety constraints. Nevertheless, it was clear to us that the primary fluvial channel lies east of or beneath the highway, and our trench T-25 was located near the margin of this system. Trench T-1 of Rockwell *et al.* (2001), which lies within 10 m to the east of T-25, also exposed this sand unit (unit 3 in their trench log).

West of the highway, unit 200 is substantially more restricted in its areal extent. North of the fault, the well-sorted sand is restricted to a narrow and shallow (<20 cm) 'feeder' channel that flowed southward across the fault. South of the fault, the sand thickens dramatically and locally reaches over 40 cm in thickness. We collected observations on the sand thickness in all exposures to provide a basis for developing an isopach map of the sand (Fig. 6). We collected these observations at a maximum spacing of 50 cm and tied all measurements to a common, surveyed system of horizontal string lines. We also measured the absolute elevations of the top and bottom of the sand relative to the string datum to provide a complete spatial reference. These data are presented later to resolve slip on the channel fill of unit 200.

The age of unit 200 is constrained to be younger than AD 1655. We dated several samples from this unit (Table 1), along with those above and below it, and use the youngest date to constrain its maximum age. Several other dates from unit 200 also fall in the same time period, but because of larger uncertainties, can only be constrained to the period from about 1490–1530 to the present (1955). All of the samples constrain the sand to the past 500 years, but sample T6-43, recovered from stratified alluvium within the channel, can be no older than AD 1655 and may be considerably younger. This, of course, assumes that this sample is in stratigraphic context, but as the deposit is well-stratified and there was no evidence otherwise,

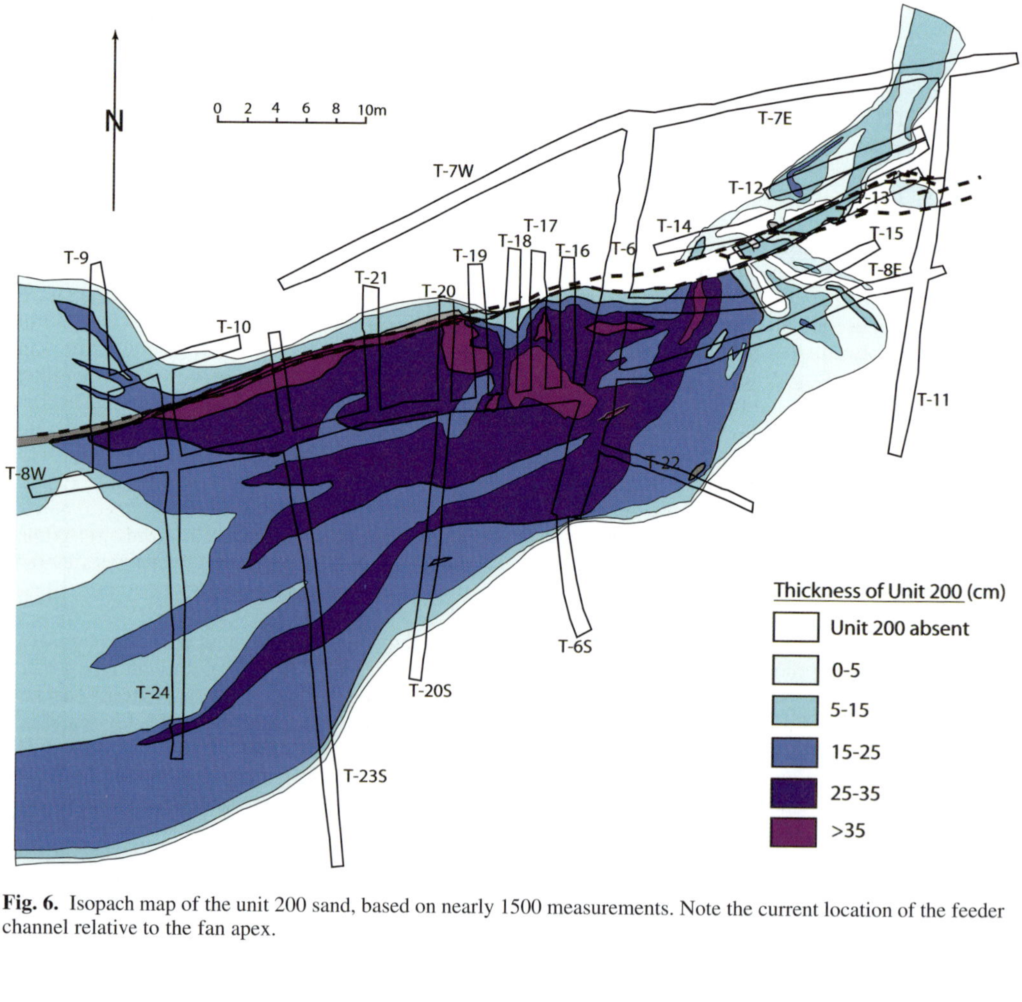

Fig. 6. Isopach map of the unit 200 sand, based on nearly 1500 measurements. Note the current location of the feeder channel relative to the fan apex.

we take this to be the case. Thus, we interpret all of the radiocarbon samples that lie above this unit and that have older apparent ages to be the result of either reworking of detrital charcoal or, more likely, the consequence of the charcoal (and its original wood) having been resident in the system for some period of time. The post-1655 date places strong constraints on which earthquakes may have produced surface rupture at this site and contributed to the observed amount of offset on unit 200, as discussed below.

The stratigraphic units above unit 200 can be designated as either sedimentation within the fault zone, possibly due to formation of a depression along the fault, or the result of overbank sedimentation by the Kavak River and its tributary channels. Units 192 to 198, recognized in the east face of T-6 (Fig. 4) and in adjacent trenches to the west, are interpreted to be a sequence of fine-grained deposits that fill a narrow trough between the unit 200 fan and a low fault scarp. Although this may be interpreted as the result of deformation along the fault, there is no direct evidence for a surface rupture in this part of the section. There is a reasonable interpretation, however, that invokes purely stratigraphic mechanisms to produce this deposit: the isopach map of the sand in Figure 6 clearly shows a fan that splays out and flows west, parallel to the fault, thereby producing a slight low along the fault. Thus, there is no need to require a faulting event to produce this low, and we prefer the non-faulting interpretation.

In contrast, units such as 110 to 150 in trench T-6 fill a depression that formed immediately after a surface rupture, as indicated by faulting and fissuring up to the base of that section. In these cases, the units may be only very locally preserved along the fault, although their significance to the interpretation of the event stratigraphy may be profound.

Units 10 to 100 are sandy to silty sediments that bury the fault scarp and are presumably derived from flood events from the Kavak River. Unit 10 is the A horizon developed in this uppermost section and is also the active plough pan in areas that are farmed, such as to the east of the highway. We dated a number of detrital charcoal samples from units 10 to 190 to provide preliminary constraints on the age of the overall section. Because unit 200 was found to be younger than AD 1655, all higher units must be as well. The radiocarbon results demonstrate a variety of dates ranging between about AD 600 and 1955 (i.e. the present), with no particular order in the section. We interpret all of these as having had a small to large component of resident age prior to their incorporation into the sediments exposed in our trenches. From this, we interpret the entire section from unit 200 to the surface as being deposited during the past 350 years or so. As this corresponds to the part of the C-14 calibration curve that cannot be resolved without very precise ages, we did not pursue further dating of this section.

Evidence for earthquakes

There was evidence observed for two large surface ruptures in nearly every trench that we excavated across the fault. In many of these, there were two deposits of well-sorted fine sand that appear to have been ejected out of the fault zone and derived from unit 200. In trench T-25 (Fig. 5), we designate the earlier of these ejecta deposits as unit 191. We also observed structural evidence for two surface ruptures, with faults and fractures extending up to a specific stratigraphic level and then being overlain by unfaulted deposits. We did not construct detailed trench logs of most exposures due to the lack of time and because our focus for many of these trenches was to map out the extent of unit 200. Nevertheless, both T-6 and T-25 record both of these events and are discussed herein.

Evidence from trench T-6. The fault in T-6 is narrow, less than 0.5 m at the base of the trench. Within this fault zone, faulting has produced liquefaction, brittle faulting, tilting, fissures, and a narrow trough into which sediment accumulated. The interpreted event horizons for each of these phenomena are coincident and correspond to the base of unit 190 and the base of unit 150.

In T-6E (east wall of trench T-6), fractures extend to the base of unit 190 and are overlain by well-bedded stratigraphy of units 160–190. Within the fault zone, a massive, well-sorted fine sand that is clearly affected by liquefaction is likely derived from unit 200, or possibly another sand below the base of the trench. This liquefaction sand is also overlain by unit 190 along both margins of the fault zone in this exposure. Finally, along the northern edge of the fault zone, a narrow depression is filled by finely laminated stratigraphy of units 160–190. We interpret the depression to be formed as a direct result of surface rupture.

The most recent event, 1912, is represented by rupture and liquefaction of units 190 up through 160 to the base of unit 150, and tilting of these units within the fault zone. Units 110 to 150 accummulated within a trough along the fault and are draped across the scarp. Unit 50 largely fills against the scarp, and units 10–30 were deposited after the scarp had been completely buried.

Evidence from trench T-25. East of the highway, we constructed detailed logs of the fault zone in trench T-25 to further constrain the timing and number of events that post-date the unit 200 sand. It should be noted that the earlier study by Rockwell *et al.* (2001) also found evidence for only two surface ruptures after deposition of their Unit 3, which is identical to unit 200 described herein. Trench T-25 lies within 10 m of Rockwell *et al.*'s (2001) trench T-1.

In the east wall of T-25, the northernmost fractures extend up through unit 200 and are overlain by another clean sand (unit 191) that we interpret as ejecta derived from unit 200. Massive clean sand fills the main fault and is also interpreted to be the result of liquefaction of unit 200. Overlying unit 191 is a sequence of bedded silt and sand units (units 160 to 190) that are not faulted by the northernmost strand of the zone. These observations all indicate a surface rupture that occurred when unit 200 was at the surface in this area.

Another set of fractures displaces all units up through 160, including unit 191 (liquefaction sand from the penultimate event). Chunks of bedded stratigraphy, composed of units 160 to 190, lie floating within the fault zone in a massive, fine sand matrix that we interpret as the result of re-liquefaction of the unit 200 sand. Unit 155a fills the trough in the fault zone and appears to be derived from unit 160, whereas unit 155b is a well-sorted sand that is identical to the unit 200 sand, has the form of a sand-blow, and is likely a second ejecta unit associated with the 1912 earthquake. All of these deposits are overlain by undeformed strata of units 100 to 130, which fill against the scarp, and units 10 to 50, which bury the scarp. The apparent scarp at the surface is evidently man-made.

The above observations indicate that there are two surface ruptures preserved in the stratigraphy, one at the contact at the base of unit 190 and one between deposition of units 130 and 155. These are identical to the relationships determined in trench T6 and we interpret these to be the same two events. Thus, at these and all other exposures that we examined in our field exercise, we noted evidence for two ruptures that produced liquefaction, surface faulting, and consequent sedimentation along the fault. Both of these events must have occurred after deposition of the channel sand of unit 200, or after AD 1655. The only two large events that may be ascribed to these surface ruptures are the large regional events of August 1766 and August 1912 (Ambraseys & Finkel 1987*a*, *b*, 1995; Ambraseys 2002*a*; Rockwell *et al.* 2001). Thus, we attribute lateral slip on unit 200 to be the cumulative result of these two earthquakes.

Determination of lateral slip

The channelized nature of unit 200 is ideal for resolving cumulative slip for the two events that post-date its deposition. From the aerial photographic analysis, it appeared that the palaeochannel containing unit 200 flowed at a high angle to the fault. We chose the area west of the highway to conduct the detailed 3D portion of this study because the area is devoid of agriculture and we were unrestricted in our ability to excavate long trenches both across and parallel to the fault zone.

In the preliminary excavations, such as trench T-6, we were not certain of the areal distribution of the unit 200 sand, so we began fault-parallel excavations to determine its extent. In all, we excavated 19 trenches and trench extensions to resolve the geometry of the unit 200 deposit. All exposures were surveyed with a Wild TC 2000 total station to provide accuracy. Further, a surveyed string line was emplaced in all trenches at the same elevation to assure accurate measurement of the sand thickness and the relative elevations of its top and base. We took over 1500 detailed measurements on the thickness of the sand, including the exact locations of the pinch-outs, to construct an isopach map of its distribution (Fig. 6).

Unit 200 is much thicker on the south side of the fault than on the north. We take this to indicate that a low scarp was present at the time the channel flowed across the fault. North of the fault, the sand is confined to a narrow channel and never exceeds 20 cm in thickness. The channel slopes to the south, towards the fault, and then thickens to over 40 cm where it crosses the fault. The overall form is that of an alluvial overflow fan, and we interpret the channel to have splayed at and beyond the fault scarp, resulting in deposition of the main fan on the south side. The fan is deflected downstream, towards the coast to the west, and its true SE side only extends south of the fault for 5–10 m. As seen in trench T-6, the fan has a convex-up profile in cross-section and is multi-lobed. The apex of the fan is exposed in trench T-15, and the fan rapidly thins to the east and pinches out in trenches T-8 and T-15. To the west, the fan is bounded on the north by the fault for a distance of about 30–35 m, and then the edge of the fan crosses the fault with 10–15 cm of deposition north of the fault. In the vicinity of T-6 and the area of the feeder channel, unit 200 is thickest away from the fault except at the fan apex.

We reconstruct the fan apex with the deepest portion of the feeder channel to resolve about 9 m of lateral slip (Fig. 7). A secondary smaller channel *c.* 8 m west of the main channel also reconstructs to a secondary fan apex, and the margins of the fan to the east of the feeder channel all realign. Furthermore, the thickest portion of the fan that ponded adjacent to the fault west of T-6 realigns to the thinner section of sand that spilled across the fault to the north, although this latter match is not tightly constrained. The uncertainty in the 9 m estimate is on the order of about 1 m, based on the

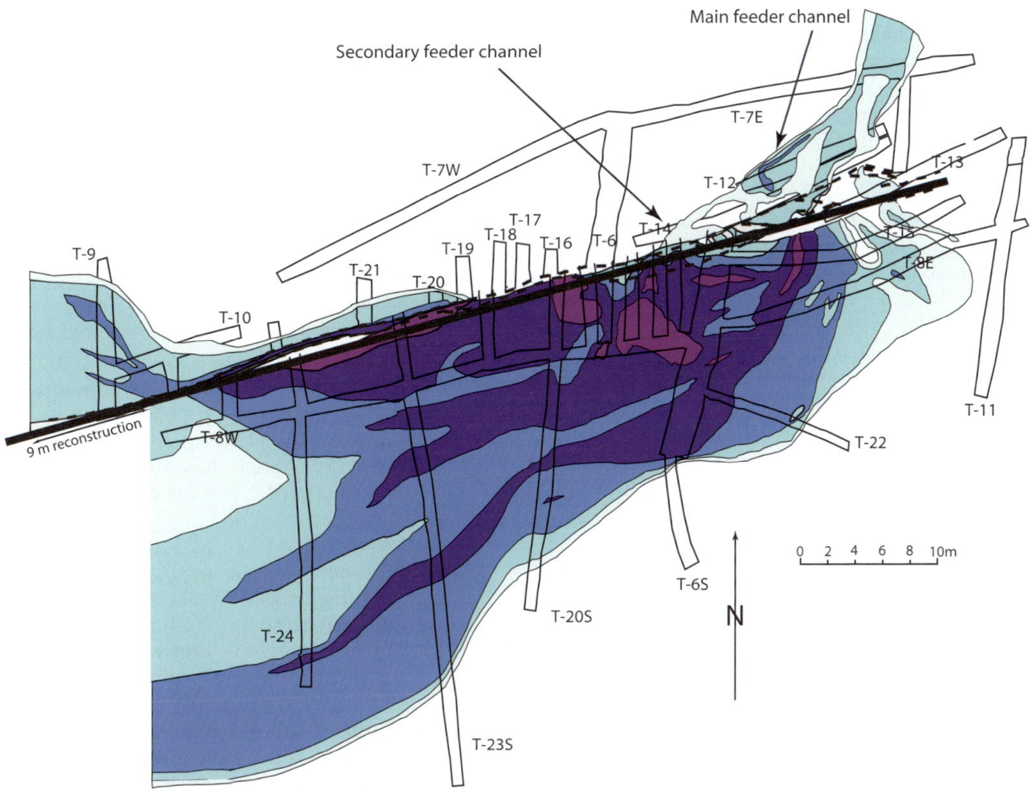

Fig. 7. Reconstruction of the fan apex with the thalweg of the feeder channel. Slip estimated at about 9 m, with about 1 m of uncertainty. Note that the eastern margin of the fan complex, along with a secondary feeder channel in T-14, also all realign with 9 m of reconstruction.

realignments and their mismatches if the reconstruction is less than 8 m or greater than 10 m. We take this 9 ± 1 m value as the cumulative slip produced by both the 1766 and 1912 earthquakes. If each earthquake produced similar slip at this site, then the *c.* 4–5 m of presumed slip agrees well with the estimated 1912 slip measured by Altunel *et al.* (2004) and Altınok *et al.* (2003) along the central and eastern onshore portions of the Ganos fault based on deflected roads and field boundaries, with their measurements located 15 to 50 km east of our study site. It is also similar to the average values for many of the large earthquakes that have ruptured the North Anatolia fault this century, and is consistent with the amount of slip predicted by Ambraseys & Finkel (1987*a*) for the 1912 earthquake, based on its inferred magnitude of about M7.4.

Timing of the past four events

The 1912 and 1766 earthquakes almost certainly produced the last two recognized surface ruptures in the Saros area. In our earlier study, we recognized evidence for two additional ruptures, but we did not have sufficient radiometric data to constrain their timing (Rockwell *et al.* 2001). In this paper, we combine all of the radiocarbon data and event evidence from both studies to better constrain the timing of the rupture history for the past 1000 years or so. For this task, we first compiled the radiocarbon dates from each trench that constrain the timing of the past four recognized events. There are clearly some dates that have substantial residency or inheritance, likely due to the burning of old wood. We therefore take only the dates from each section that did not result in stratigraphically inverted radiocarbon dates (Table 2).

We placed the radiocarbon dates in Table 2 in stratigraphic sequence in OxCal (Ramsey 2000) and analysed them for the inferred ages of the four events. Using 1912 as the most recent event and entering this as a calendar date in OxCal, the other three events are plotted as probability density functions (pdfs) of the likely age range at 2σ (Fig. 8). Event E2 plots in the appropriate timeframe for the 1766 earthquake, which we take as a

Table 2. *Radiocarbon dates used to constrain the timing of the past four Ganos fault ruptures at the Kavak–Saros site*

Sample name	Trench/unit	^{14}C date	Calibrated age range	Reference
K-T14-4S	T14/102	270 ± 40	AD 1489–1955	This study
Event E1 (1912)				
K-T13-32	T25/150	250 ± 50	AD 1489–1955	This study
Event 2 (1766?)				
K-T6-43 split	T6/200	60 ± 130	AD 1529–1955	This study
K-T6-43	T6/200	180 ± 50	AD 1655–1955	This study
K-T6-3	T6/200	260 ± 40	AD 1513–1955	This study
14C-4	T1/2b (base 200)	440 ± 60	AD 1446 (1405–1529 @ 0.73) (1542–1634 @ 0. 27)	Rockwell *et al.* (2001)
14C-5	T1/3b (*c.* 240)	530 ± 45	AD 1415 (1308–1358 @ 0.20) (1381–1451 @ 0.80)	Rockwell *et al.* (2001)
Event 3				
T5 14C-1	T5/G5	1010 ± 45	AD 1020 (965–1163)	Rockwell *et al.* (2001)
Event 4				
T5 14C-31	G5/H1	1290 ± 60	AD 755 (657–881)	Rockwell *et al.* (2001)
T5 14C-7	H1	1285 ± 45	AD 743 (664–825 @ 0.93) (833–865 @ 0.07)	Rockwell *et al.* (2001)
T5 14C-26	base H	1135 ± 45	AD 894 (791–1003)	Rockwell *et al.* (2001)

confirmation of that inference. Event E3 is more broadly constrained to the period from about AD 1000 to 1450, but fits the timeframe of the 1344 or 1354 earthquake very well. Event E4 is slightly better constrained in its timing, with a peak distribution around AD 900–1000 and a maximum calibrated range of about AD 750–1150. However, considering that the majority of radiocarbon dates displayed some residence age, either due to time in the system or burning of old wood, it is likely that the dates that constrain these older events also have some inheritance. If so, then the actual dates of the events may be younger, or at least fall in the younger portion of the distribution.

Ambraseys & Finkel (1987*b*) document large events in this region in 1063 and 1542, and a sequence in 1343–1354 (with 1344 and 1354 being to the west). The 1542 event was placed by Ambraseys & Finkel (1995) in the spurious category, and Ambraseys (2002*a*, *b*) totally ignores this event in later work, so we discount this as a likely candidate for one of the four events.

Based on the available radiocarbon data, we interpret E3 as likely being the 1344 or 1354 earthquake. The 1343–1354 sequence appears to have been a progressive rupture from the Marmara Sea to the Gulf of Saros (Ambraseys & Finkel 1991), with the 1354 being the farthest west and possibly the most destructive (and presumably the largest). Based on their tentative locations, however, E3 could correspond to either the 1344 or 1354 events. Event E4 is most likely the AD 1063 earthquake, as there are no other obvious candidates.

The average recurrence interval for ground ruptures at Saros is about 250–300 years if one takes four events in the past 1000–1200 years. A better way to frame recurrence is to take the intervals between earthquakes and take the mean, thereby also establishing the standard deviation. Using this approach, and taking the event ages as discussed, yields three intervals (146 years, 422 years (or 412 years), 281 years) with an average recurrence interval of *c.* 280 ± 110 years. Combined with the average slip per event of 4.5 m for the past two events, and assuming each of the four events displayed similarly large displacement, we estimate the slip rate at 15.9(+10/−4.5) mm/a. We emphasize, however, that this rate is strongly dependent on our assumption that the earlier two events produced similar displacements as the latter two, and this needs to be tested with future studies. Nevertheless, based on the historically recorded extent of damage, it is unlikely that the earlier events were larger than 1766 or 1912, so this rate is not likely to increase significantly.

Prior to the 1063 earthquake, there are potentially three large events reported for this area back to the fifth century AD. An event in AD 824 event is reported to have ‘devastated the Tekirdag coast ... causing considerable concern throughout Thrace’ (Ambraseys & Finkel 1987*b*), and although this may be a candidate, it may have occurred too far

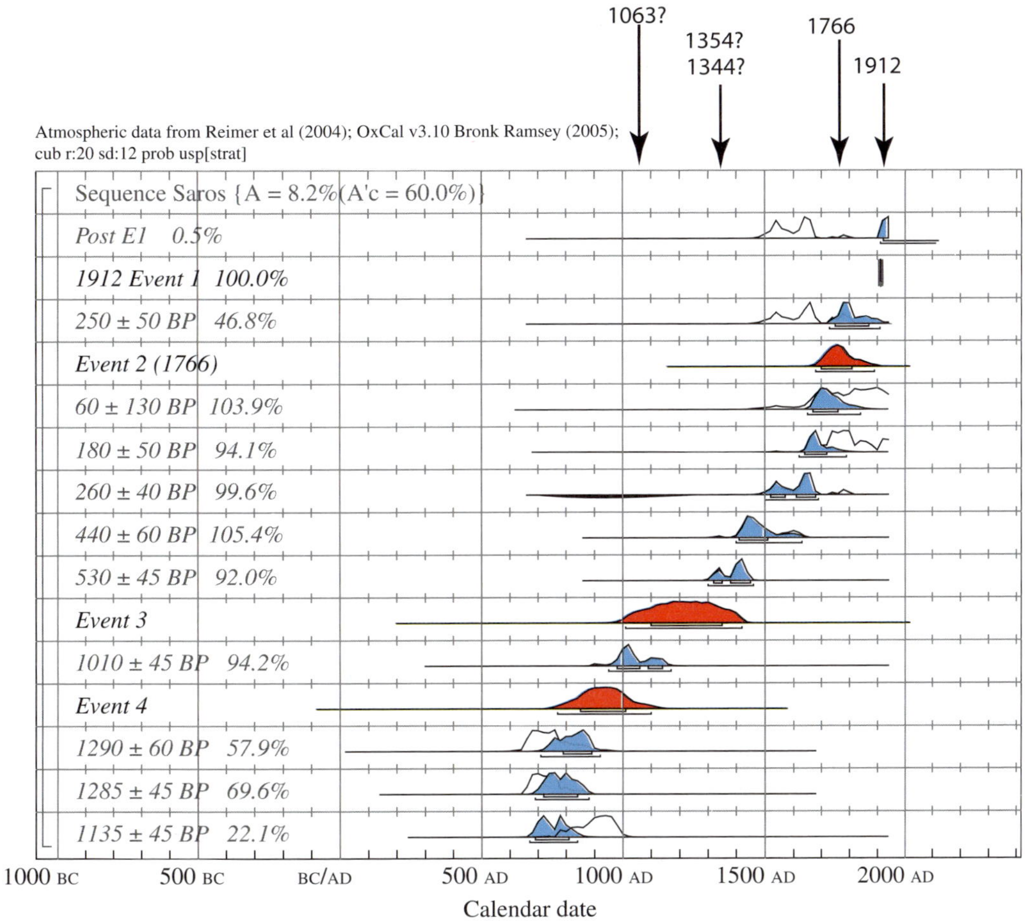

Fig. 8. OxCal plot of the probability density functions for selected (non-inverted) radiocarbon dates from the Saros (this study) and Kavak (Rockwell *et al.* 2001) studies. Note that the pdf for E2 is consistent with the 1766 earthquake. Also note that the earlier pdfs, although not well-constrained, are consistent with earthquakes in 1344 or 1354 and 1063.

to the east. Similarly, an earthquake in AD 542 is located by Ambraseys & Finkel (1991, plotted on their fig. 3) very close to their location for the 1354 earthquake near the head of the Gulf of Saros, but the damage appears to be more to the east in Constantinople (Istanbul). Ambraseys (2002*b*) later discounts this event, and suggests that this may be based on spurious reports. Finally, an event in AD 484 is reported to have been very destructive, with Gelibolu (Gallipoli) completely destroyed (Ambraseys & Finkel 1991).

If we include the 484 and the 824 earthquakes, the average recurrence interval becomes 284 $\pm$ 90 years, which is indistinguishable from the shorter-term recurrence interval. Using the average displacement of 1766 and 1912 yields a similar estimated rate of 15.8(+7.3/−3.8) mm/a for the past 1600 years if the average slip assumption of *c.* 4.5 m is used. These estimates assume that: (1) the past 1–1.6 ka is a long enough period to assess a long-term rate; (2) the historical record of large earthquakes is complete for this area which may not be the case before about AD 1000; (3) the rupture history is complete in our trenches, at least for the past 1000 years; and (4) each event has similar magnitude of displacement as we determined for the average in the past two events. This last assumption is important because we are basing our rate estimate on average displacement and average return period, not a dated feature that is offset a specific number of metres. Consequently, if some events have been very large, then we may underestimate the inferred rate using this method. However, we have likely overestimated the rate if some of these historical events were smaller than 1766 or 1912, which may have been the case for the 1344, 824 and 484 events. It is noteworthy that this rate is below the *c*. 24 mm/a (and possibly as high as 26–28 mm/a) measured by GPS for the NAF (McClusky *et al.* 2000; Reilinger *et al.* 2006), although the higher rate is allowed at the uppermost range of our estimate from the palaeoseismic data.

Palaeoseismology of the Izmit to Sapanca segment

We initiated palaeoseismic studies along the Izmit–Sapanca fault segment in October 1998 prior to the 1999 earthquakes. In our preliminary work, we focused on dating a faulted canal feature that we knew at the time to likely be the result of an effort by the Ottomans in 1591. After the earthquake, we returned to the canal to resolve how many earthquakes had affected the canal stratigraphy. We also began trenching west of the end of the canal along a small fluvial channel in the township of Köseköy with the purpose of resolving a longer record (Fig. 9). As will be shown, the records at both sites are similar and only record fault activity for the past 400 years or so.

The Ottoman canal site

Multiple periods of canal construction have been discussed in the literature (Finkel & Barka 1997), with at least two known excavation efforts. The earliest effort is pre-Roman and was intended to connect Lake Sapanca with Izmit Bay, thereby opening up commerce and access to the inland forests and other resources. A number of subsequent efforts were 'discussed' (mentioned in court records, etc.) although most of these were never undertaken. The most recent effort, for which there is direct historical documentation and known expenditures, was undertaken by the Ottomans in 1591 (Finkel & Barka 1997). In a preliminary effort in 1998, we excavated the southern margin of a large, abandoned canal at Tepetarla with the purpose of dating the construction of this prominent feature, which crosses the fault zone between Tepetarla and Köseköy. The canal extends from near Lake

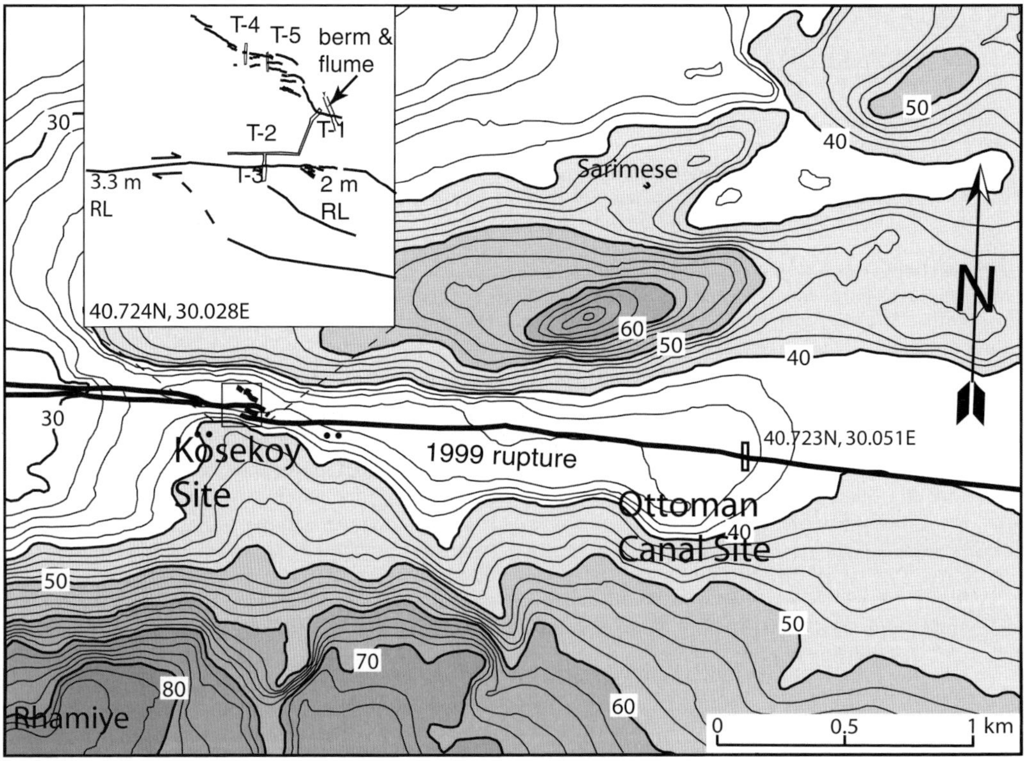

Fig. 9. Map of the Köseköy and Ottoman canal sites near Sarimeşe. The inset shows a detail of the Köseköy trenches whereas only a single trench was excavated at the canal site. The 1999 coseismic lateral displacements (Barka *et al.* 2002; Rockwell *et al.* 2002) along the fault at Köseköy are indicated in the inset.

Sapanca westward about half of the distance to Izmit Bay, consistent with historical accounts for the 1591 effort. The purpose of our initial trench was to resolve whether this was the Ottoman effort of 1591 or an earlier canal effort.

We found numerous pieces of small to large detrital charcoal, some that were associated with burn zones, that we interpret to represent cooking fires or fires to boil water for tea (as is a common practice today in Turkey). We collected eight large samples and dated four of the samples to place maximum ages on the berm construction, and therefore the canal excavation project. In that it is likely that the workers burned dead wood for their fires, and as there are numerous large trees in the area today, we surmise that all of the samples are likely older than the actual age of the berm. As it turns out, the ages of the detrital charcoal pieces range from a maximum age of 112 BC to as young as AD 1402 (Table 1). We therefore infer the berm and canal construction to post-date the youngest fourteenth century date, and to be the effort funded by the Ottomans in 1591. Thus, all of the alluvial fill within the canal must date to younger than 1591 so we did not attempt further C-14 dating in the canal fill.

In the summer of 2000, we excavated a trench across the 1999 rupture west of Lake Sapanca where the fault is entirely contained within the canal fill (Figs 1 and 9). The trench site was chosen about 2 m west of a small several-metre-long open extensional fissure resulting from a *c.* 2 m-wide releasing step-over. The trench exposed predominantly fine-grained, bedded clay-rich canal fill, although a distinctive sand was found to fill a fissure zone within the fault zone, apparently resulting from a prior rupture (Fig. 10). The fault zone is approximately 3 m wide in the trench, although the 1999 rupture zone is narrower.

The stratigraphy was differentiated into nine units, with the topmost and youngest unit (unit 1) interpreted as a plough pan. Unit 1 is a massive, organic-rich silty clay, similar in texture to several of the underlying strata. Unit 2, which is bedded and further subdivided into several subunits, is in part an alluvial fill within the fault zone. Unit 2a is a fine silty sand that is only present north of the 1999 rupture trace. Unit 2b is clayey silt that fills a

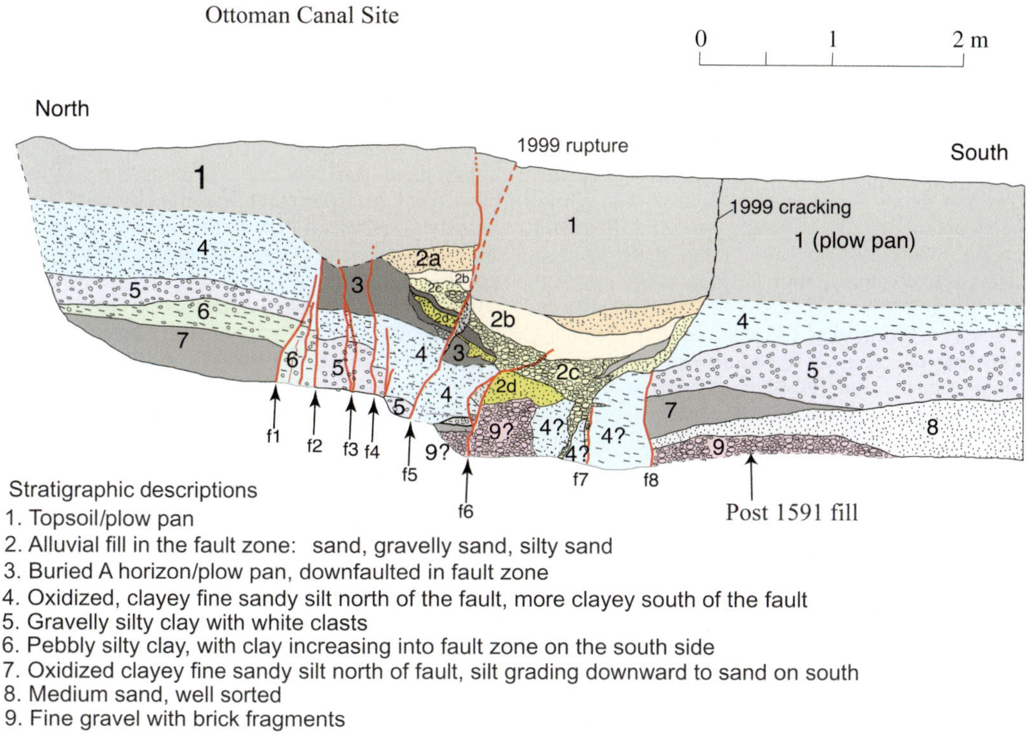

Fig. 10. Trench log of the Ottoman canal site near Tepetarla. The 1999 rupture produced slip in a narrow zone, with minor cracking over a 2 m width. All deposits exposed in the trench post-date the excavation age of the canal in AD 1591.

depression within the fault zone. In contrast, unit 2c is a coarse gravelly sand that not only fills the depression in the fault, but also extends downward in the fractures to the base of the trench. This unit grades upward into sand at its top. Unit 2d is well-sorted fine-grained sand, and we interpret this unit to be the result of liquefaction; it is likely derived from a clean sand below the base of the trench. Units 2a to 2d may constitute a fill sequence in a fissure within the fault zone following an earthquake.

Unit 3 appears to be an organic-rich, buried topsoil unit that was incorporated into the fault zone and is bounded by fault strands from an earlier event. Unit 4 is a massive, pebbly clay to clayey sandy silt (varies laterally) that grades down to the pebbly silty clay strata of units 5 and 6. These units are interpreted as quiet-water canal-fill alluvium, although the presence of scattered pebble clasts may alternatively suggest a debris flow origin. Unit 7 is an oxidized, finely bedded silt that grades downward to sand, whereas unit 8 is a well-sorted sand. Unit 9 is a sandy gravel of probable fluvial origin, and units 9 to 7 apparently represent a fining-upward fluvial sequence. The unit designated as '9?' within the fault zone is lithologically similar to both unit 9 and unit 2c and may be part of the section that liquefied or was mobilized during liquefaction of unit 9.

We did not directly date any of the strata within this trench using radiocarbon, although detrital charcoal was abundant and we collected over 50 samples from this trench. However, as the base of the canal was not encountered (the base should be greater than 5 m depth as the site is about 4–5 m above the elevation of Lake Sapanca), we infer the entire section to post-date AD 1591. In that many or most samples likely have some residence age (growth plus burning prior to burial), and because of problems with calibrating C-14 dates after about AD 1600, we did not see the utility in spending the effort to further date the section. Nevertheless, all earthquakes recorded at this site must also post-date AD 1591, which is fairly well-recorded in the history for this region.

Evidence for prior earthquakes

There is clear structural and stratigraphic evidence for at least one and probably two events prior to that of 1999. For discussion purposes, we have numbered the individual fault cracks as f1 to f8, from north to south. In a couple of cases, minor faults are grouped with more major ones. Several of these faults moved in at least one prior event, whereas only two appear to have been reactivated in 1999, which is referred to as event E1.

The 1999 (E1) rupture localized along a narrow crack, fault f5, near the centre of the trench. E1 displaces all strata up through unit 1 to the surface. A second surface crack aligns with fault f8, and this fault becomes the edge of a 2-m-wide pullapart that produced a narrow sag in 1999 only 2 m east of this trench face. In the trench wall, this fault appears to have only cracked, and no evidence of any significant displacement could be resolved. The base of the soil may be offset by a couple of centimetres, but that was not clear. Below the cracked soil, however, fault f8 defines the southern edge of the fault zone and was clearly active and a major fault in an earlier event. Along fault f5, the principal 1999 displacement surface, different units are juxtaposed and similar units have significant variations in thickness across the fault. We attribute these relationships to the *c.* 3 m of lateral slip recorded for the 1999 earthquake in this vicinity (Rockwell *et al.* 2002).

At least one earlier event, E3, is strongly indicated by the occurrence of a number of fault strands that break units 3–9 but are overlain by the unbroken soil of unit 1. Fault f2 drops an older topsoil horizon, unit 3, down against units 4 and 5, and the mismatch in unit thicknesses across this fault suggests substantial lateral slip. Faults f1, f3 and f4 also cut units 3 to 7 but are overlain by unit 1. There is no indication that any of these faults were activated in 1999. Faults f6, f7 and f8 also appear to have activated in this earlier event, which apparently resulted in an open fissure at the surface that was subsequently filled by unit 2. Unit 2d is a fine-grained sand that may be the result of a sand blow. Unit 2c, in contrast, is a gravelly sand that is intrusive downward into the fault zone and is very similar to the sandy gravel of unit 9. We observed gravelly sand mobilized during the 1999 earthquake, resulting from liquefaction and lateral spreading on the lake shoreline at Sapanca, so we surmise that this gravel may also be a consequence of liquefaction, in spite of its coarse grain size. In any case, the occurrence of the unit 2 deposits precisely within the fault zone virtually requires that an open fissure was present after deposition of unit 4 and the development of the soil of unit 3. We attribute the presence of the unit 2 deposits to this earlier faulting event.

A third event that is intermediate in age between events E1 and E3 is suggested by the breakage of the unit 2 fill by fault f6, which juxtaposes units 2d and 4, and causes a significant mismatch in the thickness of unit 2c. The upward termination of fault f6 appears to be in unit 2b, and we could not trace any evidence of this fracture upward to the modern ground surface. Based on the mismatch in stratigraphic thicknesses and the juxtaposition of dissimilar units, we infer that this fault must have significant lateral slip. The problem with a surface rupture interpretation at this stratigraphic level is that unit 2b apparently fills the depression left by event E3 and it is difficult to believe that the depression lasted for too long after the surface rupture. Thus, event E2 must have either occurred soon after event E3 or the site was closed to significant deposition for some extended period of time after event E3, which is possible considering the site's presence within the canal.

An alternative explanation is that fault f6 deformation is absorbed in the fine-grained fill of units 2b and 1 and that this fault was activated in 1999. Support for this idea is weak, but is based on the inference that unit 2b is only slightly younger in age than unit 2c, which appears to have directly resulted from the earlier event. However, the evidence for lateral slip along fault f6 is strong and would require a significant amount of strain being absorbed in unit 2b.

A third possibility is that unit 2c is a fluvial deposit that filled/eroded along the fault after event E3 and was faulted by both faults f6 and f7 during event E2. This possibility might be more attractive than the first explanation because it would permit more time between events E3 and E2. The fissuring of unit 2c downward into the fault could be explained by this mechanism and therefore would not require liquefaction of the sandy gravel of unit 2c.

We attribute event E3 to the large event of 1719 that had descriptions of damage that closely parallel those of 1999. It is the first large event after 1591 for this area and was apparently as large as 1999 (Ambraseys & Finkel 1995), consistent with the trench observations that indicate event E3 was a major surface rupture. Considering the case that event E2 occurred soon after E3, we attribute that deformation to either afterslip or possibly the 1754 earthquake, which is known to have produced damage in this region but for which the source is unknown. Later earthquakes, such as 1878 or 1894, appear to have occurred too long after event E3, and the surface soil of unit 1 would almost certainly have developed by that time.

The Köseköy site

The Köseköy site lies along a section of rupture in the SE corner of the township of Köseköy (Fig. 9), SE of Izmit and west of the end of the Ottoman canal. Rupture in this area in 1999 included about 2 m of slip along the primary fault strand, and secondary rupture along faults that splay off to the north from the main strand. The site is also in an area where the fault makes a small releasing step-over, and a strand to the south of our trench becomes the main strand farther east (Fig. 9). Rockwell *et al.* (2002) report offsets of surveyed trees adjacent to this site to be on the order of 1.8–2.25 m, consistent with slip values reported from rupture mapping after the earthquake (Barka *et al.* 2002). To both the east and west of the Köseköy area, slip along the Izmit–Sapanca rupture segment generally exceeded 3 m in 1999, and slip values as high as 3.8 m based on surveyed data were reported by Rockwell *et al.* (2002). We interpret this to mean that slip is distributed across multiple strands in the Köseköy area. We chose this site in part because the secondary faulting could be demonstrated to be principally dip-slip, making recognition and reconstructions of past earthquakes easier. Furthermore, palaeoearthquakes are often more easily recognized where multiple fault splays are present, as some secondary faults may rupture in only one or two events. Finally, part of our group conducted radar profiles at this site and identified a buried palaeochannel that is apparently right-laterally offset about 6.6 m, roughly three times the 1999 slip (Ferry *et al.* 2004).

We excavated five trenches at this site (Fig. 9), with only trench T-3 crossing the main, northern strand of the fault. The active drainage generally runs parallel to the fault zone and lies immediately south of the end of trench 3 by about 2 m. Hence, we could not extend this trench to cross the southern main strand, which had lost most of its slip in this area. Trench T-1 was excavated across a purely dip-slip fault that experienced about 60 cm of slip in 1999. A nearby small berm and associated concrete flume displayed no evidence of lateral slip (Fig. 9), thereby confirming the normal slip inference for the fault in trench T-1. Thus, no out-of-plane transfer of sediments is expected in the sediments of trench T-1 and past events should be recognized by their dip separations.

The stratigraphy in trenches T-1 and T-3 is composed of a sequence of coarse- and fine-grained strata that are interpreted as fluvial channel and overbank deposits (Figs 11 and 12). We physically traced most units between trenches T-1 and T-3 via a connecting fault-parallel trench (trench T-2, not logged in detail). Most units display some degree of lateral variability, becoming slightly coarser or finer from one trench to the other. Nevertheless, we believe that the unit designations apply consistently for both trench exposures.

Unit 1 is a dark brown plough pan (topsoil A horizon) unit that was tilled frequently. Units 2 and 3 are bedded channel deposits, with unit 2 being a distinctive sandy gravel in trench T-1, grading to a massive cobbly gravel in trench T-3. Similarly, unit 3 is a stratified coarse- to fine-grained sand in trench T-1 that grades to cobbly gravel in trench T-3. These units pinch out to the north across a palaeoscarp in trench T-1 (Fig. 12) and locally

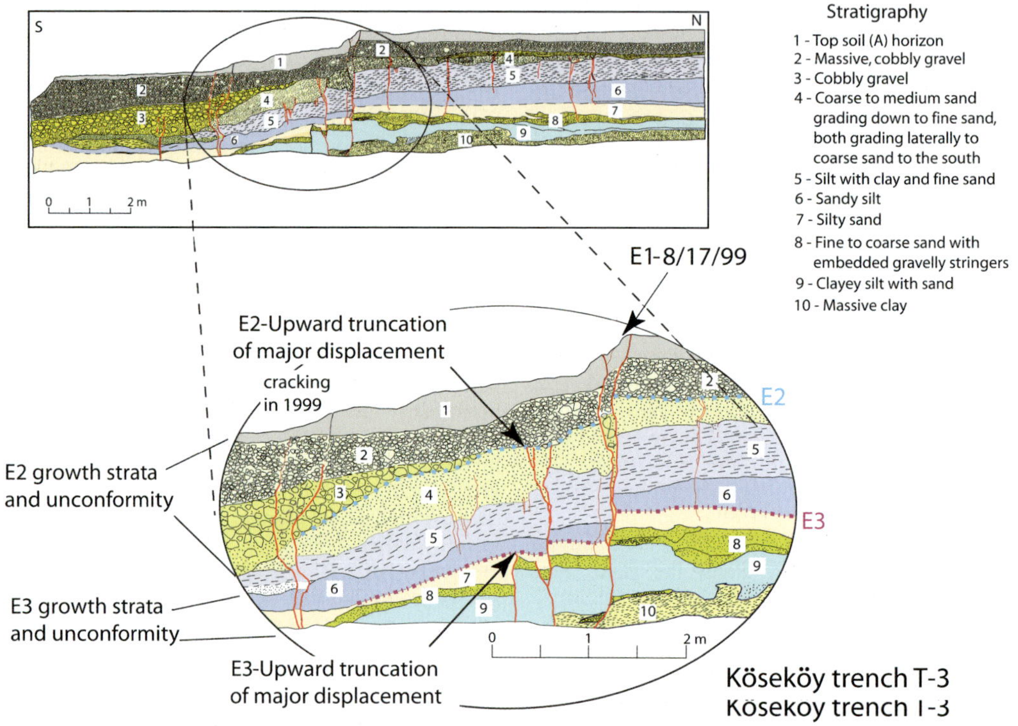

Fig. 11. Log of trench 3 at the Köseköy site. The 1999 Izmit surface rupture is indicated as E1, with two older ruptures interpreted as E2 and E3.

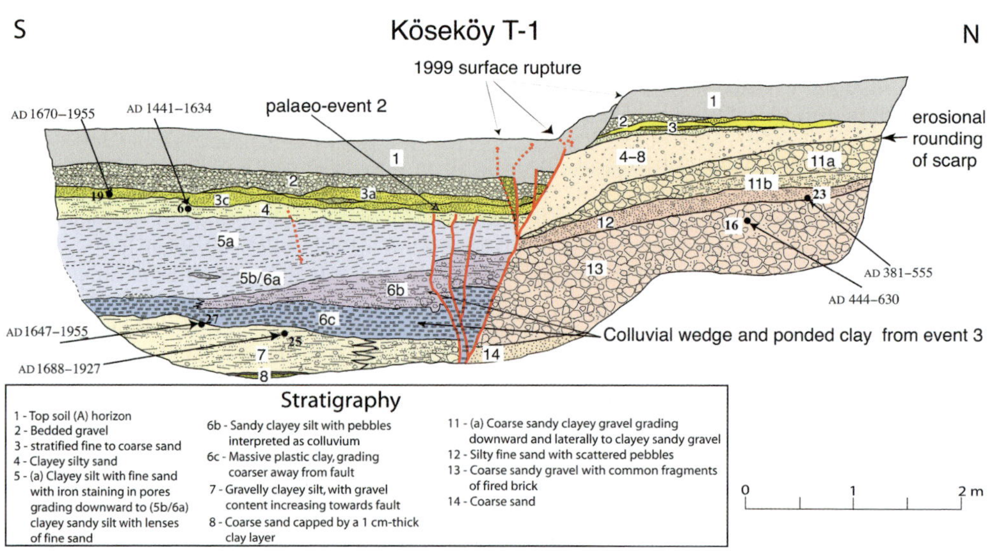

Fig. 12. Log of the Köseköy trench T-1. Unit designations stratigraphically match those given in Figure 11, hence there is no unit 9 or 10. Slip in 1999 was purely dip-slip, based on the lack of lateral offset of an adjacent flume and berm. Note the older fractures that did not rupture in 1999. Black dots with numbers are detrital charcoal sample locations for dates provided here and in Table 1.

scour into the underlying unit 4. We interpret this part of the section to represent a period of sedimentation prior to the current incision of a drainage located a few metres south of the southern end of trench T-3.

Unit 4 is a dark brown, massive, clayey silty sand (loam) that we interpret as a buried A horizon or topsoil in trench T-1, and this unit also coarsens towards the active drainage, becoming a coarse- to medium-grained sand in trench T-3. This unit was easily mappable south of the fault in T-1, but north of it, unit 4 appeared to become less distinct so we grouped it with underlying units as unit 4-8. Unit 5 is a weakly stratified clayey silt with sand that grades downward to a pebbly, sandy clayey silt. This unit is similar in all exposures and probably represents overbank sedimentation. North of the fault in trench T-1, the equivalent unit, unit 4-8 is a massive pebbly clayey silt that we interpret as predominantly of colluvial origin.

Unit 6 is a very distinctive plastic clay that was only exposed south of the fault in trench T-1 and appears to thicken towards the fault. We interpret the lower part of unit 6 as an overbank deposit, although it may be clay derived from overbank sedimentation ponded within the fault zone. The upper part of this unit thickens to the fault and may be a colluvial wedge shed from the scarp. Unit 6 coarsens southward to trench T-3, where it is of sandy silt composition.

Unit 7 is a gravelly clayey silt that appears very similar to unit 5 in trench T-1 but grades to a silty sand in trench T-3. In trench T-1, the gravel content is sparse away from the fault but increases towards the fault, suggesting a colluvial origin for part of this unit as well.

Finally, south of the fault in T-1, unit 8 is a distinctive coarse sand interpreted to be fluvial in origin. This sand was traced laterally towards trench T-3 and likely forms the channel deposit imaged by Ferry *et al.* (2004) in their radar survey. In trench T-3, the sand of unit 3 also contains stringers of fine gravel.

North of the fault, trench T-1 exposed several older units below unit 4–8. Unit 11 is a bedded coarse-grained sandy, clayey gravel interpreted to be of fluvial origin. There was no equivalent for this unit exposed south of the fault, although it is the probable source for the gravelly colluvium of unit 7. Unit 12 is a distinctive silty fine sand with scattered gravel, unit 13 is a coarsely bedded gravel containing abundant pottery and tile shards, and unit 14 is a distinctive coarse sand. Collectively, units 11 to 14 are interpreted as a fluvial sequence of strata preserved on the northern upthrown side of the fault zone.

Age control for the stratigraphy is provided by dating of individual detrital charcoal samples in trench T-1. Charcoal was abundant in our exposures, and we collected over 200 samples from the Köseköy site. In T-1 alone, nearly 40 samples were collected and we dated six. Four samples were dated from units 3 to 7 and two from the older units north of the fault.

All four samples from units 4–7 yielded modern or nearly modern results. The sample from unit 7 yielded a calibrated age of AD 1688–1927, requiring that all overlying units are also no older than 1688. Thus, most of the exposed section south of the fault was deposited during the past 335 years or so.

North of the fault, two samples were dated from units 10 and 11, with the lower sample from unit 11 yielding a calibrated date range of AD 444–630. As both sample ages are indistinguishable at 2σ, and as they both place the age of these units at about the fifth to sixth century AD, we accept these dates as the approximate age of this older fluvial section. Thus, there is a 1000-year-long hiatus in deposition on the northern side of the fault, although much of the record may be preserved at depth below the current base of T-1 to the south.

Interpretation of past earthquakes

We recognize evidence for three events recorded in the stratigraphy exposed in trenches T-1 and T-3. The most recent event is the August 1999 Izmit earthquake, and is designated as event E1. Earlier events are recognized in trench T-3 across the main fault by upward terminations of secondary splays and by tilting and growth strata. In trench T-1, earlier events are recognized on secondary splays and by the production of colluvial wedges that thin from the fault.

For instance, event E2 is indicated in the main zone by rupture of a fault splay up through unit 4, which is apparently capped by unit 2. However, unit 2 is gravel and it could be argued that slip on this fault strand is distributed in the gravel, so we rely more heavily on the dip-slip expressed from this event in trench T-1 (Fig. 13). There is apparently also tilting of units 4 and older below the inferred event E2 level in trench T-3, and there is a buttress unconformity and deposition of units 2 and 3 above this unconformity, all of which provide secondary evidence of event E2. In trench T-1, several secondary splays terminate upward at the top of unit 4, and units below this event horizon are thinner in the fault zone. This is especially evident in the reconstructions shown in Figure 13, where 1999 dip-slip was removed. In that the 1999 earthquake produced no lateral slip at trench T-1, we reconstruct the pre-1999 section assuming no out-of-plane motion (Fig. 13a), which required about 60 cm of dip reconstruction.

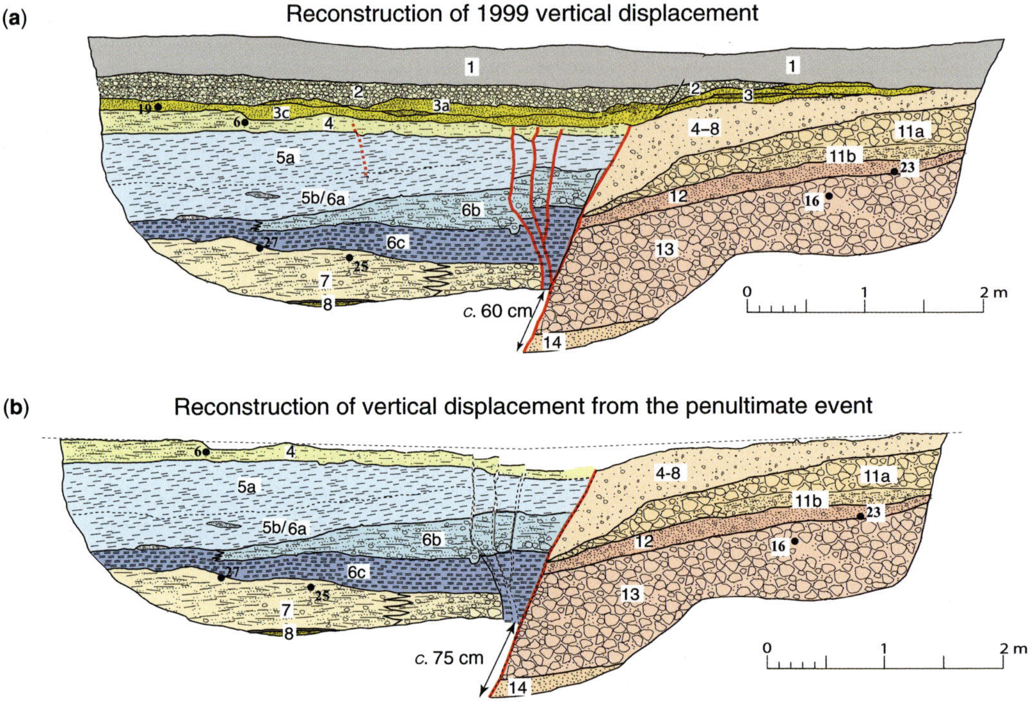

Fig. 13. Reconstruction of stratigraphy in trench T-1 to pre-1999 (**a**) and pre-penultimate event (**b**). Note the dip-slip was similar in each event. Refer to text for discussion of event interpretations.

In the log of trench T-1 in Figure 12, unit 3 pinches out north of the fault, indicating the presence of a scarp at that time. In the reconstruction of Figure 13a, these relationships are even more obvious, with units 2 and 3 pinching across a buried fault scarp produced by the penultimate event. Units 6 to 4 are faulted by the secondary strands and overlain by units 1–3. A degraded scarp that involves units 4 to 8 is also present. From these observations, we interpret event E2 to have occurred after the deposition of unit 4 and prior to the deposition of unit 3. Based on the ages of these units (Table 1), E2 must have occurred after AD 1688.

We attempted to remove the deformation of event E2 with further reconstruction of the units faulted in E2 (Fig. 13b). We reconstructed units 4 to 6 across the secondary strands, and restored or rematched units 4, 5 and 6 south of the fault to unit 4–8 north of it. This required an additional 15 cm of reconstruction, or about one-quarter of that required to remove the 1999 deformation. This reconstruction resulted in an apparent depression along the fault, part of which is explained by the uplift and subsequent erosion of unit 4 during and after event E2. The balance is explained by erosion of unit 4–8 on the north after the formation of the scarp. The volume of soil represented by this depression was evidently removed by erosion and subsequent deposition of the unit 3 channel deposits.

With the reconstruction shown in Figure 13b, we observe that unit 6c fills a depression in the fault zone (fissure fill?) that is capped by unit 6b. Unit 6b itself has a wedge-shaped geometry that we interpret as a colluvial wedge shed from the fault scarp. Based on this interpretation, event E3 occurred during deposition of unit 6. The amount of vertical slip for event E3 is similar to or more than that of 1999 if units 6b and 6c are deposited against the scarp formed in that event (the wedge thickness at the fault would be less than the actual amount of dip-slip during the causative earthquake).

There is also direct evidence for event E3 in trench T-3 across the main fault zone. Two fault splays about 1 m south of the main 1999 break rupture up through unit 7 and the base of unit 6, and are capped by additional unit 6 deposits. A radiocarbon sample recovered from unit 7 in trench T-1 post-dates AD 1688, indicating that all three of the recognized events occurred in the past 320 years.

Another observation that is consistent with the occurrence of three events after deposition of unit 8 is based on a radar survey conducted by Ferry *et al.* (2004). Prior to the trenching, they ran radar surveys parallel to the 1999 rupture and imaged a buried channel on each side of the fault. North of the main rupture, we encountered this channel in our trench 2. South of the fault, the interpreted correlative channel is laterally displaced about 6.7–7.4 m westward. We attempted to trench to the depth of the channel on the south side of the fault. However, saturated conditions and a collapsing trench wall precluded a direct look at this deposit south of the fault. If the radar correlation and estimate of slip is valid, approximately three times the amount of 1999 slip as occurred on the main rupture would be required to restore the lateral offset of the unit 8 channel. This is consistent with our inference that three events are required to explain all of the relationships observed in trench T-1, although not all were necessarily the same size.

Event E1 is the 1999 Izmit rupture. Events E2 and E3 must post-date 1688, and one is almost certainly the large 1719 earthquake that had a very similar damage distribution to 1999 (Ambraseys & Finkel 1995). Considering that 1719 is only a few decades after the bounding radiocarbon control, it is most likely that E3 corresponds to the 1719 surface rupture.

Our observations also suggest that vertical slip on the secondary normal fault in event E2 was smaller than that of 1999, although this may not reflect the amount of lateral displacement on the main fault. It is also possible that this secondary fault displays variable amounts of slip in each earthquake, possibly as a consequence of rupture direction. In any case, it is possible that event E2 was smaller than 1719 or 1999, and is possibly either 1754, 1878 or 1894. The 1754 source zone is not well known but is reported to have produced more damage in Istanbul (Ambraseys & Finkel 1995) than that which occurred in 1999. Damage from 1719 was reported from Istanbul to Bolu, whereas damage in 1754 was more strongly focused to the west. From those observations, 1754 was either larger or farther west than 1999. If the main shock of 1754 was in the Gulf of Izmit, as suggested by Ambraseys & Finkel (1991, 1995), then it is also possible that some slip propagated onto the Izmit–Sapanca segment that had ruptured only a few decades earlier. The 1878 event, in contrast, is believed to have occurred farther east than 1754 based on damage reports, but there is also much less known. Ambraseys & Finkel (1991) and Ambraseys (2000) report this as a locally destructive event between Eşme, Sapanca and Adapazarı, with damage extending to Akyazı, Izmit and Bursa. Taken at face value, the damage zone is considerably smaller than 1719 or 1754, although an east-directed rupture of smaller magnitude may be consistent with the 1878 damage effects. However, the damage zone for 1878 may be too small to be consistent with a major surface-rupturing event. In the final analysis, more historical reports on damage are needed to determine whether 1878 is a viable candidate earthquake for event E2.

The M7.3 1894 earthquake is also suggested to have ruptured from Sapanca to Izmit Bay based on historical reports of damage (Ambraseys 2001), and this earthquake would likely have resulted in surface rupture at both the Ottoman Canal and Köseköy sites if it indeed ruptured east of Izmit. (Klinger *et al.* 2003 did not see evidence for this event at their Gölcük trench site, however.) Ambraseys (2001) reports this as a large earthquake that was only slightly smaller than 1999. If correct, and if event E2 corresponds to this earthquake, then the observed smaller vertical displacement associated with E2 on the secondary strand probably reflects inherent variability in vertical displacement associated with secondary faults. Thus, there are three plausible candidates for event E2 between Izmit and Sapanca, with the potential size ranging from relatively small (1878) to relatively large (1894).

The record is too short to develop a meaningful slip rate at the Köseköy site. Nevertheless, the return period for large earthquakes in the past 350 years appears to be in the range of 140 years (1719, 1754 or 1878 or 1894, and 1999). However, based on the amount of displacement on the secondary fault exposed in T-1 at Köseköy, it is not clear if all events have similar displacement at the Köseköy trench site. In 1999, slip across the entire fault zone near our Köseköy trench site was measured from offset tree lines at *c.* 3–3.8 m (Rockwell *et al.* 2002), although we only measured about 2 m on the main strand at trench T-3. If Ferry *et al.* (2004) have correctly defined lateral displacement for all three events, then each may have been large in the Köseköy area and a rate as high as *c.* 24 mm/a is estimated, which is close to the GPS rate of 24 ± 1 mm/a (McClusky *et al.* 2000) to 28 ± 0.3 mm/a (Reilinger *et al.* 2006). Alternatively, if some events are smaller, such as suggested by the vertical component of event E2, then this likely represents a maximum rate for the past several hundred years.

Discussion of results

There are three primary conclusions from our work. First, it appears that 7–9 m of slip has accrued on the NAF both east and west of the Marmara Sea in the past 300 years. During this period, only the

April 1766 earthquake has released significant stored strain in the central Marmara itself (Ambraseys & Jackson 2000), leading to the conclusion that the Marmara is ripe for a large earthquake, consistent with interpretations by Parsons *et al.* (2000). Armijo *et al.* (2005) suggest that the 1912 earthquake may have ruptured well eastward into the Marmara based on the freshness of scarps on the sea floor. However, it is not clear how long such scarps can be preserved at the bottom of these deep basins and some of these may have survived since 1766. Further, the isoseisms for 1912 are much stronger to the west, with less damage at Tekirdag (Fig. 1) than farther west. Damage observations argue that either the 1912 rupture terminated well before the central Marmara or that the rupture propagated westward, resulting in substantial directivity of energy and concentration of damage on the Gallipoli peninsula. In either case, the central Marmara segment closest to Istanbul has apparently not failed since 1766 (or possibly 1509; Ambraseys & Jackson 2000), so hazard is high when the behaviour of adjacent segments is considered.

A second general observation is that the NAF slip rate over the past millennium is apparently lower than that indicated by GPS west of the Marmara Sea. At Saros, we determined a return period of about 283 ± 113 years, which, when combined with average displacement for the past two surface ruptures (and assuming that this average displacement is applicable for earlier events), yields a millennial rate of *c.* 16 mm/a. Reliinger *et al.* (2006) interpret GPS data to represent 26–28 mm/a of loading in the Gallipoli peninsular region, which is considerably higher than what has apparently been released in the past thousand years or so. There are three obvious explanations for this apparent discrepancy. First, it is likely that we do not record the full amount of displacement in our trenches. We determined 3D slip at only one site whereas slip can vary substantially along a rupture, as demonstrated by Rockwell *et al.* (2002) for the 1999 Izmit earthquake. In their study, they found an average of 15% distributed near-field deformation represented by warping into the fault. As this deformation extends out for several metres in areas of deep alluvium, and such is likely the case for the Saros study site, we may have determined only the minimum average displacement for the 1912 and 1766 earthquakes. Further, slip varies laterally along a rupture by 20% or more over relatively short distances, so we may have trenched in a low slip area. However, to increase displacement to that required to satisfy the GPS rate would require that the average displacement per event increase to about 6.7 m. Although this is possible, Altunel *et al.* (2004) suggest that the average far-field value for the 1912 surface rupture is closer to 4–5 m, consistent with our trench results.

For a similar length of record, Klinger *et al.* (2003) studied the rupture history of the NAF at Gölcük, where they determined that the fault had ruptured only three times since the fifteenth century. They attributed the surface ruptures described in their trenches to the large events of 1509, 1719 and 1999, which suggests a return period for large earthquakes for that segment of the NAF of about 250 years. Maximum slip at Gölcük in 1999 was about 4.5–5 m (Barka *et al.* 2002), which when combined with the return period, suggests a short-term rate of 18–20 mm/a. Farther east, along the 1944 rupture, Okumura *et al.* (2004), Rockwell *et al.* (2004) and Kondo *et al.* (2004) have determined about 25 m of slip for the past 1500 years, which also yields a rate of *c.* 16 mm/a. In summary, the NAF appears to have a surface slip rate for the past 500–1200 years, and possibly longer, that is significantly lower than that interpreted from GPS data, and this observation appears to be true both east and west of the Marmara Sea. It is possible that the past 1000 to 1500 years is too short an interval to assess the rate, and that strain release is variable over the several-thousand-year timeframe. In this scenario, the past thousand years have simply seen a relative lull in seismic slip. Alternatively, the GPS rate may be too high, at least in terms of slip on the main NAF. This can be explained by either a lower long-term GPS rate with the current observations affected by the >1000 km sequence of ruptures this past century, or by more broadly distributed slip across the NAF zone, with secondary faults accommodating up to 20% of the deformation.

A third conclusion from our trench study between Izmit and Sapanca is that this segment apparently ruptures more frequently than the section to the west (Klinger *et al.* 2003). Slip in 1999 was lower, averaging about 3–3.5 m near Izmit versus 4–5 m to the east and west (Barka *et al.* 2002). This *c.* 30-km-long segment is bound by relatively small step-overs at Lake Sapanca (1–2 km) and Gölcük (1–2 km) (Lettis *et al.* 2002), and thus appears to be weaker than the adjacent sections and can apparently fail on its own as indicated by the shorter return period and smaller amount of slip inferred for event E2. If E2 was, in fact, the result of the 1754 rupture, then it appears that re-rupture of this segment can occur after only a few decades. This is contrary to the general idea that the Izmit to Sapanca area is safe from a near-future rupture because it just sustained slip in 1999. In contrast, if E2 was 1878 or 1894, then the Izmit–Sapanca segment appears to have more periodic behaviour. Future work along this segment should concentrate on better dating of the penultimate rupture, as there are clear implications for future hazard and rupture behaviour.

This work was funded by the USGS NEHRP programme, award no. 00HQGR0034, and by Pacific Gas and Electric Company (PG & E). We are indebted to the many landowners who allowed access, the mayors of Kavakköy and Köseköy for their support, and to Lloyd Cluff and Woody Savage of PG & E who supported the field effort, allowing the students from SDSU and ITU to be involved in this work. We sincerely appreciate the reviews by Drs James Dolan and Alessandro Michetti who contributed to improvements in the presentation of this paper. Finally, none of this work would have been possible without Aykut Barka. Although a coauthor, he deserves special attention and thanks, and his death in February 2002 was a profound loss to the earthquake science community in Turkey as well as the world. He will be sorely missed.

References

ALTINOK, Y., ALPAR, B. & YALTIRAK, C. 2003. Şarköy–Mürefte 1912 Earthquake's Tsunami, extension of the associated faulting in the Marmara Sea, Turkey. *Journal of Seismology*, **7**(3), 329–346.

ALTUNEL, E., MEGHRAOUI, M., AKYUZ, S. & DIKBAS, A. 2004. Characteristics of the 1912 co-seismic rupture along the North Anatolian Fault Zone (Turkey): Implications for the expected Marmara earthquake. *Terra Nova*, **16**(4), 198–204.

AMBRASEYS, N. N. 2000. The seismicity of the Marmara Sea area, 1800–1899. *Journal of Earthquake Engineering*, **4**(3), 377–401.

AMBRASEYS, N. 2001. The earthquake of 10 July 1894 in the Gulf of Izmit (Turkey) and its relation to the earthquake of 17 August 1999. *Journal of Seismology*, **5**, 117–128.

AMBRASEYS, N. N. 2002*a*. The seismic activity of the Marmara Sea region over the last 2000 years. *Bulletin of Seismological Society of America*, **92**(1), 1–18.

AMBRASEYS, N. N. 2002*b*. Seismic sea-waves in the Marmara Sea region during the last 20 centuries. *Journal of Seismology*, **6**(4), 571–578.

AMBRASEYS, N. N. & FINKEL, C. F. 1987*a*. The Saros-Marmara earthquake of 9 August, 1912. *Earthquake Engineering and Structural Dynamics*, **15**, 189–211.

AMBRASEYS, N. N. & FINKEL, C. F. 1987*b*. Seismicity of Turkey and neighboring regions, 1899–1915. *Annales Geophysicae*, **5B**, 701–726.

AMBRASEYS, N. N. & FINKEL, C. F. 1991. Long-term seismicity of Istanbul and the Marmara Sea region. *Terra Nova*, **3**, 527–539.

AMBRASEYS, N. N. & FINKEL, C. F. 1995. *The Seismicity of Turkey and Adjacent Areas, A Historical Review, 1500–1800*. Eren Yayıncılık, Istanbul.

AMBRASEYS, N. N. & JACKSON, J. A. 2000. Seismicity of the Sea of Marmara (Turkey) since 1500. *Geophysical Journal International*, **141**, F1–F6.

ARMIJO, R., PONDARD, N. *ET AL.* 2005. Submarine fault scarps in the Sea of Marmara pull-apart (North Anatolian Fault): Implications for seismic hazard in Istanbul. *Geochemistry, Geophysics, Geosystems*, **6**, 1–29.

BARKA, A. A. 1992. The North Anatolian fault zone. *Annales Tectonicae*, **6**, 164–195.

BARKA, A., AKYUZ, H. S. *ET AL.* 2002. The surface rupture and slip distribution of the 17 August 1999 Izmit earthquake (M7.4), North Anatolian fault. *Bulletin of the Seismological Society of America* (Special Issue on the 1999 Izmit and Düzce, Turkey, Earthquakes, N. Toksoz (ed.)), **92**(1), 43–60.

BRONK RAMSEY, C. 2005. OxCal Program v3.10. www.rlaha.ox.ac.uk/oxcal/oxcal.htm.

FERRY, M., MEGHRAOUI, M., GIRAD, J.-F., ROCKWELL, T. K., KOZACI, A., AKYUZ, S. & BARKA, A. 2004. Ground-penetrating radar investigations along the North Anatolian fault near Izmit, Turkey: Constraints on the right-lateral movement and slip history. *Geology*, **32**(1), 85–88.

FINKEL, C. & BARKA, A. 1997. The Sakarya River-Lake Sapanca-Izmit Bay canal project: A reappraisal of the historical record in the light of new morphological evidence. *Istanbuler Mitteilungen*, **47**, 429–442.

KLINGER, Y., SIEH, K. *ET AL.* 2003. Paleoseismic evidence of characteristic slip on the western segment of the North Anatolian fault, Turkey. *Bulletin of the Seismological Society of America*, **93**(6), 2317–2332.

KONDO, H., OZAKSOY, V., YILDRIM, C., AWATA, Y., EMRE, O. & OKUMURA, K. 2004. 3D trenching survey at Demir Tepe site on the 1944 earthquake rupture, North Anatolian fault system, Turkey. *Annual Report on Active Fault and Paleoearthquake Research*, **4**, 231–242.

LETTIS, W., BACHHUBER, J., WITTER, R., BRANKMAN, C., RANDOLPH, C. E., BARKA, A., PAGE, W. D. & KAYA, A. 2002. Influence of releasing step-overs on surface rupture and fault segmentation: Examples from the 17 August 1999 Izmit earthquake on the North Anatolian fault, Turkey. *Bulletin of the Seismological Society of America*, **92**(1), 19–42.

MCCLUSKY, S., BALASSANIAN, S. *ET AL.* 2000. Global positioning system constraints on plate kinematics and dynamics in the eastern Mediterranean and Caucasus. *Journal of Geophysical Research*, **105**, 5695–5719.

OKUMURA, K., KONDO, H. *ET AL.* 2004. *Slip history of the 1944 Segment of the North Anatolian Fault to Quantify Irregularity of the Recurrence*. Geological Society of America Annual Meeting, **04-103**.

PARSONS, T., TODA, S., STEIN, R., BARKA, A. & DIETRICH, J. H. 2000. Heightened odds of large earthquakes near Istanbul: An interaction-based probability calculation. *Science*, **288**, 661–665.

RAMSEY, C. B. 2000. *OxCal Program Ver. 3.5*. Radiocarbon Accelerator Unit, University of Oxford. Available at: http://units.ox.ac.uk/departments/rlaha/orau/06_01.htm

REILINGER, R. E., MCCKUSKY, S. C. *ET AL.* 1997. Global positioning system measurements of present-day crustal movements in the Arabian-African-Eurasian plate collision zone. *Journal of Geophysical Research*, **102**, 9983–9999.

REILINGER, R., MCCLUSKY, S., VERNANT, P. *ET AL.* 2006. GPS constraints on continental deformation in the Africa-Arabia-Eurasia continental collision zone and implications for the dynamics of plate interactions. *Journal of Geophysical Research*, **111**, B05411, doi:10.1029/2005JB004051.

REIMER, P. J., BAILLIE, M. G. L., BARD, E. *ET AL.* 2004. IntCal04 Terrestrial Radiocarbon Age Calibration, 0–26 cal kyr BP. *Radiocarbon*, **46**, 1029–1058.

ROCKWELL, T., BARKA, A., DAWSON, T., THORUP, K. & AKYUZ, S. 2001. Paleoseismology of the Gaziköy-Saros segment of the North Anatolia fault, northwestern Turkey: Comparison of the historical and paleoseismic records, implications of regional seismic hazard, and models of earthquake recurrence. *Journal of Seismology*, **5**(3), 433–448.

ROCKWELL, T. K., LINDVALL, S., DAWSON, T., LANGRIDGE, R., LETTIS, W. & KLINGER, Y. 2002. Lateral offsets on surveyed cultural features resulting from the 1999 Izmit and Düzce earthquakes, Turkey. *Bulletin of the Seismological Society of America*, **92**(1), 79–94.

ROCKWELL, T. K., OKUMURA, K. *ET AL.* 2004. Paleoseismology of the 1912, 1944 and 1999 ruptures on the North Anatolian Fault: implications for long-term patterns of strain release. *Geological Society of America, Abstracts with Programs*, **36**(5), 51.

SIEH, K. 1996. The repetition of large-earthquake ruptures. *Proceedings of the National Academy of Sciences*, **93**, 3764–3771.

STRAUB, C. & KAHLE, H. 1995. Active crustal deformation in the Marmara Sea region, NW Anatolia, inferred from GPS measurements. *Geophysical Research Letters*, **22**(18), 2533–2536.

STRAUB, C. S. 1996. *Recent crustal deformation and strain accumulation in the Marmara Sea region, N.W. Anatolia, inferred from GPS measurements.* PhD dissertation, Swiss Federal Institute of Technology at Zurich.

Application of INQUA Environmental Seismic Intensity Scale to recent earthquakes in Japan and Taiwan

YOKO OTA[1]*, TAKASHI AZUMA[2] & YU-NONG NINA LIN[3]

[1]*Yokohama National University, Japan*

[2]*Active Fault Research Center, AIST, Japan*

[3]*Institute of Geosciences, National Taiwan University, Taiwan*

**Corresponding author (e-mail: yokoota@ntu.edu.tw)*

Abstract: The INQUA Environmental Seismic intensity scale (ESI 2007 scale) is a new seismic intensity scale proposed by the Subcommission on Palaeoseismology, INQUA, based on seismically induced ground effects. This intensity scale is expected to be useful for evaluation of detailed areal distribution of seismic intensity and also for the evaluation of intensity of palaeoearthquakes. We selected four great earthquakes to map ESI 2007 scale distribution: the 1995 Kobe; the 2004 Chuetsu, Japan; the 1935 Hsinchu-Taichung; and the 1999 Chichi, Taiwan. Proposed ESI 2007 scale maps from these areas are the mesh maps with a grid of about 1 km^2, showing more detailed intensity patterns than those previously provided by the Japan Meteorological Agency and the Central Weather Bureau for the four areas. Different responses of ground effects to the earthquakes, depending on local differences of geological materials near the surface and morphological condition of each site, are more clearly expressed by the ESI 2007 scale map, because of the large number of observed sites by the evaluation of ESI 2007 scale. Calibration exercise also reveals, however, that the classification of ESI 2007 scale needs some improvement.

The INQUA Environmental Seismic intensity scale (ESI 2007 scale) is a newly proposed intensity for the evaluation of ground effects caused by earthquakes, based on the size and scale of various environmental effects such as surface faulting, landslide, liquefaction and so forth. This intensity scale was discussed by the Working Group under the INQUA Subcommission on Palaeoseismology (2004) under the name of INQUA Earthquake Environmental Effect (EEE) seismic intensity and a simple table for this intensity classification was provided by Guerrieri *et al.* (2006). After lots of discussion, the final report was published and its name was changed at the XVII Congress of INQUA at Cairns (Guerrieri & Vittori 2007). This intensity scale is meant to be applied mainly to earthquakes (mostly palaeoearthquakes) that have no documentary records of damage and thereby allows the estimation of affected area, intensity distribution and even earthquake magnitude. The INQUA Subcommission on Palaeoseismology is now collecting ESI 2007 scale data from the world in hopes of building a database for the evaluation of palaeoseismicity.

We are cooperating in this programme, and as a first step, we use major earthquakes from seismically active Japan and Taiwan, which are located on the plate convergence boundary with many active faults as well as earthquake faults (coseismic surface faults or ruptures; hereafter, 'surface faulting'; Guerrieri *et al.* 2006) (Figs 1 and 2). We have selected four major earthquakes from Japan (the 1995 Kobe and the 2004 Chuetsu) and Taiwan (the 1935 Hsinchu-Taichung and the 1999 Chichi) (Table 1). For each of these earthquakes, we summarize the obtained results on the nature and effects of the earthquakes and provide an ESI 2007 scale map, and then we discuss some results and open issues. The surface faulting is one of the criteria for high ESI 2007 scale. Among the four earthquakes mentioned, three were associated with conspicuous surface faulting, except for the 2004 Chuetsu which had very minor surface faulting.

Since the destructive earthquakes of 1995 Kobe (Japan) and 1999 Chichi (Taiwan), it is more strongly accepted that coseismic surface faulting follows the pre-existing active fault trace. After these two events, studies on palaeoearthquakes, such as the identification of active fault traces based on geomorphologic methods and the timing of palaeoearthquakes based on trench excavation, have shown great progress. The 1995 Kobe and the 1999 Chichi earthquakes mark turning points for palaeoseismic research in both Japan and Taiwan (e.g. Ota 2000; Y.-G. Chen *et al.* 2001; Ota *et al.* 2005). In Japan, the number of trenching studies dramatically increased after the 1995 earthquake, supported by government funding and by

From: REICHERTER, K., MICHETTI, A. M. & SILVA, P. G. (eds) *Palaeoseismology: Historical and Prehistorical Records of Earthquake Ground Effects for Seismic Hazard Assessment*. The Geological Society, London, Special Publications, **316**, 55–71. DOI: 10.1144/SP316.4 0305-8719/09/$15.00

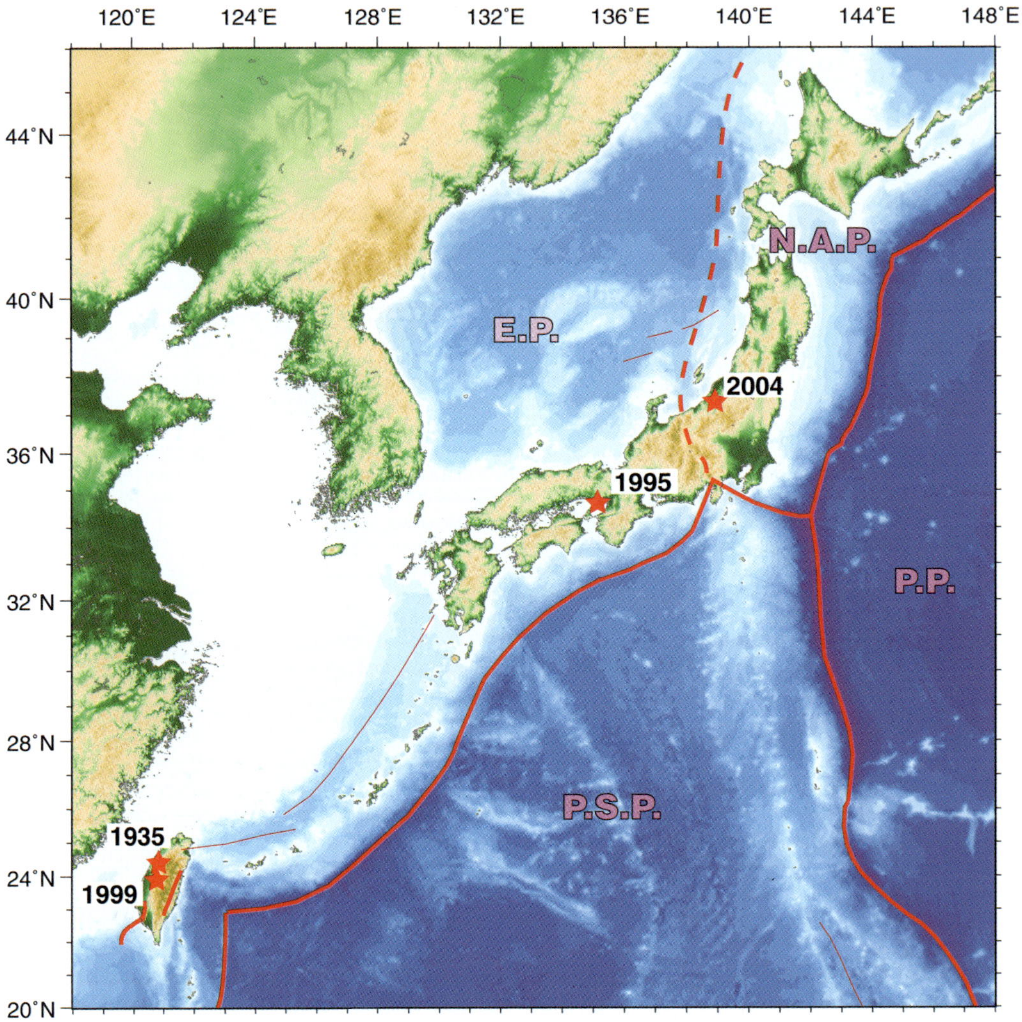

Fig. 1. Map of plate boundaries in and around Japan and Taiwan. Epicentres for four earthquakes discussed in the text are marked by star symbols. Tectonic plates: E.P., Eurasia Plate; N.A.P., North America Plate; P.P., Pacific Plate; P.S.P., Philippine Sea Plate.

cooperation of local governments, and a lot of palaeoseismic data have been obtained (Fig. 3). In Taiwan, trenching studies started just after the Chichi earthquake, providing rather short recurrence intervals of several hundred years or less along most of the Chelungpu Fault (e.g. W.-S. Chen *et al.* 2001, 2004, 2007), except for the northern part, where the penultimate event occurred at 1400–2000 yr BP (Ota *et al.* 2007). Trenching of other active faults such as the Longitudinal Valley Fault in eastern Taiwan has also been in progress (Yen *et al.* 2008). Although trenching studies contribute to the reconstruction of palaeoearthquakes, especially to their style, timing and amount of offset, they do not provide intensity distribution maps of the affected area and the possible magnitude of the specified seismic event. Thus, in addition to fault studies, the ESI 2007 scale map can provide better constraint for the estimation of the degree of shaking over wide areas affected by related earthquakes, including palaeoearthquakes.

Methods

We estimated the ESI 2007 scale, following Table 2, which is slightly modified from Guerrieri *et al.* (2006), as mentioned below. Among many seismically induced natural features, we mainly use the surface faulting, landslide and liquefaction data, because these data are usually clearly recorded.

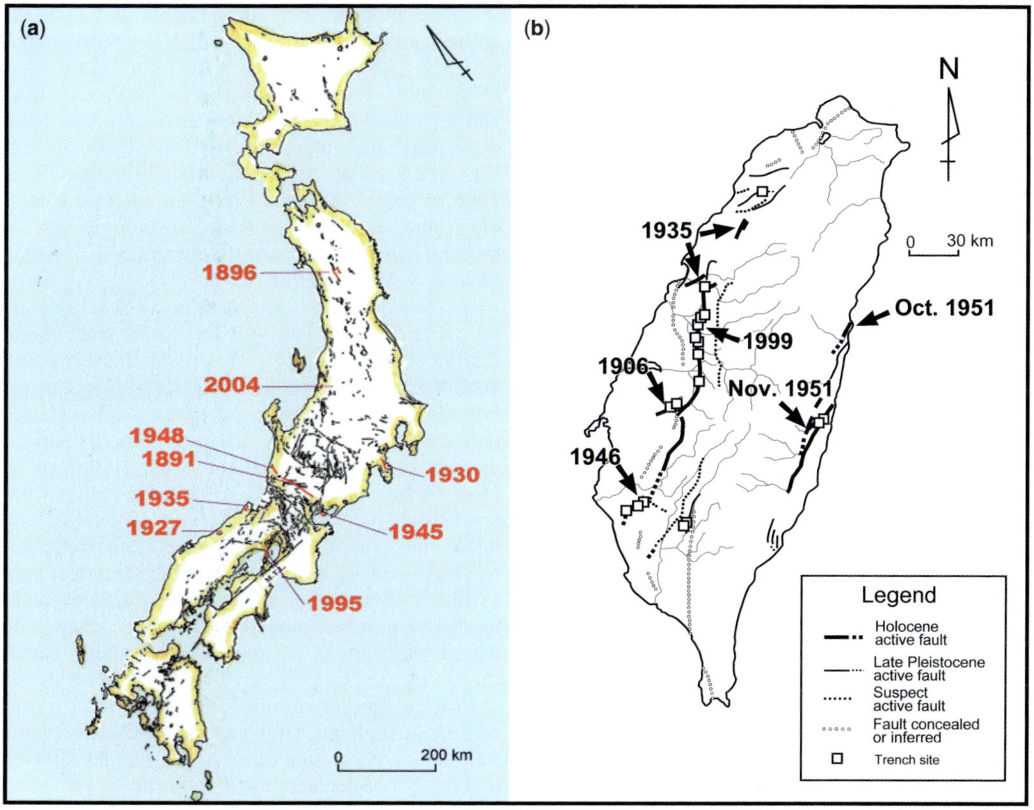

Fig. 2. Active faults and surface earthquake faults in Japan (**a**) and Taiwan (**b**). (a) Simplified from The Research Group for Active Faults of Japan (1992); (b) after Lin *et al.* (2000). Open squares are trench sites mostly on the Chelungpu Fault.

Table 2 is a summarized list for the classification of ESI 2007 scale on the basis of surface faulting, landslides and liquefaction. Since we aim to provide a mesh map to illustrate the ESI 2007 scale distribution, the length of the fault and volume of landslides, which are used as criteria by Guerrieri *et al.* (2006), are not considered. This is the major difference of our criteria from those of Guerrieri *et al.* (2006). The surface faulting is classified into four intensity degrees (VIII–XI) based on the amount of slip, and landslides are classified into four (Table 2, Fig. 4) based on the size, location and density etc. We use the liquefaction data when the diameters of affected areas are available.

We use the same method for four areas affected by the 1935, 1995, 1999 and 2004 earthquakes as follows. We draw approximately 1-km^2 grids ($10'' \times 15''$) or exact 1-km^2 grids on topographical maps of 1/25 000 scale. We evaluated the intensity using the surface faulting, landslide and liquefaction data in each grid, following the criteria shown in Table 2. Thus each grid may have two or three different ESI 2007 scale values. Finally we compiled them into one mesh map adapting the highest intensity from each grid. We also compared the ESI 2007 scale map with Japan Meteorological Agency (JMA) intensity (Table 3) and Central Weather Bureau (CWB) intensity (Table 4); both are based on the Peak Ground Acceleration (PGA) value.

Nature of four earthquake, ESI 2007 scale maps and interpretation

ESI 2007 scale map for the 1995 Kobe earthquake, southwestern Japan

General nature of the earthquake and major surface deformation. The 1995 Hyogoken-nanbu (Kobe) earthquake of M_{JMA} 7.3 occurred on 17 January. Its epicentre is located offshore between Kobe and Awaji Island. On the northwestern part of Awaji Island, close to the epicentre, surface faulting (Fig. 5a, b; right lateral-slip fault with high-angle

Table 1. *Summary of four major earthquakes discussed in the text*

Earthquake						Surface faulting		
Year/name (area)	Date/local time	M_{JMA}	M_L	M_W	Depth (km)	Name	Length (km)	Maximum offset (m)/sense
1995 Kobe (SW Japan)	17 Jan./05:46	7.3	–	6.9	16	Nojima Fault	10.5	R = 2.5, V = 1.3 (NE, reverse)
2004 Chuetsu (NE Japan)	23 Oct./17:56	6.8	–	6.6	13	Obiro Fault	1	V = 0.15 (W, reverse)
1935 Hsinshu-Taiching (NW Taiwan)	21 April/06:02	7.0	–	–	3	Chihhu Fault*	16	R = 0.6, V = 3.0 (W, reverse)
						Tuntzuchiao Fault	12	R = 2.0, V = 0.7 (W)
1999 Chichi (central Taiwan)	21 Sept./01:47	–	7.3	7.6	8	Chelungpu Fault	100	R = 7, V = 10 (E, reverse)

Major component of displacement: L, left lateral; R, right lateral; V, vertical (upthrown side and sense).
*Resulted from the major aftershock.

reverse component; see Table 1) appeared along the known active Nojima Fault, which is a right lateral-slip fault with average slip rate of 1.0 m/ka (Mizuno *et al.* 1990). The occurrence of the surface faulting and its coincidence with the active Nojima Fault were paid much attention by many scientists (e.g. Lin *et al.* 1995; Nakata *et al.* 1995; Ota *et al.* 1995; Awata & Mizuno 1998). On the east side of the island, a short (*c.* 1.6 km) and minor surface faulting with 10–15 cm of right lateral slip occurred (Awata & Mizuno 1998).

In the Kobe area across Seto Inland Sea, liquefaction (Fig. 5c) and landslide or cracks were major ground effects. A narrow zone in the densely populated Kobe area (red zone in Fig. 6a), located on the northeastward extension of the Nojima Fault, recorded JMA Intensity VII and received serious damage on buildings and roads etc., although no clear surface faulting was recognized there. It has been discussed whether there is a concealed fault under the ‘zone of seismic disaster’ characterized by JMA intensity VII, or if it is the difference in subsurface structure (e.g. difference in facies or thickness of unconsolidated deposits) that resulted in severe damage, or if some surface deformation may even exist (e.g. Watanabe & Suzuki 2000).

Strong ground motion of more than 1 g was recorded near Kobe. Highway, railroad and concrete buildings were broken down by earthquake shaking and many houses were lost due to fire.

Landslides, which include slope failures on the artificially modified land for residential use, occurred on the steep slopes around Kobe. Liquefaction was mostly observed on the artificially reclaimed island in front of Kobe downtown (Fig. 5c, *c.* 20 km distant from the epicentre) and river mouth area in Osaka plain (40–50 km distant from the epicentre). More than half of the artificially reclaimed island was damaged by sand blows and lateral spreads.

ESI 2007 scale map: interpretation and problems. Figure 6a is an ESI 2007 scale map for the 1995 earthquake using several different sources (for example, surface faulting data are mainly by Nakata *et al.* (1995) and Ota *et al.* (1995); landslide data and liquefaction data by Kokusai Kougyo Co. Ltd (1995)). ESI 2007 scale intensities range from V to X. The highest ESI 2007 scale intensities (IX and X) are identified from the surface faulting of the Nojima Fault, Awaji Island, and locally in the eastern coast.

Other high ESI 2007 scale (VII and VIII, sometimes IX) resulted from mainly liquefaction on the artificial reclaimed island near Kobe. The zone of JMA intensity VII does not correspond to the high ESI 2007 scale intensity, because this zone is estimated from the collapse of the buildings, not from

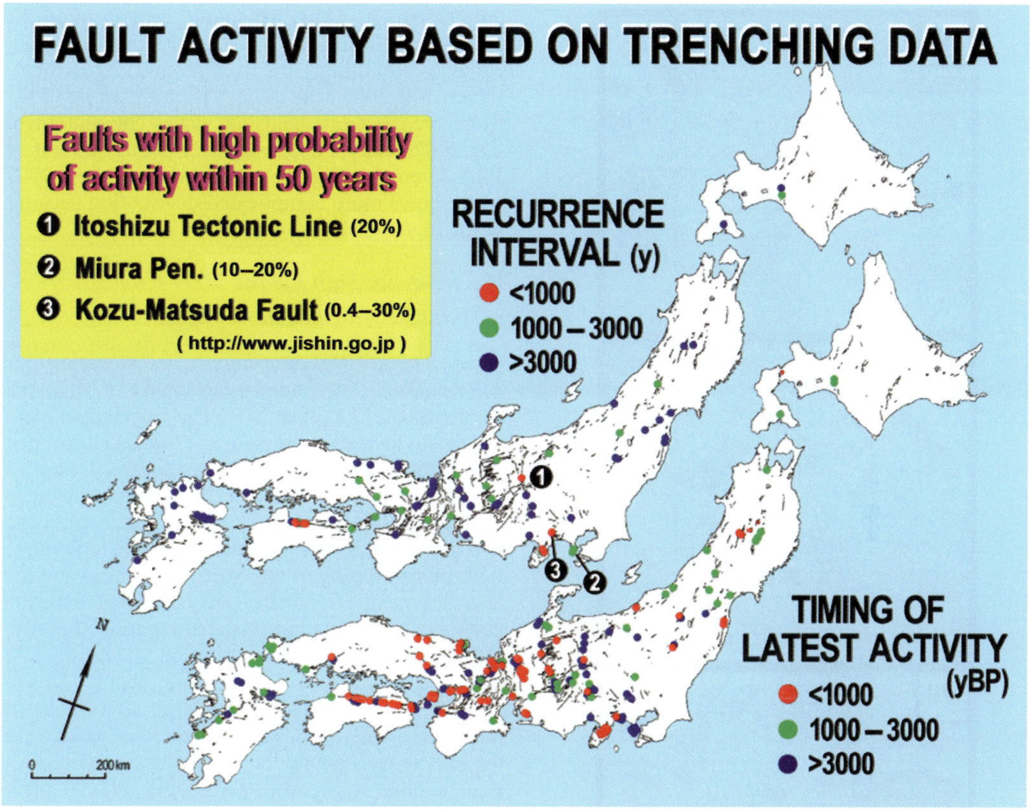

Fig. 3. Summary of average recurrence interval (above) and timing of most recent earthquakes (below) on onshore surface faulting in Japan based on trenching data (compiled by Ota in 2005). Yellow box gives earthquake probability computed for three active faults (1, Itoigawa-Shizuoka Tectonic Line Fault; 2, active faults in the Miura Peninsula; 3, Kouzu-Matsuda Fault) that are regarded to have the highest probability within 50 years.

Table 2. *Classification of ESI 2007 scale based on the surface faulting, landslides and liquefaction*

ESI 2007 scale	Surface faulting	Landslide	Liquefaction
V	–	Slope failures on steep slopes, rare	–
VI	–	Slope failures on steep slopes, frequent	–
VII	–	Collapses on gentle slopes, rare	Rare (1–4 sites in a mesh)
VIII	Offset < 0.1 m (L < 100 m)	Collapses on gentle slopes, frequent	Frequent (>5 sites in a mesh)
IX	Offset = 0.1–0.5 m (L < 1 km)	Large landslides, widespread	Subsicdence >30 cm
X	Offset = 0.5–2 m (L < 10 km)	Large landslides and rock-fall, frequent	Subsidence >1 m
XI	Offset > 2 m (L < 100 km)	–	–
XII	Offset > 10 m (L > 100 km)	–	–

Slightly modified from Guerrieri *et al.* (2006). See text for detail.

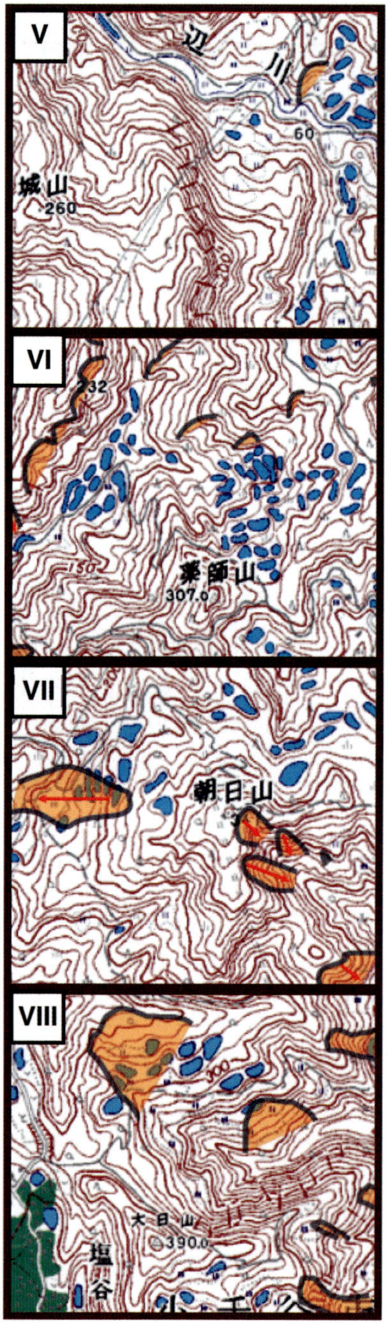

Fig. 4. Example of classification of landslides by ESI 2007 scale. Examples are from Chuetsu area. Intensity scale is modified from Guerrieri *et al.* (2006) (see text for detail). V, A few slope failures on steep slope; VI, landslides even on gentle slope; VII, frequent occurrence of slope failures on gentle slope; VIII, large landslides. Many small ponds (grey) are fish ponds unrelated to the earthquake. Each map corresponds to 1 grid, about 1 km^2.

ground effects. Thus, we should bear in mind the significant differences between the intensity from collapse of buildings and from ground effects themselves. ESI 2007 scale V to VII on the slope of the Kobe area is from landslide data. The pattern of ESI 2007 scale is approximately similar to that of JMA intensity (Fig. 6b). However, the ESI 2007 scale shows more complicated distribution because of the many observed points.

IES intensity map for the 2004 Chuetsu earthquake, northeastern Japan

General nature of the earthquake and major surface deformation. The Chuetsu earthquake of M_{JMA} 6.8 occurred on 23 October 2004. The epicentre of main shock was located in the Neogene fold and thrust belt that trends NNE–SSW in the southern part of Niigata Prefecture, NE Japan. The focal mechanism solution of the main shock shows this earthquake was generated by reverse faulting on the high-angle fault plane dipping to the west. Its solution and the distribution of aftershocks are consistent with the geological structures in this area mentioned above. JMA intensity VII was recorded at Kawaguchi near the epicentre of the main shock. Severe shaking was accompanied by large aftershocks of magnitude $>M_{JMA}$ 6. Many landslides occurred in the hills in and around the epicentral area (Fig. 7a); liquefactions were observed on the alluvial plain and fluvial terraces along the Shinano River in Ojiya and Nagaoka in the area west of the epicentre.

Very small surface faulting appeared in the eastern margin of the epicentral area. It strikes north–south and is only about 1 km long with less than 15 cm vertical offset (location is shown in Fig. 8; Maruyama *et al.* 2005). This surface faulting occurred along a trace of the pre-existing active fault, identified by recent trenching, but its trend was oblique to the major geological structure in this area. In addition to the small amount of offset and the oblique strike of this fault, the location far from the epicentre may also suggest that this fault may not be the surface expression of the seismogenic fault. Leveling data after the 2004 earthquake indicate that the western side of the epicentral area was uplifted about 0.7 m (Geographical Survey Institute of Japan 2006), but no surface faulting was recognized along this survey line. Thus, it is likely that there was no clear surface faulting of the seismogenic fault at the time of the Chuetsu earthquake, although the ground effects were very severe and caused tremendous disasters.

The area affected by this earthquake is one of the dense landslide disaster areas in Japan, because hilly land in this area is underlain by sedimentary rocks that were deposited during the Plio-Pleistocene period. These rocks were strongly folded and

Table 3. *Intensity scale used by Japan Meteorological Agency (JMA) since 1948*

JMA Intensity (before 1996)		JMA Intensity (after 1996)		
Classification	Measured intensity	Classification		PGA(gal)
0	0–0.4	0	No feeling	<0.8
I	0.5–1.4	I	Slight	0.8–2.5
II	1.5–2.4	II	Weak	2.5–8.0
III	2.5–3.4	III	Rather strong	8.0–25.0
IV	3.5–4.4	IV	Strong	25.0–80.0
V lower	4.5–4.9	V	Very strong	80.0–250.0
V upper	5.0–5.4			
VI lower	5.5–5.9	VI	Disastrous	250.0–400.0
VI upper	6.0–6.4			
VII	>6.5	VII	Very disastrous	>400

The intensity of the 1935 earthquake used the JMA scale, which was classified into seven, from 0 to VI, at that time. Intensity VII was introduced after the 1947 Fukui earthquake in Japan.

faulted and are therefore subject to mass movement. In addition to the coseismically triggered landslides of the 2004 earthquake, many old and large-scale landslides also exist (Fig. 7a). This indicates that landslides have occurred repeatedly, probably related to major earthquakes in the past, but at present we have no data to estimate the timing of landslide occurrence. Some of the large landslides were the result of bedding slip (Fig. 7b), and a huge amount of landslide debris created dammed ponds.

Liquefaction, with sand blows and up-floating buried structures, were observed in the fluvial terraces or lowland along the Shinano River. Sand blows with diameter >3 m occurred in the paddy field on the alluvial plain. Gravels were injected on the fluvial terrace about 50 m higher than the current riverbed. Manholes and buried pipes floated up (Fig. 7c) in many places up to 1.2 m in the downtown area of Ojiya City.

ESI 2007 scale map: interpretation and problems. Figure 8 is a combined IES intensity map with some observed intensity data. The major data source is the seismic disaster maps which include landslides, ground cracks and liquefaction (Geographical Survey Institute of Japan 2006). The highest intensity of IX is only locally present in the eastern margin of Figure 8, identified by the minor surface faulting. The higher ESI 2007 scale intensity (VIII) exists to the north of the epicentre, in approximately a north–south direction, but ESI 2007 scale intensity of VIII does not continue southward from the epicentre, although the geological structure there is nearly the same as to the north. Many landslides destroyed the hilly lands, especially at Yamakoshi area, north of the epicentre. Some of landslides were formed by large aftershocks. Most of the relatively low ESI 2007 scale intensities, V and VI, are identified from the presence of small-scale slope failure or scattered occurrence of small-scale landslides on the hilly lands. Generally, higher intensity occurred in the north than in the south, even though JMA intensity was mostly VI or VI+ in both areas (Fig. 8).

Table 4. *PGA-based Central Weather Bureau (CWB) intensity scale in Taiwan*

CWB Intensity	PGA (gal = cm/s^2)	PGV (cm/s)
0	<0.8	<0.22
I	0.8–2.5	0.22–0.65
II	2.5–8.0	0.65–1.9
III	8.0–25.0	1.9–5.7
IV	25.0–80.0	5.7–17
V	80.0–250.0	17–49
VI	250.0–400.0	49–75
VII	>400	>75

After Wu *et al.* (2003). The intensity of the 1935 earthquake used the JMA scale.

JMA intensity of VII is known just south of the epicentre, but ESI 2007 scale intensity is only VI there. Judging from the photo interpretation by the authors and also Geographical Survey Institute of Japan (2006), many landslides reoccupied the pre-existing landslides as mentioned before (Fig. 7a). Therefore the repeated activities of landslides are generally accepted, although some of the pre-existing huge landslides did not reactivate during this earthquake.

High ESI 2007 scale intensities of VII and VIII along the Shinano River resulted from liquefaction, which was seen even on the fluvial terrace.

ESI 2007 scale map for the 1935 Hsinchu-Taichung earthquake, northwestern Taiwan

General nature of the earthquake and surface deformation. The area affected by the 1935 earthquake

Fig. 5. Major ground effects caused by the 1995 Kobe earthquake. (**a**) Southern part of surface faulting on the Nojima Fault, cutting the artificially flattened area (photo by Sankei News Press Company, 1995, cited in Nakata & Okada 1999). (**b**) Surface faulting on the northern part of Nojima Fault. Right-lateral slip with uplift on the east shows the maximum amount of offset on the Nojima Fault (photo by Ota, January 1995). (**c**) Sand blows erupted along fissures associated with liquefaction and lateral spreading at reclaimed land in Kobe (photo by Okada, cited in Nakata & Okada 1999).

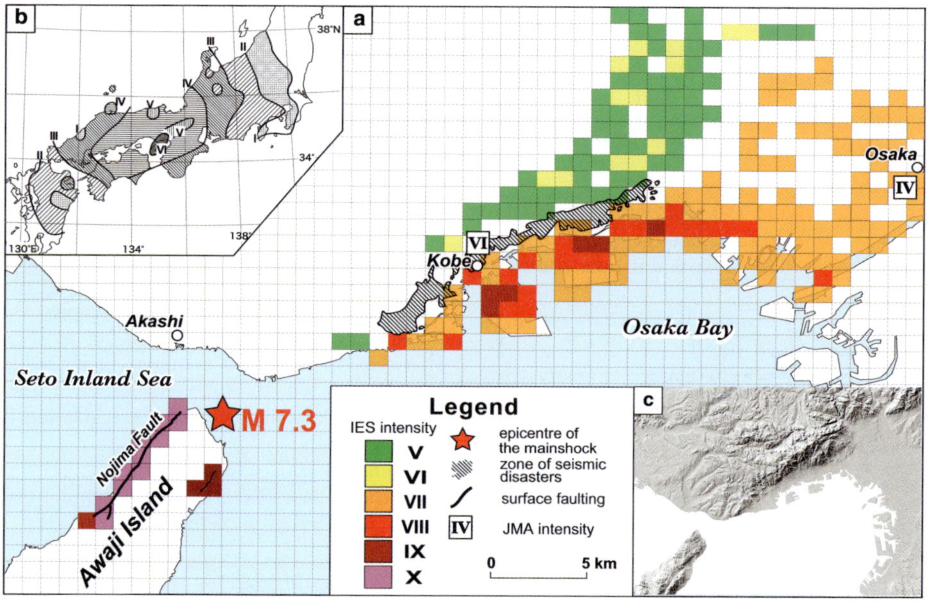

Fig. 6. ESI 2007 scale map of the Kobe area and Awaji Island. (**a**) Zone of JMA intensity VII is shown in red (based on the degree of collapse of buildings). (**b**) Generalized JMA intensity map, from Usami (2003). (**c**) Digital elevation model showing topographic relief of the area shown in (a).

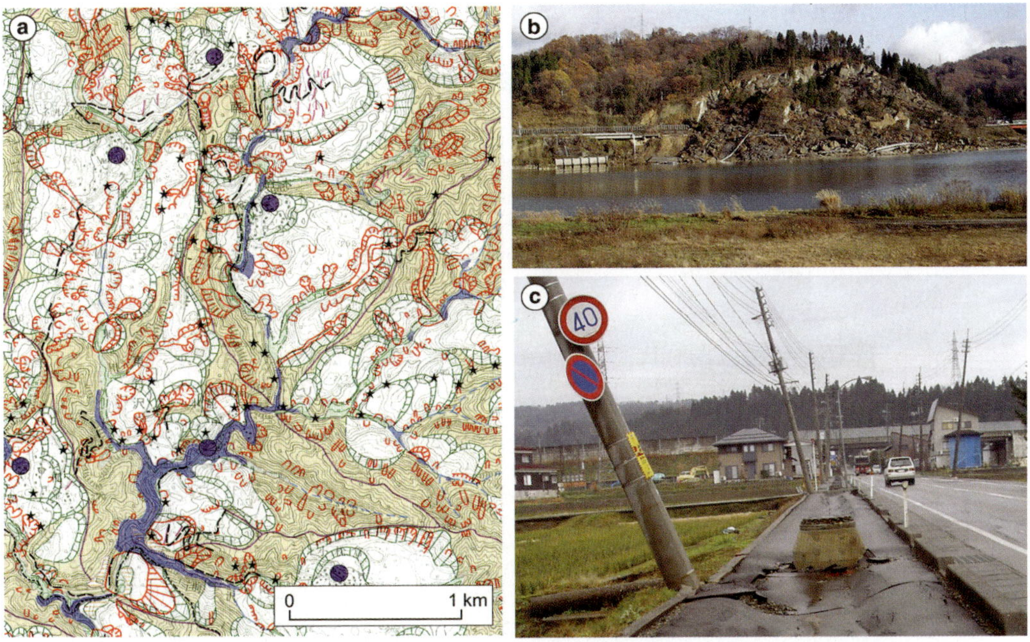

Fig. 7. Major landslides and liquefaction caused by the 2004 Chuetsu earthquake. (**a**) Example of abundant landslides (Geographical Survey Institute of Japan 2006). Colour key: red: coseismically triggered landslides by the 2004 earthquake; green, pre-existing landslides; blue, lakes dammed up by landslide debris from 2004 earthquake. (**b**) Large rock fall along the bedding plane of Tertiary rocks. A car passing at the time of the earthquake was buried and a little boy rescued three days after the earthquake (photos by Ota, November 2004). (**c**) Floating up of manhole (*c.* 0.5 m) at Ojiya City (photo by Ota, November 2004).

is in the 'Miaoli mature collision zone' (Shyu *et al.* 2005), dominated by dense thrusts and folds. The main shock (M = 7.0) on 21 April, 1935 (epicentre shown in Fig. 9) was associated with surface faulting (Tuntzuchiao Fault). Certain landslides were reported with this main shock. The main aftershock occurred further north, but there has been discussion about which one was the important aftershock (Lin 2005). The second surface faulting (the Chihhu Fault) and many small-scale landslides occurred in association with one of the major aftershocks. Those features are mapped by Otuka (1936) in detail. Other disastrous effects on the buildings, railways and so on are also remarkable, but in this paper we mostly use the maps and description of natural effects shown by Otuka (1936).

The Tuntzuchiao Fault (Figs 9, 10) is approximately 12 km long and strikes nearly N60°E from the Taan River in the north to the Tachia River in the south. Prominent features of the Tuntzuchiao Fault include a small fault scarp, up to 0.7 m high, cutting the fluvial terrace, and changes in the sense of vertical offset along the fault trace: from the west side up in the northernmost part to the east side up in the southern part. Another feature is mole tracks or *en echelon* cracks, indicating the right-lateral motion on the Tuntzuchiao Fault. The maximum amount of right-lateral offset is estimated to be 2 m. This fault truncates the late Quaternary terraces, while the proof of accumulated activity on this fault was unclear.

The Chihhu Fault (Figs 9, 10) strikes approximately N30°E, following the strike of the Neogene strata, and is approximately 16 km long. This fault is a high-angle reverse fault with up-thrown side in the west, cutting the western slope of hills, and truncates the west-flowing streams. Because the west side of the fault is uplifted, this surface deformation is characterized by a range-facing fault scarp (see inset to Fig. 9), up to 3 m high, accompanied by the formation of marshy lowland or temporal lakes in the east (down-thrown side). There is a series of saddles that cut through the western lower mountain slope, and a series of spurs that form the flank of long, dissected ridges to the west. Height difference between the saddles and spurs reaches up to 70 m, which could be the apparent accumulated offset of the Chihhu Fault. Thus repeated activity of the Chihhu Fault is very likely, although palaeoseismological study has not been carried out on this fault. Landslides are widely scattered, but the sizes are rather small.

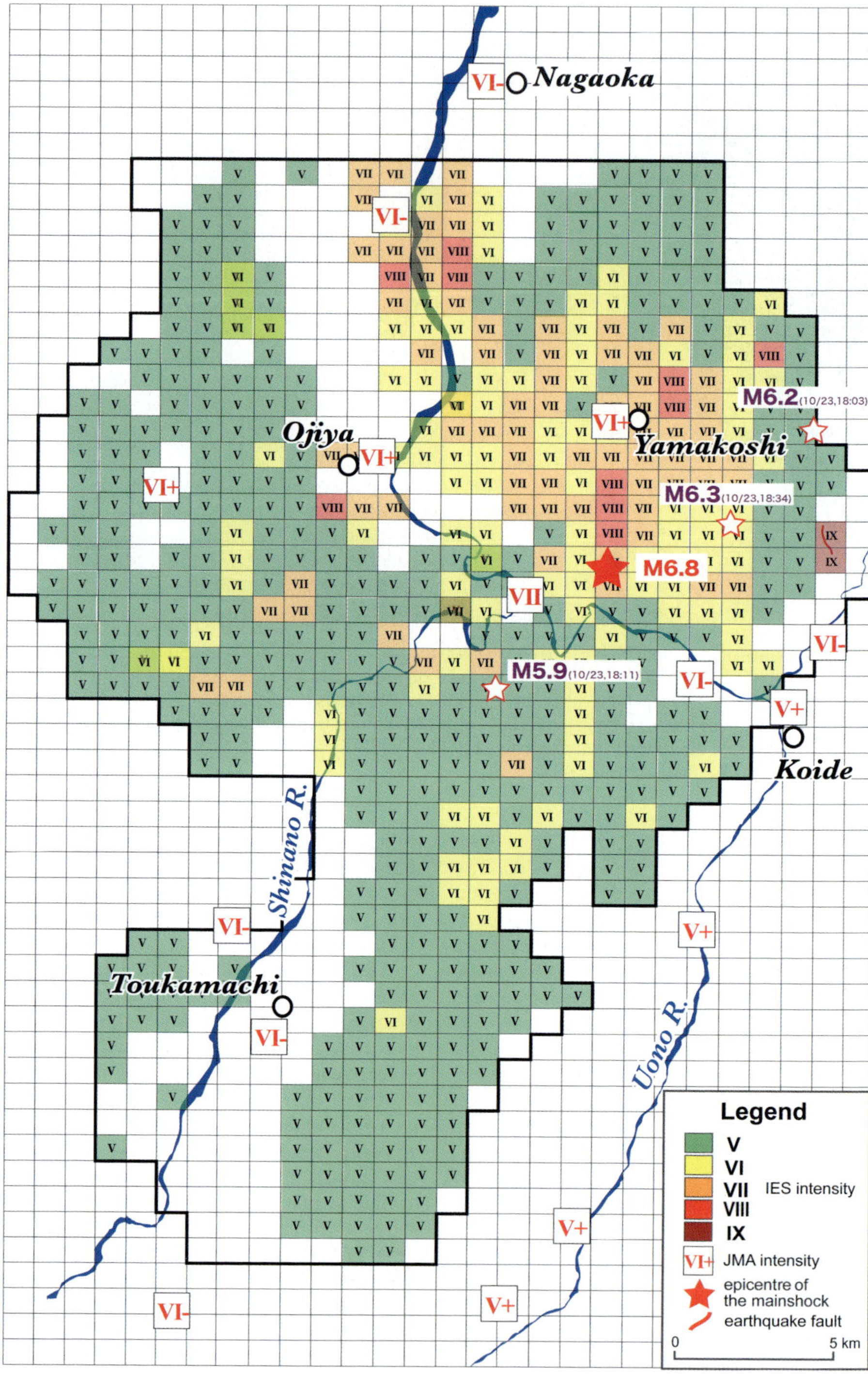

Fig. 8. ESI 2007 scale map of the 2004 Chuetsu earthquake (data are mainly based on Geographical Survey Institute of Japan 2006). Epicentres: large star, main shock; small stars, large aftershocks. Numbers in squares are observed JMA intensity. A very short surface faulting is on the right of this figure.

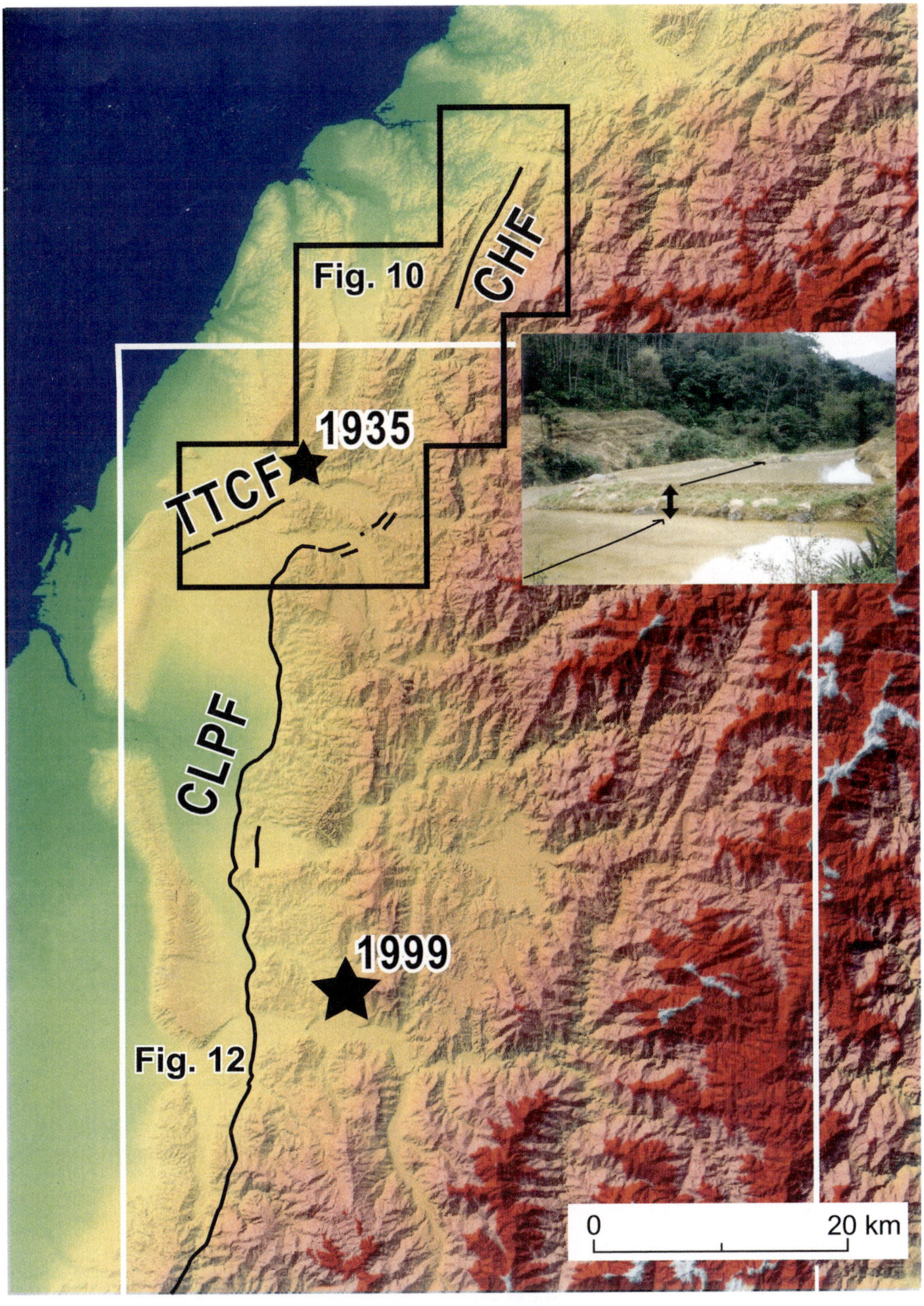

Fig. 9. Digital elevation model showing topographic relief. CHF, Chihhu Fault; CLPF, Chelungpu Fault; TTCF, Tuntzuchiao Fault. Large stars represent location of epicentre for the main shock of the 1935 and the 1999 earthquakes. Inset shows a range-facing fault scarp (thick arrow) of the CHF (stream flows upper right direction on the photo, shown by thin arrows) by the 1935 earthquake (photo by Ota, 1984).

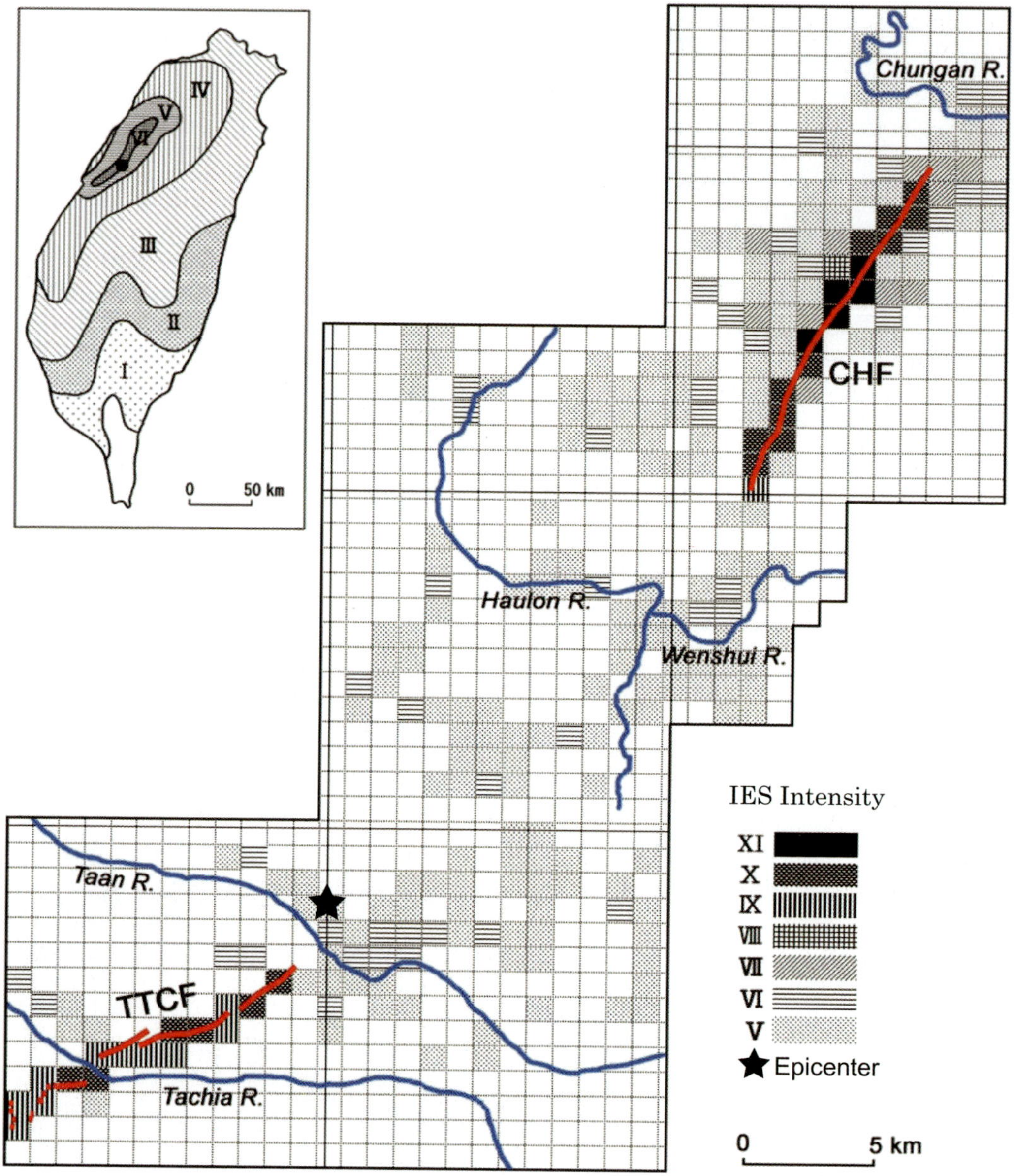

Fig. 10. ESI 2007 scale map for the 1935 earthquake. Data are from Otuka (1936). Epicentres for the main shock are relocated by Lin (2005). Inset is generalized intensity distribution by the 1935 main shock (Taipei Meteorological Observatory 1936).

ESI 2007 scale map and its interpretation. Figure 10 is a combined IES intensity map for the 1935 earthquake. IES intensities range from V to XI. High intensities (X, XI) are located on the Tuntzuchiao Fault and the Chihhu Fault. The Chihhu Fault, related to the main aftershock, shows the highest intensity. The Tuntzuchiao Fault, associated with the main shock, has a maximum intensity of only IX. The area west of the Chihhu Fault is mostly mountainous with bedrock highly folded and faulted by minor fractures. Compared with the Chihhu Fault, the area where the Tuntzuchiao went through is mostly flat fluvial terraces and is only mildly folded. The differences in geomorphological and geological settings may be the reason why the ESI 2007 scale intensity near the main shock is smaller than the intensity near the aftershock.

Higher intensity is widely observed on the west side of the Chihhu Fault. This can be explained by the fact that the western hanging wall has been more strongly shaken than the eastern footwall area. However, the contrast between both sides of the Tuntzuchiao Fault probably reflects different responses of the original landform, because the eastern part is mostly small hills while the western part is basically a terraced area. The relatively high intensity of VI corresponds to the high and steep erosional scarps cut by the Taan and Tachia rivers.

Compared to the intensity map published just after the earthquake (see inset to Fig. 10; Taipei Meteorological Observatory 1936), the areal pattern of ESI 2007 scale map is more complicated, reflecting local structure and landforms.

ESI 2007 scale map and discussion for the 1999 Chichi earthquake in central Taiwan

General nature of the Chichi earthquake and surface deformation. The Chichi earthquake ($M_w = 7.6$) of 21 September 1999 was the seismic event that caused the most serious damage and casualties in Taiwan during the past century. The epicentre was in Chichi, close to the southern part of the surface faulting and was about 10 km east of the surface faulting (Fig. 9). This remarkable 100-km-long surface faulting was caused by a low-angle thrust fault (the Chelungpu Fault) with uplift in the east. The resulting fault scarps are often characterized by flexural scarp with up-thrusting of the hanging wall over the footwall, clearly showing the shortening of the crust (Fig. 11). Vertical offset is up to 10 m. Most part of this surface faulting exactly follows the geomorphologically identified pre-existing active Chelungpu Fault (e.g. Chen *et al.* 2002; Ota *et al.* 2004), indicating repeated activity on the same fault trace during the late Quaternary. Large-scale landslides occurred on the mountains in the hanging wall, mostly taking place along major rivers where hydraulic erosion might have reduced the strength of rock. These landslides have also been repeatedly active, judging from the aerial photographs before and after the 1999 earthquake. In addition, many small landslides occurred on the eastern hanging wall. A huge ground crack was reported 30 km east of the surface faulting in Puli Basin, where strong shaking might have occurred due to basin

Fig. 11. Major surface deformation on the Chelungpu Fault by the 1999 earthquake (photos by Ota, September 1999). (**a**) Formation of a waterfall *c.* 8 m high across the Tachia River. (**b**) Flexural scarp deforming paddy field, south of the Tachia River. Red arrows indicate the base of scarp. (**c**) Flexural scarp *c.* 2 m high with convex profile on a sports ground at the central part of the Chelungpu Fault. Hanging wall overlies the footwall (blue arrow).

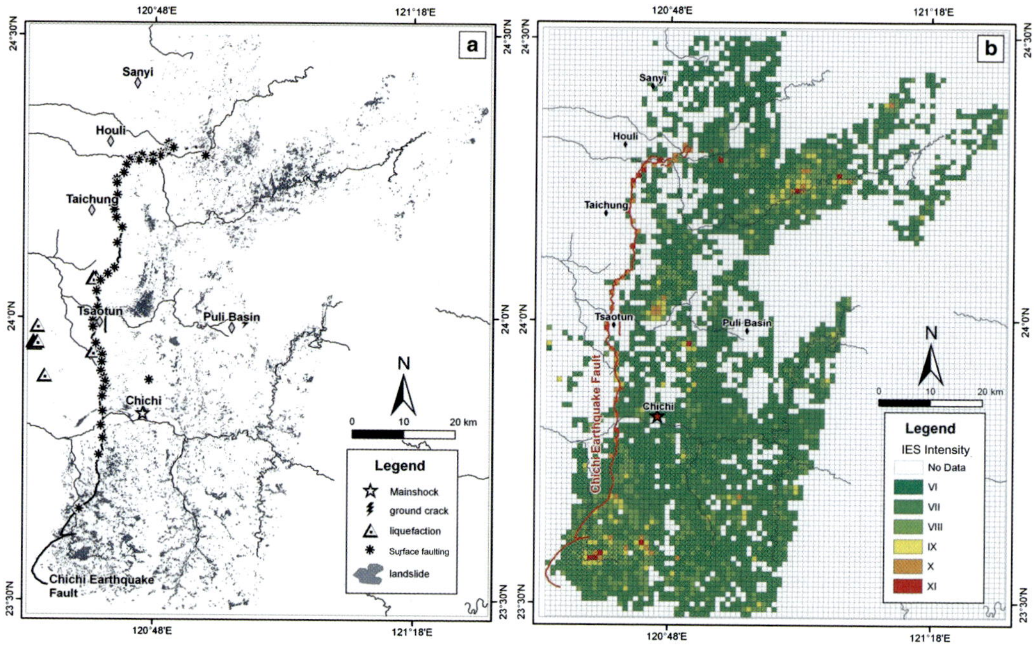

Fig. 12. ESI 2007 scale map of the 1999 Chichi earthquake. (**a**) Data used to evaluate the ESI 2007 scale of the 1999 Chichi earthquake (data source: Central Geological Survey 2000, 2002*a*, *b*, 2003*a*, *b*, 2004*a*, *b*). (**b**) ESI 2007 scale map of the Chelungpu Fault and its vicinity. Location is shown in Figure 9. Red line is the surface fault of Chelungpu Fault.

effect. Liquefaction also occurred locally in the footwall (Fig. 12a).

ESI 2007 scale map and its interpretation. Figure 12b is an IES intensity map based on data shown in Figure 12a, including the surface faulting data by Lin *et al.* (2000), a huge amount of landslide data (Central Geological Survey 2002*a*, *b*, 2003*a*, *b*, 2004*a*, *b*), one ground crack (Puli Basin) and liquefaction data in the footwall area (National Center for Research on Earthquake Engineering 1999). As we compare the original data and ESI 2007 scale map in Figure 12b, we realize that high ESI 2007 scale values are concentrated on the fault trace and on some large-scale landslides on the hanging wall, while IES intensities of VI and VII cover most of the other areas. Subtle differences of environmental effects, such as scales and sizes of landslides, liquefaction and ruptures, are revealed by this ESI 2007 scale map.

We have also prepared the distribution of PGA and PGA-based Central Weather Bureau (CWB) intensity maps (Table 4, Fig. 13a, b). The PGA records come from 650 stations in the Taiwan Strong-Motion Instrumentation Program (TSMIP) and 82 stations in the Telemetered Strong-Motion Stations (TREIRS9) (Wu *et al.* 2003). Here the PGA map shows a relatively complicated pattern, but the PGA-based CWB intensities are rather simple. The CWB intensity of VII covers most of the surface faulting and hanging wall area. It hence gives no direct indication to the rupture location and other detailed structures. It is also impossible to tell where the epicentre is through the CWB intensity map. Even in the PGA map, the concentric area with PGA value >800 Gal is not right on the epicentre. When we turn to the IES intensity map, we noticed that it shows no direct relationship with epicentral distance. It is even more difficult to construct an isoseismal map or to tell the locality of the epicentre according to the distribution of IES intensity, which is very different from what is shown by the Irpinia earthquake in the Southern Apennines of Italy (Michetti *et al.* 2004). There could be multiple reasons, one of which could be the extensively triggered strong environmental effects, especially landslides, in the hanging wall of the Chelungpu Fault. These landslides may have taken place repeatedly due to the joint result of weak geology/lithology, local topographic relief, strong river erosion, heavy precipitation and previous activities on the Chelungpu Fault or many other structures. Therefore, it is possible that the distribution of landslides after the Chichi earthquake should not be fully attributed to the effect of the earthquake itself. The landslides may simply indicate places where the threshold of failure was about to be achieved before the

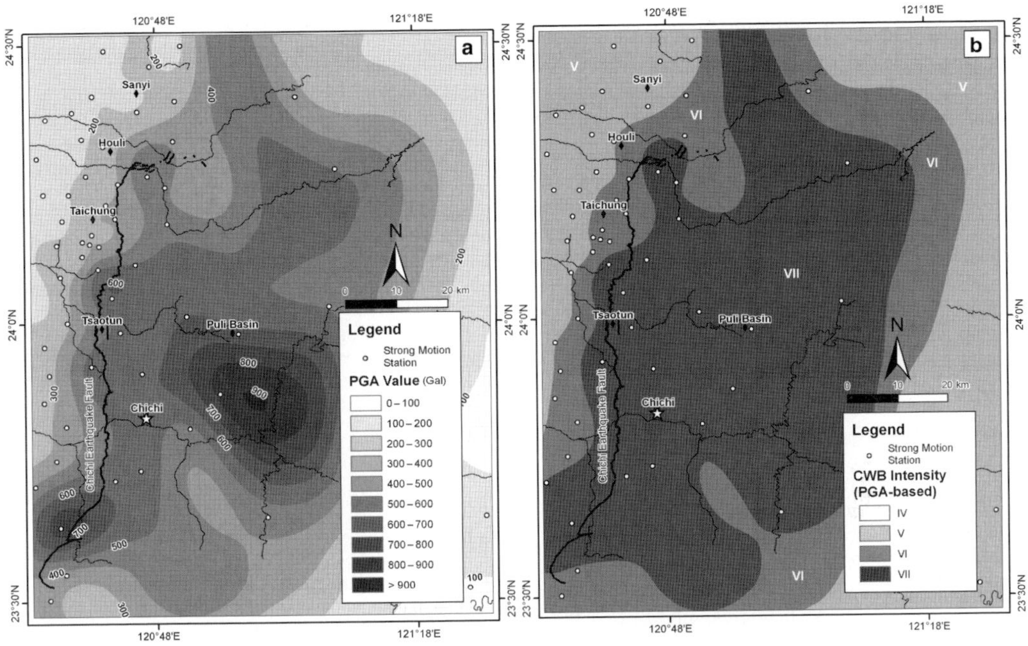

Fig. 13. PGA distribution (**a**) and PGA-based CWB intensity distribution (**b**). Areas of both figures are same as Figure 12. PGA map was made from more than 700 strong motion station records.

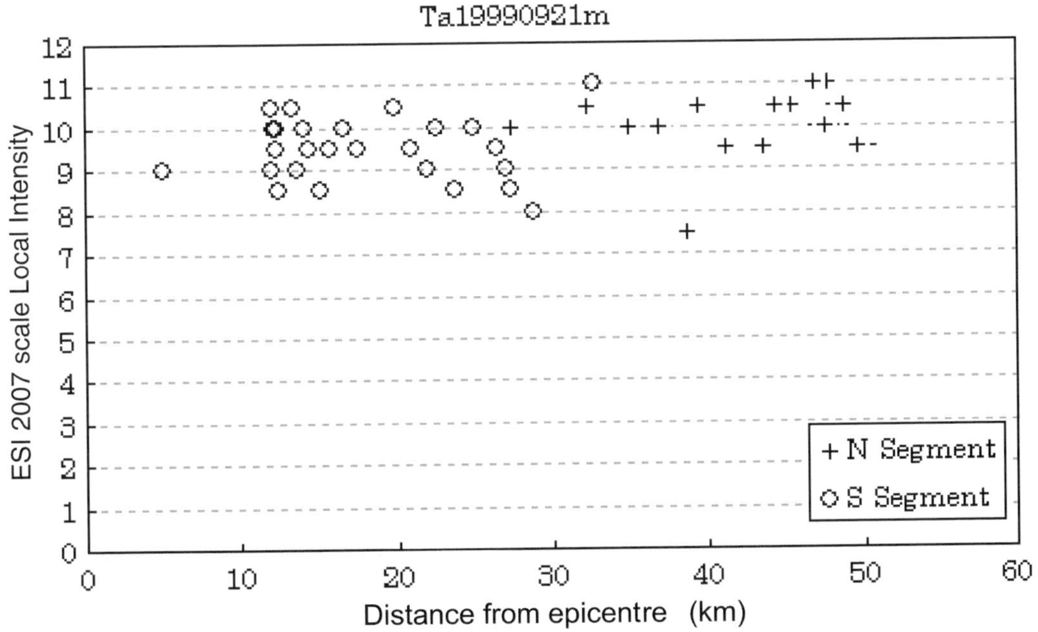

Fig. 14. Relationship between distance from the epicentre and ESI 2007 scale intensity along the Chelungpu Fault.

earthquake. This argument is verified in that there is hardly any spatial relationship between the epicentre and the original landslide data (Fig. 12a).

In Figure 14, ESI 2007 scale intensity is discordant with the epicentral distance, and shows an almost random distribution. On the other hand, the ESI 2007 scale map (Fig. 12b) shows a good relationship between locations with surface faulting and the places with strong environmental effects. Although our data do not cover all the area on the hanging wall, especially some blank zone NE of Puli, IES intensity may better reflect the local structure of geology and geomorphology than the CWB intensity.

Concluding remarks

In this paper we have presented ESI 2007 scale maps for four major earthquakes from Japan and Taiwan. These maps generally provide more detailed areal distribution of the ground effects caused by large earthquakes and reflect the local geological and geomorphological conditions than PGA-based intensity maps, because the ESI 2007 scale maps include a large number of investigated sites. Therefore, this method could be a useful key to the evaluation of intensity of palaeoearthquakes.

However, we have some more issues to consider. For example, the criterion of surface faulting offset, especially for under-thrusting tectonics, may need to be discussed, since it sets no difference between 2-m offset and 5-m offset, which in fact represents very different energy regimes. Another issue is how to deal with the different criteria. In our methodology we simply took the highest intensity value from all our evaluated criteria, yet one may raise another idea about weighting the different criteria with different factors and then summing the weighted values. Should we leave open how the ESI 2007 scale should be applied, or should we come to a common idea about how it should be done? The final issue that this study brings up is about the decision of epicentral intensity. In the case of the 1999 Chichi earthquake, the environmental effects were strongly affected by local geology and hence no concentric isoseismal contour could be drawn according to the ESI 2007 scale map. Is it a unique case, or is it also observed elsewhere in thrusting tectonics?

So far we tend to leave all the above-mentioned issues open. It is only our first step towards the application of ESI 2007 scale in Japan and Taiwan, and we hope in the near future we will have more data to nourish and improve the application of this intensity scale.

Parts of this paper were presented at the meeting of European Geoscience Union in April 2006 and at the Workshop on the Conduct of Seismic Hazard Analysis for Critical Facilities, held at International Centre for Theoretical Physics, Trieste, in May 2006. For constructive comments we thank participants at these two meetings and at the Trieste gathering of the INQUA Subcommission on Palaeoseismology. We very much appreciate especially Prof. Alessandro Michetti, who led this subcommission very actively and provided financial support of travel expense for Ota. We also thank Central Geological Survey in Taiwan for providing digital files of landslide mapping from aerial photographs, Prof. Ching-Weei Lin in National Chen Kung University for landslide mapping from SPOT image, and the National Center for Research on Earthquake Engineering for providing digital file of liquefaction. We very much appreciate Dr Brian Atwater of US Geological Survey for critical reading of the manuscript and improving the English. Finally, we thank Dr Guerrieri and Dr Shyu for their critical reading and constructive comments for this paper.

References

AWATA, Y. & MIZUNO, K. 1998. *Strip map of the surface fault features associated with the 1995 Hyogoken Nanbu earthquake, central Japan and explanatory text.* Geological Survey of Japan [in Japanese with English abstract].

CENTRAL GEOLOGICAL SURVEY. 2000. *Investigation report of 921 earthquake geology and map of surface ruptures along the Chelungpu Fault during the 1999 Chi-Chi earthquake* [in Chinese].

CENTRAL GEOLOGICAL SURVEY. 2002*a*. *Annual report of Central Geological survey*. 110–117 [in Chinese].

CENTRAL GEOLOGICAL SURVEY. 2002*b*. *Landslide survey and crisis analysis—Landslide potential analysis (1/3)* [in Chinese].

CENTRAL GEOLOGICAL SURVEY. 2003*a*. *Annual report of Central Geological survey*. 137–154 [in Chinese].

CENTRAL GEOLOGICAL SURVEY. 2003*b*. *Landslide survey and crisis analysis—Landslide potential analysis (2/3)* [in Chinese].

CENTRAL GEOLOGICAL SURVEY. 2004*a*. *Annual report of Central Geological survey*. 116–126 [in Chinese].

CENTRAL GEOLOGICAL SURVEY. 2004*b*. *Landslide survey and crisis analysis—Landslide potential analysis (3/3)* [in Chinese].

CHEN, W.-S., CHEN, Y.-G. & CHENG, H.-C. 2001. Paleoseismic study of the Chelungpu fault in the Mingjian area. *Western Pacific Earth Sciences*, **1**, 43–72.

CHEN, W.-S., LEE, K.-J. *ET AL*. 2004. Slip rate and recurrence interval of the Chelungpu fault during the past 1900 years. *Quaternary International*, **115/166**, 167–176.

CHEN, W.-S., YANG, C.-C. *ET AL*. 2007. Late Holocene paleoseismicity of the southern part of the Chelungpu Fault in central Taiwan: Evidence from the Chushan excavation site. *Bulletin of the Seismological Society of America*, **97**, 1–13.

CHEN, Y.-G., CHEN, W.-S., LEE, J.-C., LEE, Y.-H., LEE, C.-T., CHANG, H.-C. & LO, C.-H. 2001. Surface rupture of 1999 Chi-Chi earthquake yields insights on active tectonics of central Taiwan. *Bulletin of the Seismological Society of America*, **91**, 977–985.

CHEN, Y.-G., CHEN, W.-S., WANG, Y., LO, P.-W., LEE, J.-C. & LIU, T.-K. 2002. Geomorphic evidence for prior earthquakes: lessons from the 1999 Chichi earthquake in central Taiwan. *Geology*, **30**, 171–174.

GEOGRAPHICAL SURVEY INSTITUTE OF JAPAN. 2006. *1:25,000 Seismic disaster maps by 2004 Chuetsu earthquake, "Tookamachi", "Ojiya", "Yamakoshi"*.

GUERRIERI, L., VITTORI, E., ESPOSITO, E., MICHETTI, A., PORIFIDO, S., TATEVOSSIAN, R., SERVA, L. & INQUA SCALE WORKING GROUP. 2006. *The INQUA EEE intensity scale*. APAT.

GUERRIERI, L., VITTORI, E. (eds) 2007. *Environmental Seismic Intensity Scale 2007, ESI 2007*. Memorie Descrittive della Carta Geologica d'Italia, **74**. Servicio Geologica d'Italia, APAT, Roma.

KOKUSAI KOGYO CO. LTD. 1995. *Disaster maps of great Hanshin earthquake*. Tokyo.

LIN, A., IMIYA, H. *ET AL*. 1995. Investigation of the Nojima Earthquake Fault occurred on Awaji Island in the southern Hyogo Prefecture Earthquake, 1995. *Journal of Geography*, **104**, 113–126 [in Japanese with English abstract].

LIN, C.-H., CHANG, H.-C., LU, S.-T., SHIH, T.-S. & HUANG, W.-J. 2000. *An introduction to the active faults of Taiwan (second edition): explanatory text of the active fault map of Taiwan, scale 1:500,000*. Central Geological Survey Special Publication, **13** [in Chinese].

LIN, Y.-N. 2005. *Surface deformation and seismogenic structure model of the 1935 Hsinchu-Taichung earthquake (MGR = 7.1) in Miaoli, northwestern Taiwan*. Master thesis of Institute of Geosciences, National Taiwan University.

MARUYAMA, T., FUSEJIMA, Y., YOSHIOKA, T., AWATA, Y. & MATSU'URA, T. 2005. Characteristics of the surface rupture associated with the 2004 Mid Niigata Prefecture earthquake, central Japan and their seismotectonic implications. *Earth Planets Space*, **57**, 521–526.

MICHETTI, A. M., ESPOSITO, E. *ET AL*. 2004. The INQUA Scale. An innovative approach for assessing earthquake intensities based on seismically-induced ground effects in natural environment. *In*: VITTORI, E., COMERCI, V. (eds) *Special paper, APAT, ESI 2007*. Memorie Descrittive della Carta Geologica d'Italia, **74**. Servicio Geologic d'Italia, APAT, Roma.

MIZUNO, K., SANGAWA, A. & TAKAHASHI, Y. 1990. *1:50,000 Geological Map series, "Akashi" and explanatory text*. Geological Survey of Japan [in Japanese].

NAKATA, T. & OKADA, A. (eds) 1999. *Nojima Fault: Pictorial Record and Explanatory Text: Surface ruptures associated with the 1995 Hyogo-ken Nanbu Earthquake*. University of Tokyo Press [in Japanese with English figure captions].

NAKATA, T., YOMOGIDA, K., ODAKA, J., SAKAMOTO, T., ASAHI, K. & CHIDA, N. 1995. Surface fault ruptures associated with the 1995 Hyogoken-Nanbu Earthquake. *Journal of Geography*, **104**, 127–142 [in Japanese with English abstract].

NATIONAL CENTER FOR RESEARCH ON EARTHQUAKE ENGINEERING. 1999. *921 Chichi earthquake hazard survey—geotechnical engineering hazard survey* [in Chinese].

OTA, Y. 2000. Changes in paleoseismic studies of active faults in Japan since the 1995 Kobe earthquake. *Journal of Geodynamics*, **29**, 425–443.

OTA, Y., HORINO, M. & SURVEY GROUP ON DISASTERS, GEOGRAPHICAL SURVEY INSTITUTE. 1995. Notes on the earthquake fault on Awaji Island, associated with the 1995 Hyogoken-nanbu Earthquake and related disasters. *Journal of Geography*, **104**, 143–155 [in Japanese with English abstract].

OTA, Y., WATANABE, M., SUZUKI, Y. & SAWA, H. 2004. Geomorphological identification of pre-existing active Chelungpu Fault in central Taiwan, especially its relation to the location of the surface rupture by the 1999 Chichi earthquake. *Quaternary International*, **115–116**, 155–166.

OTA, Y., CHEN, Y.-G. & CHEN, B.-S. 2005. Review of paleoseismological and active fault studies in the light of the Chichi earthquake of September, 21, 1999. *Tectonophysics*, **408**, 63–77.

OTA, Y., SHISHIKURA, M. *ET AL*. 2007. Low-angle reverse faulting during two earthquakes on the northern part of the Chelungpu Fault, deduced from the Fengyuan trench, central Taiwan. *Terrestrial, Atmospheric and Oceanic Sciences*, **18**, 55–66.

OTUKA, Y. 1936. The earthquake of central Taiwan (Formosa), April 21, 1935, and earthquake faults. *Bulletin of the Earthquake Research Institute*, Suppl. **3**, 22–74 [in Japanese with English abstract].

RESEARCH GROUP FOR ACTIVE FAULTS OF JAPAN. 1992. *Maps of active faults in Japan with an explanatory text*. University of Tokyo Press.

SHYU, J.B.H., SIEH, K., CHEN, Y.-G. & LIU, C.-S. 2005. The neotectonic architecture of Taiwan and its implication for future large earthquakes. *Journal of Geophysical Research*, **110**. DOI: 10, 1029-2004JB003251, B08402.

TAIPEI METEOROLOGICAL OBSERVATORY. 1936. *The report on the April 21st, 1935 Hsinchu-Taicyung Earthquake*. Taipei Meteorological Observatory [in Japanese].

USAMI, T. 2003. *Materials for Comprehensive List of Destructive Earthquakes in Japan*. University of Tokyo Press [in Japanese].

WATANABE, M. & SUZUKI, Y. 2000. Active faults around Kobe and surface deformation during the 1995 Hyogoken-nanbu earthquake. *Chikyu Monthly* (extra edition), **28**, 62–68.

WORKING GROUP UNDER THE INQUA SUBCOMMISSION ON PALEOSEISMICITY. 2004. The INQUA scale: An innovative approach for assessing earthquake intensities based on seismically-induced ground effects in natural environment. *Memorie Descrittive Della Carta Geologica D'Italia*, **LXVII**.

WU, Y.-M., TENG, T.-L., SHIN, T.-C. & HSAIO, N.-G. 2003. Relationship between peak ground acceleration, peak ground velocity and intensity in Taiwan. *Bulletin of the Seismological Society of America*, **93**, 386–390.

YEN, I.-C., CHEN, W.-S., CHEN, C.-C., YANG, B., YANG, HUANG, N.-W. & LIN, C.-W. 2008. Paleoseismology of the Rueisuei Segment of the Longitudinal Valley Fault, Eastern Taiwan. *Bulletin of the Seismological Society of America*, **98**, 1737–1749.

Earthquake intensity assessment based on environmental effects: principles and case studies

R. E. TATEVOSSIAN*, E. A. ROGOZHIN, S. S. AREFIEV & A. N. OVSYUCHENKO

Institute of Physics of the Earth, RAS, ul. B. Gruzinskaya 10, Moscow 123995, Russia

**Corresponding author (e-mail: ruben@ifz.ru)*

Abstract: The comparison of intensity assessments based on macroseismic data and Earthquake Environmental Effects (EEE) is presented. Specific problems faced when assessing intensities using different types of scales are discussed. Two case studies of recent earthquakes with magnitudes $M_S = 7.4$ (Altai, 2003, and Neftegorsk, 1995) are used to illustrate the applicability of the INQUA EEE scale. The Altai earthquake was accompanied by surface faulting of *c.* 70 km length and up to 2 m of horizontal and 70 cm of vertical offset; secondary EEE were observed over 3000 km^2. The dominant type of surface faulting during the Neftegorsk earthquake was strike-slip. The length of surface faulting was up to 46 km, maximum horizontal offset was 8.1 m, and average offset coherent with seismic moment was 3.9 m; secondary EEE were observed occasionally at considerable distance from the epicentre on wet seashore sands. Application of the INQUA scale shows the epicentral intensity of the Altai earthquake to be X degrees. Most consistent with all types of data (rupture length, maximum and average offsets) intensity assessment for the Neftegorsk earthquake which is within the X–XI degree range. Taking into account environmental effects in intensity scales is an essential requirement: it follows from the complex nature of an earthquake impact, which spans a very broad frequency range, including static deformations. The case studies illustrate that the intensity assessment of an earthquake, based only on damage to buildings, will be essentially incomplete.

The INQUA scale is designed to calibrate earthquake intensity based on environmental effects (Michetti *et al.* 2004). A revised version of the scale was published in 2007 (Guerrieri & Vittori 2007); a new name for the scale was adopted, the Environmental Seismic Intensity scale (ESI 2007). The importance of Earthquake Environmental Effects (EEE) as a tool to measure earthquake intensity is being increasingly acknowledged (e.g. Dengler & McPherson 1993). The reasons for this are as follows.

(1) *The EEE size is the most reliable parameter for the intensity assessment of strong earthquakes.* For major earthquakes (intensity equal to or more than X) the effects on the built environment suffer from saturation (man-made structures are nearly completely destroyed), therefore intensity assessments based only on the damage caused to buildings cannot work properly.

(2) *The use of EEE ensures comparability of earthquake intensity assessments worldwide.* The EEE are free from influence of cultural and technological features, which can have essential differences in various parts of the world. Moreover, earthquake-prone areas can be located completely or partially in deserted places, where only the effects on the natural environment are available for the estimate of intensity.

(3) *Intensity based on EEE is the best tool to compare recent, historic and pre-historic earthquakes.* As the impact of an earthquake on the artificial environment depends on the distribution of urbanized areas, it is difficult to compare two seismic events that occurred in the same area, but distant in time. This approach extends the time coverage of earthquake catalogues to prehistoric times.

Immediately after publication of the ESI 2007 scale, studies began that had the goal of testing it. Guerrieri *et al.* (2007) proposed rules of the scale application. A classification of effect types was suggested: it recognizes primary and secondary environmental effects. The primary effects (surface faulting) are linked more or less directly to earthquake source; the secondary effects (slope movements, ground cracks, liquefaction, etc.) represent the result of a complex interaction of source features, seismic wave propagation paths and local conditions. Guerrieri *et al.* (2007) formulated conceptual differences between intensity ranges, which can be assigned to a single environmental effect and intensity in a natural locality. They proposed how to move from the size of a single effect toward intensity assessment at a locality (for explanations see Fig. 1b). Tatevossian (2007) gives a

From: REICHERTER, K., MICHETTI, A. M. & SILVA, P. G. (eds) *Palaeoseismology: Historical and Prehistorical Records of Earthquake Ground Effects for Seismic Hazard Assessment*. The Geological Society, London, Special Publications, **316**, 73–91. DOI: 10.1144/SP316.5 0305-8719/09/$15.00

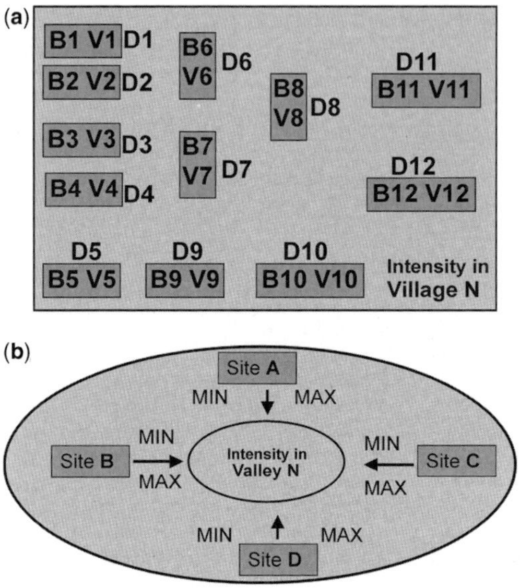

Fig. 1. Intensity assessment in localities. (**a**) Inhabited locality (village N). B1 to B12 are buildings characterized by a certain vulnerability class (V1 to V12). D1 to D12 are assessed damage degrees of each inspected building. (**b**) Natural locality (valley N). Site A to Site D are single EEE sites in the valley. MIN, MAX are intensity ranges, which have to be ascribed to each site.

detailed analysis of the scale application to historic earthquakes. It was demonstrated that the ESI 2007 scale is a very useful tool to gain a comprehensive image of a historic earthquake. In this paper, which is part of a worldwide testing of the ESI 2007 scale, results of the scale application to intensity calibration of two recent earthquakes are presented. When addressing recent earthquakes we are able to analyse relationships between the earthquake parameters based on instrumental and geological data. This can give useful hints on how to develop the ESI 2007 scale further.

The ESI 2007 scale is designed not to replace but to overpass some limitations of the macroseismic scales based mostly on damage caused to the buildings, and ignoring environmental effects (like EMS 98; Grünthal 1998). To keep the door open for future incorporation of intensity scales of different types, the ESI 2007 scale has to be logically consistent with them. Assessment of macroseismic intensity in a inhabited locality is explained in Figure 1a. During expert inspection of a village, damage degree and vulnerability class are assessed for each building (in practice, for a representative sample of the buildings) (Table 1). According to the EMS 98 scale, damage is classified in five degrees (V is for complete collapse). Vulnerability

Table 1. *Illustration of damage degree assessment to buildings based on macroseismic effects*

Building	Vulnerability	Observed damage	Damage degree
B1	V1 = C	Hair cracks in plaster	D1= I
B12	V12 = A	Some of partition walls fall, no damage to bearing walls	D12 = II

assessment is based on the building type (brick, concrete, wood, etc.) and takes into account a number of factors (quality of design and construction works, actual state, etc.), which are able to alter the basic vulnerability class. All buildings are classified into six groups from A to F (A is the most vulnerable). Intensity in the locality is assessed based on damage degree statistics, addressed in a qualitative way (few, some, most). This approach can be easily adopted for local intensity assessment based on EEE. In the case of a natural locality (e.g. a valley), instead of buildings single EEE sites (i.e. landslide, rockfall, ground cracks, etc.) can be used (Fig. 1b). Intensity range is assessed for each site analogous to damage degree of buildings in the EMS scale (Table 2). Intensity in the natural locality has to be assessed based on this range according to expert evaluation (Guerrieri *et al.* 2007). For example, according to the information in Table 2, the most consistent assessment for the valley N in Figure 1 will be VIII degree.

It is more difficult to maintain the same approach when locating the epicentre and assessing epicentral intensity (I_0) in the case of macroseismic and environmental effects (Fig. 2). In the case of macroseismic effects the epicentre is a point, the location of which is deduced from the spatial

Table 2. *Illustration of intensity range assessment to natural locality based on EEE*

Site	Effect type	Observed damage	Intensity range
A	Ground cracks	Fractures up to 60 cm wide observed in loose alluvial deposits	VIII–IX
D	Landslide	Landslide at steep slope up to 2×10^4 m^3	VII–VIII

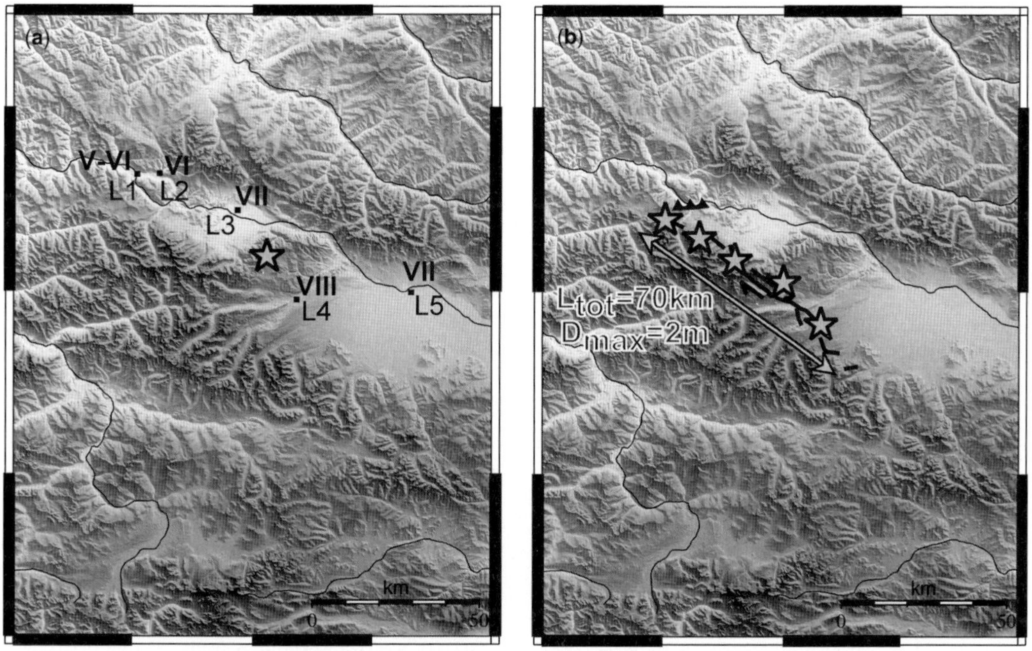

Fig. 2. Imaginary cases of epicentre location and epicentral intensity assessment based on macroseismic (**a**) and environmental (**b**) effects. L1 to L5 are five localities in the earthquake-prone area, for which macroseismic intensities are assessed using the scheme in Figure 1a. Star marks position of epicentre. Thick black lines are mapped surface faulting segments. Total rupture length and maximum offset are reported.

distribution of observed intensities in localities; I_0 degree is an extrapolation from observed ones (Fig. 2a). We have to answer the following question: What intensity would be observed in that point (star in Fig. 2a) if there were a village? Some seismologists avoid the extrapolation procedure and prefer to manipulate only with maximum observed intensity, but still assigned to the epicentre. So, we have a distribution of points (localities) over a certain area, from which the position of another point (the earthquake epicentre) has to be derived. But a dimensionless point can represent surface faulting of several tens or hundreds of kilometres only conventionally. Where should the epicentre be placed in Figure 2b? It is not easy to answer this question, especially if we recall that often the maximum offset is not in the middle of the surface faulting. Distribution of surface faulting allows depiction of an epicentral area, rather than a dimensionless point. When the source is not exposed on the surface it is possible to locate the epicentre on the basis of the distribution of secondary effects. Therefore, I_0 becomes the earthquake's characteristic parameter, like seismic moment M_0, which is proportional to the product of rupture area and average displacement over it.

In this paper we will address two case studies to compare macroseismic and EEE intensities, and to highlight the specific problems faced in assessing intensities and how reliable they are.

Case studies are two recent earthquakes (Altai, 2003, and Neftegorsk, 1995) that occurred in different seismotectonic conditions (Fig. 3). Both earthquakes had the same magnitude ($M_S = 7.4$; all magnitudes according to International Seismological Centre 2003), so they are roughly equal from an instrumental point of view. They were shallow source earthquakes, and both manifested surface faulting. But the macroseismic, environmental and social impacts of these seismic events are essentially different, which makes them interesting objects for detailed study.

The Altai earthquake, 27 September 2003

The epicentral area of the Altai earthquake ($M_S = 7.4$) is situated in Russia, close to the frontier with China, Mongolia and Kazakhstan (Fig. 4). The regional topography is sharp; the mountains reach 4500 m a.s.l, separated by depressions at 1500 m a.s.l. The earthquake was felt over a large territory

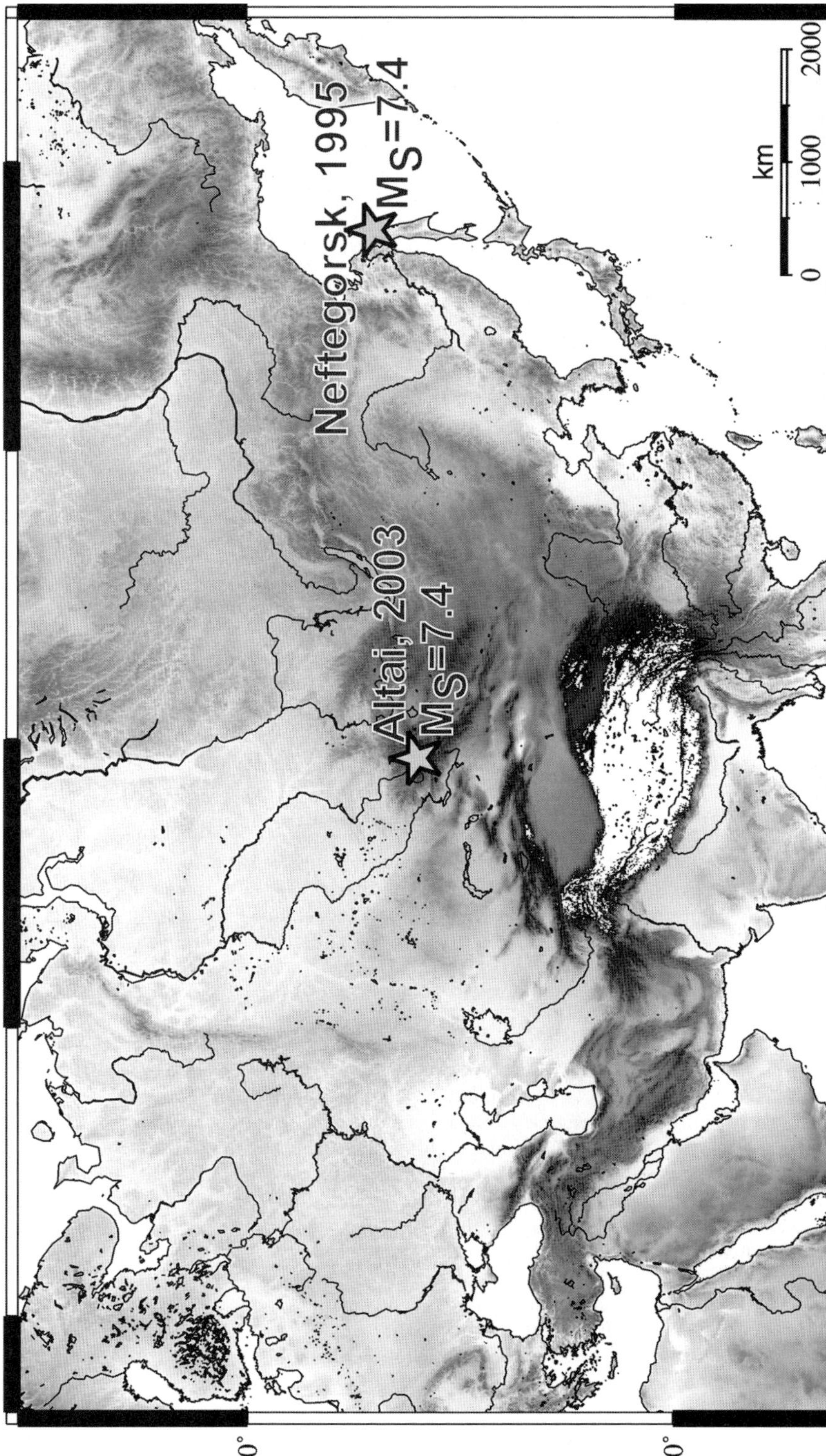

Fig. 3. Location of epicentres of the Altai 2003 and the Neftegorsk 1995 earthquakes.

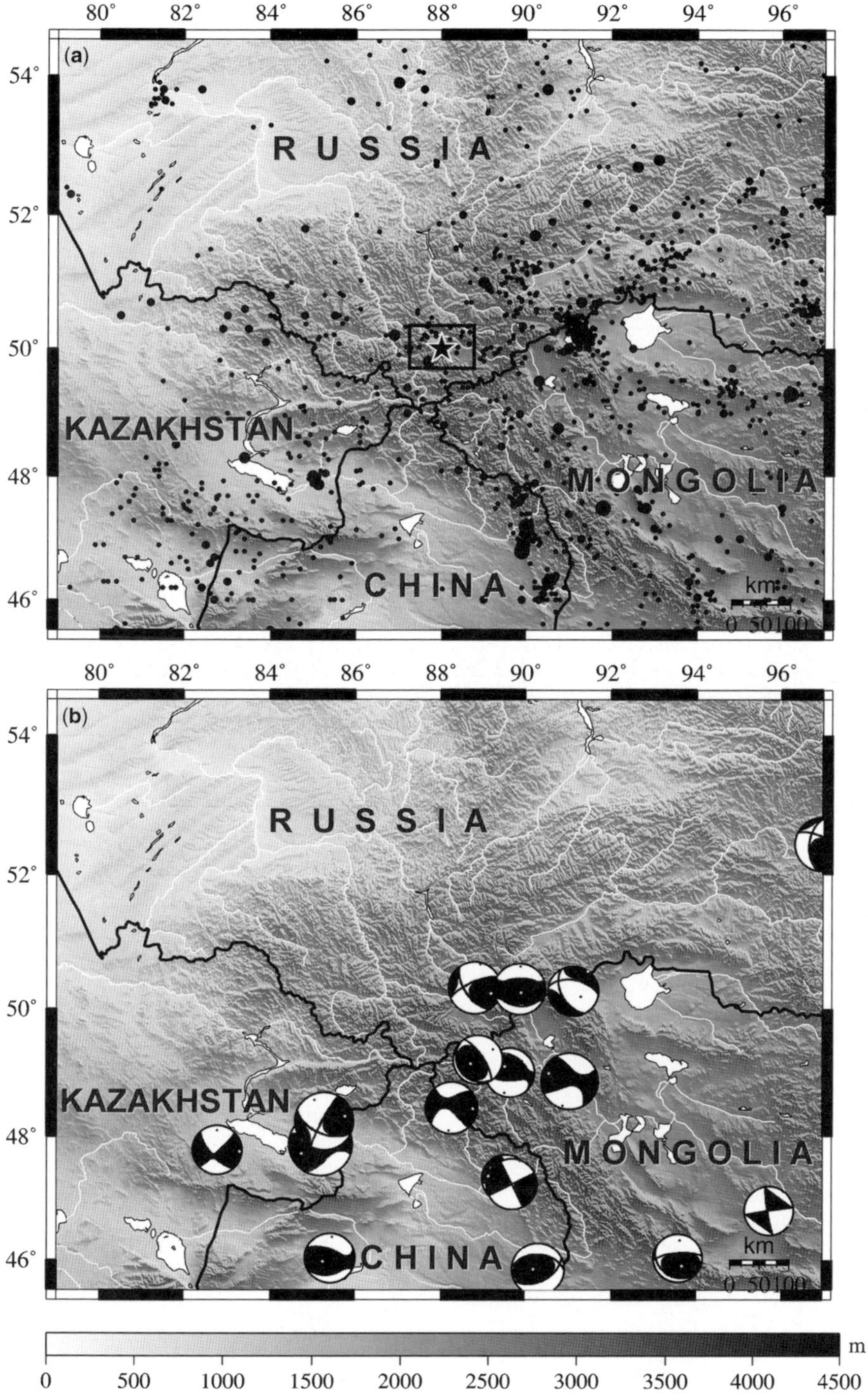

Fig. 4. Geographical and seismotectonic settings of the epicentral area of the Altai 2003 earthquake. Plotted area roughly corresponds to the area over which the earthquake was felt. (**a**) Epicentre map (black dots) for historical and instrumental times. Star shows position of the mainshock; rectangle outlines the area where environmental effects of the earthquake were observed. (**b**) CMT solution plots.

(up to 2000 km of distance); the area of its environmental effects is outlined by a rectangle in Figure 4a. The earthquake did not occur in an aseismic place; however, all previously known earthquakes in its epicentral area were of considerably smaller magnitudes (the largest was M = 6.0). The seismic history is short: although the first earthquake mentioned in the catalogue is in 1761, regular information starts in the 1850s. The epicentre of the 2003 earthquake is within the Altai mountain system, which crosses the Russian border and continues into Mongolia and China. The dominant type of faulting in the region is strike-slip with reverse and normal components (Fig. 4b) (CMT catalog 2008). According to the same source (CMT catalog 2008) the mainshock in 2003 was a strike-slip event with a small thrust component.

The aftershock sequence is unusual. Within three days earthquakes with magnitudes $M_S = 7.4$ (the mainshock), 6.6 and 7.0 occurred. This is very far from what is expected according to Båth's statistical law, which predicts the magnitude of the largest aftershock to be one unit smaller than the mainshock magnitude (Båth 1965). These three shocks have similar centroid moment tensor (CMT) solutions, in which the predominant movement is strike-slip (Fig. 5a). One of the nodal planes strikes NW–SE, which is perfectly consistent with the general orientation of surface faulting and extension of secondary EEE effects. Possibly some of the EEE in the northern part of the epicentral area occurred due to the $M_S = 7.0$ aftershock. But because field survey (Rogozhin *et al.* 2007) started two weeks after the mainshock, this hypothesis can not be verified.

The Altai earthquake aftershock map is shown in Figure 5b. Due to the epicentral observations with dense temporary network, the location accuracy is high (within 1–3 km) (Arefiev *et al.* 2006). The whole length of the surface faulting is pronounced in the aftershock area.

Both primary (Fig. 6) and secondary (Fig. 7) EEE types were associated with the Altai earthquake. The dominant type of surface faulting is strike-slip with some vertical component, which is consistent with CMT solution. In the northern termination normal faulting was observed, oblique to the general orientation of the system of strike-slip faults (Fig. 8a). It is not clear if it occurred during the mainshock, or the magnitude 7.0 aftershock. It is not unusual for strike-slip faults to terminate with normal or reverse faulting, which helps to accommodate strains accumulated within the system of strike-slip faults. Various secondary effects were observed: ground cracks, liquefactions, rockfalls and landslides. Faults of vibrational origin occurred parallel to the surface faulting. There is almost no offset along them (or it is negligible). We consider them as secondary EEE type.

The total length of surface faulting is *c.* 70 km, maximum horizontal offset reaches 2 m, and maximum vertical offset is up to 70 cm. Applying the ESI 2007 scale, the epicentral intensity is X degrees based on surface faulting parameters (close to the limit between X and XI degrees). The same epicentral intensity degree follows from the total area (*c.* 3000 km^2) affected by secondary EEE. Consistency between epicentral intensity assessments based on different effect types and sizes (total length and maximum offset of surface faulting, and total area of secondary EEE) makes $I_0 = X$ a reasonably accurate evaluation. Thanks to detailed field observations, data on fault segment parameters are available; this makes it possible to differentiate at least two zones within the epicentral area (Fig. 8a). In zone A horizontal offsets of different segments vary from 1 to 2 m; in zone B offsets are in the order of some tens of centimetres (but not reaching 50 cm). A valley where the surface faulting has a different strike from the general orientation of the fault system, separates these two zones. The instrumental epicentre of the $M_S = 6.6$ aftershock, that occurred 6 h 10 min after the mainshock, coincides with this place. In zone B four aftershocks occurred with $M_S > 4.5$, the largest one with $M_S = 7.0$. Intensity X can be assessed in the central and southern parts (zone A) and intensity IX in the northern branch (zone B), based on details of slip distribution along the surface faulting system.

Figure 8a gives information on macroseismic intensities in villages located in the epicentral area. The Altai earthquake was not destructive, but it caused damage (Gol'din *et al.* 2004). The highest observed intensity was VIII degree assigned to Beltir. There is a certain inconsistency: the village of Beltir is *c.* 7 km distant from the fault segment to which intensity X has been evaluated above (zone A). Therefore, macroseismic intensity in Beltir is less than what could be expected taking into account distance from the fault to the locality and surface faulting parameters. From the panoramic view of Beltir (Fig. 8b) it seems nothing serious happened there. But Figure 8c and d show extensive EEE in the village: ground collapse and liquefaction. The fact that buildings were not destroyed can probably be explained by the special behaviour of wooden constructions, which keep their integrity even under high seismic loadings.

A couple of years before the earthquake an expedition had been sent to the Altai region to find evidence of palaeoearthquakes (Rogozhin & Platonova 2002). The idea to check this region for palaeoearthquakes follows from the short-term seismic history and the inconsistency of the maximum observed earthquake magnitudes in the Russian and Mongolian parts of the Altai

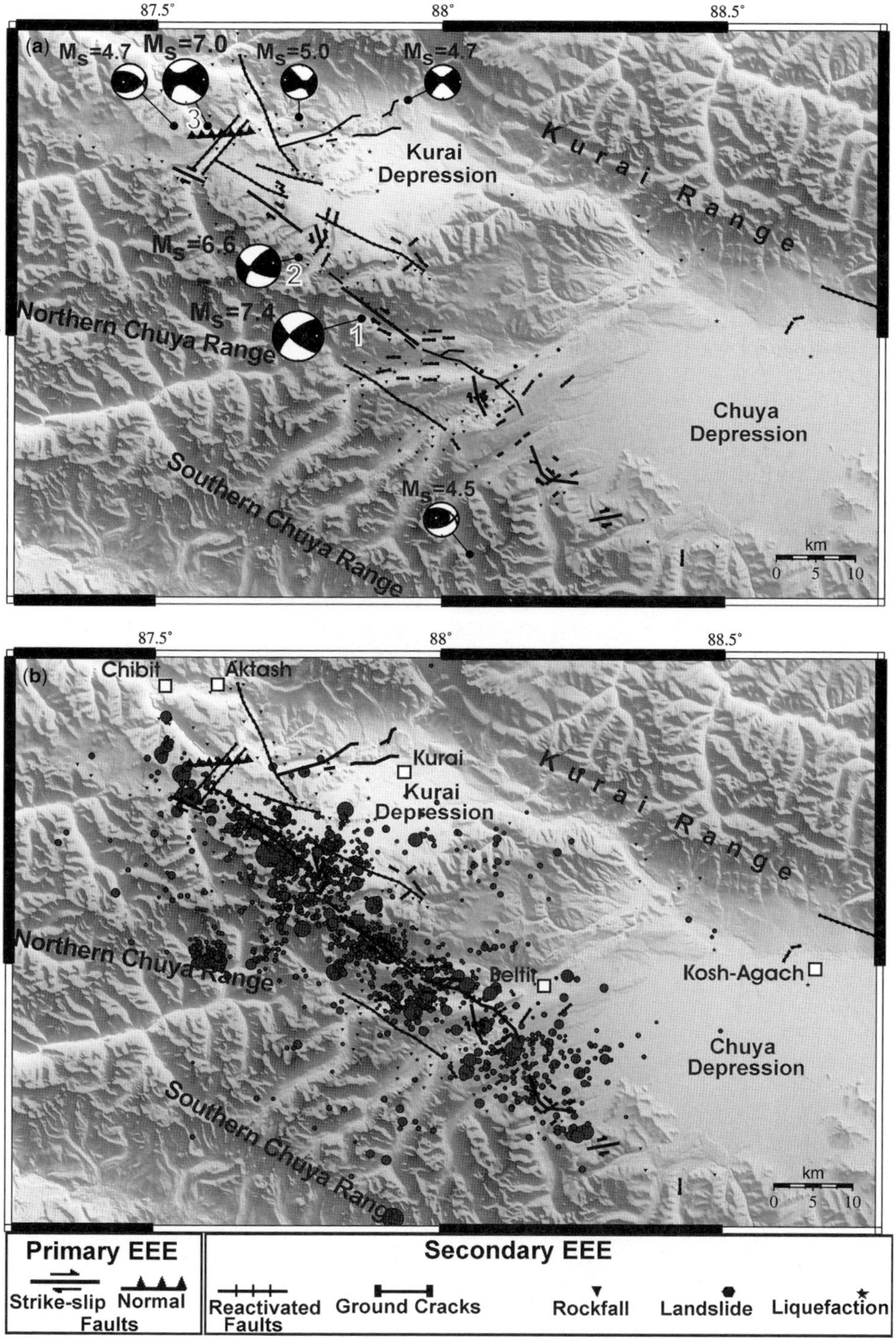

Fig. 5. Environmental effects and aftershocks of the Altai earthquake. (**a**) CMT solutions of largest aftershocks: numbers give sequential order of earthquakes (1 is the mainshock). (**b**) Epicentre map.

Fig. 6. Photos of surface faulting associated with the Altai earthquake.

Fig. 7. Photos of secondary EEE associated with the Altai earthquake: (**a**) rockfall, (**b**, **d**) landslides, (**c**) ground collapse.

Mountains. In the Mongolian part, earthquakes with $M > 8$ are known, while in the Russian part maximum reported magnitudes were 6 or less. A wealth of data were collected, which allowed ^{14}C dating. Some additional data were collected during observations in the epicentral area of the Altai earthquake. The map in Figure 9 shows locations of all the sites, where dating gives *c.* 1000 years BP. From this area, we can deduce the palaeoearthquake magnitude to be in the range 7.0–7.5 taking into account uncertainties of relationships for magnitude evaluation and accuracy of dating (we cannot exclude the possibility that not all samples are linked with one earthquake).

The Neftegorsk earthquake, 27 May 1995

The Neftegorsk earthquake ($M_S = 7.4$) occurred in lowlands of Sakhalin Island. Topography is flat, with the highest point in the epicentral area at *c.* 120 m a.s.l. The earthquake almost completely destroyed the town of Neftegorsk. The seismic history of the region is very short: it started practically together with instrumental seismology (the

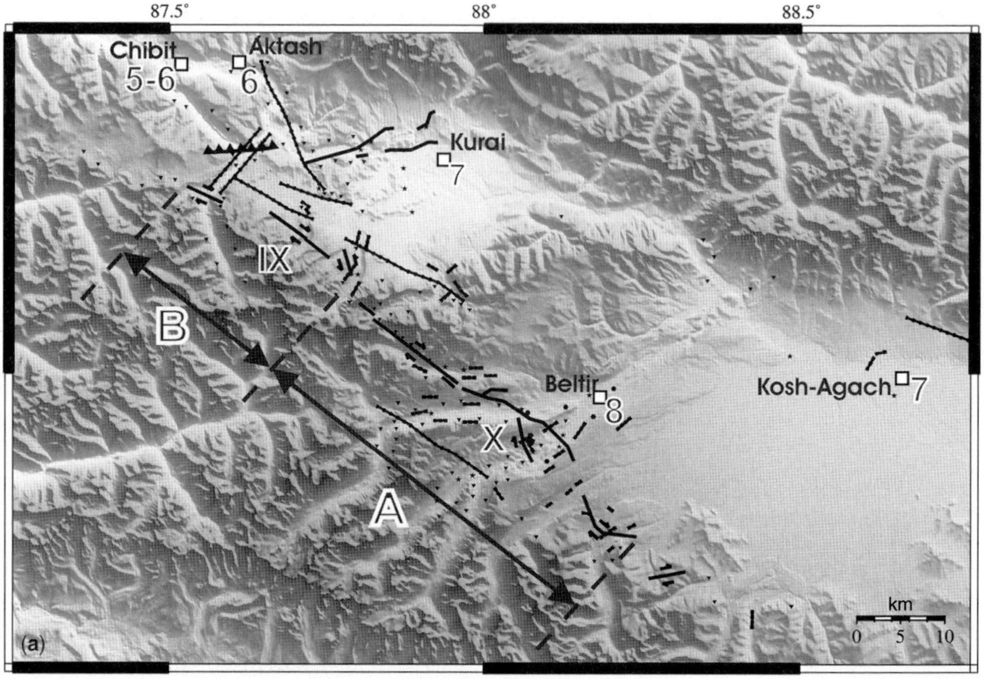

Fig. 8. Macroseismic and environmental effects of the Altai earthquake. (**a**) Macroseismic (Arabic numerals; Gol'din *et al.* 2004) and EEE (Roman numerals) intensities. A and B are two zones with considerably different surface faulting parameters. (**b**) Panoramic view of Beltir; (**c**) ground collapse and (**d**) liquefaction in Beltir.

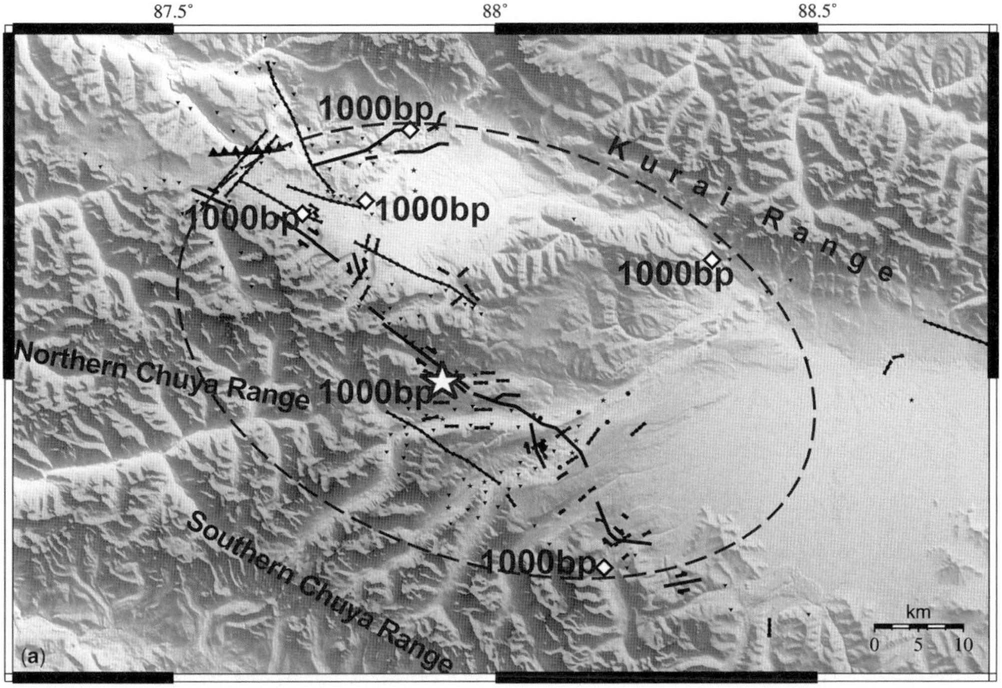

Fig. 9. Palaeoearthquake in the epicentral area of the Altai earthquake. (**a**) Site map, where palaeosoil dating gives 1000 BP (marked by diamonds). Star marks the site of photo in (b). (**b**) Segment of the surface faulting. Arrow points to the place where the sample of palaeosoil for dating was taken.

first earthquake in the catalogue is dated to 1895). The Neftegorsk earthquake occurred in a place where magnitudes known in the vicinities of the mainshock prior to 1995 are 2 units smaller. Larger earthquakes with magnitudes greater than 7 are concentrated between Sakhalin and Hokkaido islands (Fig. 10a). Most of the CMT solutions correspond to strike-slip faults, some to thrust faults (Fig. 10b).

The aftershock sequence of the Neftegorsk earthquake is also unusual. But, in contrast to the Altai earthquake sequence, the magnitude of the largest aftershock is much smaller than could be expected from Båth's law: within six months after

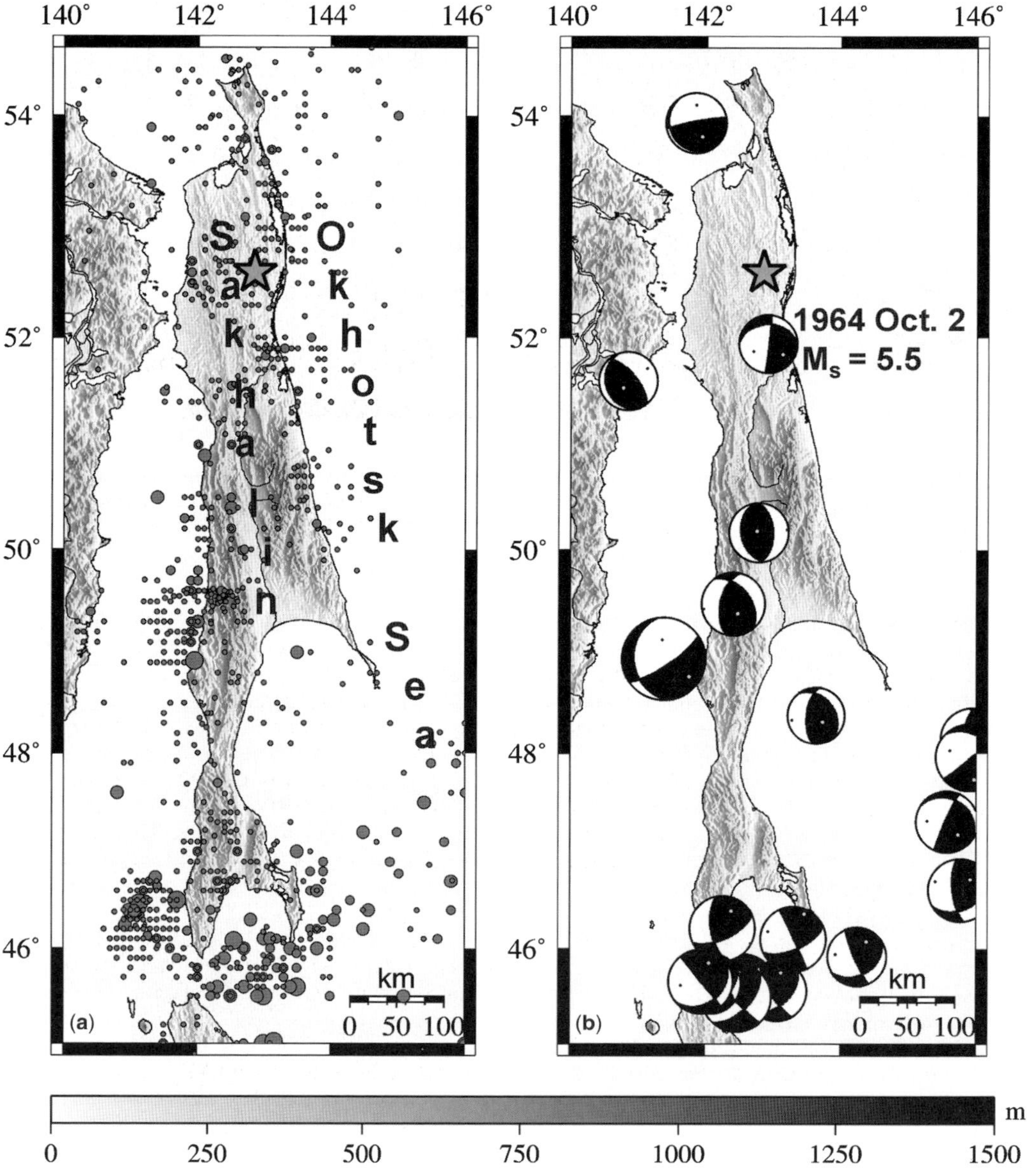

Fig. 10. Geographical and seismotectonic settings of the epicentral area of the Neftegorsk earthquake. (**a**) Epicentre map (black dots) for historical and instrumental times. Star shows position of the mainshock. (**b**) CMT solution plots. The Nogliki 1964 earthquake ($M_S = 5.5$) is the largest that occurred in the vicinity of the Neftegorsk earthquake (at *c*. 100 km distance).

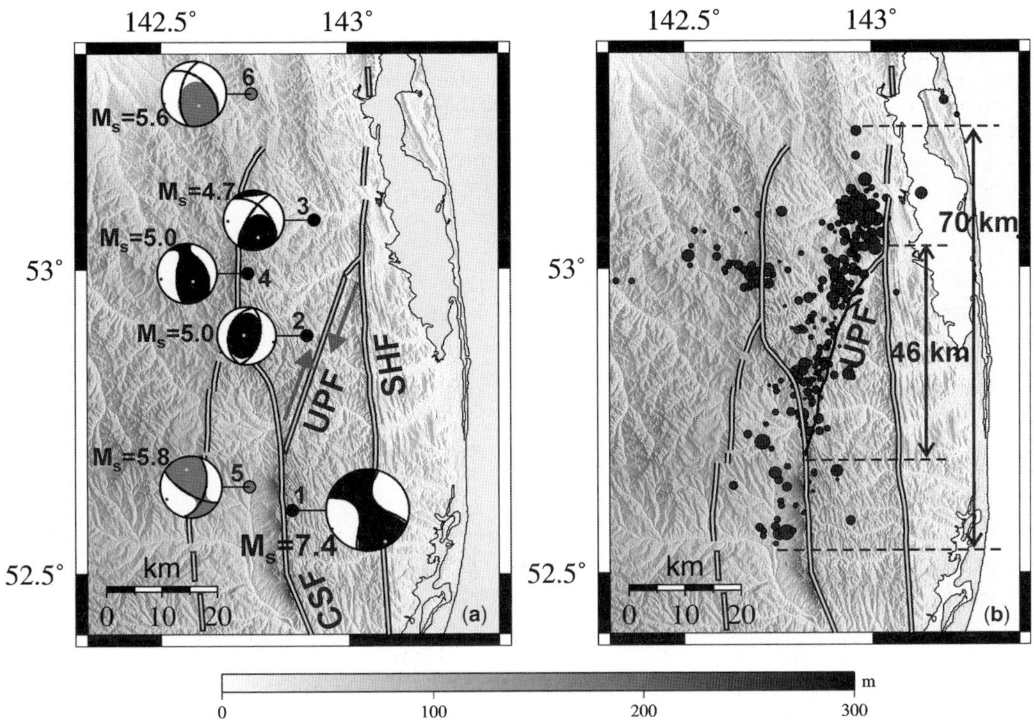

Fig. 11. Aftershocks and surface faulting of the Neftegorsk earthquake. (**a**) CMT solutions of largest aftershocks. Numbers give sequential order of earthquakes (1 is the mainshock). Aftershocks 5 and 6 occurred nine months after the mainshock. Abbreviations: CSF, Central Sakhalin Fault; SHF, Sakhalin-Hokkaido Fault; UPF, Upper Piltun Fault. (**b**) Epicentre map. Solid black line marks trace of the surface fault.

the mainshock ($M_S = 7.4$) the biggest aftershock was 5.0 (Fig. 11a). Later on, two shocks with $M_S > 5$ occurred in December 1995 and January 1996 but, because of the long time gap after the mainshock, they can be considered as independent earthquakes. Neither the mainshock nor any of the aftershocks are associated with the global scale Sakhalin-Hokkaido Fault, extending over 2000 km (Rogozhin 1995). Surface faulting (see Fig. 12 for details) shows that the source is associated with the low-rank Upper Piltun Fault. Aftershock cloud length is approximately 70 km, which is much more than the surface rupture total length (46 km). Therefore, only part (*c.* 65%) of the source rupture was exposed in surface faulting (Fig. 11b). This situation is not exceptional: rare aftershock cloud length coincides with surface rupture length. But even if this difference is remarkable, the corresponding EEE intensity is often the same or differs not more than one degree.

The dominant type of faulting is strike-slip (Fig. 12). Considerable thrusting occurs only at fault terminals. Surface rupture is continuous, except a small piece near Neftegorsk. But slip distribution along the rupture is very inhomogeneous. The central part (C), where practically pure strike-slip movements were observed, shows offset from 4 to 8 m (maximum 8.1 m). In parts B and D the vertical and horizontal components of slip are of the same order (though the horizontal is everywhere larger); neither component exceeds 4 m (Fig. 12a, b) (Kozhurin & Strel'tsov 1995; Shimamoto *et al.* 1995). The solid vertical bar in Figure 12b is plotted at 3.9 m. This is the slip amount consistent with CMT solution. The fault segment A is the only piece separated from the main rupture; horizontal and vertical offsets are just a few centimetres. Secondary EEE appeared occasionally, mostly in seashore sands, in the form of liquefactions and ground collapse. They cannot be used for I_0 assessments because, as shown in Tatevossian (2007), secondary effects that occurred out of the area of maximum EEE are not representative for epicentral intensity.

In this case epicentral intensity assessment based on EEE meets some difficulties. There is a certain inconsistency between surface faulting length (46 km) and maximum offset (8.1 m). They give epicentral intensity X and XI respectively. One of the reasons for such a discrepancy can be the fact,

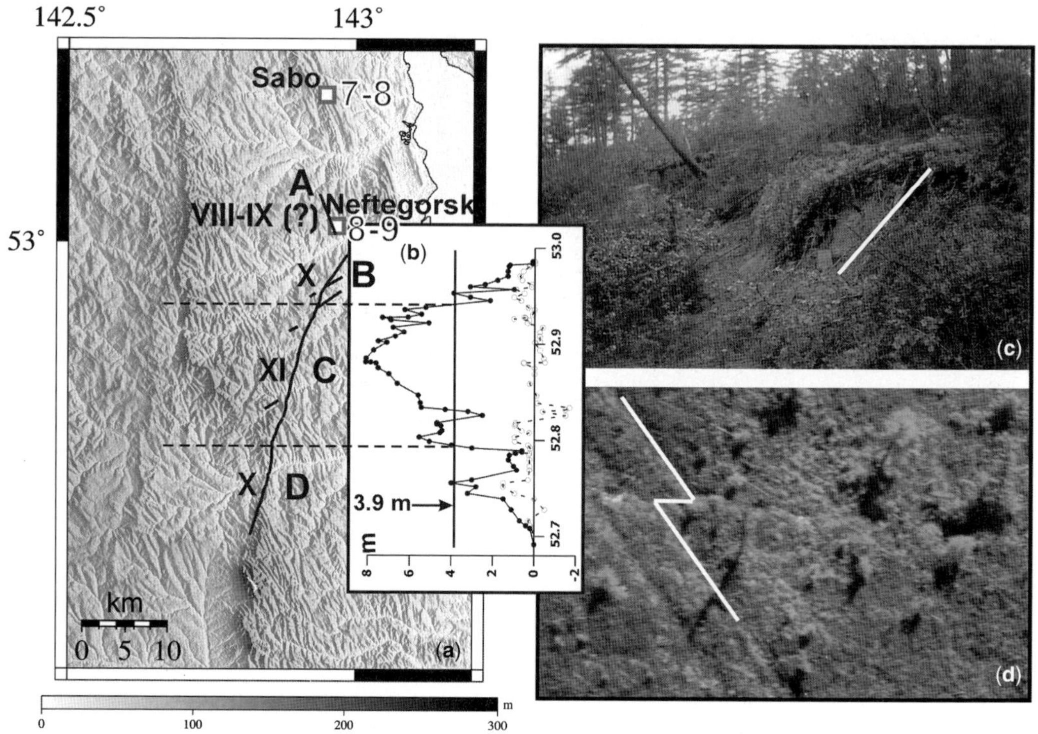

Fig. 12. Surface faulting parameters and images. (**a**) Fault map. A, B, C, D are four segments with considerably different slip parameters. In Arabic numerals are macroseismic intensities from Ivashchenko *et al.* (1995); in Roman numerals are EEE intensities. (**b**) Vertical (dashed line) and horizontal (solid) offsets along the fault. (**c**) Photo of thrust offset. (**d**) The strike-slip segment of the fault, viewed from a helicopter.

already mentioned, that the surface fault length is only 65% of the aftershock cloud length, so it does not represent the whole source. For a 70 km total length the discrepancy would be less dramatic. The other reason is the anomalously large offset for a magnitude 7.4 seismic event likely caused by local conditions. The epicentral intensity in the range X–XI gives the most consistent assessment both with surface faulting total length and maximum slip.

Slip distribution over rupture length enables differences in intensity to be recognized in the central and marginal parts of surface faulting. Corresponding intensities for zones A–D are marked in Figure 12a.

Macroseismic and EEE intensities can be compared for only one locality, Neftegorsk, which is the only place where EEE were observed in close vicinity to an inhabited locality. In principle, macroseismic and EEE intensities are mutually consistent: both give VIII–IX degree. But there was some uncertainty in macroseismic intensity assessment due to the problem of the accurate evaluation of vulnerability of constructions (Koff *et al.* 1995). This problem is illustrated in Figure 13a and Table 3. We get very incoherent intensity assessments based on different building types. Compared with the problems of intensity assessment based on damage to buildings, one degree of uncertainty in EEE epicentral intensity assessment is acceptable.

Gaps in spatial distribution of macroseismic data are evident from Figure 13b. In densely populated Europe it may be hard to imagine how large can be the gaps in sparsely populated areas in other parts of the world.

At the same distances, macroseismic intensities north of the rupture zone are higher than to the south. Probably we are dealing with some directivity effect. This was also observed in the case of one of the largest aftershocks, for which we have strong motion records (Fig. 14). At two stations (NFG and SNFG), located in the same source-station azimuth, maximum acceleration of ground motion (a_{max}) recorded during the $M_S = 4.7$ aftershock is practically the same, although the SNFG is twice as far from the epicentre. Meanwhile, the a_{max} at station SAB, located at the same distance from the epicentre as SNFG but in the opposite direction, is almost twice as high.

Fig. 13. Macroseismic effects of the Neftegorsk earthquake. (**a**) Panoramic view of Neftegorsk town. (**b**) Map of localities that felt intensity IV and higher (from Ivashchenko *et al.* 1995). Dashed area shows the gap in locality distribution.

In the fault zone of the Neftegorsk earthquake several trenches were cut. Figure 15 shows evidence of a palaeoearthquake. Carbon-14 dating of buried palaeosoil cut by a liquefaction channel gives the age 1800 BP (Rogozhin 1995).

Conclusions

Table 4 summarizes a comparison of the results of the Altai and the Neftegorsk earthquake studies.

The seismicity pattern in the regions of both earthquakes (low seismic activity in historic and instrumental times together with evidence of strong palaeoearthquakes in prehistoric times) stresses the importance of the EEE-based intensity scale: it is the only scale able to extend to the past our knowledge of seismic history. Another reason supporting the importance of the ESI 2007 scale is the sparse distribution of localities in the epicentral areas of both earthquakes. The nearest locality in the case of the Altai earthquake is approximately 7 km distant from the surface fault segment. Neftegorsk town, the closest locality to the surface fault, is situated at its terminus, where offset is only a few centimetres, while maximum observed slip reaches 8.1 m. As a result, maximum observed intensities based on damage in inhabited localities have a considerably lower degree than epicentral intensities. Both earthquakes again raise the problem of the correct evaluation of the vulnerability of constructions. For example, the intensity VIII assessed for Beltir (Altai earthquake) may follow from the specific behaviour of wooden buildings: they are able to keep the integrity of the construction under heavy seismic loading. Looking at ground collapse in Beltir it is easy to imagine what would happen there if there were concrete five-storey buildings, which were completely ruined in Neftegorsk town. But of course, all problems cannot be reduced only to 'external' factors, like construction quality: the strong azimuthal dependence of a_{max} found in records for one of the largest aftershocks of the Neftegorsk earthquake suggests that a specific mainshock source or local structural properties can be responsible for the observed asymmetry of the spatial distribution of damage.

Testing the ESI 2007 scale shows that one of the sources for inaccuracy in intensity assessment can be related to strong aftershock(s) creating new, or extending an existing segment of surface faulting that occurred during the mainshock. But this is not a problem of only EEE-based scale: macroseismic effect also can have a cumulative nature, when a strong aftershock follows the mainshock. Moreover,

Table 3. *Intensity assessment in Neftegorsk based on damage to buildings*

Building type	Intensity
Complete collapse of 17 five-storey buildings (type A or A1)	IX–X
Damage degree 3–4 to buildings of type B (5 buildings)	VIII
Damage degree 1 to type C7 (4 buildings)	VII
Final assessment for the town	VIII–IX

Based on Koff *et al.* (1995).

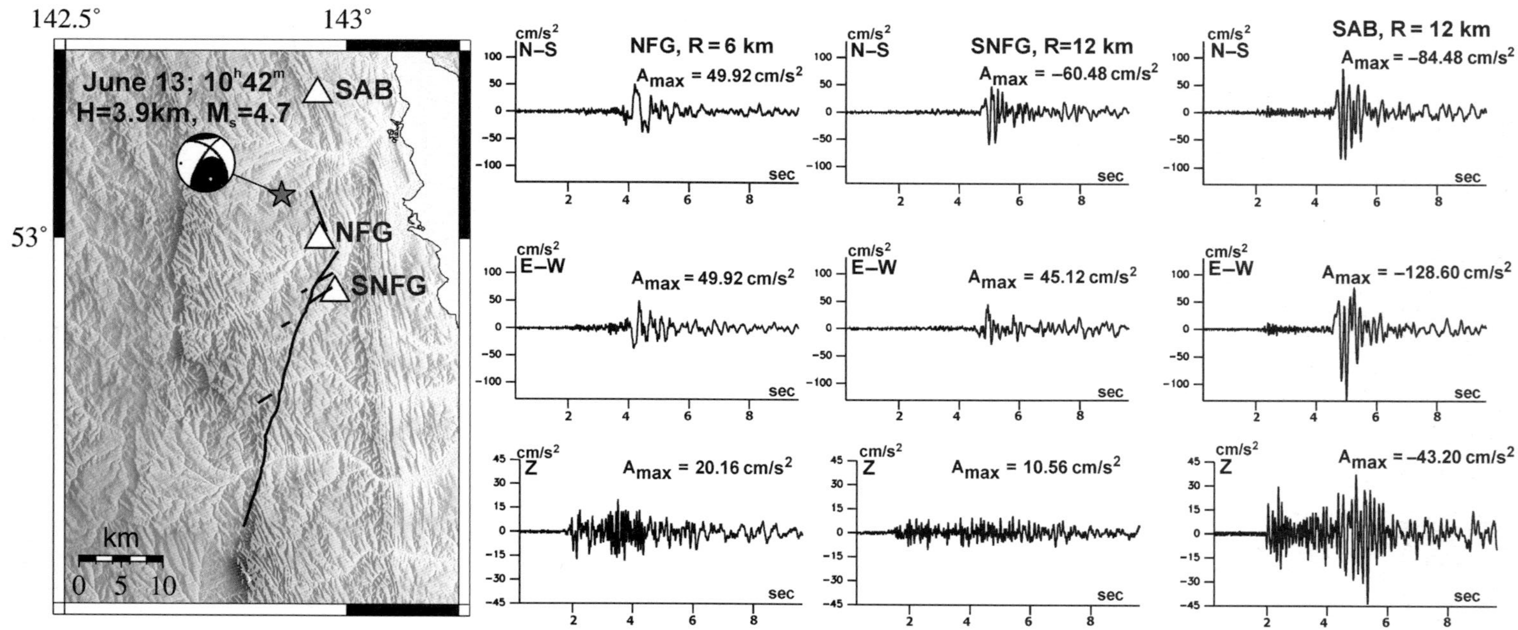

Fig. 14. Station map and strong motion records from $M_S = 4.7$ aftershock of the Neftegorsk earthquake at stations SAB, NFG and SNFG. The star marks epicentre position.

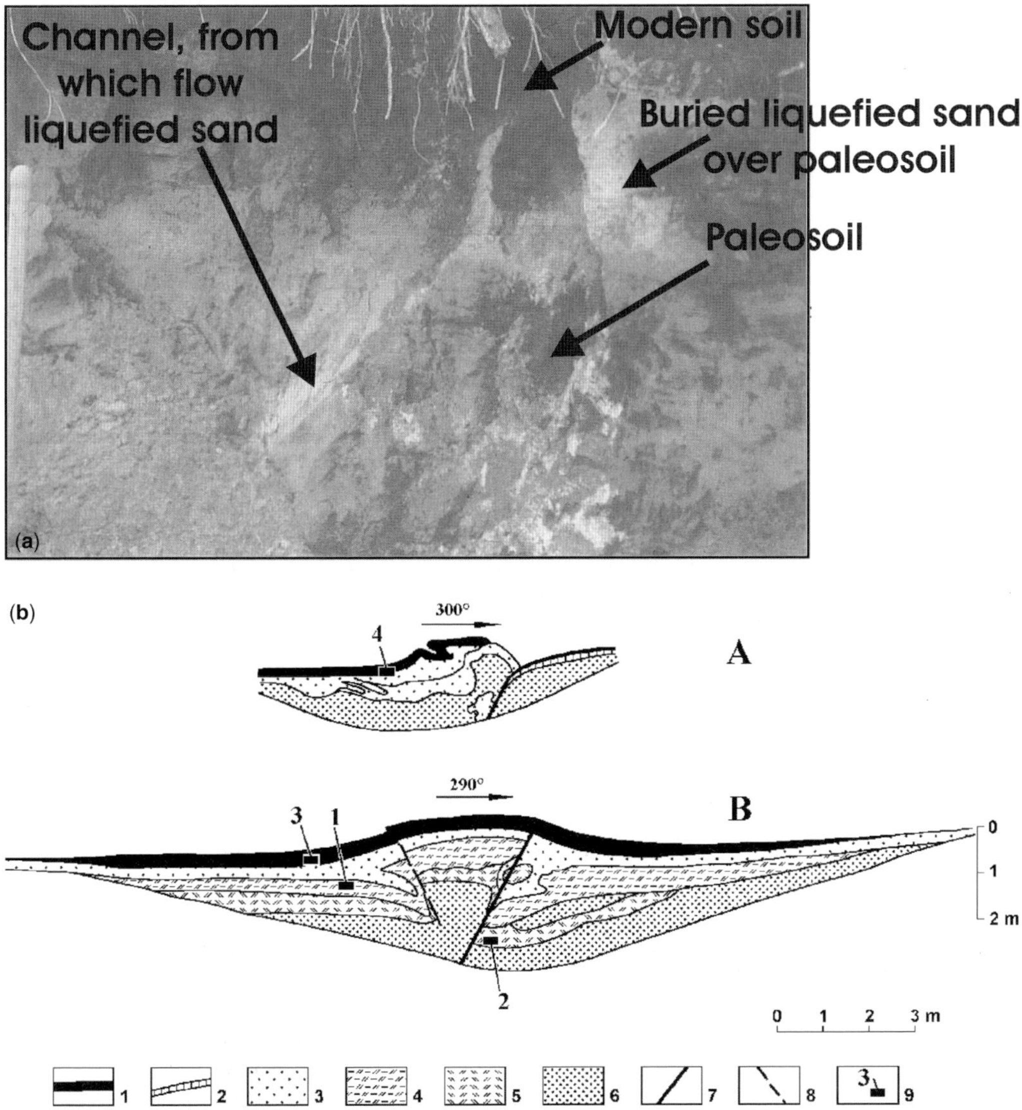

Fig. 15. Evidence of palaeoearthquake in the fault zone of the Neftegorsk earthquake (after Rogozhin 1995). (**a**) Photo of the trench. (**b**) Schematic cross-section Legend: 1, modern soil; 2, loam; 3, white quartz sand; 4, buried palaeosoil (hard); 5, buried palaeosoil (loose); 6, yellow sand; 7, main fault plane; 8, auxiliary fault plane; 9, places from which palaeosoil samples were taken for ^{14}C dating.

the cumulative effects can be stronger in the case of damage to buildings, as aftershock affects buildings already weakened by the seismic loading from the mainshock.

When assessing I_0, it should be remembered that the whole source rupture length is not always manifested in surface faulting. This can bring considerable inconsistency between the surface faulting length and maximum offset along it. The most coherent I_0 assessment in such cases would be to give a range of intensity grades, corresponding to each of the parameters. Probably the weighting should be introduced giving higher impact to the offset value.

The total area of secondary EEE type can be used to evaluate I_0 but this assessment can be reliable only if the conditions are favourable for their occurrence. For example, in the case of the Altai earthquake, intensity assessments based on primary and secondary EEE are coherent, while

Table 4. *Comparison of Neftegorsk and Altai earthquakes*

Feature	Altai, 2003	Neftegorsk, 1995
1. Seismicity pattern	The earthquake magnitude is 1.3 units larger than ever observed in its epicentral area. In the same seismotectonic zone, at much larger distances, very big earthquakes are known (gigantic seismic events in Mongolian Altai).	The earthquake magnitude is 1.9 units larger than maximum ever observed in its epicentral area. In the same seismotectonic zone, at much larger distances, very big earthquakes are known ($M \geq 7$ seismic events between Sakhalin and Hokkaido islands).
2. Palaeoearthquakes	Seismic history of the region is short. Addressing palaeoearthquake data completely changes our evaluation of the seismicity pattern. Evidence of $M \approx 7.5$ earthquakes found 1000 BP.	Seismic history of the region is short. Addressing palaeoearthquake data completely changes our evaluation of the seismicity pattern. Evidence of $M \approx 7.5$ earthquakes is found 1800 BP.
3. Aftershock sequence	Anomalous aftershock sequence: high magnitude swarm (3 events with M_S 6.6–7.4 in 3 days).	Anomalous aftershock sequence: very large (2.4 units) gap between mainshock and largest aftershock magnitudes (in 6 months).
4. Surface faulting parameters	Surface faulting length is 70 km. Dominant type is strike-slip (maximum slip 2 m), maximum vertical offset is 60 cm. According to the aftershock distrnhibution the whole source exposed in surface faulting.	Surface faulting length is 46 km. Almost pure strike-slip (maximum slip 8.1 m), vertical offsets are observed only at surface fault terminals (maximum 2 m). According to the aftershock distribution *c.* 65% of the source exposed in surface faulting.
5. Epicentral intensity assessment	Intensive secondary effects (landslides, rockfalls, ground cracks) accompany the zone of surface faulting. Their total area can be used to assess I_0. It is consistent with assessment based on surface faulting length and offset – X degrees.	Secondary EEE were observed occasionally (mostly liquefactions along seashore). They cannot be used for I_0 assessment. I_0 based on surface faulting length is X, based on maximum offset – XI. Offset, consistent with M_0, gives $I_0 = X$.
6. Macroseismic and EEE intensity relationships	Macroseismic intensity assessment faced considerable problems in correct evaluation of vulnerability of wooden buildings.	Intensity assessment in Neftegorsk demonstrates problems to get coherent intensity assessment for different construction types. The range of intensities spans from VII to IX–X.

for the Neftegorsk earthquake secondary EEE (mostly liquefactions) occurred occasionally in seashore sands and give very little information for I_0 assessment.

The main reason why environmental effects were excluded from intensity assessments (e.g. in EMS 98) is their instability: EEE can be found in a very wide range of intensities. But this is true mostly for secondary EEE types. Moreover, this instability comes from a lack of statistical approach. Indeed, when the intensity assessment is deduced from a single EEE (a single landslide) its volume can be strongly affected by vulnerability of the slope. But when different levels of spatial generalization of information are introduced (Fig. 1b) the assessments become more stable. On the other hand, the dependence of intensity assessments on vulnerability is inherent to macroseismic effects: intensity assessment in Neftegorsk ranges within four degrees, depending on building types taken into account for that purpose (Fig. 13, Table 3).

Finally, if the environmental effects were ignored we get $I_0 =$ VIII–IX for the Neftegorsk earthquake: this for an earthquake that produced 8.1 m of slip at the surface. What engineering construction can stand such an offset, being built just over the surface fault? And what should we expect (how many metres of surface fault offset) from I_0 X, or XI, if intensity VIII–IX is ascribed to an earthquake generating 8.1 m of offset? By excluding the environmental effects, especially primary EEE, we not only miss a valuable piece of information, sometimes the only one available in sparsely populated areas, we also miss the low frequency (static) content of an earthquake impact. Any macroseismic scale based only on degree of damage to man-made constructions emphasizes the vibrational part (ground acceleration) of the earthquake impact.

This simplification helps to construct a coherent intensity scale. In the far field, where static deformations are negligible, such an approach can be considered reasonable. But in the epicentral area, where the static offset reaches the order of several metres, intensity assessments ignoring this component are useless. Therefore, taking into account environmental effects in intensity scales is an essential requirement in densely populated areas also: it follows from the complex nature of an earthquake impact, which spans a very broad frequency range, including static deformations.

This work is partly supported by RFBR grants 07-05-00702a, 08-05-00103a and 08-05-00598a. The authors would like to express gratitude to Leonello Serva and Vladimir Trifonov for careful revision of the manuscript and helpful remarks.

References

AREFIEV, S. S., APTEKMAN, Z. Y. *ET AL.* 2006. The source and aftershocks of the Altai (Chuya) earthquake of 2003. *Izvestiya, Physics of the Solid Earth*, **42**(2), 167–177.

BÅTH, M. 1965. Lateral inhomogeneities of the upper mantle. *Tectonophysics*, **2**(6), 483–514.

CMT CATALOG. 2008. Global Centroid Moment Tensor database (formerly Harvard CMT catalog). Available at: http://www.globalcmt.org

DENGLER, L. & MCPHERSON, R. 1993. The 17 August 1991 Honeydaw earthquake North Coast California: a case for revising the Modified Mercalli Scale in sparsely populated areas. *BSSA*, **83**, 1081–1094.

GOL'DIN, S. V., SELEZNEV, V. S. *ET AL.* 2004. The Chuya (Altai) earthquake in 2003: Materials of seismological survey. *In*: GLIKO, A. O. (ed.) *Sil'noye zemletryaseniye na Altaye*. IFZ RAN, Moscow, 55–60 [in Russian].

GRÜNTHAL, G. (ed.) 1998. *European Macroseismic Scale 1998*. Cahiers du Centre Europeen de Geodynamique et de Seismologie, **15**, Luxembourg.

GUERRIERI, L. & VITTORI, E. (eds) 2007. Intensity scale ESI2007. *Memorie descritive della carta geologica d'Italia*, **LXXIV**.

GUERRIERI, L., TATEVOSSIAN, R. *ET AL.* 2007. Earthquake environmental effects (EEE) and intensity assessment: the INQUA scale project. *Bollettino della Società Geologica Italiana*, **126**(2) 375–386.

INTERNATIONAL SEISMOLOGICAL CENTRE. 2003. *On-line Bulletin*. Available at: http://www.isc.ac.uk/ Bull. ISC, Thatcham, UK.

IVASHCHENKO, A. I., KUZNETSOV, D. P. *ET AL.* 1995. The Neftegorsk 27(28) May, 1995, earthquake in Sakhalin. *In*: *Federal'naya sistema seismologicheskikh nablyudeniy i prognoza zemletryaseniy*. Neftegorskoye zemletryaseniye 27 (28).05.1995, Moskva, 48–67 [in Russian].

KOFF, G. L., KOTLOV, V. F., TEN SU MUN, LADONTSEV, E. A. & SHAKHRAMAN'YAN, M. A. 1995. Engineering analyses of macroseismic consequences of the Neftegorsk 27(28).05.95 earthquake. *In*: *Federal'naya sistema seismologicheskikh nablyudeniy i prognoza zemletryaseniy*. Neftegorskoye zemletryaseniye 27 (28).05.1995, Moskva, 139–154 [in Russian].

KOZHURIN, A. I. & STREL'TSOV, M. I. 1995. Seismotectonic manifestations of the earthquake in May 27(28), 1995 in Northern Sakhalin. *In*: *Federal'naya sistema seismologicheskikh nablyudeniy i prognoza zemletryaseniy*. Neftegorskoye zemletryaseniye 27 (28).05.1995, Moskva, 95–100 [in Russian].

MICHETTI, A. M., ESPOSITO, E. *ET AL.* 2004. The INQUA scale. An innovative approach for assessing earthquake intensities based on seismically-induced ground effects in natural environment. *Memorie descritive della carta geologica d'Italia*, **LXVII**.

ROGOZHIN, E. A. 1995. The Neftegorsk, May 27 (28), 1995, earthquake: Geological effects and tectonic setting of the source. *In*: *Federal'naya sistema seismologicheskikh nablyudeniy i prognoza zemletryaseniy*. Neftegorskoye zemletryaseniye 27 (28).05.1995, Moskva, 80–94 [in Russian].

ROGOZHIN, E. A. & PLATONOVA, S. G. 2002. *Source zones of strong earthquakes in Altai Mountains in Holocene*. IFZ RAN, Moscow [in Russian].

ROGOZHIN, E. A., OVSYUCHENKO, A. N., MARAKHANOV, A. V. & USHANOVA, E. A. 2007. Tectonic position and geological features of the Altai, 2003, earthquake. *Geotektonika*, **2**, 3–22. [in Russian].

SHIMAMOTO, T., VATANABE, M. & SUDZUKI, Y. 1995. Surface faulting of the Neftegorsk May 27(28), 1995, earthquake. *In*: *Federal'naya sistema seismologicheskikh nablyudeniy i prognoza zemletryaseniy*. Neftegorskoye zemletryaseniye 27 (28).05.1995, Moskva, 101–116 [in Russian].

TATEVOSSIAN, R. E. 2007. The Verny, 1887, earthquake in Central Asia: Application of the INQUA scale based on coseismic environmental effects. *Quaternary International*, **173–174**, 23–29.

Surface and subsurface palaeoseismic records at the ancient Roman city of *Baelo Claudia* and the Bolonia Bay area, Cádiz (south Spain)

PABLO G. SILVA[1]*, KLAUS REICHERTER[2], CHRISTOPH GRÜTZNER[2], TERESA BARDAJÍ[3], JAVIER LARIO[4], JOSE L. GOY[5], CARI ZAZO[6] & PETER BECKER-HEIDMANN[7]

[1]*Dpto. Geología, Universidad de Salamanca, Escuela Politécnica Superior de Ávila. C/Hornos Caleros, 50 05003-Ávila, Spain*

[2]*Institut für Neotektonik und Georisiken, RWTH Aachen University, Lochnerstr. 4-20, 52056 Aachen, Germany*

[3]*Dpto. Geología, Universidad de Alcalá de Henares, 28871-Alcalá de Henares, Madrid, Spain*

[4]*Dpto. Ciencias Analíticas, Fac. Ciencias. Universidad Nacional de Educación a distancia (UNED), 28040-Madrid, Spain*

[5]*Dpto. Geología, Universidad de Salamanca, Fac. Ciencias, 37008-Salamanca, Spain*

[6]*Dpto. Geología, Museo Nacional de Ciencias Naturales (CSIC), 28006-Madrid, Spain*

[7]*Institut für Bodenkunde, Universität Hamburg, Allende-Platz 2, 20146 Hamburg, Germany*

**Corresponding author (e-mail: pgsilva@usal.es)*

Abstract: The Roman archaeological site of *Baelo Claudia* (Cádiz, south Spain) is located within the Gibraltar Arch, a region with no significant recent or historical seismicity. However, previous studies have emphasized the occurrence of repeated strong archaeoseismic damage (intensity ≥ IX MSK) at *Baelo Claudia* tentatively bracketed in this study around AD 40–60 and AD 260–290. A multidisciplinary study has been carried out including the detailed mapping of surface deformation and building damage, surface geology and geomorphology, collection of structural data, and an extensive ground penetrating radar (GPR) survey. The obtained data are not conclusive when considered separately, but evident links between archaeoseismic damage, structural and GPR data indicate that the destruction of the city was linked to seismic shaking. The analysis of the pattern and orientation of deformation clearly indicates SW–NE directed compression due to ground shaking. This analysis also focuses on localized landslides and liquefaction processes, which appear to be coeval with the earthquakes, but the poor geotechnical parameters of the clayey substratum were determinant to amplify the observed level of destruction. The application of the present Spanish seismic code (NCSE-02) indicates that intensity VIII MSK (0.24–0.26 g) can be reached in this zone for 500 year return periods.

Previous studies reported on the occurrence of repeated strong archaeoseismic damage (intensity ≥ IX MSK) in the ancient Roman city of *Baelo Claudia* (first to fourth centuries AD), located at the axial zone of the Gibraltar Strait in southern Spain (Menanteau *et al.* 1983; Goy *et al.* 1994; Sillères 1997; Silva *et al.* 2005, 2006). The archaeological stratigraphy of the city evidences two major episodes of abrupt city destruction tentatively bracketed during AD 40–60 and AD 350–395, separated by an intervening horizon of demolition for city rebuilding (Sillères 1997; Silva *et al.* 2005), elsewhere interpreted as characteristic of many earthquake-damaged archaeological sites in the Mediterranean (i.e. Stiros 1996; Marco *et al.* 1997; Hancock & Altunel 1997; Stiros & Papageourgiu 2001; Altunel *et al.* 2003).

The urban geology and general geomorphological features of the study area are described in the previous works of Borja *et al.* (1993), Alonso-Villalobos *et al.* (2003) and Silva *et al.* (2005), evidencing that the Roman city was differentially founded on plastic clayey substratum, Late Quaternary sandy and clayey materials thickening to the coast (up to 5 m thick), and a variable amount of poorly compacted Roman artificial fillings. In addition, all these materials are characterized by poor geotechnical parameters, including high swelling rates for the underlying clayey substratum (Borja *et al.* 1993). In detail, Silva *et al.* (2005)

From: REICHERTER, K., MICHETTI, A. M. & SILVA, P. G. (eds) *Palaeoseismology: Historical and Prehistorical Records of Earthquake Ground Effects for Seismic Hazard Assessment*. The Geological Society, London, Special Publications, **316**, 93–121. DOI: 10.1144/SP316.6 0305-8719/09/$15.00

applied the existing seismic code of Spain (NCSE-94 1997) concluding that it will be necessary to invoke site effect amplification to link the observed damage (over VIII–IX MSK) with local seismic sources.

On the other hand, these authors dismiss linking the observed ground deformations with the well-known AD 365 Crete tsunami event as earlier proposed by other authors (Menanteau *et al.* 1983), because no evidence of tsunami-related damage is recorded in the ruins. However, recent findings in small fluvial outlets east of *Baelo Claudia* indicate probable tsunami occurrences in cal. 2150–1825 BP (Alonso-Villalobos *et al.* 2003) and again in the fifteenth and sixteenth centuries (Becker-Heidmann *et al.* 2007) in Bolonia Bay. The tsunami event during Roman times is geologically well documented along the Spanish Atlantic coast of the Gibraltar Strait from the Doñana marshlands to Cádiz (Lario *et al.* 2001; Luque *et al.* 2002; Ruíz *et al.* 2004) and might correspond with one of the three historically documented tsunamis that occurred between 219 BC and 60 BC in this zone (Campos 1991; Luque *et al.* 2001). Nevertheless, strong seismic events occurred during the Roman period, presumably affecting southern Spain: the AD 33, AD 346 and AD 382 earthquakes are listed in the Spanish seismic catalogues (i.e. Galbis 1932; Martínez Solares & Mezcua 2002). All these events are poorly documented and supposedly related to the far-field Cape of San Vicente seismic source, responsible for the AD 1775 Lisbon event (Campos 1991; Luque *et al.* 2001). Dates of some of these historic events match with the assumed ages of AD 40–60 and AD 365–395 proposed for the two events of destruction recorded at *Baelo Claudia* (Menanteau *et al.* 1983; Silva *et al.* 2005). However, even these relatively strong earthquakes produced only ground motions of moderate intensities in the Gibraltar Strait region between VI–VII MSK (Martínez Solares *et al.* 1979) and V–VI EMS (Mártinez Solares 2001) as recorded during the AD 1775 Lisbon event. Ground motions leading to these intensities are insufficient to explain the extensive destruction of the city, making it necessary to consider and check the specific site effect amplification at *Baelo Claudia*.

In any case, an important limitation of archaeoseismological data is that they generally do not allow identification of the causative seismic source (i.e. capable fault) producing the observed architectural disruptions (Hancock & Altunel 1997; Altunel *et al.* 2003; Similox-Tohon *et al.* 2006). Additionally, in the particular case of *Baelo Claudia*, most of the observed archaeoseismological disruptions can be classified as secondary effects of ground shaking (Silva *et al.* 2005), but might have also been produced by other natural hazards (landslides, ground subsidence, soil creep, etc.) under specific climate and weather conditions. Therefore, thoughtful research is necessary to identify the true nature of the observed deformations by means of analysis of their surface and subsurface records. We tend to divide the observed archaeoseismological damage into two groups: (1) release of impulse-like high energy during seismic events and rupture-like processes; and (2) low energy events, mainly due to gravitational forces, producing 'slow' processes, like creep and landslides.

This work focused on the detailed mapping of surface deformations and architectural disruptions within the ancient urban zone of *Baelo Claudia*, in order to determine the extension, nature and structural pattern of the recorded deformations as earlier performed for other archaeological sites in the Mediterranean zone and central Europe (Korjenkov & Marzor 1999; Hinzen & Schütte 2003; Monaco & Tortorici 2004; Similox-Tohon *et al.* 2006). Measurement of structural data on fractures, cracks, shocks, and pop-up structures affecting the ancient Roman pavements and walls, directions of collapse of columns, houses and city walls, offer enough data to discuss the directivity and pattern of the ground movement triggering the observed architectural destruction. Therefore, we use these kinematic indicators preserved in the ruins in order to probe the directional nature of the deformations to separate effects caused by the eventual progressive ruin and burying of the city from those pointing to seismic shaking. Geophysical analysis of subsurface archaeological remains by means of an extensive ground penetrating radar (GPR) survey has been carried out in order to determine the depth extension of the observed surface deformations. Additionally GPR data have been probed to offer quality information about the subsurface location of probable event horizons.

Geodynamic setting of the Gibraltar Strait area

The Gibraltar Strait connects the Atlantic with the Mediterranean, above the convergent Africa–Eurasia plate boundary in the westernmost area of the Betic Cordilleras (southern Spain) and the Moroccan Rif (NW Africa). In this area, the Africa–Eurasia plate boundary is rather undefined as manifested by scattering of earthquake foci. The continental collision zone extends westwards by the displacement of nappes of the Betic Internal Zone. The stacking and thrusting mainly occurred during the early Miocene, including large-scale folding and back-thrusts in the Gibraltar Arc (Leblanc 1990; Sanz de Galdeano 1990; Weijermars 1991). Late Neogene post-collisional convergence resulted in the development of a diffuse plate boundary of more than 700 km width, in which the Africa–Eurasia convergence was distributed through a wide range of interactive NW- and SE-trending transpressive and transtensive structures linked to the broad NW–SE plate convergence (Weijermars 1991; Vázquez & Vegas 2000). The tectonic regime is consistent with regional studies for the Betic Cordilleras (Galindo-Zaldivar *et al.* 1993; Reicherter & Peters 2005) and the entire Iberian Peninsula (Herraiz *et al.* 2000). SH_{max} trajectories, based on Plio-Quaternary deformations, borehole breakouts, focal and moment tensor solutions (Herraiz *et al.* 2000; Jiménez-Munt *et al.* 2001; Stich *et al.* 2003), indicate NNW–SSE directed horizontal compressive stresses in the Gibraltar Strait area.

This compressive stress field induces mainly a transpressive setting, defined by the development of conjugate master strike-slip faults, trending NE–SW (left-lateral sense of displacement) and NW–SE (right-lateral sense of displacement; Fig. 1). The geodynamic situation is characteristic for the emerged northern sector of the Gibraltar Strait (Benkhelil 1977; Goy *et al.* 1995; Gracia *et al.* 1999), but has also been described from the African coast of Morocco (Akil *et al.* 1995), and deduced from the submarine sector of the Strait (Sandoval *et al.* 1996).

An obvious discrepancy between present plate convergence rates of 4–5 mm/a (Noomen *et al.* 1993) up to 5.6 mm/a (Jolivet *et al.* 1999) and estimated maximum uplift rates of 0.15 mm/a in this zone during the late Quaternary seems to indicate that no plane-constant volume strain occurs at this diffuse plate boundary (Zazo *et al.* 1999). Lateral east–west expulsion/extrusion of the different crustal wedges bounded by the aforementioned conjugate set of strike-slip faults at both sides of the strait edge would provide a plausible scenario. This scenario is particularly relevant in the Atlantic sector of the strait, where WNW compression has been deduced (González Lastras *et al.* 1991; Galindo-Zaldivar *et al.* 1993) from palaeostress analysis on Late Miocene and Early Pliocene sediments (Figs 1 and 2).

During the Miocene (Vejer-Barbate: Atlantic littoral) and Pliocene (Algeciras Bay: Mediterranean littoral) north–south trending marine grabens formed on both sides of the Gibraltar Strait

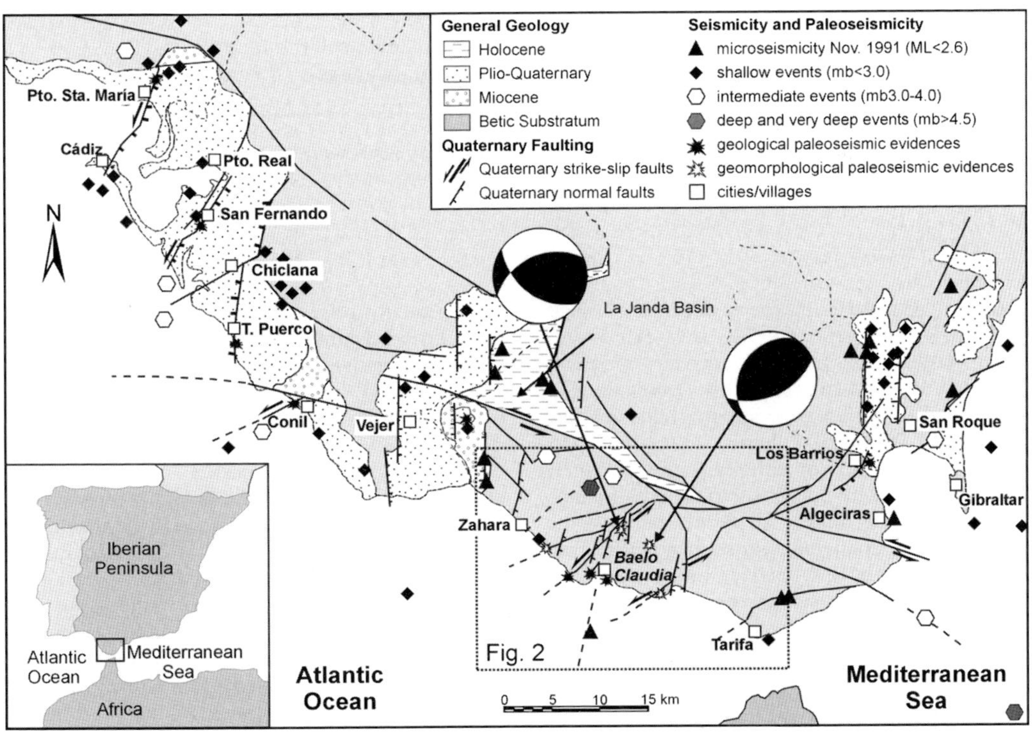

Fig. 1. Main Quaternary faults and seismic activity in the Gibraltar Strait. Data compiled from Benkhelil (1977), Goy *et al.* (1995), Gracia *et al.* (1999) and Zazo *et al.* (1999). Seismic data from the IGN database (IGN 1994) updated with data available in the IGN website. Microseismic data from Makris & Egloff (1993). Stress distribution from structural measurements on Plio-Quaternary materials in La Laja and San Bartolome ranges.

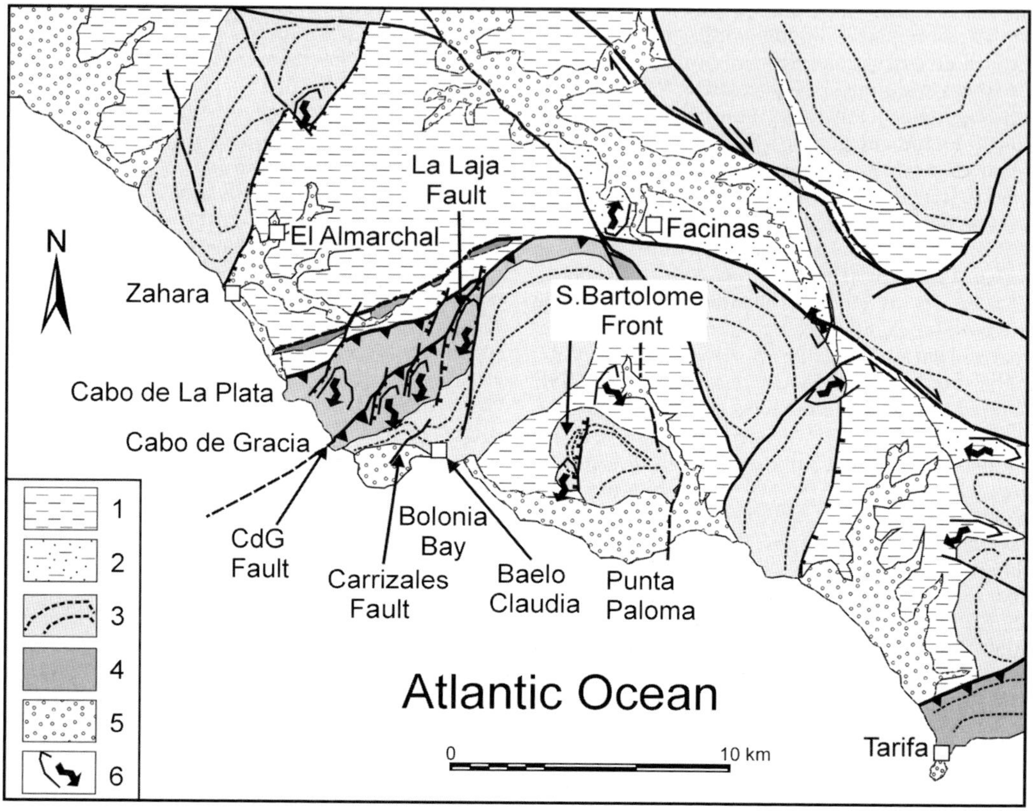

Fig. 2. Regional map of neotectonic structures and flysch units of the Gibraltar Arc in the surveyed zone. Legend: 1, El Almarchal Unit (plastic clays); 2, Facinas Unit (plastic clays); 3, El Aljibe flysch nappe (mainly sandstones): dotted lines = trace of Betic of uprighted stratification planes; 4, flysch slabs activated during the neotectonic period (Bolonia and Tarifa); 5, post-collisional Pliocene and Quaternary deposits; 6, landslide units. After Silva *et al.* (2006). For location see Figure 1.

(previously emerged). These sedimentary troughs subsided in response to an east–west extension at the uppermost crustal levels driven by crustal lateral extrusion (Zazo *et al.* 1999). During the Plio-Pleistocene, Late Neogene marine deposits were uplifted and deformed in the present-day transpressive setting (Fig. 1). In the following, the subsequent Quaternary marine sedimentation was mainly restricted to the littoral areas of the Gibraltar Strait. Recent continental Quaternary sediments are well developed only in river valleys and in the fluvio-lacustrine tectonic depression of La Janda Basin (Figs 1 and 2).

Geology of the studied area

Highly tectonized Gibraltar Flysch deposits dominate bedrock geology in the study area. These comprise the Eocene to Aquitanian Aljibe, Algeciras and Bolonia sandstone units. These Tertiary units are folded intensely and have been thrust over very plastic Cretaceous–Eocene turbiditic sandy and clayey sediments of Facinas and Almarchal units. The emplacement of the nappes took place between the Burdigalian and the Late Tortonian (*c.* 16 to 6.5 Ma) under a WNW–ESE compressive stress field (Sanz de Galdeano 1990; Weijermars 1991) during the Alpine period. This deformation process resulted in the formation of westward-plunging stacked nappes affected by recumbent folding and subsequent episodes of backthrusting and backfolding (González Lastras *et al.* 1991). The westward tectonic transport, the associated gravitational collapse, and correlative fragmentation of the upper flysch nappes (Aljibe Unit) are the primary processes responsible for the present topography of the Campo de Gibraltar. Limited clockwise rotation and dextral displacements of the fragmented nappe fronts as well as the individualization of detached blocks occurred (González Lastras *et al.* 1991); parts of those detached blocks now constitute the Cabo de Gracia and San

Bartolome ranges in the studied area (Fig. 2). These Alpine structures were deformed later in a NNW–SSE directed compressive stress field from the Late Miocene on, resulting in NW–SE and north–south trending graben systems and extensional normal faulting (e.g. Reicherter & Peters 2005).

Around *Baelo Claudia*, the Cabo de Gracia coastal range constitutes a NE–SW trending uplifted and detached slab of the Aljibe flysch nappe overriding the more plastic Cretaceous Almarchal clays and thin-bedded sandstones (Fig. 2). The flysch block is complexly affected by NNE–SSW upright folds, and overprinted by younger NE–SW strike-slip and reverse faults (Gonzalez Lastras *et al.* 1991; SECEGSA 1988), such as the Cabo de Gracia Fault (Goy *et al.* 1995). However, the youngest structures are north–south striking normal faults cross-cutting the older fault systems, like the La Laja Fault (Fig. 2). Normal faulting appears like subsidiary structures linked to apparent extensional steps along the trace of the NE–SW major strike-slip faults. In most of the cases, the trace of these normal faults is adapted to well-defined lithostructural contacts as those developed between the rigid Aljibe sandstones and the plastic claystones of the El Almarchal and Facinas formations. The most impressive topographic front is developed along La Laja Range north of *Baelo Claudia*, which shows geomorphological evidence of recent activity, such as well-developed triangular facets, hanging valleys (Fig. 3) as well as other minor geomorphic markers of tectonic activity. In most cases normal faults accommodated large mass-wasting processes occurring along the nearly vertical and bedding-parallel topographic fronts generated by thrusting and/or faulting within the Cabo de Gracia coastal range (Figs 2 and 3). Large mass-wasting processes (onshore) and mega-avalanches (offshore) have been proven to be a relevant process shaping the landscape of the Gibraltar Strait region from post-Messinian times onwards (Baraza *et al.* 1992; Sandoval *et al.* 1996; Esteras *et al.* 2000; Silva *et al.* 2006) but their relationships with local or remote seismic activity remain unclear.

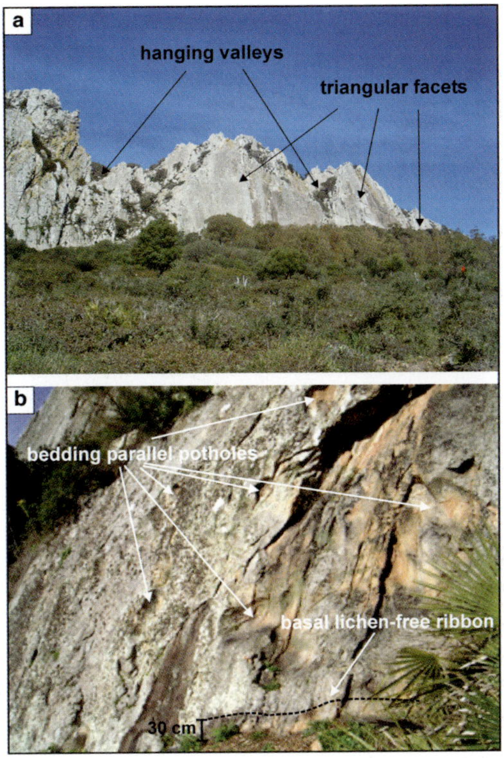

Fig. 3. La Laja range front. (**a**) Geomorphology of La Laja Range with well-developed triangular facets and hanging valleys in near-vertical bedding Aljibe Sandstone Formation (see Fig. 2 for location). (**b**) Geomorphic markers of probable accumulated coseismic displacements along the base of la Laja range front (potholes and basal lichen-free pedogenic ribbon).

Seismic activity in the studied area

Seismicity of the axial zone of the Gibraltar Strait area (6°50′W–5°14′E/35°50′N–36°25′N) is described in previous studies (Goy *et al.* 1995; Silva *et al.* 2006). Only 66 earthquakes are catalogued for the period 1750–1993 in this area, demonstrating a relatively low seismic activity. However, there is no historical seismic information prior to the year 1750 (Reicherter 2001). The archaeological remains of the ancient Roman city of *Baelo Claudia* indicate the occurrence of severe earthquake damage from the late first century AD to the late fourth century AD (Menanteau *et al.* 1983; Sillères 1997). These Roman events fall in the time-span of remote earthquakes, which probably occurred in the Gulf of Cádiz–Cape of San Vicente area between AD 33 and AD 382 (Martínez Solares & Mezcua 2002).

Seismic data illustrate the spatial relationships between the recorded epicentres and the main Quaternary faults of the coastal zone of the Gibraltar Strait (Fig. 1). Offshore, NE–SW and north–south trending faults have been extrapolated from the onshore faults. These correspond to similarly running faults in Bolonia Bay as deduced from high-resolution seismic profiling (Hübscher *et al.* 2007). Intermediate to deep (14–60 km) moderate events (3.1–4.2 mb) have been recorded in the Atlantic sector of the Gibraltar Strait in close relationship with the NE–SW trending strike-slip fault system. This observation led Goy *et al.* (1995) to catalogue this conjugate strike-slip fault system as seismic source during the late Quaternary. However, about 95% of the instrumental seismicity of the studied zone is commonly shallow (<10 km), weak (mb $\leq$ 3.0) and closely linked to the system of north–south trending normal faults segmenting the Betic nappes in the Gibraltar Strait region (Fig. 1). These faults are nicely represented within the Plio-Quaternary grabens of Barbate and Algeciras Bays (Goy *et al.* 1995; Silva *et al.* 2006), but also in the studied zone segmenting the Cabo de Gracia and San Bartolome ranges, adjacent to the ancient Roman city of *Baelo Claudia* (Fig. 2).

Instrumental records demonstrate that common ongoing seismic activity at shallower crustal levels is especially linked to the set of north–south normal faults (Fig. 1), but classified as an area of low earthquake hazard (Makris & Egloff 1993). Silva *et al.* (2005) related the archaeoseismic damage recorded at *Baelo Claudia* with the offshore activity of the NE–SW Cabo de Gracia strike-slip fault (Figs 1 and 2). However, all known local historic events never exceeded VI MSK maximum intensity (Goy *et al.* 1995) in the Gibraltar Strait area and major offshore events produced in the Gulf of Cadiz area (i.e. 1755 Lisbon earthquake) reached maximum intensities of only VI–VII MSK (Martínez Solares *et al.* 1979) or V–VI EMS (Martínez Solares & Mezcua 2002) in this zone. Moreover, relevant strong earthquakes have been recorded in the adjacent sector of the Alboran Sea east of the zone mapped in Figure 1. These events reached maximum magnitudes of 7.2 and onshore intensities of VI–VII MSK in the Málaga zone (Baraza *et al.* 1992), but no intensity data are available for the studied zone. This same situation occurs for the strongest historic events (VIII–X MSK) that occurred in this Mediterranean sector in AD 1494 and 1680 (Martínez Solares & Mezcua 2002). In addition, the catalogued, presumably strong events that occurred during the Roman period (AD 33 and AD 382) in the Gulf of Cadiz–Cape of San Vicente zone are poorly documented and do not offer macroseismic intensity data for the Gibraltar Strait region (Martínez Solares & Mezcua 2002). Therefore, the available seismic data do not support strong seismic damage in the Gibraltar Strait region, which is in disagreement with the archaeoseismological evidence in *Baelo Claudia* studied in this work.

Geomorphology, tectonics and palaeoseismic indicators around the city of *Baelo Claudia*

Outside the archaeological park of *Baelo Claudia*, the most extensive outcrops are the pervasively folded and tectonized Almarchal and Facinas formations (Fig. 2) of the Betic Flysch Zone in the Campo de Gibraltar. The Cretaceous to Miocene rocks form a clayey substratum in the ancient city of *Baelo Claudia*. These soft materials are framed horseshoe-like to the east and west by the rigid sandstones of the Aljibe Formation, which are also folded and steeply dipping (almost vertical, partly overturned). Goy *et al.* (1995) and Silva *et al.* (2005, 2006) offered several palaeoseismic indicators around the ancient Roman city focusing in the Cabo de Gracia and Carrizales faults, located between 5 and 10 km west of *Baelo Claudia*. Both vertical faults have a NE–SW orientation with sinistral strike-slip and minor normal displacement affecting Late Pleistocene or Pleistocene deposits. The Cabo de Gracia Fault displaces Late Pleistocene OIS 5 marine deposits and the overlying dunes exhibit liquefaction structures (Goy *et al.* 1994, 1995; Silva *et al.* 2005). The Carrizales Fault disrupts Pleistocene conglomerates including a well-developed clayey red palaeosol, pervasively affected by hydroplastic deformations generating pocket collapse-like structures, which are filled by dewatered Pleistocene conglomerates and sands. The latter are also affected by faulting. All these relatively young deformation structures are apparently fossilized by younger Bronze age and Roman aeolian deposits (Goy *et al.* 1995; Silva *et al.* 2006). In this work further apparent earthquake-related environmental effects of a purely geomorphic nature have been explored. Both cases are clearly linked to NNW–SSE to north–south tectonic structures located north of Baelo at the La Laja and San Bartolome range fronts (Fig. 2).

La Laja range front

The La Laja range front is developed along the vertical contact between the Miocene Aljibe sandstones and the adjacent Betic substratum following a NNW–SSW direction for about 5 km along the eastern side of the Cabo the Gracia coastal range NW of *Baelo Claudia*. Along the range front, the steeply inclined sandstone strata display well-developed triangular facets or 'flat irons' (Fig. 3a) with a spacing of 0.4–0.7 km and maximum amplitudes of more than 40 m. Hanging tributary valleys at 15 to 20 m height occur between consecutive triangular facets. These valleys developed clear bottleneck-shaped outlets, which are hanging in relation to the present range front basal knick-point, indicating relative accumulated offset (Crosby *et al.* 2007) along La Laja range front during the Pleistocene. Other minor features, such as palaeopotholes of springs, are developed on and between the vertical strata planes parallel to the bedding evidenced by pseudo-elliptical subhorizontal sapping cavities of metric scale hanging more than 2 m above the present basal knick-point of the range front. At present similar springs are developed at the basal knick-point of the range, so these hanging features should indicate its former position recording the history of differential uplift along the La Laja range front (Fig. 3b). Potholes, palaeosprings and hanging valleys occur at different elevations and may reveal relevant information about the recent tectonic uplift of La Laja range front.

The contact between the Aljibe sandstone and the Betic substratum at the basal knick-point of the range front is characterized by a centimetre-scale (10–20 cm) thin reddish basal pedogenic ribbon, visible at the present knick-point of the range continuous along more than 4 km. In contrast to the bedding planes of the Aljibe, this basal ribbon is only weakly weathered (or karstified) and shows no lichen growth (Fig. 3b). This ribbon is similar to that developed along active bedrock-fault scarps, indicating recent movements. Typically, coseismic features formed during historic and recent moderate-magnitude earthquakes ($M_W > 5.5$) in the Mediterranean region (Armijo *et al.* 1992; Michetti *et al.* 2000; Vittori *et al.* 2000; Monaco & Tortorici 2004) and may indicate historical fault activity.

Summarizing all these geomorphological features, it can be concluded that the La Laja range front clearly matches with the set of tectonic geomorphological anomalies, which are normally associated with active normal faulting. The thin ribbon or pedogenic veneer can as well be associated with ground settlement or slow active sliding. Fault-trenching analyses are planned in the basal knick-point of La Laja range in order to discriminate the nature of the observed features.

San Bartolome range front

The San Bartolome range front situated NE of *Baelo Claudia* is constituted by gently dipping weathered Aljibe sandstone underlain by softer and clayey Cretaceous flysch formations of the Betic Zone. Here, a mesa-type relief developed, elongated in a NNW–SSE direction (Fig. 2). The NW front of the Sierra San Bartolome has a NNE–SSW trending relevant scarp of about 50 m height and 2.5–3 km length mainly developed by differential erosion, but affected by normal faulting backslope. The erosive scarp is mantled by talus sheet containing a large number of big sandstone blocks. The rockfall deposits include individual blocks of a volume larger than 600 m^3, which have been transported a distance of about 400 m away from the scarp. Rockfall is an active process accumulating blocks in this zone, clear recent fallen rocks created rows of fallen trees in the present talus slope of the range, and numerous tumbled tree and block impacts are visible in the tree cover (Fig. 4a). Some of the bigger blocks contain open sets of post-Roman medieval graves (late fourth to sixth centuries AD) carved in the top of the sandstones. Some of them are clearly overturned and rotated (Fig. 4b) indicating their provenance from the top of the escarpment. The process of this block accumulation is unknown, but some studies have highlighted that anomalous

Fig. 4. San Bartolome range front. (**a**) View of the San Bartolome range front from the city of *Baelo Claudia* (SW) showing the escarpment carved in sandstones affected by faulting and multiple rockfalls. Note the row of fallen trees at the talus slope (arrow). (**b**) Tumbled block in the talus slope of the San Bartolome range front with over-turn medieval graves.

rockfall accumulations may be related to moderate to strong earthquake shaking at intensities larger than VI MSK (i.e. Burbank & Anderson 2001; Guerrieri & Vittori 2007). In particular, recent moderate earthquakes in SE Spain demonstrated that V^{+} EMS to VI MSK intensities are sufficient to trigger multiple rockfalls of large boulders (>10 m^3) along rocky escarpments developed on weathered rocks around the epicentral area (<5 km) under horizontal ground accelerations of about 0.024 g (Murphy Corella 2005; Benito *et al.* 2007). In our case, the number of tumbled blocks accumulated in the talus slope of the San Bartolome range exceeds 40, but a systematic analysis is necessary to unravel their episodic or progressive history of accumulation in order to relate it with seismic shaking.

The Carrizales Fault

The new observations on roughly north–south trending tectonic structures or erosive escarpments led to the proposal that apparent palaeoseismic evidence is not only linked to the set of NE–SW strike-slip fault systems in the Gibraltar Strait area as earlier proposed by Silva *et al.* (2005, 2006). Additionally, north–south landscape elements can be linked to normal faulting, otherwise clearly related to the current and shallower tectonic activity in this area. The Carrizales Fault first reported by Goy *et al.* (1995) appears to be a deflected prolongation of the NNE–SSW normal fault trace responsible for the occurrence of the La Laja range front, which developed in the firmer Aljibe sandstones NW of *Baelo Claudia* (Fig. 2). The structures revisited in the outcrop of the Carrizales Fault at Punta Camarinal, west of *Baelo Claudia*, indicate more outstanding recent deformation than previously reported (Goy *et al.* 1995; Silva *et al.* 2006). They form part of a complex micro-horst and graben with buried topography of metric scale developed within Pleistocene conglomerates (Fig. 5) including liquefaction of gravel and sand bodies. The idea that earth–west subsidiary extension in the Gibraltar area can account for many of the neotectonic or palaeoseismological landscape features around the Bolonia Bay area (Goy *et al.* 1995) is supported by the new data reported in this paper.

Fig. 5. Deformation on Plio-Pleistocene conglomerates affected by the Carrizales Fault (west of *Baelo Claudia*) apparently sealed by Late Holocene to Bronze Age aeolian deposits.

Surface geology and horizons of destruction within the ancient urban area

The geology of the urban area of the ancient *Baelo Claudia* is widely described in the works of Borja *et al.* (1993) and Silva *et al.* (2005). These authors differentiated three poorly compacted main geological units resting on a thick succession (up to 8 m thick) of clayey and plastic materials of the Gibraltar flysch units. Surficial deposits constitute a complex polygenic layer, in which pre-Roman, Roman and post-Roman formations are clearly differentiated in the lower sector of the city. In addition, this is the zone of *Baelo Claudia* in which ground and architectural disruptions are pervasively developed. A detailed geological and geomorphological sketch map and a cross-section across the urban area of the ancient Roman city are illustrated in Figure 6.

Pre-Roman formations are constituted by Late Pleistocene to Late Holocene beach and aeolian deposits, which grade progressively inland into fine-grained colluviums with a maximum thickness of about 2.1 m, thinning towards the coast. The Roman formation is the best developed and reaches thicknesses of 1.2–2.5 m. The geoarchaeological horizon of the city is constituted by artificial fillings for ground levelling (Sillières 1997; Silva *et al.* 2005), which embed the scarce remains and foundations of the first city building period (before AD 60). The artificial fillings are generally of silty to clayey nature, but large stone blocks, bricks and boulders of the removed soft substratum are also incorporated. In most cases this level is topped by an artificially compacted Roman mortar of *c.* 0.1 m in thickness (Silva *et al.* 2005), or by well-defined wall-collapse levels dated archaeologically to AD 40–60 (Sillères 1997). Silva *et al.* (2005) interpreted this entire formation as an anomalous 'demolition' archaeological horizon, on which the new imperial city was founded. Finally, the Roman and post-Roman formations are made up of a large colluvial and anthropogenic level that embedded the present remains and architectural disruptions of the monumental zone of the city, before its systematic archaeological excavation from the early 1970s. Numismatic and pottery findings date the main episode of destruction of *Baelo Claudia* to the second half of the third century AD and the eventual ruin of the Roman city in the late fourth century (Sillères 1997). However, the post-Roman archaeological level is not completely excavated and in most of the eastern and upper (north) sectors of the ancient city this level remains intact and has been the subject of an extensive geophysical survey. Tables 1 and 2 summarize the relationships between the different building phases of the city, destruction events and the different surface formations described by Silva *et al.* (2005).

Building damage and structural data recorded in *Baelo Claudia*

The building damage preserved in the ancient city of *Baelo Claudia* can be classified into two main groups: (a) deformation structures and wall collapse horizons preserved within the ground excavated by archaeological probe trenches; and (b) deformation structures and architectural disruptions presently at surface in the excavated sectors of the city (Table 2).

The deformation structures preserved within the ground are mainly associated with the Roman geoarchaeological horizon and can be related to the first episode of city destruction (AD 40–60). Most of them are linked to collapse horizons around the eastern sector of the city wall at the base of the Roman level. These collapse horizons are constituted by large blocks (0.5×0.3 m) of the first city wall that fell down directionally in a 'domino' style towards the west or SW. In other trenches, similar collapse levels are observed at the top of the Roman formation only covered by a thin post-Roman colluvium. In these last findings, the collapse is non-directional, sometimes towards the west, but others towards the east. These can be related to the second episode of city destruction; however, they can hardly be interpreted as a directional collapse process, like those expected from seismic wave propagation. The abandonment of the city in the period AD 365–390 may have led to the degradation of the buildings. Besides these collapse levels, only rare deformational structures are still well-preserved in the northeastern corner of the forum (Silva *et al.* 2005) and the Isis Temple area affecting the foundations of house-walls directly founded on the plastic clayey substratum (Fig. 6). New observations, detailed mapping and geophysical surveys in this part of *Baelo Claudia* led us to relate these sharply folded and disrupted architectural remains with a relatively shallow (*c.* 3 m) sliding plane of a landslide affecting a small area around 200 m^2 (Figs 6 and 7).

The deformation structures presently preserved at the surface are affecting the archaeological remains belonging to the second period of the city (after its first destruction in the first century AD.) and can be related to the second event of city destruction and eventual abandonment (AD 365–390). Most of these architectural disturbances were described in an earlier paper by Silva *et al.* (2005), comprising faulted and westward-tilted city walls, directional SW collapse of the columns of the Basilica, and pervasive pop-up arrays in the

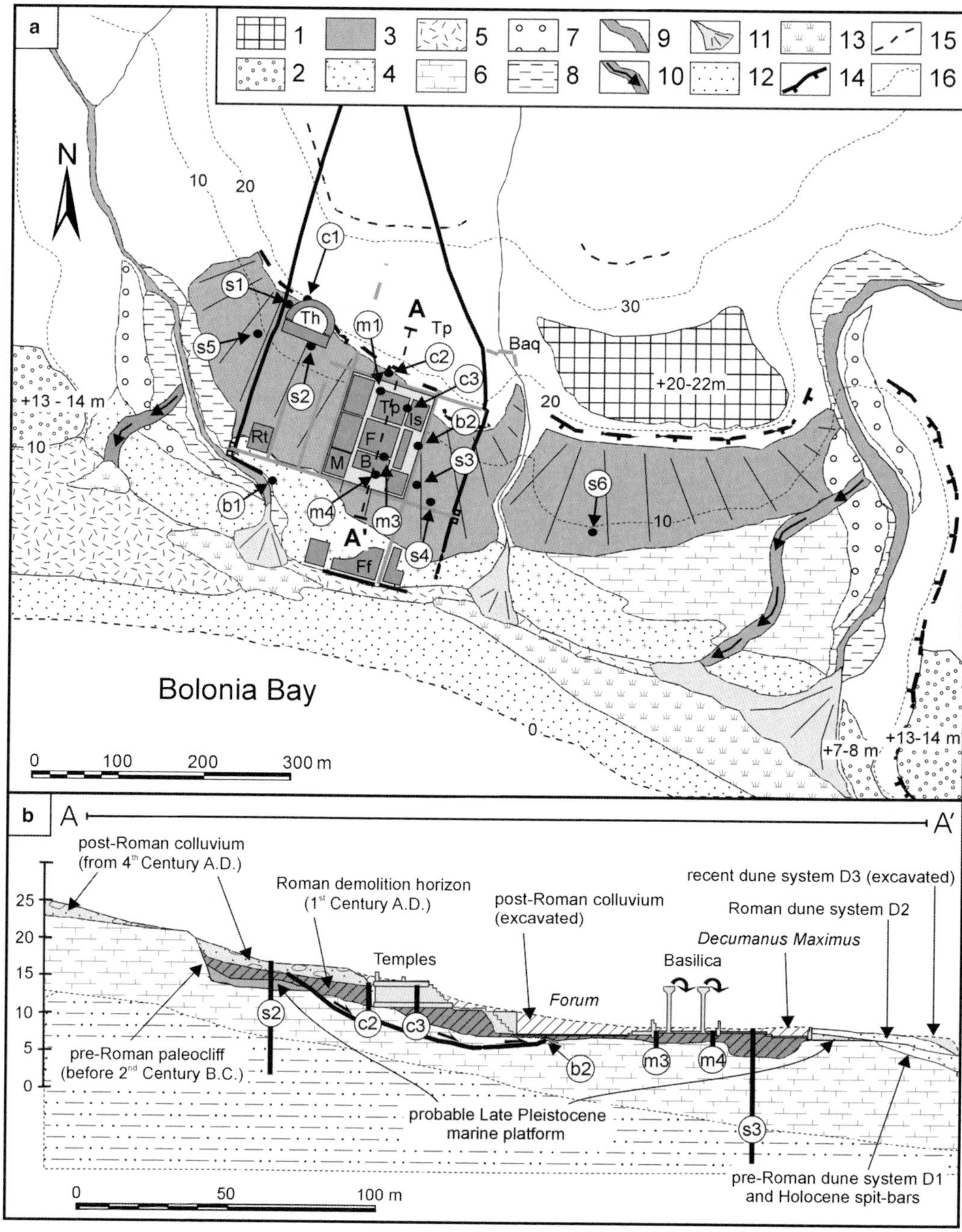

Fig. 6. Geology and geomorphology of *Baelo Claudia*. (**a**) Geomorphological map of the ancient urban area of *Baelo Claudia*. Key: Th, theatre; Tp, temples; F, forum; B, basilica; M, Macellum; Ff, fish factories; Rt, Roman thermes; Baq, broken aqueduct. Also included are the locations of the wells (s) described in Borja *et al.* (1993), and the archaeological trenches reported by (m) Menanteau *et al.* (1983), (b) Silva *et al.* (2005) and (c) Borja *et al.* (1993). Dotted grey lines illustrate the ancient harbour structures and the Roman palaeoshoreline reported by Alonso-Villalobos *et al.* (2003). (**b**) Geological cross-section displaying the relationships and arrangement of the deformations and Roman buildings. Legend: 1, marine abrasion platform; 2, Late Pleistocene marine terraces; 3, Betic substratum; 4, holocene spit-bar system including D_1 and D_2 dune system of south Spain; 5, recent D_3 dune system; 6, lagoon deposits; 7, fluvial terraces; 8, flood plains; 9, channel beds; 10, abandoned channels; 11, terminal river systems; 12, beach deposits; 13, marshes; 14, palaeocliff; 15, bedrock scarps; 16, contour levels. Modified from Silva *et al.* (2005).

Table 1. *Settlement history of* Baelo Claudia, *including geomorphic and human processes*

Dates	Historical features	Archaeological milestones	Geomorphic and human processes
Before late 2nd century BC	No urban settlement in the zone	Neolithic tombs on the cliff of Cape Camarinal; remains in a cave in Sierra de la Plata	Colluvial slopes partially bury staircase marine terraces; dune system development (D_1) on the Holocene spit-bar system
Late 2nd century BC	**First Roman settlement**	First fish factories; increasing commercial activity with Africa; *Oppidum* was situated in the Silla del Papa (N of *Baelo*)	Scarce human modification; local incorporation of archaeological artifacts to surface formations
Late 2nd century BC to middle 1st century AD	**First building phase** (lower coastal sector); first settlement becomes *Oppidum latinum* (category of Roman city); major urban reforms	Construction of the ancient square or forum, ancient capitol, temples, and first city walls (*c.* AD 10–20); amplification and improvement of first fish factories and private buildings in the typical orthogonal urban pattern	Major landscape reworking; ground digging; generation of artificial talus on colluvial slopes N of the forum square; soil beheading and alteration of surface hydrology
Middle 1st century AD (*c.* AD 40–60) PROBABLE EARTHQUAKE	**Demolition** and/or **collapse** of private and public buildings; damage of city walls	Blocky demolition horizon overlays clayey substratum, folded Roman pavement and house foundations in the lower sector of the city and old *Macellum*; ancient temples leveled up to a metre above the ancient ground surface	Probable localized landslide event NE of the forum; overall ground levelling; artificial silty and clayey filling with incorporation of large architectural elements
Late 1st century AD to early 2nd century AD	**Second building phase** maximum extension and prosperity of the city, which acquires the imperial label of '*Claudia*' at *c.* AD 48–50; improvement of the harbor south of the city	Development of the monumental zone (Basilica, temples, forum, *Curia*, capitol, *Macellum* and theatre); recycling of previous architectural elements, and partial use of former foundations; ancient forum square shortened north and south; reparation, reinforcement and thickening of the 1st century city wall; new urban orthogonal pattern, including all the pavements of the forum and streets (E–W *Decumanus* and N–S *Cardos*)	Rebuilding of the city on the '**demolition horizon**', artificially cemented on surface by *opus cementum;* creation of an artificial terrace (500 m^2) for the building of the new capitol and Isis Temple north of the forum

(*Continued*)

Table 1. *Continued*

Dates	Historical features	Archaeological milestones	Geomorphic and human processes
Late 3rd century AD (*c.* AD 260–290) PROBABLE EARTHQUAKE	**Abrupt ruin and urban depopulation;** the damaged city was never fully abandoned, but the monumental buildings were never restored or rebuilt	Pop-up-like deformations in forum and *Decumanus Maximus;* collapse of Basilica columns and *Macellum* roofs; Isis Temple partially collapsed; west tilting of most house walls and city walls; harbur and fish factories presumably abandoned	Probable localized and reactivated landslide event (*c.* 5000 m^2), affecting the artificial terrace NE of the forum; severe ground settlement and subsidence affecting the south side of the basilica; liquefaction?
Late 3rd century AD to middle 4th century AD	**Third building phase;** poor quality buildings are constructed over the ruins, following a different urban pattern	Damaged structures are artificially levelled and new poor quality buildings are constructed over the ruins; only the partially damaged remains of the *Macellum* and Basilica, Isis Temple and theatre survived; fish factories and some sectors of the theatre were used as houses; the city wall was never restored or reinforced	Intervening horizon of ruins levelling some of the damaged urban structures for city rebuilding
Late 4th century AD (*c.* 350–395 AD) EVENTUAL RUIN OF THE ROMAN CITY	**Eventual ruin** of remains of Isis Temple, *Macellum* and Basilica and definitive abandonment of the ancient Roman city	Severely damaged remains of the Isis Temple, *Macellum*, and Basilica; southern sector of the theatre progressively collapsed on the intervening horizon of ruins; Necropolis still active	Roman and post-Roman colluvial formations bury the damaged remains of the city; coastward shifting due to development of dune system D_2
Late 4th century AD to ca. 8th century AD	**Late urban development;** small palaeo-Christian to Visigothic settlements and graves on former monumental zone, with a very different urban pattern	For first time, graves and tombs are carved within the ancient monumental zone of the city; definitive abandonment shortly before the AD 711 Moors conquest of the Iberian Peninsula	Colluvial burying of the destroyed Roman city; soil swelling and slope creeping amplify and magnify pavement damage; burying of lowermost sector of the city by dunes
Late 8th century AD to 1292 AD and present	**Military garrison of the Moors;** small military sentinel and watchtower built on the ruins of the theatre	In AD 1292, Reconquista of Tarifa by the Spanish; 15th century possible tsunami hits coast AD 1870–1907 first archaeological findings; AD 1917–1921 first archaeological excavations	Slope wash sediments containing Roman rubbish (bones, teeth of animals, e.g. pigs, and seashells) and ceramics and glass

Modified from Silva *et al.* (2005).

Table 2. *Lithological, structural units and mechanical properties of those for the* Baelo Claudia *site*

Unit	Depth (thickness)	Lithology	Age	Mechanical properties
Upper Complex Polygenic Unit				
Post-Roman colluvium coastward grading to D_2 dune system	(0.8–1.0 m)	Colluvial deposits with archaeological artifacts, bones, charcoal, etc.	<4th century AD	Poorly cemented no data not relevant for this study
Roman demolition horizon	(1.5–2.3 m)	Artificial filling; stone blocks, bricks, and boulders of clays pasted in a clayey matrix*	Approx. AD 40–60	C3 soil (NCS-94); UCS < 2 kg/cm^2; SPT (N = 14 to 20); top consolidated (Roman moles): SPT peaks of N $>$ 90 in the upper 0.1 m
Pre-Roman colluvium coastward grading to D_1 dune system	(0.9–2.1 m)	Subangular gravels and boulders with clayey-sandy matrix; capped in S2 by 0.2 m thick littoral deposits at +13 m above the present sea-level*	Holocene to Upper Pleistocene	Poorly cemented no data (only relevant at the toe of the palaeocliff: area of the temples, north of the forum)
Plastic Substratum				
Upper plastic clays	2.3–4.4 m (6 m)	Variegated plastic clays and marls, with limited soil development at top; forms the natural foundations of the different Roman settlements covering the entire lower sector*	Eocene Bolonia Fm.	CH-CL soil (USCS); expansive soils; C3 Soil (NSC-94); UCS: 3.82–2.45 kg/cm^2; liquid limit: 50–61; plastic limit: 22–27; swelling potential: 0.6–1.70 kg/cm^2; locally up to 2 kg/cm^2
Basal clays	8.0–9.1 m (6–7 m)	Red clays with interbedded sandstones	Cretaceous-Eocene Betic Flysch	CH Soil (USCS)/C2 Soil (NSC-94); liquid limit: 49–61; plastic limit: 28–34

Damaged buildings founded partially or Totally in these materials.
Modified from Silva *et al.* (2005).

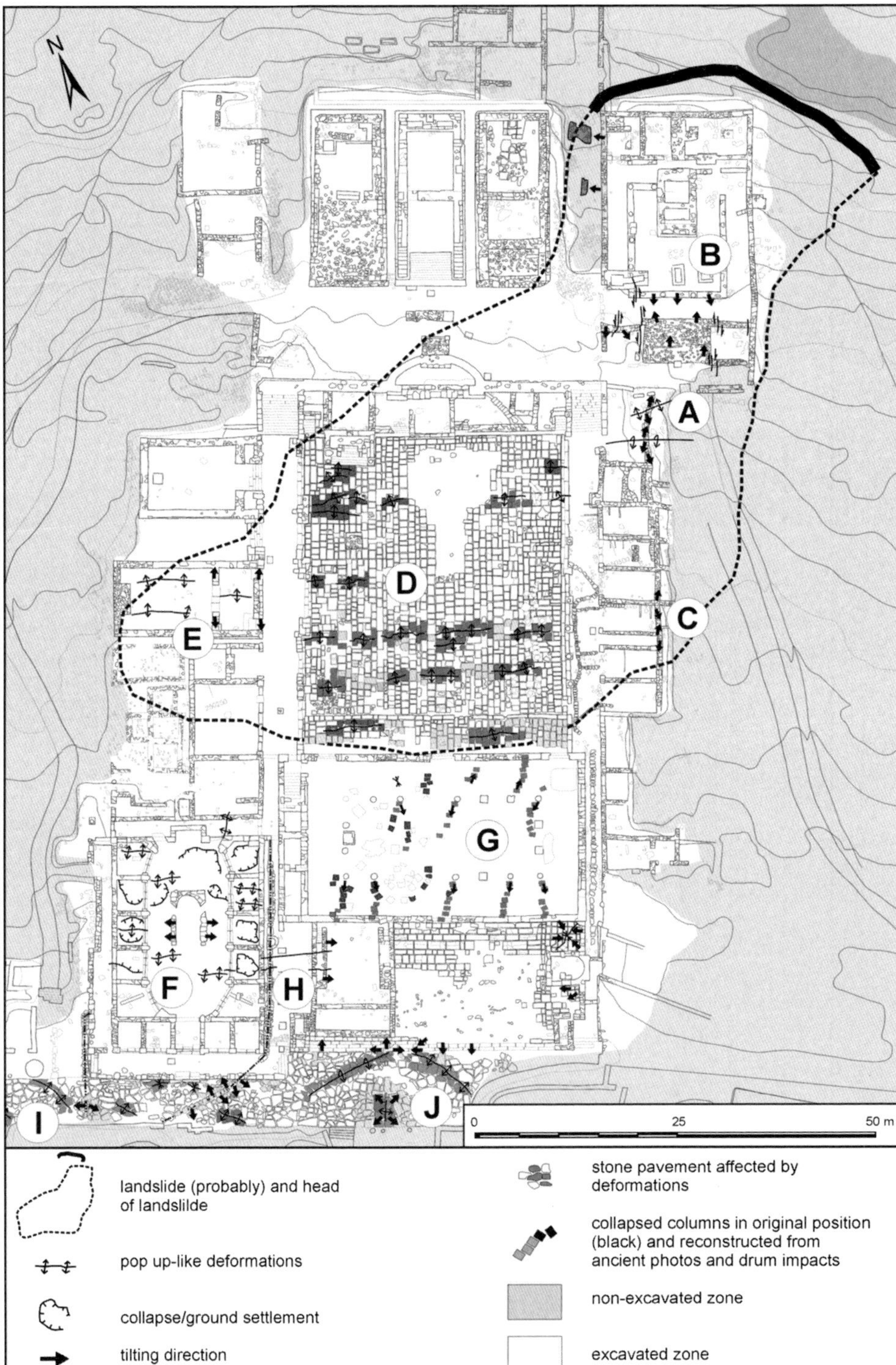

Fig. 7. Detailed map of surface deformations and architectural disruptions observable in the lower sector of the ancient Roman city of *Baelo Claudia*. All the mapped disruptions belong to the second period of city destruction (AD 260–290). Letters A–J indicate locations of sites discussed in the text.

pavements of the forum and *Decumanus Maximus*. In the present work all these structures have been carefully mapped, and detailed measurements of the orientations of the pop-up arrays, directional collapse of columns, directional tilting of walls, pavement joints, and corner break-outs of the pavement flagstones allow the direction of deformation to be quantified and characterized (Fig. 8). A total

Fig. 8. Main architectural deformations observed on the present ruins of *Baelo Claudia*. (**a**) Propagation of cracks in the walls of the ancient basilica and adjacent macellum with indication of the collapse orientations of the columns (and relative rose diagram for column collapse). (**b**) Dropped keystone in the western wall of the basilica, see pencil as scale. (**c**) Pop-up arrays affecting the eastern sector of the Decumanus Maximus and rose diagram of preferred orientations for these structures. (**d**) Shocks developed in the corners of individual flagstones and rose diagram of measured orientations. (**e**) Fractures (i.e. shear joints) in individual flagstones and rose diagram of measured orientations. (**f**) Pop-up deformation affecting multiple flagstones in the forum.

of 225 structural measurements has been done for this study, 186 of them corresponding to joint orientations (101), pop-ups and flexures (42), dip directions on disrupted flagstones (30) and shock breakouts in flagstones (13). The rest correspond to the tilting direction of the eastern perimeter of the city wall, and direction of collapse of columns of the basilica and the Isis Temple. Because the basilica columns were restored and built up, the collapse directions have been measured from old photos and aerial photos taken before the restoration, and from the direction of the impacts in the concrete floor (in Roman, Opus Dominum) of the basilica. Details of the distribution and features of architectural deformations are displayed in Figures 6 and 7. The rose diagrams of structural measurements are illustrated in Figure 8.

In detail, our new mapping results reveal evidence for a landslide in the NE corner of the ancient forum affecting the zone of the Isis Temple (B in Fig. 7; Fig. 9b). Here, walls are offset by between several decimetres to more than 1 m, and most parts of the walls are anomalously tilted towards the south and north (Fig. 9b and c). Below the artificial topographic step that was carved to build the forum, all houses situated in the east show severe

Fig. 9. Deformations recorded around the Isis Temple. (**a**) Collapsed wall onto the ruins and debris of the western side of the Isis Temple displaying the stratigraphy of destruction horizons. (**b**) Close-up view of northward tilting and offsets in the Isis Temple; white arrows indicate the tilting direction. (**c**) General view of northward tilting on foundations, pavements and walls of the Isis Temple (see Fig. 7 for location); white arrows indicate the tilting direction. (**d**) General view of severely folded walls along the eastern side of the forum downslope the Isis Temple (see Fig. 7a–c for location).

distortion and bending (A and C in Fig. 7; Fig. 9d). East–west directed up-thrusting and folding as well as the pop-up structures affecting the pavement of the forum (D, I and J in Fig. 7; Fig. 8c and f) are interpreted to be induced by discrete landslides, which affected the pavements and early foundations (Fig. 9d).

East–west trending low-amplitude folds are clearly deforming the stairs, base of house-walls and the Opus Dominum (Roman mortar) of the ancient Curia at the western zone of the forum (E in Fig. 7) and the ancient macellum (market) located immediately SW of the forum (F in Fig. 7). These architectural disruptions may also be related to the landslide event as a consequence of limited earth flow at the landslide toe as illustrated in the cross-section of Figure 6b. However, the deformation structures observed close to and within the macellum area are unidirectional. Numismatic and pottery findings indicate that the macellum had very limited use by the Romans after the earthquake from *c.* AD 260 to AD 395, when commercial activity eventually terminated (Menanteaú *et al.* 1983; Sillères 1997). Summarizing these results, we can say that most of the observed structures in this area of the city are compatible with a SSW-directed and east–west trending complex landslide event as suggested from the rose diagrams inset in Figure 8.

The area of the Isis Temple is only partly excavated (B in Fig. 7). Some of the best examples of building deformation are observable in the Isis Temple. Fallen columns (SW-directed) and wall and pillar collapses are also directed in a south to SW direction, but also scarp and slope parallel (Fig. 9b and c). A crude stratigraphy based on Roman pottery allows us to date the collapse event to the fourth century AD. We took a variety of soil samples directly below the fallen columns and walls (Fig. 9a); radiometric carbon dating is in progress. The Isis Temple was constructed on a small topographic scarp and shows other deformation features such as offsets, tilting and bending of walls and foundations (Fig. 9b and c).

Located south of the forum, the remains of the ancient basilica are one of the most outstanding examples of earthquake architectural deformation, before it was restored, as occurring in many places of historical seismic shaking in the eastern Mediterranean region (Ambraseys 1971, 2006; Stiros 1996). Most of the columns collapsed towards the SW and SSW, with the column drums in a domino style (C in Fig. 6; Fig. 8a). Drum impacts in the ancient floor of this building are numerous and there was no debris layer between the column drums and the ancient floor (Silléres 1997), indicating the sharp and sudden character of the collapse event (Silva *et al.* 2005). The directional collapse of the columns is incompatible with the proposed SSW-directed landslide event, and hence indicates a ground movement directed towards the NE to NNE (if the basilica columns were destroyed by seismic waves). In addition, in the western north–south wall of an annexed building of the basilica there are some fractures cutting through two or more adjacent blocks (Fig. 8a). Those affect most of the preserved wall and can be continued to the eastern wall of the building and to the adjacent western wall of the macellum stores (H in Fig. 7). One of these fractures is also associated with the subsidence of a keystone of an arch-window located in the western wall of the edifice annexed to the basilica (Fig. 8b). Such fractures have been associated with minimum earthquake intensities of VIII MSK by Korjenkov & Mazor (1999) and Hinzen (2005).

South of the basilica, the decumanus maximus (main east–west trending road) displays a wide variety of pop-up arrays of flagstones (I and J in Fig. 7; Fig. 8c) such as synclinal and anticlinal structures (Fig. 7). The detailed mapping carried out for this work (flagstone by flagstone) enabled those deformations due to ancient subsurface canalizations to be distinguished from others resulting from horizontal ground acceleration. Aside from the more spectacular deformations induced by the presence of subsurface Roman structures, the rest of the pop-up arrays are arranged following a main N130–125E orientation and a secondary N50–60E one (Fig. 8c). Vertical displacements of up to 30 cm and flagstone up-thrusting are common (Fig. 8f). A wide variety of joints and corner break-outs due to horizontal shocks densely fracture the set of flagstones. Structural data measured for pop-up structures, joints and shocks directions from flagstone break-outs indicate very congruent orientations. The break-outs are mostly found along the NE edges of the flagstones. Orientation of these indicators is systematically distributed, pointing to a shock from the SW, and folding in a NW–SE direction (Fig. 8). At the end of the Decumanus Maximus, just to the south of the basilica, a semi-circular dome-like structure with a radius of about 6 m affects the entire ancient pavement suggesting severe ground subsidence (J in Fig. 7) including liquefaction.

The city wall surrounds the village and was built for representative and not defensive purposes. At the ends of the Decumanus Maximus, are two main gates, the Puerto de Gades (Cádiz) in the west, and the eastern gate (Fig. 6). The wall is equipped with several watchtowers. Generally, the city walls are in a very bad condition. Parts are not excavated or are collapsed; others are missing, because of later quarry use. We studied and measured the entire remains of the city walls. The eastern wall and its towers are preserved and excavated in parts up to the groundings. During the excavation, older remains of a former city wall were encountered; this wall is topped by a 'demolition horizon' with big blocks of wall boulders. This horizon may correspond to the AD 40–60 earthquake outlined by Silva *et al.* (2005). Major damage is also observed along the eastern city wall, which is mainly tilted to the WSW between 15° and 20°, but in some cases up to 25° (Fig. 10). Tilting is accompanied by intense fracturing and rotation of individual segments against each other, often in an anti-clockwise sense. Some of those cracks in the northern bastion and close to it display offsets on the order of centimetres to metres, but commonly they are smaller than 20 cm. Partly, the cracks were already restored by Romans (Menanteau *et al.* 1983).

The eastern aqueduct outside the city walls crosses a little creek; the western part of the aqueduct collapsed downhill, and now forms seven parts. The entrance through the city walls is almost perpendicular and was excavated in 2007 (Fig. 6). Some of the arcs show rotational displacement around a horizontal axis; this might be interpreted as a slow deformational feature (low energy) originating probably from small creek-parallel landslides.

In the upper sector of the city the main archaeological feature is the theatre of the first century AD (Fig. 6) built during the first phase of the city and directly founded in the clayey substratum. It displays not only earthquake or landslide deformation, but also a lot of restoration. The Moors started to build a watchtower in the right part of the theatre.

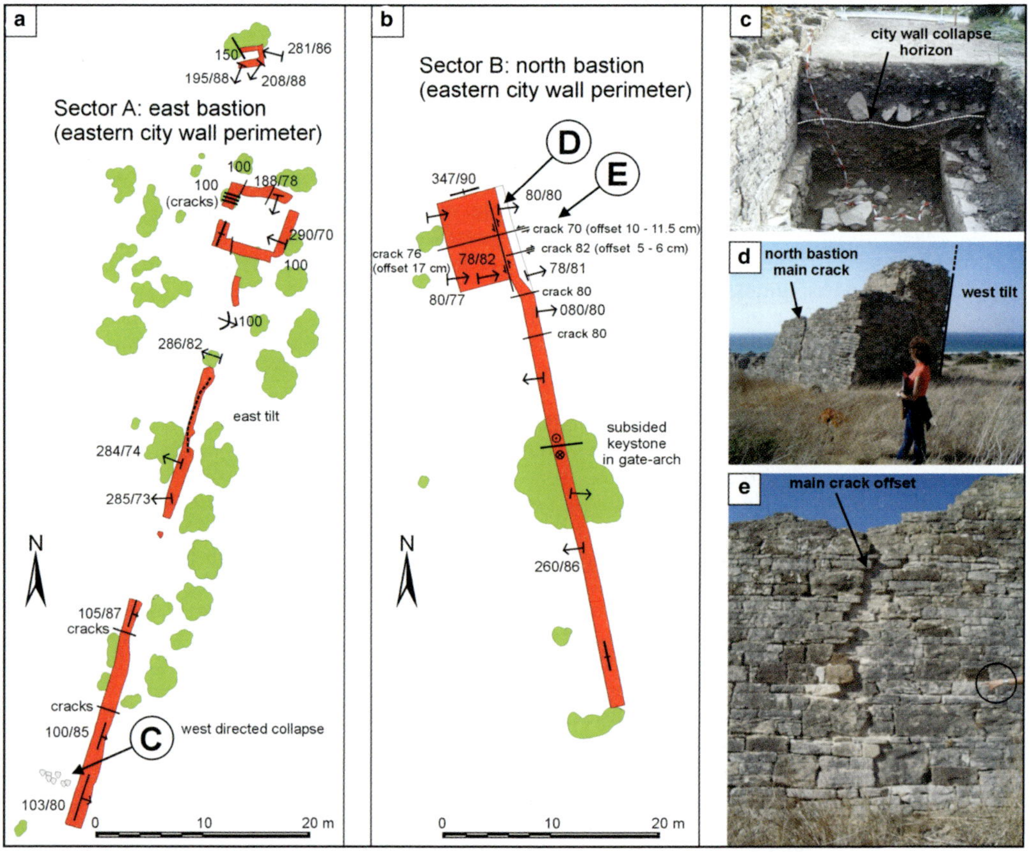

Fig. 10. Deformations recorded along the eastern city wall perimeter. (**a**) Sector A (east bastion) and (**b**) Sector B (north bastion). Conventional dip symbols represent tilting measured within the city wall; arrows represent large block occurrence indicating direction. Areas in green represent present vegetation cover. (**c**) Archaeological trench displaying a wall collapse level associated to the AD 40–60 event in the southern termination of the east bastion of the city wall (Sector A) for location see (A). (**d**) General view of the westward tilted and fractured main watchtower in the northern zone of the north bastion (Sector B) for location see (B). (**e**) Close-up view of the main crack affecting to the watchtower in (D) displaying centimetric offset (Sector B); see hand for scale. For location see (B).

The theatre served as a living place for many people in post-Roman times. Therefore, it is quite difficult to recognize any deformation related to earthquake damage. Presently, the entire inner part is completely restored and filled with concrete; however, the collapsed stairway to the loggia is still in situ, and directed towards the south. Open cracks in the walls and inclined walls are interpreted as generated by slow deformation. On the other hand, big fallen blocks of the tiers are attributed to coseismic damage. A lot of triangle-shaped cracks and breakouts are found at pillar bases of the building, which are probably related to deformation during a high-energy event.

On the contrary, in the lower coastal sector of the city, south of the Decumanus Maximus the fish factories zone shows almost no destruction due to sudden seismic shocks or landsliding (only minor cracks). Fish factories are founded in a sandy substratum, which is different than the rest of the city, and were completely buried by dunes after their abandonment. Therefore, this zone should have had much smaller site effects than those buildings founded in loose debris or clayey materials, and slow deformational processes linked to the post-Roman alluvium burying the rest of the city were absent in the sandy dunes embedding the ruins.

Geophysical data: ground penetrating radar survey

Methodology, equipment and processing

For about 15 years, ground penetrating radar (GPR) has often been employed for detecting buried faults or underground tectonic structures, sedimentary structures in all kinds of rocks, or archaeological remains. Different GPR methods have also more recently been applied to archaeological surveying with very good results (Vaughan 1986; Goodman 1994; Hruska & Fuchs 1999; Pipan *et al.* 1999; Sambueli *et al.* 1999; Basile *et al.* 2000) and an extensive review has been supplied by Conyers (2004).

Electromagnetic radiation of waves in the radio band at metre wavelength is used in GPR investigations. Radar waves can be reflected at interfaces between materials with differing dielectric permittivities (ε_r), such as between rocks and air or sand and silt. Since the velocity of the wave is known, the depth to the interface can be determined by the time it takes for the echo to return (two-way-traveltime, TWT), and the velocity of electromagnetic waves in a material is specified by its dielectric permittivity. In contrast to seismic reflection, the time-scale of TWT is on the order of nanoseconds, and hence GPR can resolve objects and inhomogeneities of less than 10 cm, depending on the frequency of the antenna applied. On the other hand, attenuation of the radar waves by conductive materials strongly limits penetration depth of the electromagnetic pulses.

The SIR 2 GPR system by GSSI, with different antenna models (45, 200, 300, 400 MHz) and with a survey wheel, has been used to study subsurface structures at *Baelo Claudia*. The applied antenna frequencies allow resolution of objects, discontinuities or strata in a decimetre to centimetre range, from *c.* 20 cm (200 MHz) to *c.* 10 cm (400 MHz). Penetration depth varies depending on the antenna frequencies and the physical properties of the ground and location of groundwater table(s). As the achievable depth is inversely related to the conductivity (σ) of the materials, clayey layers, wet soils and seawater may act as a barrier to wave propagation. In general, maximum interpretable depth with 200 MHz antenna arrangements is about 20 m in dry carbonate rocks or coarse-grained clastics (Beres *et al.* 1995; Smith & Jol 1995; Neal 2004), However, maximum penetration in this study, dealing with poorly compacted and plastic clayey materials, has been about 5–6 m.

Normally, after good parametric field settings, the raw data need to be processed. The REFLEX® software (version 4.2; Sandmeier 2006) has been used in this study to process and filter the data and to optimize imaging. Time migrations, topographic corrections and 3D modelling from 2D GPR arrays have also been implemented using the same software facilities.

GPR results of the *Baelo Claudia* area

A total of 7 km of radar profiles has been collected in the ruins over an area of 0.2 km^2 using the 300 MHz GSSI antenna and the SIR2 system (Fig. 11). Some low-frequency measurements (45 MHz) have been performed where vegetation and archaeological remains allowed the handling of those large antennae. Most of the profiling was concentrated on the damaged sector of the city area (forum, temples (A in Fig. 11) and theatre (B in Fig. 11). In these zones, NNE–SSW orientated GPR arrays were developed in order to survey the scarp connecting the upper and lower sectors of the city, still unexcavated east to the forum and south of the theatre. Some profiles were surveyed with two antenna models (Fig. 11) in order to combine the advantages of deep penetration by the 45 MHz antenna with the higher resolution at shallow depths provided by the 300 MHz system. An extensive GPR survey was developed around the perimeter of the city wall, and complementary arrays were performed near the fish factories

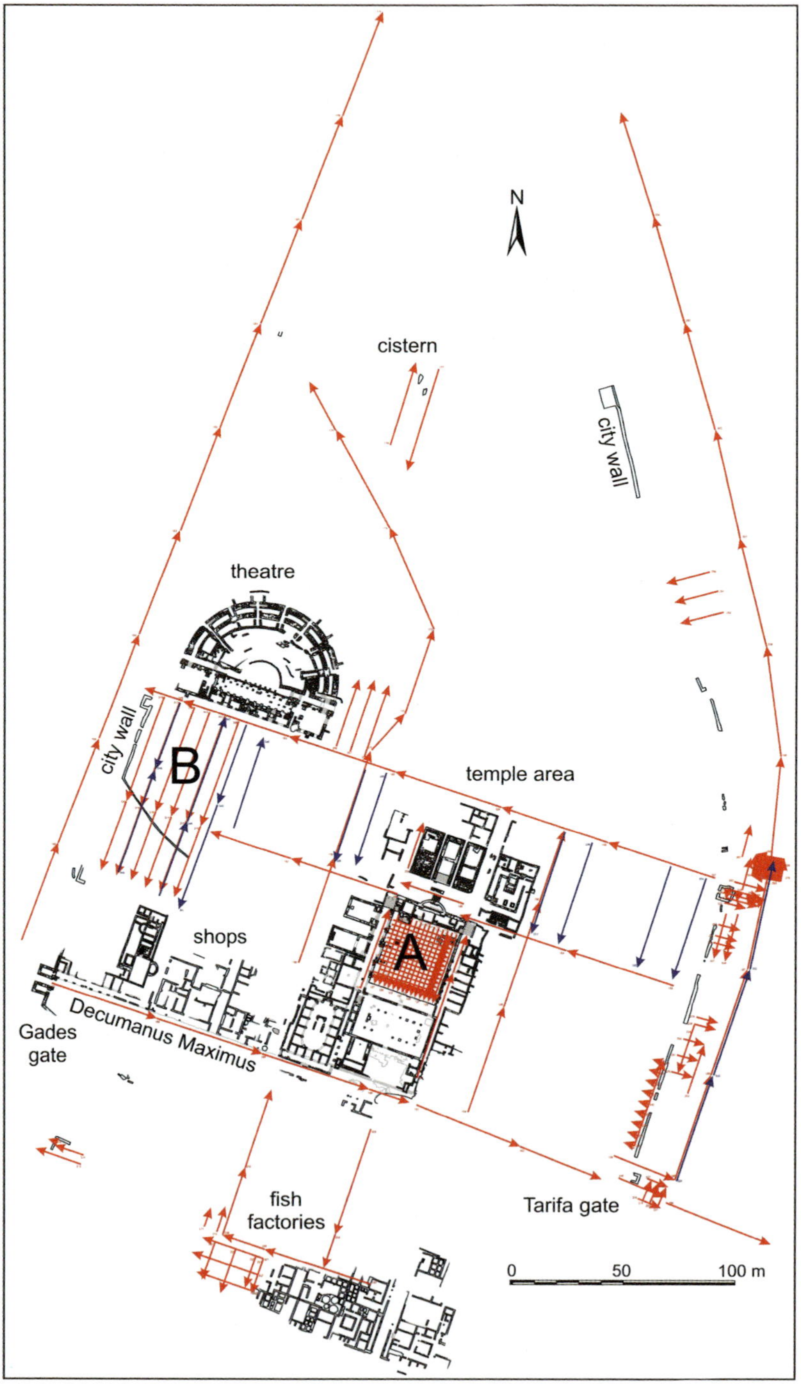

Fig. 11. Map of the radar investigations inside the ruins, 300 MHz profiles in red, 45 MHz profiles in blue. 0.5 m spaced grid at the aqueduct (red rectangle at the very east of the ruins), 10 × 12 m. A, 2 m spaced grid at the forum, 25 × 33 m. B, Detailed investigations with both antenna types at the scarp south of the theatre.

sector (Fig. 11). The present study is illustrated only with those radar profiles that clearly display surface deformations and subsurface imaging about the location of probable horizon events and damaged building remains.

3D surveys were carried out in two specific areas. The first one aimed at the damaged aqueduct (partially buried in the year 2006, now excavated) at the eastern part of the city wall. A 10 × 12 m grid with 0.50 m spacing was performed in order to create a 3D cube from single lines. On the forum, the spacing of the 25 × 33 m grid was set to 2 m, which is too distant for a real 3D analysis. Nevertheless, these data also allow a detailed interpretation of a large area.

The 300 MHz profiles show clearly detailed information of the subsurface in high resolution. In several parallel and perpendicular radargrams we found changes in the reflection pattern around 45 ns TWT (cf. 2.4 m depth), which is interpreted as an 'event horizon' (Fig. 12). Additionally, ruptured and destroyed wall remains are visible in nearly every part of the city. These architectural remains usually form small-scale reflection anomalies including diffraction hyperbolae and increased amplitudes (Fig. 12). In the case of walls, features may be delineated in parallel profiles and therefore allow the mapping of streets and houses. The 'event horizon' is characterized by flat-lying reflectors and by a marked angular unconformity above warped and distorted reflections (e.g. file 154, east of the forum, in Fig. 11 and files 253–255, at the eastern city wall in Fig. 13). This 'event horizon' may be interpreted as the last earthquake event, dated *c*. AD 260–290. The flat-lying reflections possibly represent post-event sedimentation. Southward-dipping and slope-perpendicular structures in some places support this idea.

Large-scale amplitude changes in the upper parts of the radargrams as in Figure 12 are due to near-surface conductivity inhomogeneities. The reason for this effect may be changes in humidity (even absence of the lawn) or different materials in the uppermost layer. Architectural remains also produce amplitude variations; however, these effects can be distinguished considering depth, sharply delineated boundaries and their spatial extension. The 300 MHz profiles document the occurrence of similar westward-directed buried wall collapse levels around the city wall (Fig. 13). In contrast, distortion, failure and breakdown of house walls and foundations at the NE corner of the forum and Isis Temple document the occurrence of localized landsliding directed towards the SW. Radargrams at these zones image probable detachment horizons dipping towards the SW delineating the boundaries between the archaeological level and the overlying post-Roman colluvium and the underlying clayey substratum, which can be interpreted as part of the head zone of the landslide (Fig. 12).

The 45 MHz low-frequency profiles do not reveal such detailed information of the shallow subsurface, but instead allow imaging of structures at greater depths, in this case up to 20 m. Figure 12 shows profile 346, which may count as representative for what can be seen in parallel lines. The uppermost layer clearly differs from deeper structures and its mostly parallel reflections are interpreted as post-Roman colluvial sediments. Their position and physical properties fit the section published by Silva *et al.* (2005). For this reason, GPR can be used to map these layers in adjacent areas and to determine changes in depth and thickness. The middle section is interpreted as the pre-Roman colluvial level, followed by southward-dipping marine terraces. Depending on the material above those marine units, they are not visible in every profile. A highly conductive layer at the top of the profile, seawater intrusions, and clayey or wet sediments will lead to high attenuation of the radar waves, thus any underlying structures may not be visible.

Discussion

The mapped architectural damage at *Baelo Claudia* falls within the categories of secondary and ground-shaking archaeoseismological evidence listed by Hancock & Altunel (1997). Archaeological evidence is often more ambiguous than generally assumed, because other episodic high-energy natural events like tsunamis, storm surges or landslides can also account for the observed archaeological evidence (Stiros 1996; Silva *et al.* 2005; Similox-Tohon *et al.* 2006). Therefore, in all the archaeoseismological analyses it is necessary to maintain a proper balance between tectonic, geomorphological and geotechnical factors on one hand, and historic, anthropogenic and archaeological factors on the other hand (Karcz & Kafri 1978; Galadini *et al.* 2006). The *Baelo Claudia* case remains problematic since no catalogued local historic event can be related to the deformation. On the contrary, the wide range of characteristic deformational features and architectural disturbances exhibit a confident similarity to many other ancient cases throughout the Mediterranean, and especially the occurrence of rebuilding phases separated by destruction or demolition horizons (Silva *et al.* 2005). From the available geotechnical data (Borja *et al.* 1993; Silva *et al.* 2005), it can be concluded that geomorphological and geotechnical factors in the lower sector of the city promote relevant amplification of ground motion, and that the

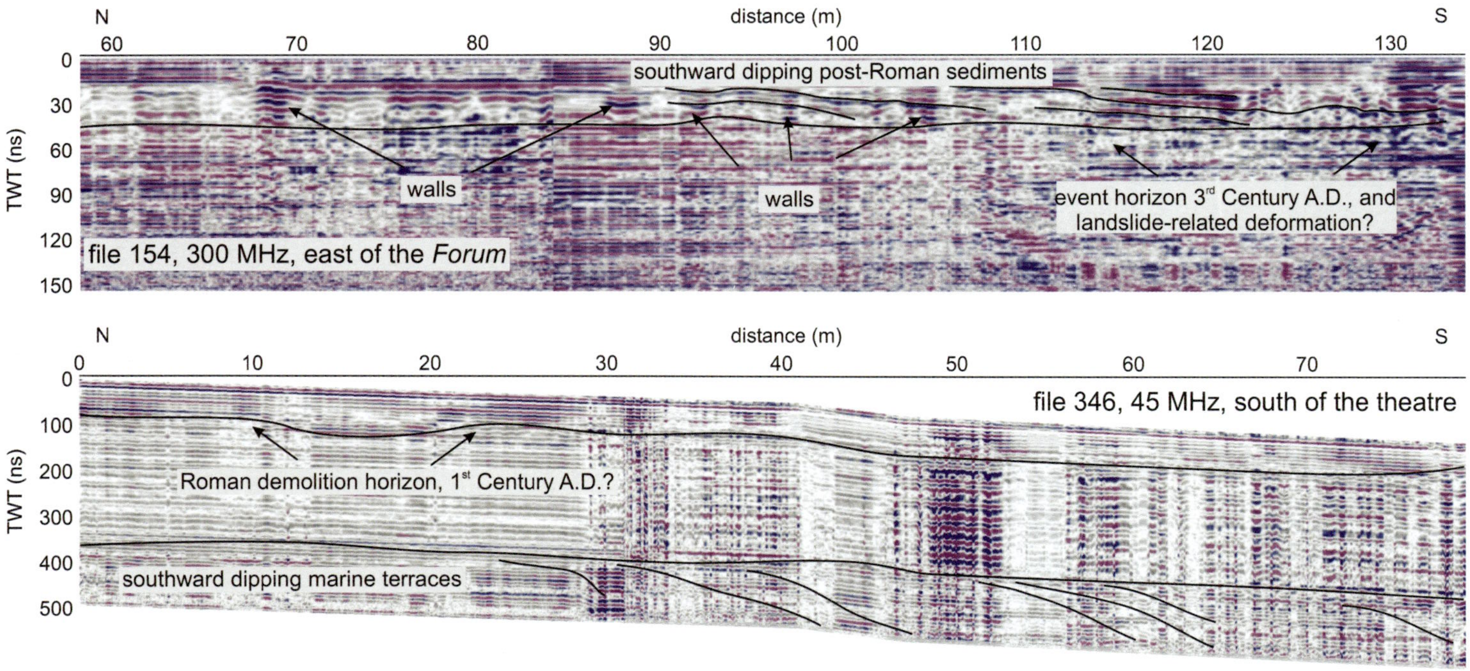

Fig. 12. Radar profiles 154 (300 MHz, east of the forum marked A in Fig. 11) and 346 (45 MHz, south of the theatre crossing the scarp, marked B in Fig. 11). 100 ns TWT represent *c*. 5 m depth, calculating with a wave velocity of about 0.1 m/ns.

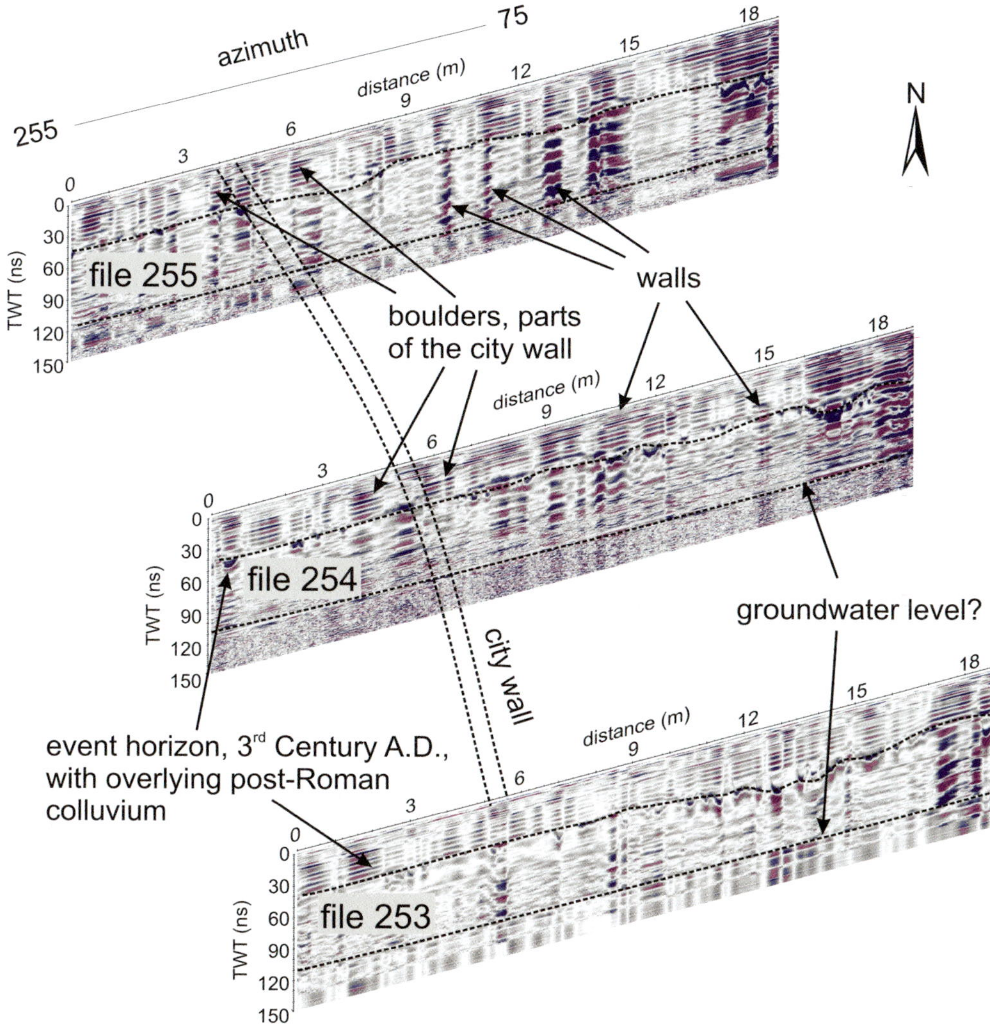

Fig. 13. Profiles 253, 254 and 255, 300 MHZ, crossing the eastern city wall. Here, fallen boulders produce characteristic reflection patterns and there are hints of the deformation of the city wall. The event horizon is clearly visible; the change of the reflection pattern at 110 ns may indicate the groundwater level.

observed deformations (specially in the pavements) were augmented during the burying process of the Roman city ruins, especially pavements located on artificial fillings (i.e. demolition horizon). However, cracks, jointing, break-outs, and pop-up structures in the ancient Roman pavements point directly to earthquake damage by SW to NE ground motion, also supported by the arrangement and the mainly unidirectional collapse of the columns of the Basilica.

On the other hand, small mass movements have been mapped and/or characterized in this study. They were supportive of the destruction of parts of the city and amplified the deformation during and after the last earthquake. In detail, most of the deformations recorded at the Isis Temple match with the head zone of a landslide and most of the deformations observed downslope at the forum, curia and macellum, match with those characteristic of a landslide toe-zone (Fig. 7). Deformations and inertial rotations of blocks in the aqueduct zone also can be linked to slower mass-movement processes on clayey steep slopes. On the contrary, those buildings totally or partially founded in the coastal aeolian materials south of the Decumanus Maximus (fish factories) do not record relevant deformations.

Ground acceleration parameters and the Spanish seismic code

The application of the former Spanish seismic code (NCSE-94 1997) led Silva *et al.* (2005) to evaluate a local site maximum ground horizontal spectral pseudo-acceleration between 0.15 and 0.21g for dominant vibration periods between 0.20 to 0.77 seconds for different seismic scenarios. These vibration periods are mainly dangerous for low buildings (high-frequency structures: Coburn & Spence 1992) such as the Roman edifices analysed in this study.

The presumed ground response under the directions of the updated Spanish seismic code NCSE-02 (2000) has also been evaluated in order to evaluate the specific site amplification at the Bolonia Bay area. The present Spanish seismic code takes into account specific site effect amplification (S) due to the geotechnical features of the ground, which was not considered in the previous NCSE-94. NCSE-02 introduces a new typified kind of soil (C-IV) corresponding to loose granular and/or cohesive soils with s-wave velocities down to 200 m/s. The properties of the poorly compacted Roman artificial fillings on which the imperial buildings were founded are within the range of the new C-IV soil class according to the available geotechnical data (Borja *et al.* 1993). Taking this into account the amplification coefficient (S) to consider for the *Baelo Claudia* site is $S = 1.44$. On the other hand, since Bolonia Bay is located close to the limit of two different ground acceleration areas codified in NCSE-02 (0.07g for Zahara county and 0.04g for Tarifa country), the resulting values for the expected ground site accelerations may vary between 0.1g (Zahara area) and 0.06g (Tarifa area) for the most conservative scenarios with lowest risk coefficients ($\rho = 1.0$). This follows the normalized NCSE-02 equation:

$$a_c = S\,\rho\,a_b$$

where S is the amplification coefficient, a_b the basic ground horizontal acceleration codified in NCSE-02, and a_c the resultant horizontal ground acceleration to consider in the calculus of the ground response elastic spectra. The risk coefficient (ρ) is a non-dimensional parameter that considers the probability of exceedence of a_c during the functional time-period (t) of the buildings, 50 years for normal buildings and 100 years for special buildings, and main engineering facilities (NCSE-02).

Under the conditions listed above, the spectral pseudo-accelerations ($S_{am} = \alpha T$) can reach maximum values from 0.24g (Tarifa-type zone) to 0.26g (Zahara-type zone) for dominant vibration periods between 0.13 to 0.57 seconds (Fig. 14). These new estimates increase the expected site spectral accelerations in the *Baelo* area from 0.24 to 0.26g for the 10% probability of exceedence in 50 years or 475 years return period. These values raise the theoretical ground movement to the lower levels generally, associated with VIII MSK intensity in firmer soils (0.25 to 0.30g; Bolt 1993). In the same way dominant vibration periods are shortened, with respect to those obtained by Silva *et al.* (2005), falling into the category of those

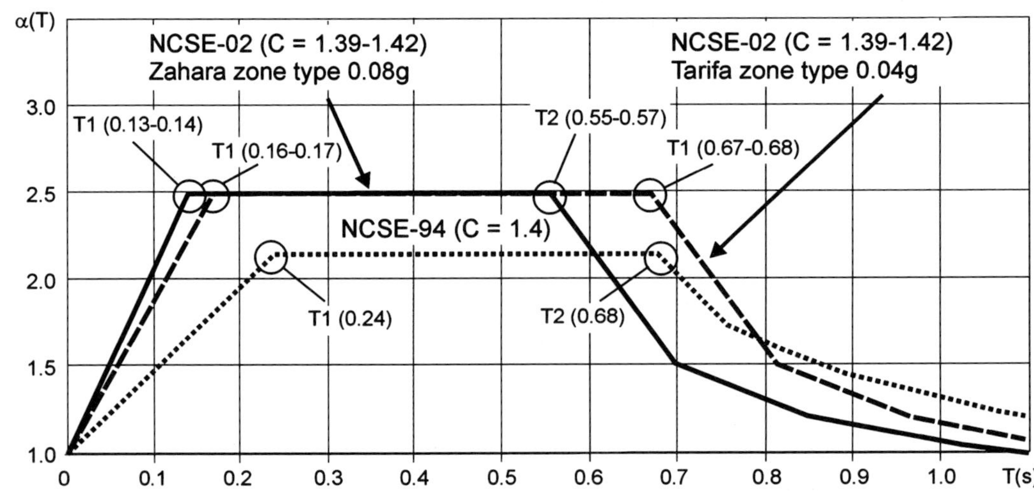

Fig. 14. Ground acceleration spectra for the *Baelo Claudia* site. Dotted line indicates spectra resulting from application of the former Spanish seismic code (NCS-94). Bold (Zahara zone type) and bold-dotted (Tarifa zone type) lines indicate spectra resulting from application of the present Spanish seismic code (NCSE-02) considering amplification values.

especially dangerous for low-rise (high-frequency structures) like those in the ancient Roman city of *Baelo Claudia*. On the other hand, due to this last consideration and as frequently stated, archaeoseismic damage is commonly associated with local seismicity rather than with the ground shaking induced by relatively remote events (Stiros 1996).

Here, we are considering repeated earthquake damage between AD 40–60 and AD 260–290. We have to consider return periods in the range of 200 to 250 years and, therefore, ground accelerations within the range of those described by NCSE-02 are applied here. However, seismic hazard at the present times (*c*. 1700 years after the last event) will be underestimated when using the present Spanish seismic code.

Probable Intensity from environmental and architectural effects recorded in Baelo Claudia

Following the directions of the recently updated ESI INQUA intensity scale for the environmental effects of earthquakes (Guerrieri & Vittori 2007), the size of the probable landslide event (*c*. 600 m^3), apparent ground settlement (10–100 cm subsidence) and widespread pop-up-like structures in paved zones caused by ground undulations match with the observed effects of intensity VIII ESI events. In detail, recent moderate earthquakes (M_S 5 to 6) in southern Italy (Michetti *et al.* 2000) and Greece (Lekkas *et al.* 1996; Mariolakos *et al.* 1998) indicate that similar pop-up features and landslides on artificial slopes can be produced from intensities of VII MSK onwards. On the other hand, jointed trespassing of two or more adjacent blocks has been attributed to minimum intensities of VIII MSK by Korjenkov & Mazor (1999). In the studied case, environmental effects of seismic shaking are highlighted by the pattern of deformations in the architectural remains directly founded on poorly compacted artificial fillings, where site effects were relevant.

The structural analysis of the deformation recorded at *Baelo Claudia*, both geomorphological and architectural evidence, can be reasonably explained by a SW–NE directed ground motion. In detail, rose diagrams of jointing, shocks, breakouts and pop-up orientations (Fig. 8) are compatible with a compressive SW to NE horizontal seismic action. In most cases, house and city walls orientated in north–south directions following the 'cardos' (the north–south Roman array of streets), display dominant tilting and/or collapse towards WSW orientations, and those orientated east–west following the Decumanus Maximus towards SW–SE orientations. Reconstruction of column collapse directions also point to dominant SSW orientations, although some of them display random collapse orientations mainly due to the eventual process of destruction of the damaged remains (Sillères 1997). Following the lessons from other historic and recent strong earthquakes (i.e. Korjenkov & Mazor 1999; Ambraseys 2006), linear architectural elements and columns are commonly tilted and/or collapsed against the sense of the surface wave propagation.

As mentioned above, kinematic markers, such as shear jointing, fracturing and shocks measured in the ancient Roman pavement along the Decumanus Maximus, also point to a broad SW–NE compressive stress field during the suspected seismogenic event. In detail, fold axes linked to the set of pop-up-like arrays developed in this same sector of the city are preferentially orientated in a NW–SE direction, pointing to a similarly orientated SW–NE compressive stress field (Fig. 7). Assuming that the deformations described here were caused by seismic wave propagation, they should be related with directional SW–NE compression waves (P-waves and/or Rayleigh waves). Therefore, the seismic source responsible for the damage of this ancient city should be located somewhere SSW offshore the city.

The uncharacteristic structural data from the Isis Temple zone do not match with the proposed SW–NE ground motion. There most of the east–west and north–south architectural elements are tilted (7–11°) upslope towards the north, and east–west walls are affected by roughly north–south cracks displaying centimetric offsets (<30 cm). However, this case is probably linked to a moderately seated (*c*. 3 m depth) landslide of about 0.6 km^3 volume (200 m^2 × 3 m) as displayed by the topographic head scarp and the GPR profiles (Figs 6 and 12). Furthermore, east–west upthrusted, folded and bent waving walls along the eastern side of the forum, and pervasive east–west pop-up folding of the ancient Roman pavement are presumably related to the body-toe contact of the landslide (Fig. 7). The size and characteristics of this event are also within the range of secondary effects for minimum intensities of VII ESI events (Guerrieri & Vittori 2007).

Conclusions

The analysis of the disturbed archaeological remains and urban geology of the ancient Roman city of *Baelo Claudia* indicates that the two recorded periods of abrupt city destruction (Table 1) can now be bracketed to AD 40–60 and approximately AD 260–290. The previously proposed date of AD 350–395 for the second event of destruction (Silva *et al.* 2005) is only related with the eventual ruin of the city after the abandonment of Roman inhabitation and can be deduced from the work of Sillères (1997) and new archaeological findings (Table 2).

The bracketing ages can be roughly related to poorly documented historic events (AD 33–382) catalogued for the Gulf of Cadiz–Cape of San Vicente (Martínez Solares & Mezcua 2002) far away from the studied Roman city (*c.* 400 km to the west). However, as commonly stated, archaeoseismic damage normally records local events affecting low-rise, high-frequency buildings (Stiros 1996) and closer seismic sources should be considered to explain the level of destruction (≥VIII MSK) recorded at *Baelo Claudia* (Silva *et al.* 2005).

The pattern of archaeological damage mapped during this study and the structural measurements on joints, fractures, shocks, pop-ups developed in the Roman pavements, and tilting and collapse orientations of house walls, city walls and columns point to a SW towards NE sense of compressional ground motion, or to a roughly NNW–SSE compressive stress field, which agree with the present S_{Hmax} direction promoted by the Africa–Eurasia convergence at this zone (Herraiz *et al.* 2000; Stich *et al.* 2003). Considering that directed ground motion was associated with seismic wave propagation, all the analysed structural data point directly to SW to NE P-waves or surface Rayleigh waves, locating the presumed seismic source SSW offshore of *Baelo Claudia*. At this area the only known tectonic structures are the offshore prolongation of the Cabo de Gracia Fault in the SSW and some discrete north–south to NNE–SSW faults immediately south of Baelo revealed in recent seismic profiling (Hübscher *et al.* 2007).

New geomorphological and palaeoseismological data based in the list of environmental earthquake effects quoted in the ESI INQUA scale (Guerrieri & Vittori 2007) indicate that NNE–SSE normal faults occurring around *Baelo Claudia* could have experienced palaeoseismic activity during the Late Quaternary to prehistoric times (i.e. La Laja and San Bartolome range fronts). Therefore, the list of possible seismic sources that can produce architectural damage at the studied Roman city is increased with respect to those proposed by Silva *et al.* (2005), which listed only NE–SW strike-slip faults. In fact, north–south normal faults in the axial zone of the Gibraltar Strait axial area account for most of the 95% of the seismic activity instrumentally recorded in this zone, otherwise shallow (<10 km deep) and normally of low magnitude (<3.5 mb). In addition, the suspected palaeoseismological features associated with the analysed onshore segments of these faults are commonly associated with minimum intensity levels of VII ESI scale (Guerrieri & Vittori 2007). On the other hand, recorded building damage clearly points to intensity levels of VIII MSK when compared with the data reported for other, better known archaeoseismological sites (i.e. Korjenkov & Mazor 1999; Similox-Tohon *et al.* 2006; Galadini *et al.* 2006).

The application of the updated seismic code of Spain (NCSE-02) indicates that amplification factors considered in this new code can attain maximum spectral pseudo-accelerations of 0.24g to 0.26g for a 475-year return period. These theoretical values of acceleration are in turn normally associated with minimum intensities of VIII MSK (Bolt 1993). These theoretical results are valid for seismic engineering purposes and lack scientific value for larger/shorter return periods, but in the studied case they help to constrain that site amplification can attain intensity levels within the range of those deduced from environmental effects and building damage. However, this study has also determined that localized landsliding at the Isis Temple zone and NE corner of the forum, and ground settlement anomalies in the southern sector of the basilica surely amplified the observed deformations during and after the last seismic event. GPR data help to constrain the approximate landslide volume to about 600 m^3.

At the moment, the available data compiled for the different building phases of the city, stratigraphy and structural measurements, strongly point to recurrent earthquake damage during the mid-first century AD, and especially for the late third century AD, coming from a still unknown offshore seismic source south of *Baelo Claudia* and inducing a minimum intensity of VIII MSK, but locally amplified by site effects. However, the Baelo case is still an open debate and radiocarbon dating (in progress) is necessary for a refined assessment of the dates of destruction. In the same way, further research on seismological modelling of architectural remains by means of the collection of more structural data, acquisition of detailed geotechnical parameters of constructive stones and underlying soils, full interpretation of GPR surveys and additional geo-electrical tomography will help to interpret the nature and pattern of the observed damage.

This work has been supported by the Spanish–German Acciones Integradas Program HA2004-0098, by the Spanish Research Projects CGL2005-04655/BTE (USAL), CGL2005-01336/BTE (CSIC) and by the Deutsche Forschungsgemeinschaft project Re 1361/9. The authors are grateful to the archaeologists Angel Muñoz and Iván García who helped and facilitated the fieldwork within the Conjunto Arqueológico de *Baelo Claudia* (Junta de Andalucía). The comments of Klaus H. Hinzen (University of Köln, Germany) and Miguel A. Rodríguez-Pascua (IGME, Spain) helped to improve the quality of this work.

References

AKIL, M., CHALOUAN, A., ISTIQUAM, H., EL FAHSSI, A. & MAMAD, A. 1995. Les formations quaternaires de la Rive Sud du Detroit de Gibraltar: Stratigraphie et déformation. *In*: ESTERAS, M. (ed.) *El Enlace Fijo del Estrecho de Gibraltar*. SECEG, Madrid, Vol. 2, 71–79.

ALONSO-VILLALOBOS, F. J., GRACIA-PRIETO, F. J., MÉNANTEAU, L., OJEDA, R., BENAVENTE, J. & MARTÍNEZ, J. A. 2003. Paléogeographie de l'anse de Bolonia (Tarifa, Espagne) à l'époque romaine. *In*: *The Mediterranean World Environment and History*. Elsevier, Amsterdam, 407–417.

ALTUNEL, E., STEWART, I., BARKA, A. & PICCARDI, L. 2003. Earthquake faulting at the Ancient Cnidus, SW Turkey. *Turkish Journal of Earth Sciences*, **12**, 137–121.

AMBRASEYS, N. N. 1971. Value of historical records of earthquakes. *Nature*, **232**, 375–379.

AMBRASEYS, N. N. 2006. Earthquakes and archaeology. *Journal of Archaeological Science*, **20**, 1–9.

ARMIJO, R., LYON-CAEN, H. & PAPANASTASSIOU, P. 1992. East–west extension and Holocene fault scarps in the Hellenic arc. *Geology*, **20**, 491–494.

BARAZA, J., ERCILLA, G. & LEE, J. 1992. Geotechnical properties and preliminary assessment of sediment stability on the continental slope of the Northwestern Alboran Sea. *Geo-Marine Letters*, **12**, 150–156.

BASILE, V., CARROZZO, M. T., NEGRI, S., NUZZO, L., QUARTA, T. & VILLANI, A. V. 2000. A ground-penetrating radar survey for archaeological investigations in an urban area Lecce, Italy. *Journal of Applied Geophysics*, **44**, 15–32.

BECKER-HEIDMANN, P., REICHERTER, K. & SILVA, P. G. 2007. ^{14}C dated charcoal and sediment drilling cores as first evidence of Holocene tsunami at the Southern Spanish coast. *Radiocarbon*, **49**(2), 827–835.

BENITO, B., CAPOTE, R., MURPHY-CORELLA, P. *ET AL.* 2007. An overview of the damaging and low magnitude La Paca Earthquake (Mw 4.8) on January 29, 2005: Context, seismotectonics and seismic risk implications for Southeast Spain. *Bulletin of the Seismological Society of America*, **97**, 671–690.

BENKHELIL, J. 1977. *Etude néotectonique de la terminaison occidentale des Cordillères Bètiques (Espagne)*. Thèse d'Etat, Université de Nice, France.

BERES, M., GREEN, A. G., HUGGENBERGER, P. & HORSTMEYER, H. 1995. Mapping the architecture of glaciofluvial sediments with three-dimensional georadar. *Geology*, **23**, 1087–1090.

BOLT, B. A. 1993. *Earthquakes*, (3rd edition). Freeman, New York.

BORJA, F., MARTÍNEZ, A. & TROYA, A. 1993. El proceso histórico de destrucción de la ciudad Romana de *Baelo Claudia* (SW Cádiz): Caracterización geotécnica y análisis de formaciones superficiales. *Arqueología Espacial*, **16/17**, 297–308.

BURBANK, D. W. & ANDERSON, R. S. 2001. *Tectonic Geomorphology*. Blackwell Science, Oxford.

CAMPOS, M. J. 1991. Tsunami hazard in Spanish coast of the Iberian Peninsula. *Science of Tsunami Hazards*, **9**(1), 83–90.

COBURN, A. & SPENCE, R. 1992. Site selection and seismic hazard assessment. *Earthquake Protection*, **187**, 187–213.

CONYERS, L. B. 2004. *Ground-Penetrating Radar for Archeology*. Altamira Press, Walnut Creek.

CROSBY, B. T., WHIPPLE, K. X., GASPARINI, N. M. & WOBUS, C. W. 2007. Formation of fluvial hanging valleys: Theory and simulation. *Journal of Geophysical Research*, **112**, F03S10. DOI: 10.1029/2006JF000566.

ESTERAS, M., IZQUIERDO, J., SANDOVAL, N. G. & BAHMAD, A. 2000. Evolución morfológica y estratigráfica Plio-Cuaternaria del Umbral de Camarinal (Estrecho de Gibraltar) basada en sondeos marinos. *Revista de la Sociedad Geológica de España*, **13**(3–4), 539–550.

GALADINI, F., HINZEN, K. G. & STIROS, S. 2006. Archaeoseismology: methodological issues and procedure. *Journal of Seismology*, **10**(4), 395–414.

GALBIS, R. J. 1932. *Catálogo sísmico de la zona comprendida entre los meridianos 5° E y 20° W de Greenwich y los paralelos 45° y 25°N*. Dirección General del Instituto Geográfico, Catastral y de Estadística, Madrid.

GALINDO-ZALDÍVAR, J., GONZÁLEZ-LODEIRO, F. & JABALOY, A. 1993. Stress and palaeostress in the Betic-Rif cordilleras (Miocene to the present). *Tectonophysics*, **227**, 105–126.

GONZÁLEZ LASTRAS, J., GARCÍA DE DOMINGO, A., HERNAIZ, P. P., ZAZO, C. & GOY, J. L. 1991. *Cartografía geológica y memoria de la hoja 13-47 (Tarifa)*. Mapa Geológico de España, Escala 1:50.000. Plan MAGNA, 2ª Serie, IGME, Madrid.

GOODMAN, D. 1994. Ground-penetrating radar simulation in engineering and archaeology. *Geophysics*, **59**(2), 224–232.

GOY, J. L., ZAZO, C. *ET AL.* 1994. Pop up-like deformation of a Roman floor and liquefaction structures in SW Spain as possible paleoseismic indicators. *Bulletin of the INQUA Neotectonics Comisión*, **17**, 42–44.

GOY, J. L., ZAZO, C., SILVA, P. G., LARIO, J., BARDAJÍ, T. & SOMOZA, L. 1995. Evaluación geomorfológica del comportamiento neotectónico del Estrecho de Gibraltar durante el Cuaternario. *In*: ESTERAS, M. (ed.) *El Enlace Fijo del Estrecho de Gibraltar*. SECEGSA, Madrid, Vol. 2, 51–69.

GRACIA, F. J., RODRÍGUEZ VIDAL, J., BENAVENTE, J., CÁCERES, L. & LÓPEZ AGUAYO, F. 1999. Tectónica Cuaternaria en la Bahia de Cádiz. *In*: PALLÍ, L. & ROQUÉ, C. (eds) *Avances en el estudio del Cuaternario español*. AEQUA-UdG, Girona, 67–74.

GUERRIERI, L. & VITTORI, E. 2007. *Intensity Scale ESI 2007*. Servizio Geologico d'Italia, APAT, Rome, Memoria Descritiva de la Carta Geologica d'Italia, **74**.

HANCOCK, P. L. & ALTUNEL, E. 1997. Faulted archaeological relicts at Hierapolis (Pamukkale), Turkey. *Journal of Geodynamics*, **24**, 21–36.

HERRAIZ, M., DE VICENTE, G. *ET AL.* 2000. The recent (upper Miocene to Quaternary) and present tectonic stress distributions in the Iberian Peninsula. *Tectonics*, **19**, 762–786.

HINZEN, K. G. 2005. The use of engineering seismological models to interpret archaeoseismological findings in Tolbiacum, Germany: A case study. *Bulletin of the Seismological Society of America*, **95**(2), 521–539.

HINZEN, K. G. & SCHÜTTE, S. 2003. Evidence for earthquake damage on Roman building in Cologne, Germany. *Seismological Research Letters*, **74**(2), 124–139.

HRUSKA, J. & FUCHS, G. 1999. GPR prospection in ancient Ephesos. *Journal of Applied Geophysics*, **41**, 293–312.

HÜBSCHER, C., BETZLER, C., REICHERTER, K. & LÜDMANN, T. 2007. *The Meteor M69/1 cruise report, CARBMED.* Report, University of Hamburg.

IGN. 1994. *Banco de Datos Sísmicos, Servicio Nacional de Sismología.* Instituto Geográfico Nacional (IGN), Madrid, Petición de datos no. 94/0001 (data-file).

JIMÉNEZ-MUNT, I., FERNÁNDEZ, M., TORNE, M. & BIRD, P. 2001. The transition from linear to diffuse plate boundary in the Azores – Gibraltar Region: results from a thin sheet model. *Earth and Planetary Science letters*, **192**, 175–189.

JOLIVET, L., FRIZON DE LAMOTTE, D., HORVÁTH, F., MASCLE, A. & SÉRRANE, M. 1999. The Mediterranean basins: Tertiary extension within the Alpine Orogen – an introduction. *In*: DURAND, B., JOLIVET, L., HORVÁTH, F. & SÉRRANE, M. (eds) *The Mediterranean Basins: Tertiary Extension within the Alpine Orogen.* Geological Society, London, Special Publications, **156**, 1–14.

KARCZ, I. & KAFRI, U. 1978. Evaluation of supposed archaeoseismic damage in Israel. *Journal of Archaeological Science*, **5**, 237–253.

KORJENKOV, A. M. & MAZOR, E. 1999. Earthquake characteristics reconstructed from archaeological damage patterns: Shivita, the Negev Desert, Israel. *Natural Hazards*, **18**, 193–226.

LARIO, J., ZAZO, C. *ET AL.* 2001. Particle size and magnetic properties of Holocene estuarine deposits from the Doñana National Park (SW Iberia): evidence of gradual and abrupt coastal sedimentation. *Zeitschrift für Geomorphologie*, **45**, 33–54.

LEBLANC, D. 1990. Tectonic adaptation of the external zones around the curved core of an orogen: the Gibraltar Arc. *Journal of Structural Geology*, **12**, 1013–1018.

LEKKAS, E., LOZIOS, S., SKOURTSOS, E. & KRANIS, H. 1996. Liquefaction, ground fissures, and coastline change during the Egio earthquake (15 June 1995: Central Greece). *Terra Nova*, **8**, 648–654.

LUQUE, L., LARIO, J., ZAZO, C., GOY, J. L., DABRIO, C. J. & SILVA, P. G. 2001. Tsunami deposits as palaeoseismic indicators: examples from the Spanish coast. *Acta Geológica Hispanica*, **36**, 197–211.

LUQUE, L., LARIO, J., CIVIS, J., SILVA, P. G., GOY, J. L., ZAZO, C. & DABRIO, C. J. 2002. Sedimentary record of a tsunami during Roman times, Bay of Cadiz, Spain. *Journal of Quaternary Science*, **17**, 623–631.

MAKRIS, J. & EGLOFF, E. 1993. *Seismic and seismicity study of the Gibraltar Strait.* Report, SECEGSA-SNED Gibraltar Strait Project, University of Hamburg.

MARCO, S., AGNON, A., ELLENBLUM, R., EIDELMAN, A., BASSON, U. & BOAS, A. 1997. 817-year old walls offset sinistrally 2.1 m by the Dead Sea Transform, Israel. *Journal of Geodynamics*, **24**, 11–20.

MARIOLAKOS, I., FOUNTOULIS, I. & MARIOLAKOS, D. 1988. Deformation structures at the Gula of Corinth, Greece, induced by the Egion earthquake of 15-6-95. *Proceedings 8th IAEG Congress*, Vancouver, Canada, 789–795.

MÁRTINEZ SOLARES, J. M. 2001. *Los efectos en España del Terremoto de Lisboa (1 de Noviembre de 1755).* Instituto Geográfico Nacional (IGN), Madrid, Monografía no. 19.

MARTÍNEZ SOLARES, J. M. & MEZCUA, J. 2002. *Catálogo Sísmico de la Península Ibérica (880 AC–1900).* Instituto Geográfico Nacional (IGN), Madrid, Monografía no. 18.

MARTÍNEZ SOLARES, J. M., LÓPEZ ARROYO, A. & MEZCUA, J. 1979. Isoseismal map of the 1755 Lisbon Earthquake obtained from Spanish data. *Tectonophysics*, **53**, 301–313.

MENANTEAU, L., VANNEY, J. R. & ZAZO, C. 1983. *Belo II : Belo et son environment (Detroit de Gibraltar), Etude physique d'un site antique.* Casa de Velazquez, Serie Archeologie **4**, Broccard, París.

MICHETTI, M., FERRELI, L. *ET AL.* 2000. Ground effects during the 9 September 1988, Mw = 5.6, Lauria Earthquake and the seismic potential of the "Aseismic" Pollino region in southern Italy. *Seismological Research Letters*, **71**, 31–46.

MONACO, C. & TORTORICI, L. 2004. Faulting and effects of earthquakes on Minoan archaeological sites in Crete (Greece). *Tectonophysics*, **382**, 103–116.

MURPHY CORELLA, P. 2005. *Field Report for the January 29th 2005 earthquakes in Lorca, Spain.* Available at: http://www.proteccioncivil-andalucia.org/Documentos/SismoLorca.htm

NCSE-94. 1997. Norma de Construcción Sismo-Resistente Española. *In*: *Normativa de Acciones en la Edificación A Serie Normativas.* Ministerio de Fomento, Madrid.

NCSE-02. 2000. Norma de Construcción Sismo-Resistente Española. *In*: *Normativa de Acciones en la Edificación AE-02 (NBE) Serie Normativas*, Ministerio de Fomento, Madrid.

NEAL, A. 2004. Ground-penetrating radar and its use in sedimentology: principles, problems and progress. *Earth Science Reviews*, **66**, 261–330.

NOOMEN, R., AMBROSIOUS, B. & WAKKER, K. 1993. Crustal motions in the Mediterranean region determined from Laser ranging to Lageos. *In*: SMITH, D. E. & TRUCCOTTE, D. L. (eds) *Contribution of Space Geodesy to Geodynamics.* AGU, Washington, Earth Dynamics, Geodynamics Series, **23**, 331–346.

PIPAN, M., BARADELLO, L., FORTE, E., PRIZZON, A. & FINETTI, I. 1999. 2-D and 3-D processing and interpretation of multi-fold ground penetrating radar data: a case history from an archaeological site. *Journal of Applied Geophysics*, **41**, 271–292.

REICHERTER, K. R. 2001. Paleoseismologic advances in the Granada Basin (Betic Cordilleras, southern Spain). *Acta Geológica Hispánica*, **36**, 267–282.

REICHERTER, K. R. & PETERS, G. 2005. Neotectonic evolution of the Central Betic Cordilleras (southern Spain). *Tectonophysics*, **405**, 191–212.

RUÍZ, F., RODRÍGUEZ-RAMÍREZ, A. *ET AL.* 2004. Late Holocene evolution of the southwestern Doñana National Park (Guadalquivir Estuary, SW Spain): a multivariate approach. *Palaeogeography, Palaeoclimatology, Palaeoecology*, **204**, 47–64.

SAMBUELI, L., SOCCO, L. V. & BRECCIAROLI, L. 1999. Acquisition and processing of electric, magnetic and GPR data on a Roman site *Victimulae*, Salussola, Biella. *Journal of Applied Geophysics*, **41**, 189–204.

SANDMEIER, K. H. 2006. *Reflex-Win Version 4.2 radar processing and interpretation software package*. Sandmeier Scientific Software, Karlsruhe.

SANDOVAL, N. G., SANZ, J. L. & IZQUIERDO, F. J. 1996. Fisiografía y Geología del Umbral del Estrecho de Gibraltar. *Geogaceta*, **20**(2), 343–346.

SANZ DE GALDEANO, C. 1990. Geologic evolution of the Betic Cordilleras in the Western Mediterranean, Miocene to present. *Tectonophysics*, **172**, 107–119.

SECEGSA 1988. *Mapa Geológico del Estrecho de Gibraltar -Orilla Norte- Sc. 1:100.000*. SECEGSA, Madrid.

SILLIÈRES, P. 1997. *Baelo Claudia: Una ciudad Romana de la Bética*. Junta de Andalucía, Casa de Velázquez, Madrid.

SILVA, P. G., BORJA, F. ET AL. 2005. Archaeoseismic record at the ancient Roman city *of Baelo Claudia* (Cádiz, South Spain). *Tectonophysics*, **408**, 129–146.

SILVA, P. G., GOY, J. L. ET AL. 2006. Neotectonic fault mapping at the Gibraltar Strait Tunel area, Bolonia Bay (South Spain). *Engineering Geology*, **84**, 31–47.

SIMILOX-TOHON, D., SINTUBIN, M., FERNÁNDEZ, M., MUCHEZ, P. & WAELKENS, M. 2006. Identification o fan active fault by a multidisciplinary study at the archaeological site of Sagalassos (SW Turkey). *Tectonophysics*, **420**, 371–387.

SMITH, D. G. & JOL, H. M. 1995. Ground-penetrating radar: antenna frequencies and maximum probable depths of penetration in Quaternary sediments. *Journal of Applied Geophysics*, **33**, 93–100.

STICH, D., AMMON, C. J. & MORALES, J. 2003. Moment tensor solutions for small and moderate earthquakes in the Ibero-Maghreb region. *Journal of Geophysical Research (Solid Earth)*, **108**(B3) 2148, doi: 10.1029/2002JB002057.

STIROS, S. 1996. Identification of earthquakes from archaeological data: methodology, criteria and limitations. *In*: STIROS, S. & JONES, R. E. (eds) *Archaeoseismology Occasional paper no. 7 of the Fitch Laboratory*. British School at Athens, Athens, 129–152.

STIROS, S. & PAPAGEOURGIU, S. 2001. Seismicity of western Crete and the destruction of the town of Kisamos at AD 365: Archaeological evidence. *Journal of Seismology*, **5**, 381–397.

VAUGHAN, C. J. 1986. Ground-penetrating radar surveys used in archaeological investigation. *Geophysics*, **51**(3), 595–604.

VÁZQUEZ, J. T. & VEGAS, R. 2000. Acomodación de la convergencia entre África y la Península Ibérica, Golfo de Cádiz, Mar de Alborán, a partir del análisis de terremotos. *Geogaceta*, **27**, 171–174.

VITTORI, E., DEIANA, G. ET AL. 2000. Ground effects and surface faulting in the September-October 1997 Umbria-Marche (Central Italy) seismic sequence. *Journal of Geodynamics*, **29**, 535–564.

WEIJERMARS, R. 1991. Geology and tectonics of the Betic Zone, SE Spain. *Earth Science Reviews*, **31**, 153–236.

ZAZO, C., SILVA, P. G., GOY, J. L., HILLAIRE-MARCEL, C., LARIO, J., BARDAJÍ, T. & GONZÁLEZ, A. 1999. Coastal uplift in continental collision plate boundaries: Data from the Last interglacial marine terraces of the Gibraltar Strait area (South Spain). *Tectonophysics*, **301**, 95–119.

Ground effects of the 18 October 1992, Murindo earthquake (NW Colombia), using the Environmental Seismic Intensity Scale (ESI 2007) for the assessment of intensity

S. MOSQUERA-MACHADO[1]*, C. LALINDE-PULIDO[2], E. SALCEDO-HURTADO[3] & A. M. MICHETTI[4]

[1]*Am Risc, LP, 20405 SH 249, Suite 430, Houston, TX 77070, USA*

[2]*Carrera 83A, N° 34A–23, Apto. Edificio Villa Laureles, Medellin, Colombia*

[3]*Departamento de Geografia, Universidad del Valle, Edificio 384, Ciudad Universitaria Melendez, Cali, Colombia*

[4]*Dipartimento di Scienze Chimiche e Ambientali, Università dell'Insubria, Via Valleggio, 11, 22100, Como, Italy*

**Corresponding author (e-mail: smosqueramachado@gmail.com)*

Abstract: The macroseismic intensity of the 18 October 1992 Murindo-Atrato earthquake that affected the northwestern states of Colombia (Chocó and Antioquia) is reassessed using the newly developed INQUA Environmental Seismic Intensity Scale (ESI 2007) which is based on the evaluation of earthquake environmental effects. To generate the ESI 2007 isoseismal map of northwestern Colombia, a geographical information system was used. Unifying the available information on the seismological and active tectonics framework including historical seismicity, hypocentral depths, foreshocks, aftershocks, focal mechanism, macroseismic data under the same GIS and the map of Quaternary faults allowed us to reinterpret the geological and environmental effects of the 1992 earthquakes sequence. A total of 24 sites from the areas of Quibdó, Bojayá, Rio Sucio, Murindo, Vigía del Fuerte and Turbo were evaluated. A systematic comparison among evaluated intensities (Modified Mercalli and ESI scale) revealed differences from one to two degrees. According to the ESI 2007 scale, the epicentral intensity I_o is XI. This represents one degree higher than the epicentral intensity obtained using MM and Medveded Sponhauer Karnik (MSK) intensity scales, probably due to the lack of suitable observations on building damage in this poorly populated and developed region. This information is also useful in order to shed some light on the persistent question of the exact location and dimension of the main rupture zone associated with the earthquake. The isoseismal map derived from the integration of the whole set of environmental effects with other macroseismic data strongly suggests that the causative tectonic structure is the Murindo fault. However, the rupture length derived from the distribution of ground effects is greater than the Murindo fault length, implying that other nearby fault segments were activated during the 1992 event. The new isoseismal map resulting from this work is relevant for the assessment of future seismic risk in the northwestern region of Colombia. Overall, the application of the ESI 2007 scale to the 18 October 1992 earthquake, and to similar strong events in the region, can be useful for disaster management and planning, estimation of damage, and post-earthquake recovery efforts.

On 17 and 18 October 1992 a disastrous sequence of two strong shallow crustal earthquakes ($M_s = 6.6$ and $M_s = 7.3$) occurred in Atrato Valley near Murindo in northwestern Colombia (Fig. 1). The focal depth of the 18 October main shock was about 10 km according to Arvidsson *et al.* (2002, and references therein). This was the largest earthquake to strike northwestern Colombia during the modern seismological period (last 30 years). This earthquake caused enormous destruction in Murindo, Rio Sucio and Bojayá, and a wide spectrum of environmental effects ranging from small cracks to extensive landslides, liquefaction, and mud volcanoes that occurred at a distance of 150 km from the epicentre. According to Martinez *et al.* (1994), in places as far as the city of Medellin (more than 130 km away from the epicentral area), nearly 243 buildings were damaged and ten people were killed.

Governmental and scientific institutions (Colombian Geological Survey, Ingeominas; Environmental Protection Agency of Chocó State,

From: REICHERTER, K., MICHETTI, A. M. & SILVA, P. G. (eds) *Palaeoseismology: Historical and Prehistorical Records of Earthquake Ground Effects for Seismic Hazard Assessment*. The Geological Society, London, Special Publications, **316**, 123–144. DOI: 10.1144/SP316.7 0305-8719/09/$15.00

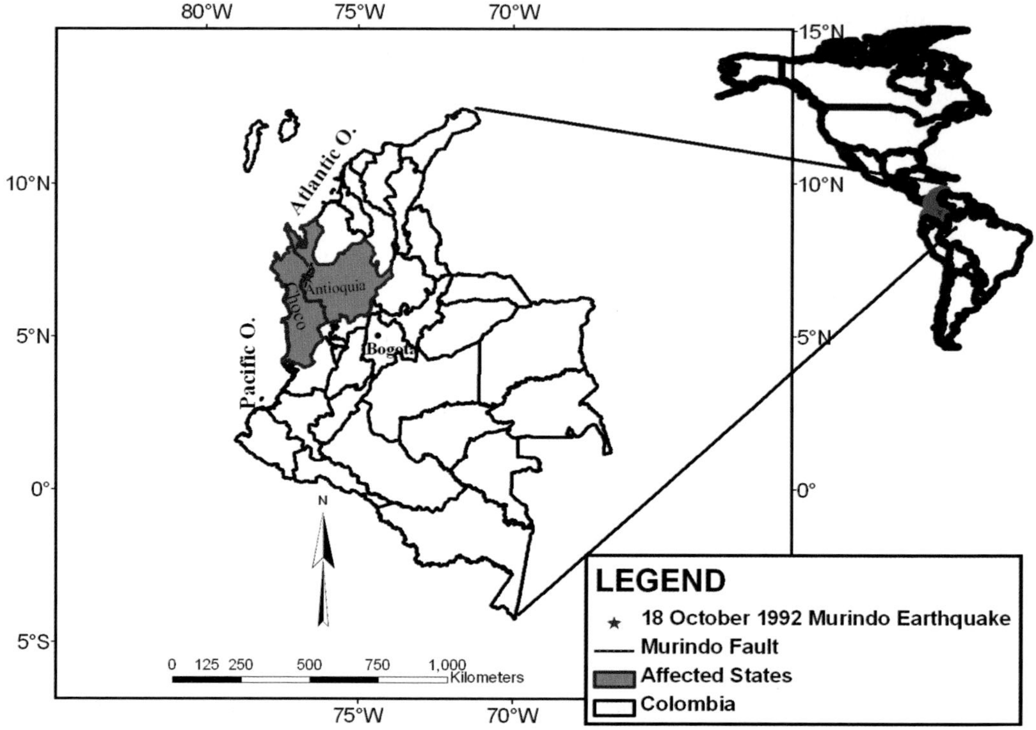

Fig. 1. Geographic location of the area affected by the 18 October 1992 Murindo earthquake (Colombia).

Codechoco) deployed teams to the area in the aftermath of the events to evaluate the earthquake effects as well as advance the understanding of earthquake emergency management processes in the region.

Due to the shallow hypocentral depth and large magnitude of the earthquake, surface faulting probably occurred. There was great uncertainty about the exact location and rupture length of the causative faults for the earthquake sequence, because the tropical fluvial setting and difficult access in the epicentral area do not allow reasonably accurate identification of earthquake fault scarps. No evidence of surface faulting has been reported in the literature. Martinez *et al.* (1994) and Paris *et al.* (2000) ascribe the October 1992 rupture to the Murindo fault, a well-known NNW trending fault with mostly left-lateral strike-slip kinematics. On the basis of geological and macroseismic data Paris *et al.* (2000) interpret the coseismic rupture as affecting the whole length of the fault, *c.* 75 km long. Conversely, Ardvisson *et al.* (2002), based on foreshocks and aftershocks joint-hypocentre relocations, propose a NNE trending source not clearly related to any of the major faults known in the region; their estimated rupture length is *c.* 90 km. According to Li & Toksöz (1993), the study of the source time functions for the mainshock indicates instead a NE trending, mostly reverse fault with a total length of *c.* 140 km.

In this paper, we show an effective approach using the newly developed Environmental Seismic Intensity Scale (ESI 2007) based on earthquake environmental effects (Michetti *et al.* 2007) in an attempt to shed some light on the persistent question of the source parameters of the main rupture zone associated with the earthquake. More than 60 scientific publications were compiled. A geographical information system was used to unify different kinds of information such as seismological framework, historical seismicity of the area, hypocentral depth, foreshocks, aftershocks, focal mechanisms, macroseismic data, Quaternary fault maps, plate tectonic setting, slip rates and data from previous studies on active tectonics and palaeoseismology of the area, with the reinterpretation of geological and environmental effects of the earthquake. Primary and secondary effects have been identified by field survey (Coral & Salcedo 1992; Mosquera-Machado *et al.* 1992; Mosquera-Machado 1994*a*, *b*, 2002; Parra 2002, 2003) and by the analysis and reinterpretation of official reports written at different times. We assessed the Environmental Seismic Intensity at selected localities, bearing in mind the criteria of completeness and detail of description of environmental effects. For 24 localities in the municipalities of Quibdó, Bojaya, Rio Sucio (Chocó State) and Vigía del Fuerte, Murindo and Turbo (Antioquia State) we obtained well constrained new intensities of earthquake environmental effects (EEE), suitable for integration with the existing macroseismic and instrumental data.

Geological and seismological framework

Regional tectonic setting

Because of their position in the northwestern corner of Colombia, the Chocó and Antioquia states are tectonically controlled by the interaction of three main plates (the Caribbean, South American and Nazca plates) and the assemblage of microcontinental blocks and fragments bounded by high-strain suture zones, wide and narrow mobile belts and transcurrent fault systems (Pennington 1981; Kellogg *et al.* 1985; Duque-Caro 1990*a*; Freymueller *et al.* 1993; Taboada *et al.* 2000).

The tectonic complexity of the region was modelled by Taboada *et al.* (2000) using local seismological, tectonic and global tomographic data. The model suggests the existence of a confluent environment of four plates (Fig. 2): the North Andes block as part of the South American Plate, the Panama block, the Caribbean and the Nazca plates. The result of these interactions is three main tectonic features in the region: the Colombian Nazca–Pacific subduction, the North Andes block and the Southern Caribbean plate boundary zone (Freymueller *et al.* 1993). The North Andes block is moving NNE relative to stable South America and compressed in the east–west direction, whereas in the north it is converging with the Caribbean Plate (Pennington 1981). The rates calculated with GPS are for a northwesterly convergence of Panama and North Andes at about 21 mm/a (Kellogg & Vega 1995). The velocity of the Nazca Plate relative to the stable South America plate measured with GPS is about 6 mm/a (Trenkamp *et al.* 2002).

The Caribbean Plate has been and is currently being obliquely subducted beneath the northern continental margin of South America (Burke *et al.* 1984). The measured rates of convergence along northwestern Colombia are 17 mm/a (Kellog & Bonini 1982), 10 mm/a (Freymueller *et al.* 1993), 1.3 ± 0.3 cm/a (Van der Hilst & Mann 1994) and 20 ± 2 mm/a (Trenkamp *et al.* 2002). Norabuena *et al.* (1998), using space geodetic observations, concluded that the Nazca Plate is moving eastward with respect to stable South America at a rate of 60 mm/a.

Geological and geomorphological setting of the study area

Almost all the seismic effects of the 18 October 1992 event were found within the microplate called the Chocó block. According to Duque-Caro (1990*b*), the Chocó block is constituted by a tectonic mélange of materials, especially at its eastern margin, in which disrupted strata and inclusions of Upper Cretaceous–Palaeocene, Eocene–Oligocene, and Miocene exotic blocks are dispersed in a pelitic matrix of middle Miocene age. Within the Chocó block five major structural features are defined: the Uramita, Atrato, Murindo, and Baudo fault zones, and the Itsmina Deformed Zone (Duque-Caro 1990*b*; Paris *et al.* 2000).

The Uramita Fault Zone is the suture between the Chocó block and the Cordillera Occidental in NW South America and delineates the eastern boundary of the Chocó block, and the Itsmina Deformed Zone to the south (Case *et al.* 1971).

Irving (1971, 1975) described the Atrato Fault as 'an extraordinary rift zone along the eastern margin of the Atrato River Valley and the western margin of the western Cordillera. The fault extends for several hundred kilometers south from the Gulf of Uraba.' Haffer (1967) had already described and discussed a fault called Uraba, apparently with the same characteristics as Irving's Atrato Fault. According to Kellogg *et al.* (1989), the northernmost land extension of the Atrato Fault coincides with the trace of the Uramita Fault Zone.

Geomorphological evidence of the Murindo fault was observed in 1979 by Woodward-Clyde Consultants and has been document in the work of Page (1986). Paris & Romero (1994) and Paris *et al.* (2000) compiled all the available data of Murindo and concluded that the Murindo fault is a left-lateral fault bounding the western margin of the Atrato region. It extends next to the western slope of the Cordillera Occidental of Colombia, from the Rio Arquia in the south to the Rio Sucio and the basin of the Rio Atrato in the north. The Murindo fault places Cretaceous volcanic rocks against Tertiary turbidites, and cross-cuts Tertiary quartz-diorite and granodiorite.

The Mutata fault is located near the junction of the Nazca, Caribbean and South American plates (Paris *et al.* 2000), between the Rio Penderisco and the Caribbean Sea. The Mutata fault places Cretaceous intrusive rocks and greenstones (to the east) in contact with sedimentary Tertiary rocks (to the west).

The Unguia fault is a 155 km long reverse, dextral (right-lateral) fault located in the Darien area of northwestern Colombia. It has an irregular arcuate trace in map view, but it has a general north tendency (Paris *et al.* 2000).

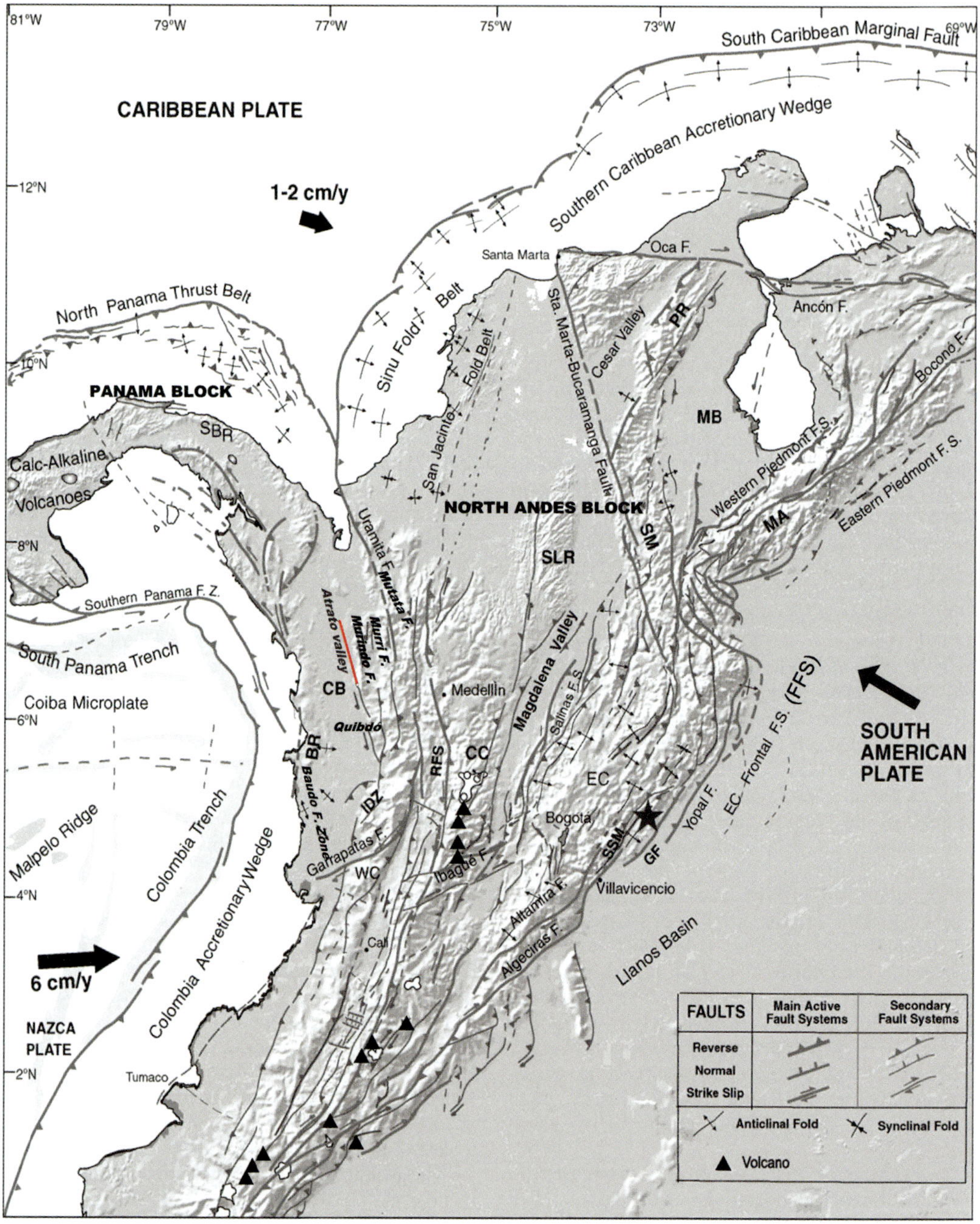

Fig. 2. Neotectonic map of Colombia with the main fault systems. Dashed lines show trends of faults and plate boundaries (after Taboada *et al.* 2000; Dimate *et al.* 2003). The solid arrows show the movement of the Nazca and Caribbean plates relative to the South American Plate. Solid triangles indicate the location of volcanoes. Abbreviations: CB, Panama-Choco Block; CC, Central Cordillera; EC, Eastern Cordillera; IDZ, Itsmina Deformed Zone; RFS, Romeral Fault; WC, Western Cordillera. The trace of the Murindo Fault, in red, is modified from Paris *et al.* (2000).

The Murri fault is 88 km long located in the western flank of the Cordillera Occidental of Colombia (Paris *et al.* 2000). The fault puts Cretaceous mafic igneous rock to the east in contact with Tertiary marine sedimentary rocks to the west.

The Baudo Fault represents the suture of the Dabeiba Arch, and the western margins of the Baudo Arch (Haffer 1967).

The Istmina Deformed Zone (IDZ in Fig. 2) separates the Atrato and San Juan basins and marks the southern boundary of the Chocó block, which trends N60°E (Nygren 1950; Bueno & Govea 1976).

Geomorphology

In the zone affected by the 1992 Murindo earthquake, three main regions can be identified: the Uraba plain, the Andean part of Antioquia and Chocó, and the Atrato plain (Fig. 3).

The Urabá plain includes the affected municipalities of Apartadó Turbo, Necoclí, San Juan de Urabá and Mutatá, and is characterized by low, flat relief surrounded by the Eastern Cordillera in the east and south. It comprises recent soft alluvial sediments that cover the lowlands around the Uraba Gulf like low energy alluvial fans. The Andean part of Antioquia corresponds to the western Andean zone of the state. It includes the municipalities of Dabeiba, Frontino, Cañas Gordas, Santa Fe de Antioquia, Urrao and Concordia. The Atrato plain is 80 km wide and extends from the south of Quibdó to the Urabá Gulf. It includes the affected municipalities of Murindo, Bojaya, Vigia del Fuerte, Rio Sucio and Quibdó. The topography of the plain is characterized by flat alluvial flood plain and meandering streams. The absence of topographic relief in this depositional plain has resulted in poor drainage, extensive swamps, multiple fluvial channels and small shallows lakes.

Historical seismicity

Historical accounts of the seismicity of Colombia indicate that in the past the NW sector of the country has also been the epicentre of many earthquakes with macroseismic characteristics very similar to the two events that occurred in 1992 (Table 1). The epicentre distribution of historical earthquakes clearly shows a high level of seismic activity along the Atrato–Murindo fault system (Fig. 4). Analyses of the historical seismicity of the area reveal that several destructive earthquakes have occurred in the region within the past 500 years (Ramirez 1975; Goberna 1985). The activity of the first 30 years of the twentieth century has been characterized by the concentration of a

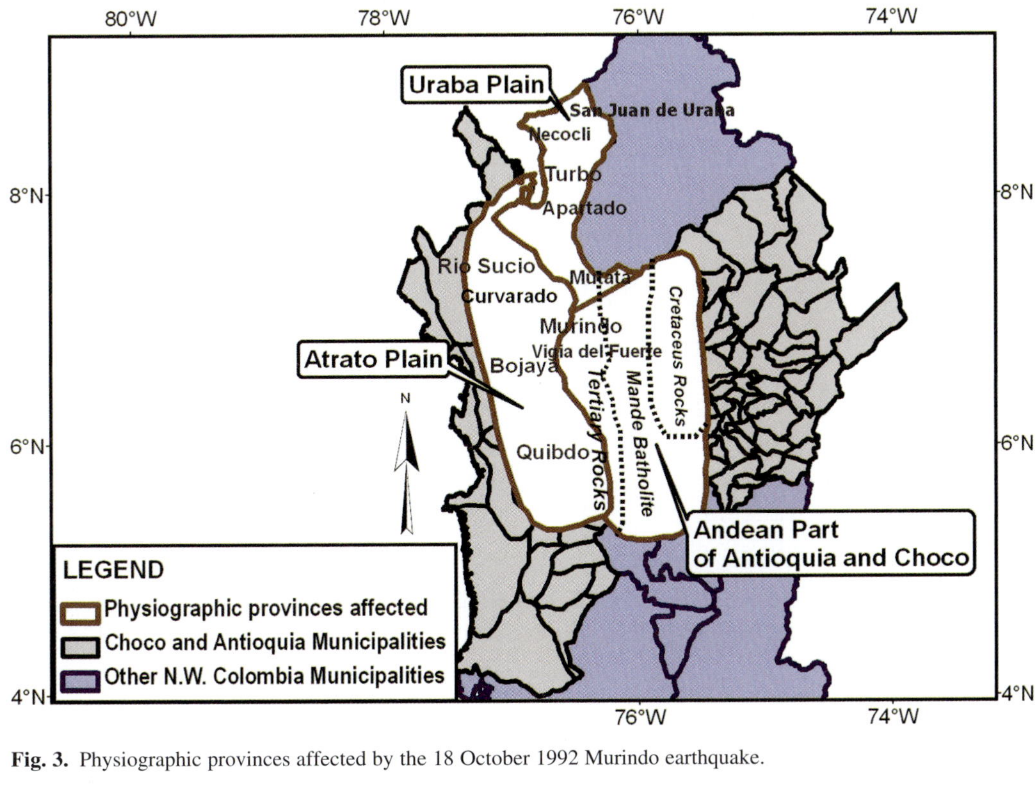

Fig. 3. Physiographic provinces affected by the 18 October 1992 Murindo earthquake.

Table 1. *Important earthquakes in Northwestern Colombia*

No	Date			Coordinates			Magnitude M_S	Intensity* I_{MAX} MM Scale	Location	Source†
	Day	Month	Year	Lat. °N	Long. °W	Depth (km)				
1	07	09	1882	76.2	8.7	–	–	IX	Turbo-Antioquia	1
2	08	03	1883	76.9	7.4	–	–	VII	Ríosucio–Chocó	1, 2
3	01	12	1903	76.4	6.4	–	–	VII	Frontino Antioquia	1, 2
4	14	02	1952	76.4	7.5	44	6.7	VIII	Pavarandocito–Antioquia	1, 2
5	13	03	1960	77.0	7.5	60	6.1	VIII	Riosucio–Chocó	1
6	01	09	1960	77.0	6.6	56	–	V	Bojayá–Chocó	1
7	22	09	1960	77.7	6.9	56	–	VI	Coredó–Chocó	1
8	16	08	1965	77.5	5.2	28	5.2	VI	Purrichá–Chocó	1
9	21	01	1966	77.4	5.2	04	4.7	V	Purrichá–Chocó	1
10	22	01	1966	77.4	5.2	–	–	–	Purrichá–Chocó	1
11	23	01	1966	77.4	5.2	–	–	–	Purrichá–Chocó	1
12	04	02	1966	76.3	5.6	–	–	–	Bagadó–Chocó	1
13	25	02	1966	77.3	5.3	34	4.4	V	Purrichá–Chocó	1
14	06	05	1967	77.5	6.9	23	4.0	V	Coredó–Chocó	1
15	31	10	1968	76.5	6.5	24	4.7	–	SE de Murrí–Antioquia	1
16	14	02	1969	76.6	6.0	15	4.0	–	Bebaramá–Chocó	1
17	08	04	1970	76.4	6.5	43	3.9	–	Urrao–Antioquia	1
18	20	05	1970	77.5	5.7	33	4.9	V	Pacific coast-W. Colombia	1
19	05	08	1970	76.2	5.7	6	4.6	–	Guaduas–Chocó	1
20	26	12	1971	77.3	6.4	9	4.8	V	Pacific Coast Chocó Col.	1
21	30	12	1971	77.7	5.6	43	4.9	V	Costa. Chocó Col.	1
22	15	04	1972	76.9	6.9	42	4.9	–	La Isla–Chocó	1
23	16	06	1972	78.1	5.2	33	4.7	–	Chocó Col.	1
24	18	06	1972	77.2	5.6	33	4.7	–	North-Catrú-Chocó	1
25	17	09	1972	77.6	5.7	22	5.4	VI	Pacific Coast Chocó	1
26	08	11	1972	77.3	6.4	33	4.6	–	Pacific Coast Chocó	1
27	26	11	1972	77.4	5.0	48	4.8	V	Pacific Coast Chocó Col.	1
28	17	10	1973	77.2	7.5	15	4.9	V	Ríosucio–Chocó	1
29	02	12	1973	77.4	6.7	76	4.1	–	Pacific Coast Chocó	1
30	17	10	1992	76.8	6.6	<15	6.6	–	Murindó–Chocó-Antioquia	3
31	18	10	1992	77.0	6.9	<15	7.3	–	Murindó–Chocó-Antioquia	3

*Intensity data taken from Ingeominas (1999).

†1, Ramírez (1975); 2, Arango-López & Velásquez (1993); 3, NEIC, (2007).

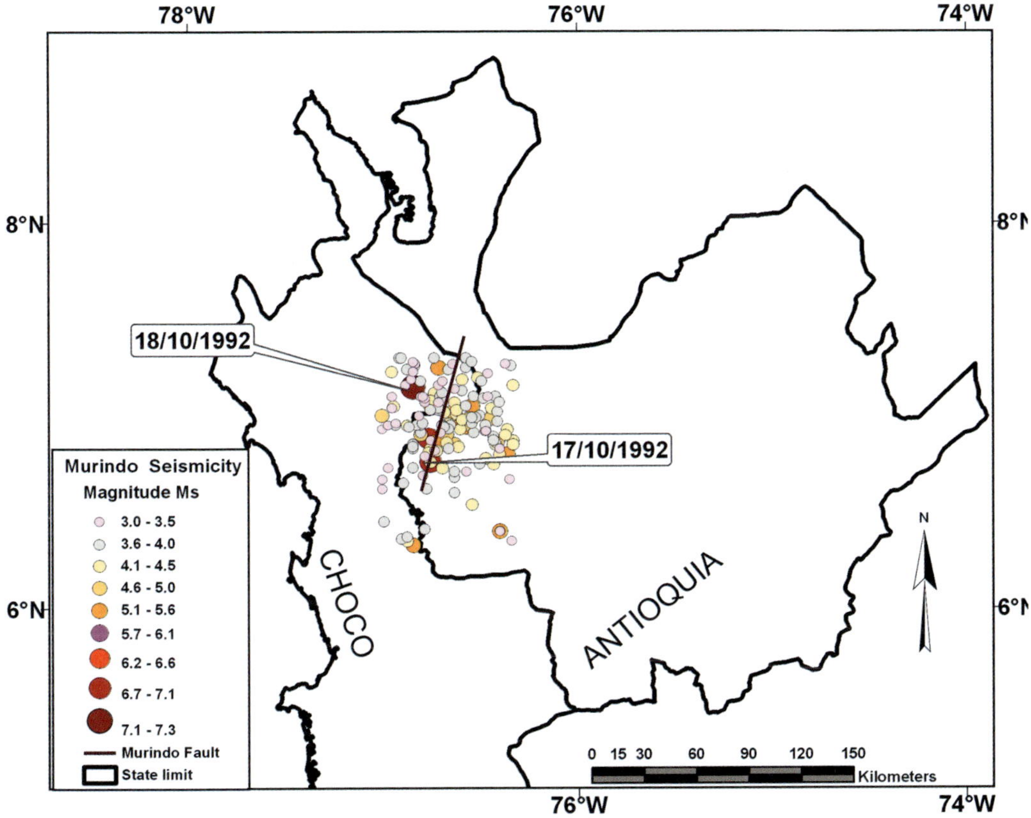

Fig. 4. Historic seismicity of the Murindo area (magnitude Ms ≥ 3.0). Location of epicentres from the National Seismologic Network of Colombia (NSRC).

higher level of seismicity along the Pacific coast (Ramirez 1975). To the west and south of Atrato, seismicity was relatively high in the 1970s. A pair of earthquakes in 1970, $M_S = 6.6$ and $M_S = 6.5$, and the earthquake in 1975, $M_S = 6.5$, occurred close to the Pacific coast (Garcia *et al.* 1984). The most important historical earthquakes of this region occurred in the nineteenth century as described below.

On 7 September 1882 at 02:56 local time, a large earthquake occurred in the north of Panama. This event affected all the territory of the Isthmus of Panama and a large part of the Chocó and Antioquia states in Colombia. A crater near Riosucio (Chocó) was formed producing an eruption of sands and ashes. Perrey (1858) described that in Turbo (Uraba Gulf), a thermal source flooded all the streets of the city, causing major losses.

On 8 March 1883, a large earthquake occurred in Pavarandó (Chocó state). This event caused liquefaction characterized by the presence of small volcanoes, as indicated in the report published in *El Periódico La Voz de Antioquia* by the engineer J. H. White in 1883. He describes specifically the phenomenon: 'Simultaneously with the earth movement the last March 8, appeared several volcanoes in the valleys of river Leon and Sucio, within the limits of Antioquia department. The number and area of these volcanoes is unknown, but according the reports from the Caribbean coast, they extend until the Gulf of Darien. Between Pavarandó and the Leon River, there are several centers of eruptions, which throw a big amount of mud and hot water; with those materials, they caused great damage in an extended zone and had also caused the obstruction of the Leon River. There is a great probability that during the winter, the water on top of the volcanoes will trigger important landslides.'

On 1 December 1903, an earthquake struck the village of Frontino, Antioquia, and extensive damage to dwellings was reported.

The 18 October 1992 Murindo earthquake

On 18 October 1992 at 15 hours 12 minutes and 9.80 seconds (Harvard Centroid Moment Tensor (CMT

2007), an earthquake of magnitude M_s 7.3 (USGS-NEIC) struck northwestern Colombia in the Atrato Valley area, killing ten people and causing severe damage in 33 municipalities (Martinez *et al.* 1994). The seismic period began on 15 October with a M_s 4.5 shock. Particularly relevant was the M_s 6.6 foreshock of 17 October located *c.* 20 km south of the mainshock (Li & Toksöz 1993; Ammon *et al.* 1994). The aftershocks lasted until April 1993. After a short period of inactivity, the area was affected by moderate shocks in October 1994 and in 1995.

Previous studies

Mosquera-Machado (Mosquera-Machado *et al.* 1992) led a commission sent by Codechoco for evaluation of the damage, as well as for site selection for the relocation of three villages of Bojaya and Rio Sucio municipalities (Chocó state). During this field trip, all municipalities affected by the earthquake were covered and a detailed study of the environmental effects was performed. Coordinates and measures of the liquefaction zones were captured using a GPS.

Coral & Salcedo (1992) conducted field investigations in the region in the aftermath of the earthquake to evaluate the macroseismic field of the event using the Modified Mercalli (MM) 1956 scale (Fig. 5).

A detailed description of the earthquake effects and the tectonic emplacement and historic seismicity of the area was compiled by Martinez *et al.* (1994). Ramirez & Bustamante (1996) depicted the effects of the earthquake and analysed the institutional Disaster Risk Management framework that was in place to manage the emergency.

Escallón (2000) described the 17 and 18 October 1992, Murindo earthquake sequence as one of the most important events in the seismicity of Colombia in the last century.

Lalinde *et al.* (2004) and Lalinde & Sanchez (2007) made the first attempt to assess the environmental seismic intensity (ESI) of the most devastating recent Colombian earthquakes. They concluded that more detailed surveys of the environmental effects of the earthquakes are needed to assign a unique ESI intensity value for the studied events.

The fault complexity of the 17 and 18 October Murindo earthquakes was studied by Wallace &

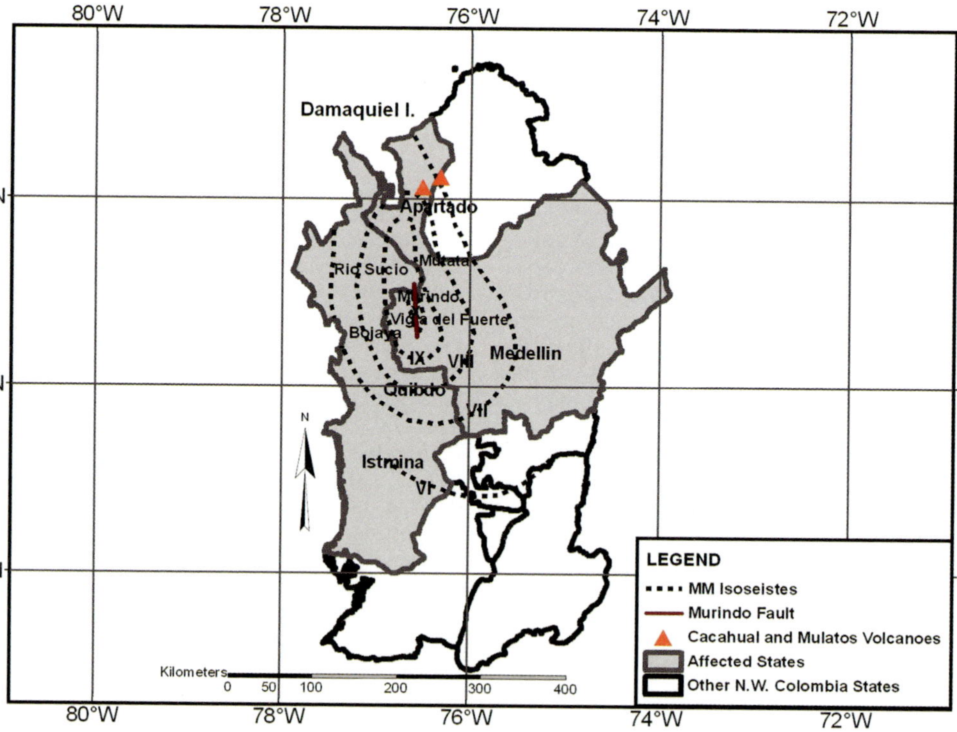

Fig. 5. Modified Mercalli 1956 intensity map of the 18 October 1992 Murindo earthquake after Coral & Salcedo (1992); solid triangles show the location of mud volcanoes that erupted in Uraba Municipality a few minutes after the main event; trace of the Murindo fault modified from Paris *et al.* (2000).

Table 2. *The 17 and 18 October 1992 Murindo earthquakes Source Parameter (NEIC)*

Date	Time (UTC)	Epicentre		Moment		
		Latitude (deg N)	Longitude (deg W)	Depth (km)	M_w	M_o (10^{19} Nm)
17/10/1992	08:32:40	6.845	76.806	8.0	6.6	0.77
18/10/1992	15:11:59	7.075	76.862	11.0	7.2	8.4
13/09/1994	10:01:32	7.054	76.678	30.0	6.0	0.13

Beck (1993), Li & Toksöz (1993), Ammon *et al.* (1994) and Arvidsson *et al.* (2002). They estimated the length of the fault zone to range from 90 to 140 km.

The focal mechanism and source parameters

The two main events of the 17 and 18 October 1992 earthquake sequence were reported by the NEIC with similar focal parameters and Ms magnitude of 6.6 and 7.3, respectively (Table 2).

For the main event (18 October), the CMT solution shows a nodal plane in direction N21°E concordant with a left-lateral fault plane, that in turn coincides with the new event of 13 September 1994 in the same area with similar focal mechanism solution (Fig. 6 and Tables 2, 3). However, for the largest foreshock (17 October), considered as a premonitory of the seismic sequence (Arvidsson *et al.* 2002), the focal mechanism indicates a NNE trending reverse fault plane. The initial polarity of the emergent waveform of the 18 October event is opposite to that of 17 October, as shown by

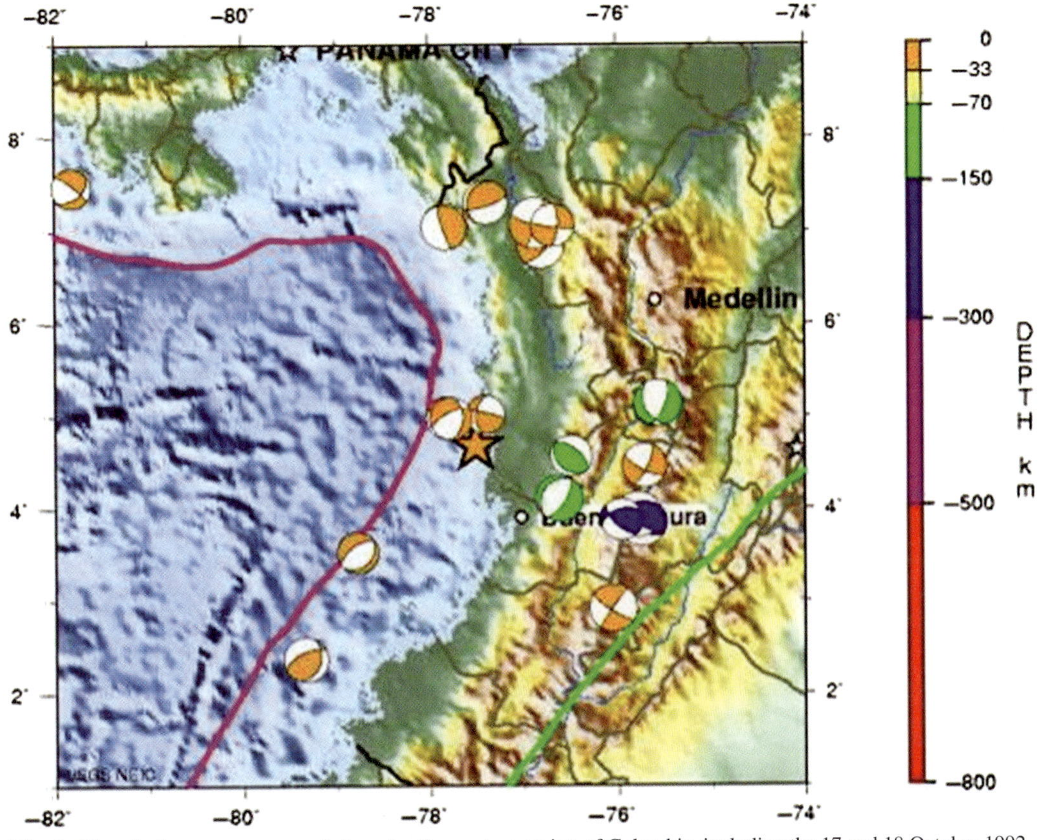

Fig. 6. Historical moment tensor solutions for the western region of Colombia, including the 17 and 18 October 1992 events that occurred in the northwestern region of Colombia.

Table 3. *Focal mechanism solutions for the 17 and 18 October 1992 Murindo earthquakes (CMT)*

Date	Principle axes				Nodal planes					
	P		T		1			2		
	Azimuth	Plunge	Azimuth	Plunge	Strike	Dip	Slip	Strike	Dip	Slip
17/10/1992	153	16	266	54	280	40	143	39	67	56
18/10/1992	336	8	66	3	111	82	−176	21	86	−8
13/09/1994	323	20	125	20	219	25	−84	33	65	−93

Ammon *et al.* (1994). Based on the source radiation directivity Li & Toksöz (1993) demonstrated that the main shock was a complex event composed by two main subevents that occurred 1 second apart. They calculated the rupture lengths for these subevents to be around 50 and 90 km, respectively. Their suggested rupture direction for both events is S50°W.

Wallace & Beck's (1993) 'best' source mechanism of the foreshock is a NE trending thrust event (strike = 43°, dip = 41°, rake = 31°), while for the mainshock the 'best' faulting mechanism is a north-trending left-lateral strike-slip fault (strike = 5°, dip = 92°, rake = 31°).

The spatial distribution of aftershocks reveals a complicated faulting geometry in the rupture area. Although the focal solutions obtained by different authors did not agree completely, they all show the subevent nature of the main shock rupture, and suggest an approximately north-trending tectonic causative structure.

Earthquake environmental effects

Induced effects associated with the 18 October 1992 Murindo earthquake were very common and widespread because conditions were given for either slope failure in tropical forested areas of rough topography, or liquefaction and lateral spread in low-lying young alluvial plains (Table 4). Permanent ground deformations were observed, not only near the Murindo fault zone, but also at distant sites affected by induced effects like soil liquefaction, grounds cracks, lateral spreads, mud volcanoes eruption and slides. According to Coral & Salcedo (1992) and Martinez *et al.* (1994) the distribution of ground effects produced by the foreshock of 17 October was similar to that of the main shock. When it has been possible to discriminate between the effects of the two events, macroseismic intensities assessed for the strong foreshock were typically one degree lower than those induced by the main shock. The 17 October foreshock occurred very close (*c.* 20 km south) to the 18 October mainshock. Taking into account that the mainshock was significantly stronger than the foreshock, in the following we consider the environmental effects as related to the mainshock only.

Ground cracks. Cracks were observed along the Atrato, Murindo and Jiguamiando rivers in the Atrato Medio Region from Quibdó to Murindo in a distance of 90 km (Fig. 7). They cut through soft young alluvial and loose material and vary in nature from place to place. They appear almost continuously, mainly parallel and perpendicular to the Atrato River, in lengths of a few metres up to 20 m. They varied in width from millimetres up to 1 m in Bojaya and Riosucio municipalities. To the west, ground cracks with millimetre opening have been reported as far away as Itsmina, some 220 km SSW of the epicentre. To the east, these effects are observed only up to the surrounds of Medellin, about 130 km away from the epicentre. Cracks were observed also on the northern part of Atrato River in Turbo Municipality.

The distribution range of the ground cracks along the Atrato River varied from a few metres up to 30 m. Other ground cracks caused the loss of the river talus by creating small islands up to $5 \times 3\ m^2$, which floated in the river and finally disappeared helped by the torrential rain that followed the earthquake.

Landslides. The areas affected by the 18 October Murindo earthquake are mainly alluvial plains and less than 17% correspond to smaller ranges of the eastern part of the Western Cordillera. The mountainous areas near the epicentre were extensively affected by sliding (Fig. 8). The hilly parts of the Murindo, Jiguamiando and Rio Sucio rivers were affected by slope failures (Fig. 9) covering an area of *c.* 480 km^2 (Martinez *et al.* 1994). Most landslides occurred along the Murindo fault (Parra 2002). All the vegetation and soils that covered the terrain along the Murindo fault were completely destroyed. In sum, between 30 and 40% of the vegetation cover of the area was lost. The main road was damaged by landslides, rock slides and rock fall in various levels at several places. About 40% of

Table 4. *Seismically induced ground effects per locality with EEE intensity values; last column shows the MM intensity values for comparison*

ID	Locality	Lat.	Long.	Effects	EEE Intensity	MM 1956 Intensity
1	San Juan de Uraba	8.77	−76.50	Mud ejection from Mulatos Volcano was reported by inhabitants of Mulatos village. The mud flow path was observed by the surveyor scientists. There was evidence that Damaquiel Island with an approximate area of 7.5 km^2 emerged in the coast near Damaquiel town.	7	6
2	San Pedro de Uraba	8.33	−76.38	50 000 m^3 of mud were expelled during the eruption of Cahahual Volcano a few minutes after the earthquake. The eruption was followed by gases that by sudden ignition killed 7 persons and injured 20 others.	7	6
3	Turbo	8.13	−76.74	Liquefaction and small ground fissures with ejection of sands were observed parallel to Atrato River in the North West districts	8	7
4	Barranquillita	7.58	−76.72	Widespread liquefaction. Cracks filled with water and mud were observed. Some cases of sand ejection from fissures.	10	9
5	Pavarandocito	7.39	−76.60	Ground fissures with ejection of mud and sand boils were observed. Several cases of sand ejection from cracks were reported. Significant landslide and rock fall were observed.	10	9
6	Mutatá	7.26	−76.56	Extensive landslides were observed as well as widespread liquefaction. 20% of vegetation cover was lost.	10	8
7	La Isla	6.97	−76.78	Widespread liquefaction. Regime change of the river. Grounds cracks perpendicular to Murindo River were observed.	10	10
8	San Jose de la Calle	6.71	−76.92	Widespread liquefaction and subsidence (more than 1 m) were observed. Cracks parallel to the river up to 80 cm wide. Collapse of the talus of the river *c.* 7 m. Total loss of the soil cohesion. This locality was relocated.	10	8
9	Vigia del Fuerte	6.55	−76.84	Liquefaction and cracks with ejection of sand were observed in the SE part of the city. The cracks that crossed the south part of the district parallel to Atrato River caused the lost of several metres of river bank. Islands (1 × 1 m were floating in Atrato River after the collapse of the river bank.	10	8

(Continued)

Table 4. *Continued*

ID	Locality	Lat.	Long.	Effects	EEE Intensity	MM 1956 Intensity
10	Bella Vista	6.57	−76.90	Liquefaction and cracks with ejection of mud were observed in Pueblo Nuevo District in the northern part of the city. The cracks that crossed the entire district parallel to the river (up to 10 m from the river bank) caused the loss of more than 5 m of river bank. Islands (approximately 2 × 2 m were floating in Atrato River after the collapse of the river bank.	10	8
11	Medellin	6.24	−75.61	Few isolated landslides involving very few materials in poorly consolidated soils in the district of Brisas and of very small size were reported in unconsolidated soils in las Brisas de Robledo and Villatina.	7	7
12	Quibdó	5.83	−76.50	Small ground fissures were observed parallel to Atrato River shore in the districts of Kennedy and San Vicente. Cracks were reported in several constructions of the districts, Medrano, Alameda, La yesquita, La primera, EL Jardin and el Silencio.	7	7
13	Istmina	5.18	−77.71	Very small and few isolated cracks were observed in the shore of San Juan River in the Camellon district	5	6
14	Buchadó	6.33	−76.80	Ground fissures up to 80 m wide with ejection of mud and water were observed along all the river shore up to 25 m from the river bank. The river bank collapsed and formed floating island up to 1 × 1.5 m. Widespread liquefaction.	9	8
15	Tagachí	7.63	−78.71	Ground fissures (10–30 cm) parallel to the river shore half filled with water were observed.	8	8
16	Opogadó	6.82	−76.91	Widespread liquefaction and cracks parallel to river.	10	8
17	Rio Sucio	7.34	−76.97	Widespread liquefaction. River bank collapsed and formed small island up to 2 m. Ground fissures filled with water along the river shore up to 10 m were observed. Ground fissures with ejection of mud and sand boils were observed. Some landslides were observed in the area, vegetation cover was diminished.	10	8
18	La Grande	7.12	−76.90	Ground fissures with ejection of mud and sand were observed. Large landslides and loss of forest cover were observed. Liquefaction and subsidence up to 1 m was observed.	10	9

(*Continued*)

Table 4. *Continued*

ID	Locality	Lat.	Long.	Effects	EEE Intensity	MM 1956 Intensity
19	Curvaradó	6.38	−76.29	Ground fissures with ejection of mud and sand were observed. Also cracks on the pavement parallel to the shoreline were reported. Large landslides were observed in the mountainous part of this village.	10	9
20	Llano Rico	7.09	−76.71	Liquefaction and cracks parallel to Curvarado River. Landslides were also observed in small dimensions. Liquefaction accompanied with sand eruptions from fissures up to 80 cm wide.	10	8
21	Puerto Conto	6.54	−76.87	Ground fissures half filled with liquid up to 1 m wide and 50–80 cm deep were observed parallel to the river. River bank collapsed (around 2 m). Liquefaction in the northern part of the village was observed.	9	8
22	La Boba	6.53	−76.86	Ground cracks parallel to the river up to 30 cm wide and up to 80 cm deep were observed. Also cracks on the foundation of houses were observed.	9	8
23	La Loma	6.61	−76.93	Ground fissures with ejection of mud and sand boils were observed. Also cracks on the pavement parallel to the shoreline were reported.	9	8
24	Murindó	6.98	−76.78	Extensive liquefaction (widespread liquefaction), subsidence. Rock and landslides widespread in an area *c.* 480 km^2 along the Murindo Fault. Two east–west oval sectors on each side of the fault, uplift in the west were sand and ground water were ejected, and subsidence to the east. Changes on the river channel were observed. Landslide and rock fall were observed in the southeastern, north and central part of the epicentre area. Ground crack open up to 1 m large were observed in some districts of the town. Vigorous shaking of trees was observed. Many trees fell even in flat places repressing Murindo river. Uplift of the water table. Changes in the hydrological regime were observed. The total area affected by liquefaction was estimated *c.* 187.418 km^2 from an ellipse of 353 × 169 km.	11	10

Fig. 7. Ground cracks induced by the Murindo events of October 1992 in Bojaya, (photo: S. Mosquera-Machado). For municipality locations, see Figure 3.

Fig. 8. Extensive landslides in hilly areas and along the river banks of Atrato River: (**a**) extensive landslide in the mountains with destruction of vegetal cover (photo: E. Parra); (**b** and **c**) collapse of Atrato river banks (photo: S. Mosquera-Machado); (**d**) obstruction of the Murindo River caused by landslides (photo: E. Parra).

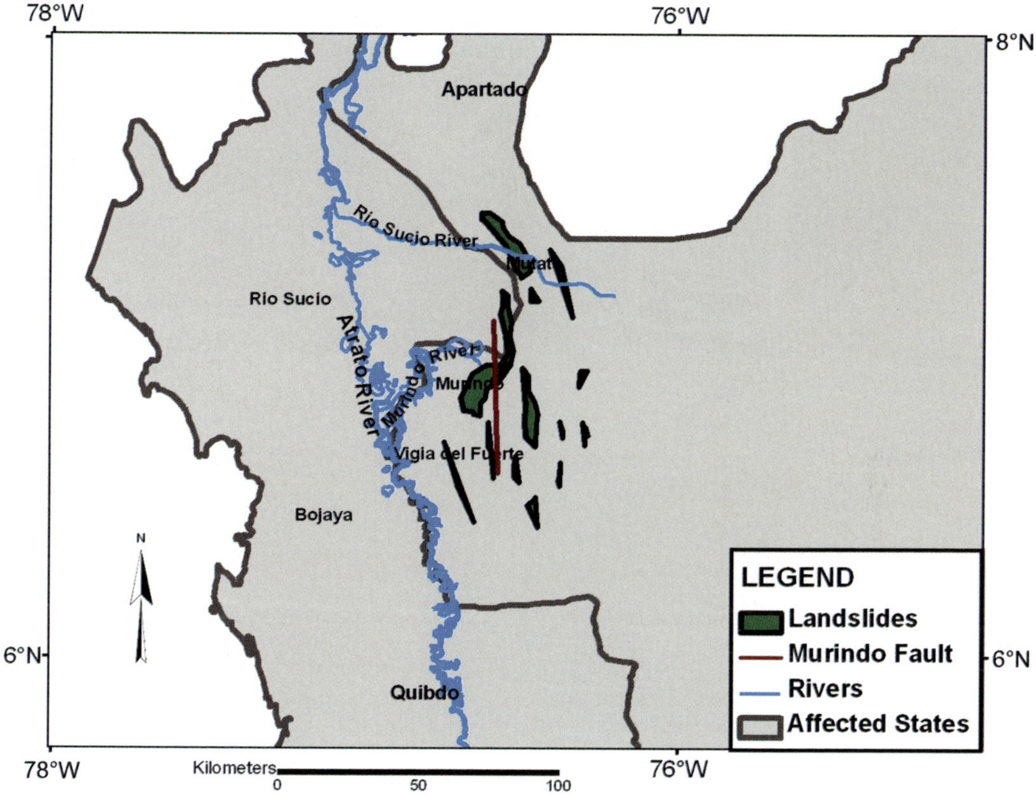

Fig. 9. Areas affected by landslides; trace of the Murindo fault modified from Paris *et al.* (2000).

vegetal coverage along the Murindo fault was destroyed by landslides, causing an obstruction of the Murindo River for more than 12 km in length. Failure of slopes was also observed along the banks of the Atrato River between Vigia del Fuerte and Rio Sucio, and was associated with open cracks around the margins (Mosquera-Machado *et al.* 1992). Landslides of small magnitude were also reported in the hilly Villatina village in Medellin, which had been recently built on unconsolidated soils. As a consequence of the landslides and vegetal coverage loss, several rivers (Rio Sucio, Murindo and Jiguamiando) had their channels obstructed and their capacity for sediment transport severely diminished.

Liquefaction and lateral spreading. The most frequent liquefaction features reported in association with the Murindo 1992 earthquake were sand blows and sand-vent fractures, related to lateral spreading. Liquefaction and water upsurge were present in areas from the confluence of Arquia and Atrato rivers up to the Uraba Gulf. The water of the Atrato and Murindo rivers became muddy.

Vertical subsidence of 1.5 m was observed in la Grande (Rio Sucio) and 50 cm in San Jose de la Calle (Mosquera-Machado *et al.* 1992; Mosquera-Machado 1994*b*).

Large fissures, 30 cm to 2 m wide, were observed along the affected zones beginning from Vigia del Fuerte and Bojaya. Almost all lateral spreads did show venting.

Sand blows were observed between Quibdó and Rio Sucio in an area with elliptical shape.

All reported liquefaction features are in active alluvial plains along the Atrato, Murindo, Curvarado and Rio Sucio rivers, as well in the soils of Uraba Plain. In San Jose de la Calle (Bojaya) and La Grande (Rio Sucio), south and north of the epicentral area, the pressure of the escaping water–sand mixture was high enough to uproot fully grown trees, and dislocate houses (Fig. 10). At both sites, the soil structure after the earthquake was not adequate for construction (Mosquera-Machado *et al.* 1992), and villages had to be relocated.

Beginning from the village of Buchado (Vigia del Fuerte), almost all riverbanks were damaged

Fig. 10. Liquefaction with water upsurge in la Grande, Rio Sucio (photo: S. Mosquera-Machado).

by lateral spreading, opening deep cracks that paralleled the rivers. It was common to observe blocks of riverbank that slid down to, toppled, or laterally moved into the river bottom, regardless of bank height, such as along the Atrato River between Bojaya and Rio Sucio municipalities. Lateral spreading also damaged some schools in more high areas such as La Loma in Bojayá. The farthest evidence of lateral spreading was found on the El Tigre village, south of Quibdó city on the left bank of the Atrato River, more than 120 km south of the epicentral area.

Within the town of Murindo, the liquefaction was extensive and vent fractures and widespread lateral spread were observed ubiquitously (Fig. 11).

GPS were utilized to get the exact location of liquefaction sites (minimum ten points in each locality). Less accessible points were captured from a helicopter again using GPS. Then points imported into a GIS were used to create the ellipsoidal shape of the liquefaction zone (Fig. 12). The total area of liquefaction was estimated as *c.* 50 000 km^2.

Other effects. Three associated effects have been mentioned and precisely described by locals and witnessed by field investigators: the eruption of the Cacahual mud volcano in San Pedro de Uraba and Mulatos volcano in the village with the same name (Turbo Municipality); the change of the

Fig. 11. Extensive liquefaction and lateral spreading in Murindo (photo: E. Parra).

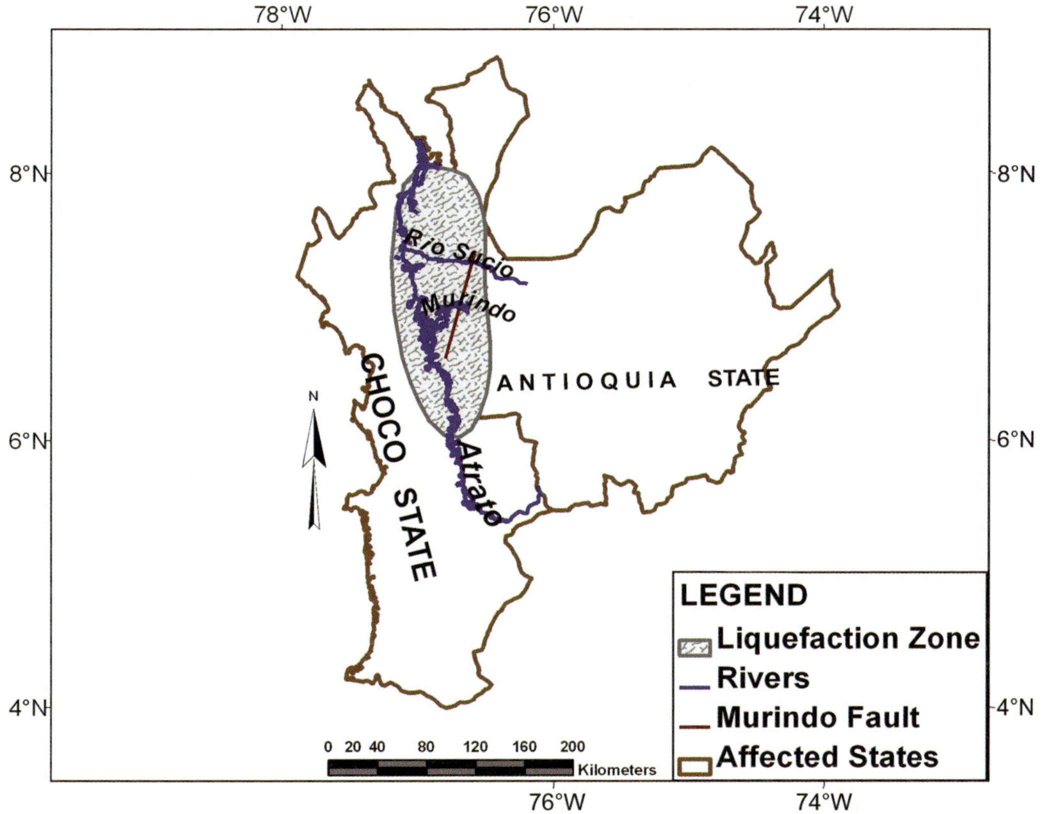

Fig. 12. Total area affected by liquefaction (*c*. 50 000 km^2).

Murindo river channel; and the emergence of Damaquiel Island near the Uraba Gulf. The Cacahual volcano erupted around 50 000 m^3 of mud and sands (Fig. 13) followed by the ignition of gases that caused the death of seven people and injured 20 others (Fig. 14). The emergence of Damaquiel Island attracted our attention as a possible coseismic effect, although it appeared in the area periodically (each five to seven years) as reported by inhabitants (Fig. 15). Along the right bank of Murindo River in Murindo, the river shore gained around 10 m into the village. The change of channel direction may

Fig. 13. Eruption of Cacahual mud volcano a few minutes after the 18 October 1992 Murindo earthquake (photo: E. Parra).

Fig. 14. Cacahual mud volcano fire followed by ignition of hydrocarbon and gas, a few minutes after the 18 October 1992 Murindo earthquake (photo: E. Parra).

Fig. 15. Damaquiel Island emerged in the Gulf of Uraba after the 18 October 1992 Murindo earthquake (photo: E. Parra).

be observed in aerial photography before and after the earthquake.

Environmental Seismic Intensity Scale (ESI 2007) assessment

The newly proposed Environmental Seismic Intensity Scale (ESI 2007) (Michetti *et al.* 2007), relying solely on modifications to the geological environment, provides a potentially powerful new tool for the evaluation of the strength of the earthquake in terms of its associated natural effects.

All the available information, including observations made immediately after the earthquake sequence and later, was revised in order to apply the ESI 2007 scale. All papers with description of the earthquakes and seismicity of the area were analysed to add description of environmental effects to the observed localities and for comparison purposes. A total of 24 localities were retained for macroseismic and environmental description and analysis. An intensity degree was attributed to each site according to the scale (Table 3), and the maximum degree was assigned to the locality. The ESI isoseismal map shown in Figure 16 integrates all these data. We derive an epicentral intensity I_0 of XI, covering an ellipsoid around the Murindo fault. This value of I_0 is in good agreement with the length of the ruptured fault calculated using the foreshock and aftershock of the main event, that is in the order of 90 to 140 km (according to the different interpretations of instrumental data; e.g. Li & Toksöz 1993; Wallace & Beck 1993; Arvidsson *et al.* 2002).

The effects described in the ESI 2007 scale intensity XI (c, d, e and f; see Appendix in Reicherter *et al.* 2009) (Michetti *et al.* 2007) clearly characterize the Murindo earthquakes'

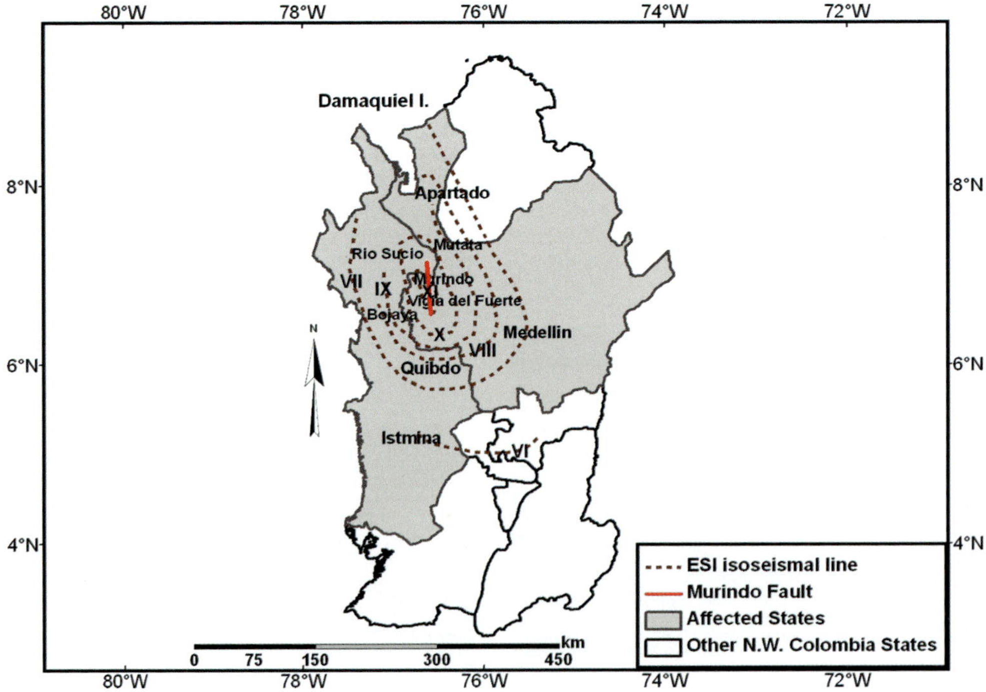

Fig. 16. ESI 2007 isoseismal map of the 18 October 1992 Murindo earthquake; trace of the Murindo fault modified from Paris *et al.* (2000).

epicentral area. A comparison between the ESI 2007 isoseismal map and the Modified Mercalli scale MM-1956 isoseismal map by Coral & Salcedo (1992) shows a difference from one to two degrees. The ESI 2007 I_o value of XI is one degree higher than using the MM-1956 scale. The final intensity map was also drawn from the combination of the MM-56 and ESI 2007 intensities value (Fig. 17). To construct this, when both damage and environmental effects were present in a certain locality, the worst effect was retained. In such an area characterized by low density of population, usually the environmental effects were worst and gave a more realistic picture of the severity of the earthquake. To this end, in our opinion the final map presented in Figure 17 is a proper description of the intensity of the earthquake, as intended in the formal definition of the most widely used 12-degree intensity scales worldwide (MCS, MM-1931, MM-1956, MSK; Michetti *et al.* 2004).

Discussion and conclusions

Hazard mapping and risk assessment form the foundation of the risk management decision-making process by providing information essential to understanding the nature and characteristics of the community's risk. The first step in seismic risk assessment is the development of hazard maps, which may be in the form of probabilistic or intensity maps. These maps are useful for operational risk management and for disaster mitigation. In earthquake-prone areas estimation of both the intensity of the earthquake and the extent of the affected area are essential. The approach presented in this study deals with estimation of both, drawing especial attention to the definition of the intensity in the near field. In particular, the use of the environmental earthquake effects only to decipher the source effects of the earthquake has the advantage of using the available recent or palaeoseismic data to estimate the maximum intensity. In particular, we reinterpreted the available data on coseismic effects but also integrated the seismic, tectonic, palaeoseismic and focal mechanism data in a geographical information system, allowing comparison with traditional macroseismic intensity assessment and the characterization of source parameters. According to this reinterpretation, it was possible to estimate that for the 1992 Murindo earthquake $I_o = XI$ and not X, as derived from observations on the effects on the environment. In fact,

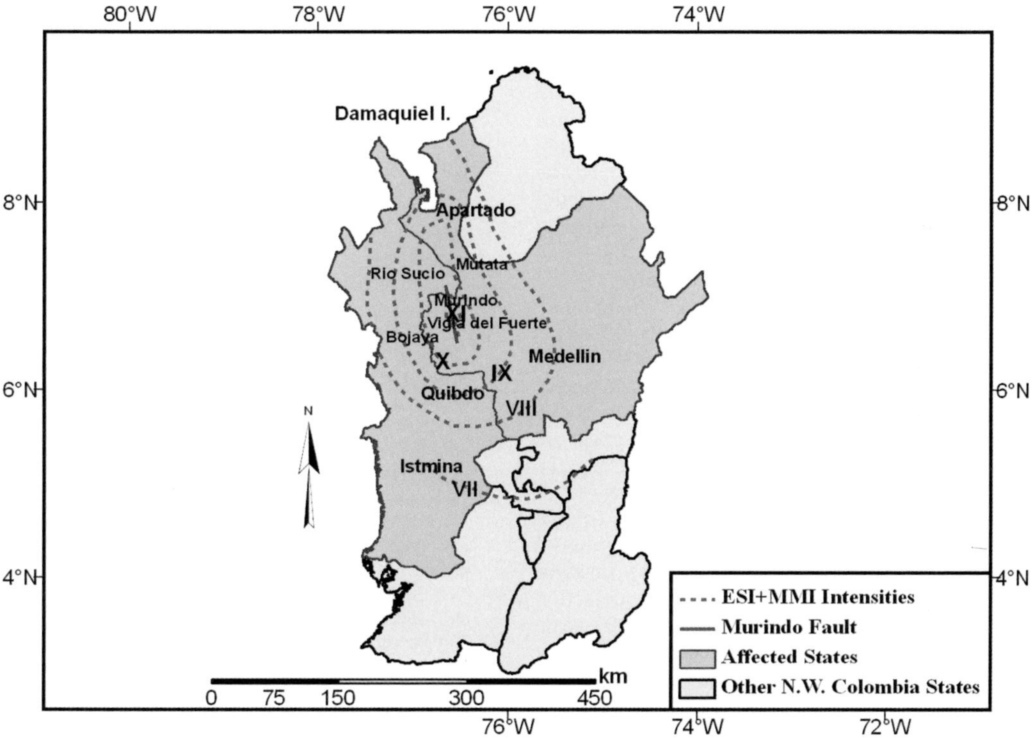

Fig. 17. Final isoseismal map of 18 October 1992 Murindo earthquake; trace of the Murindo fault modified from Paris *et al.* (2000).

MM- 1956 intensity underestimates the effects of the Murindo event because the area is sparsely populated and poorly developed in terms of economy, transportation network, and engineering facilities. According to the ESI scale, for intensity XI, the end-to-end surface rupture length is in the order of 100 km, as suggested by instrumental data (Li & Toksöz 1993; Arvidsson *et al.* 2002); and the source of the 18 October 1992 event is *c.* north–south trending, in agreement with Wallace & Beck (1993). The location of the fault rupture obtained here is in agreement with the Murindo fault trace as mapped by Paris *et al.* (2000). However, the rupture length indicated by Paris *et al.* (2000) is *c.* 70 km, much lower than that suggested by the ESI scale epicentral intensity XI and by the seismological data. We suggest the rupture of the 18 October 1992 earthquake involved the reactivation of the whole trace of the Murindo fault as mapped by Paris *et al.* (2000), and also the reactivation of other secondary fault segments associated with the main Murindo fault, for a total rupture length of more than 100 km.

When comparing through plotting on a distance/intensity magnitude graph, the main and total areas affected by slope failure during the 1992 earthquake do fit quite well against the mean regression line for worldwide earthquake data compiled by Keefer (1984), but they do not fit the total area of environmental effects as described in the ESI scale intensity XI (total area in the order of 10 000 km^2). The difference between the expected dimension of the area affected by slope failures for the ESI intensity XI and the total slope failure area calculated from field observation can be explained by the morphology of the epicentral region, mainly formed by alluvial plains and only around 17% corresponding to hilly areas belonging to the western part of the western cordillera.

Conversely, the total area affected by liquefaction, *c.* 50 000 km^2, fits very well with the definition of intensity XI given in the ESI 2007 scale. Liquefaction of saturated deposits caused by earthquakes continues to be a major cause of earthquake-related damage. That is why liquefaction hazard maps are increasingly being incorporated into earthquake risk mitigation practices. This is why the use of the environmental effects to predict, assess or give a relationship between the liquefaction potential of the earthquakes is a factor that would improve liquefaction hazard mapping. The liquefaction area that was calculated from the ESI 2007 scale assessment is georeferenced and constituted a very important input that would enhance the preparation of a more complete hazard map for the affected area.

The final intensity map issued from the combination of the MM and ESI intensity maps of the 18 October 1992 Murindo earthquake drawn in this study clearly shows three interesting aspects: (1) the intensity in the epicentral area I_o is one degree greater than intensity calculated with the MM-1956 scale alone; (2) the area with intensity XI stretches over 60 km in length; and (3) the region affected by liquefaction is very broad, being in coherence with the XI degree of the ESI scale, thus giving a clear indication of the zones where a very detailed liquefaction analysis is needed to define the suitability of soil reconstruction practices and future development planning.

The procedure followed in the present study for the assessment of intensity in 24 sites of NW Colombia in terms of intensity using the earthquake environmental effects has several advantages: (1) a more realistic value of the intensity in the epicentre was found, which is independent of damage to the built environment by the use of all known and the most reliable seismological, tectonic and geological data of the region; (2) the source parameters were defined even though no surface fault was visible because of the dense vegetation cover in the area, according to the criteria defined in the ESI scale for the identification of macroseismically derived surface faulting (e.g. Shebalin 1972; Branno *et al.* 1986; Serva *et al.* 2007).

The unconditional collaboration of Dr Eduardo Parra from Colombian Geophysical Survey Ingeominas, who gave us a great data set of field observations of the 1992 Murindo earthquakes, is greatly acknowledged. Reviews by Drs Eliana Porfido and Frank Audemard greatly improved the manuscript.

References

AMMON, C. J., LAY, T., VELASCO, A. A. & VIDALE, J. E. 1994. Routine estimation of earthquake source complexity; the 18 October 1992 Colombian earthquake. *Bulletin of the Seismological Society of America*, **84**, 1266–1271.

ARANGO-LÓPEZ, M. & VELÁSQUEZ, A. 1993. *Catálogo Histórico Sísmico para Medellín. Estudio de Riesgo Sísmico para Medellín.* Alcaldía de Medellín.

ARVIDSSON, R., TORAL-BOUTET, J. & KULHANEK, O. 2002. Foreshocks and aftershocks of the Mw = 7.1, earthquake in the Atrato region, Colombia. *Journal of Seismology*, **6**, 1–11.

BRANNO, A., ESPOSITO, E., LUONGO, G., MARTURANO, A., PORFIDO, S. & RINOLDIS, V. 1986. The largest earthquake of Apennines, southern Italy. *AIG-IAEGI*, **4**, 3–14.

BUENO, R. & GOVEA, C. 1976. Potential for exploration and development of hydrocarbons in Atrato Valley and Pacific coastal and shelf basins of Colombia. *In*: HALBOUTY, M. T., MAHER, J. C. & LIAN, H. M. (eds) *Circum-Pacific Energy and Mineral Resources*. American Association of Petroleum Geologists, Memoir, **25**, 318–327.

BURKE, K., COOPER, C., DEWEY, J., MANN, P. & PINDELL, J. 1984. Caribbean tectonics and relative plate motions. *In*: BONINI, W. E., HARDGRAVE, R. B. & SHAGAM, R. (eds) *The Caribbean–South American Plate Boundary and Regional Tectonics*. Geological Society of America, Memoir, **162**, 31–64.

CASE, J. E., DURAN, L. G., LOPEZ, R. A. & MOORE, W. R. 1971. Tectonic investigations in western Colombia and eastern Panama. *Geological Society of America Bulletin*, **82**, 2685–2711.

CMT. 2007. The Harvard Centroid Moment Tensor (CMT) database. Available at: http://www.globalcmt.org/ 8 October 2007.

CORAL, C. & SALCEDO, E. 1992. *Mapa preliminar de isosistas del terremoto del 18 de octubre de 1992*. Reporte interno Ingeominas.

DIMATE, C., RIVERA, L. *ET AL*. 2003. The 19 January 1995 Tauramena (Colombia) earthquake: geometry and stress regime. *Tectonophysics*, **363**, 159–180.

DUQUE-CARO, H. 1990*a*. Neogene stratigraphy, paleoceanography and paleobiology in northwest South America and the evolution of the Panama Seaway. *Palaeogeography, Palaeoclimatology and Palaeoecology*, **77**, 203–234.

DUQUE-CARO, H. 1990*b*. The Choco Block in the northwestern corner of South America: structural, tectonostratigraphic, and paleogeographic implications. *Journal of South American Earth Sciences*, **3**, 71–84.

ESCALLÓN, J. 2000. La sismicidad Colombiana en la última década del siglo XX. *In*: *Revista Red Sismológica Regional del Eje Cafetero – Viejo Caldas y Tolima*. INGEOMINAS, Bogota, 5–11.

FREYMUELLER, J. T., KELLOGG, J. N. & VEGA, V. 1993. Motions at the Andean region. *Journal of Geophysical Research*, **98**, 21853–21863.

GARCIA, R., SARRIA, M., ESPINOSA, S. A., BERNAL, L. & PUCCINI, L. M. 1984. *Estudio general del riesgo sísmico de Colombia*. Asociación Colombiana de Ingeniería Sísmica.

GOBERNA, J. R. 1985. Informe Nacional de Colombia, en Catálogo de Terremotos para America del Sur, CERESIS, Descripción del Catálogo e Informes Nacionales, Programa para la mitigación de los efectos de los terremotos en la región andina. *In*: ASKEW, B. L. & ALGERMISSEN, S. T. (eds) *Earthquake Mitigation Program in the Andean Region (Project SISRA)*. Ceresis, Lima.

HAFFER, J. 1967. *On the Geology of the Uraba and northern Choco regions, northwestern Colombia*. Ecopetro, Open File Report, **809** (GR-357).

INGEOMINAS. 1999. *Actualización Catálogo de sismos de Colombia para estudios de amenaza sísmica 1566–1998. Datos de Hipocentros e Intensidades*. Proyecto Estudio de la Amenaza Sísmica de Colombia, Alvarado, C., Bogotá.

IRVING, E. M. 1971. La evolución de los Andes más septentionales de Colombia. *Ingeominas, Boletín Geológico*, **19**(2), 1–90.

IRVING, E. M. 1975. *Structural evolution of the northwestern Andes, Colombia*. U.S. Geological Survey, Professional Paper, **846**.

KEEFER, D. 1984. Landslides caused by earthquakes. *Geological Society of America Bulletin*, **95**, 406–421.

KELLOGG, J. N. & BONINI, W. E. 1982. Subduction of the Caribbean Plate and basement uplifts in the overriding South American Plate. *Tectonics*, **1**, 251–276.

KELLOGG, J. N. & VEGA, V. 1995. Tectonic development of Panama, Costa Rica, and the Colombian Andes: Constrains from Global Positioning System geodetic studies and gravity. *In*: MANN, P. (ed.) *Geologic and Tectonic Development of the Caribbean Plate Boundary in Southern Central America*. Geological Society of America, Special Papers, **295**, 75–90.

KELLOGG, J. N., OGUJIOFOR, I. J. & KANSAKA, D. R. 1985. Cenozoic tectonics of the Panama and North Andes blocks. *Memorias-Congreso Latinoamericano de Geologia*, **6**, 34–49.

KELLOGG, J. N., DIXON, T. & NEILAND, R. 1989. Central and South American GPS Geodesy. *EOS Transactions*, **70**, 649–656.

LALINDE, C. & SANCHEZ, J. 2007. Earthquake Environmental Effects in Colombia during the Past 35 Years: INQUA Scale Project. *Bulletin of the Seismological Society of America*, **97**(2), 646–654.

LALINDE, C., ESTRADA, B. & FARBIARZ, J. 2004. *Preliminary Application of the INQUA Scale to Recent Colombian Earthquakes*. Available at: http://www.apat.gov.it/site/_Files/Inqua/Colombia_WG_Report.pdf

LI, Y. & TOKSÖZ, N. 1993. Study of the source process of the 1992 Ms = 7.3 earthquake with the empirical Green's function method. *Geophysical Research Letters*, **20**, 1087–1090.

MARTÍNEZ, J. M., PARRA, E., PARÍS, G., FORERO, C., BUSTAMANTE, M., CARDONA, O. D. & JARAMILLO, J. D. 1994. Los sismos del Atrato Medio 17 y 18 de Octubre de 1992 Noroccidente de Colombia. *Revista Ingeominas*, **4**, 35–76.

MICHETTI, A. M., ESPOSITO, E. *ET AL*. 2004. *The INQUA Scale an Innovative Approach for Assessing Earthquake Intensities Based on Seismically Induced Ground Effects in Natural Environment*. Memorie Descrittive della Carta Geologica D'Italia, Special Paper, **LXVII**.

MICHETTI, A. M., AUDEMARD, F. *ET AL*. 2007. *Environmental Seismic Intensity Scale 2007-ESI 2007*. Memorie Descrittive della Carta Geologica d'Italia, **74**, Servizio Geologico d'Italia-Dipartimento Difesa del Suolo, APAT, Roma, Italy.

MOSQUERA MACHADO, S. C. 1994*a*. *Evaluacion Técnica de los daños ocasionados por el desbordamiento de los rios Andágueda, Atrato, San Juan, Baudó y Sipí*. Codechoco, Quibdó.

MOSQUERA MACHADO, S. C. 1994*b*. *Evaluation tecnica de las pérdidas de los sismos del 17 et 18 octubre 1992 y estado actual de la reconstrucción*. Informe interno de Codechoco.

MOSQUERA-MACHADO, S. C. 2002. *Analyse multi-aléas et risques naturels dans le Département du Chocó, Nord- Ouest de Colombie*. Doctoral dissertation, University of Genèva, Vol. 37.

MOSQUERA-MACHADO, S. C., LUNA, P. & GARCIA, N. 1992. *Informe técnico de la evaluación de las zonas afectadas por los sismos del 17 y 18 de octubre de 1992 y posibles sitios de reubicación de los pueblos evacuados*. Informe Interno de Codechoco, Quibdo.

NEIC. 2007. National Earthquake Information Center. Available at: http://earthquake.usgs.gov/regional/neic/.

NORABUENA, E. O., LEFFLER-GRIFFIN, L. *ET AL.* 1998. Space geodetic observations of Nazca-South American convergence across the central Andes. *Science*, **279**, 358–362.

NYGREN, W. E. 1950. The Bolivar geosyncline in northwestern South America. *Bulletin of the American Association of Petroleum Geologists*, **34**, 1998–2006.

PAGE, W. D. 1986. *Seismic geology and seismicity of Northwestern Colombia: San Francisco, California*. Woodward-Clyde Consultants Report for ISA and Integral Ltda., Medellín.

PARIS, G. & ROMERO, R. 1994. Fallas Activas en Colombia: Ingeominas, Santafé de Bogotá. *Boletín Geológico*, **34**(2–3), 3–26.

PARÍS, G., MACHETTE, M. N., DART, R. L. & HALLER, K. M. 2000. *Map and Database of Quaternary Faults and Folds in Colombia and its Offshore Regions*. US Geological Survey, Open File Report **00–0284**.

PARRA, E. 2002. *Estudio Geológico y Geomorfológico para la reubicación de la cabecera municipal de Murindó*. Ingeominas.

PARRA, E. 2003. Informe de la visita practicada por comisión de verificación, las comunidades del consejo comunitario de Jiguamiando y las nueve comunidades de Curbarado. Ingeominas.

PENNINGTON, W. D. 1981. Subduction of the eastern Panama Basin and seismotectonics of northwestern South America. *Journal of Geophysical Research*, **86**, 10753–10770.

PERREY, A. 1858. Documents sur les tremblements de Terrea du Pérou, dans la Colombie et dans le Bassin de l'Amazone. Présentés a la séance du 7 november 1857. *Bulletin de l'Academia Royale de Belgique, Bruxelles*, **VII**, 134.

RAMÍREZ, J. E. 1975. *Historia de los Terremotos en Colombia* (2nd edn). Instituto Geográfico Agustín Codazzi, Bogotá.

RAMÍREZ, F. & BUSTAMANTE, M. 1996. Los sismos de octubre de 1992 en el Atrato Medio, Colombia. *In*: MASKREY, A. (ed.) *Terremotos en el Trópico Húmedo. La Gestión de Desastres del Alto Mayo, Perú (1990–1991); Limón, Costa Rica (1991); y Atrato Medio, Colombia (1992)*. La Red, Santafé de Bogotá. Available at: http://www.desenredando.org/public/libros/1996/tth/TerremotosTropicoHumedo-1.0.0.*pdf*

REICHERTER, K., MICHETTI, A. M. & SILVA, P. G. 2009. Palaeseismology: historical and prehistorical records of earthquake ground effects for seismic hazard assessment. *In*: REICHERTER, K., MICHETTI, A. M. & SILVA, P. G. (eds) *Palaeoseismology: Historical and Prehistorical Records of Earthquake Ground Effects for Seismic Hazard Assessment*. Geological Society, London, Special Publications, **316**, 1–10.

SERVA, L., ESPOSITO, E., GUERRIERI, L., PORFIDO, S., VITTORI, E. & COMERCI, V. 2007. Environmental effects from some historical earthquakes in Southern Apennines (Italy) and macroseismic intensity assessment: contribution to INQUA EEE Scale Project. *Quaternary International*. DOI: 10.1016/J. quaint. 2007.03.015.

SHEBALIN, N. R. 1972. Macroseismic data of information on source parameters of large earthquakes. *Physics of the Earth and Planetary Interiors*, **6**, 614–323.

TABOADA, A., RIVERA, L. A. *ET AL.* 2000. Geodynamics in northern Andes: Subductions and intracontinental deformations (Colombia). *Tectonics*, **19**, 787–813.

TRENKAMP, R., KELLOGG, J. N., FREYMUELLER, J. T. & MORA, H. 2002. Wide plate margin deformation, southern Central America and northwestern South America, CASA GPS observations. *Journal of South American Earth Sciences*, **15**(2), 157–171.

VAN DER HILST, R. & MANN, P. 1994. Tectonic implications of tomographic images of subducted lithosphere beneath northwestern South America. *Geology*, **22**, 451–454.

VOZ DE ANTIOQUIA PERIÓDICO. 1883. Año I, Trim. II, No. 14, pag. 84.

WALLACE, T. & BECK, S. L. 1993. The Oct. 17–18, 1992 Colombian earthquakes; slip partitioning or faulting complexity. *Seismological Research Letters*, **64**, 29–35.

WOODWARD-CLYDE CONSULTANTS. 1979. *Report for ISA and Integral Ltda*. Medellín, Colombia.

Prehistoric seismicity-induced liquefaction along the western segment of the strike-slip Kunlun fault, northern Tibet

AIMING LIN[1*] & JIANMING GUO[2]

[1]*Institute of Geosciences, Faculty of Science, Shizuoka University, Shizuoka 422-8529, Japan*

[2]*Graduate School of Science and Technology, Shizuoka University, Ohya 836, Shizuoka 422-8529, Japan*

**Corresponding author (e-mail: slin@ipc.shizuoka.ac.jp)*

Abstract: The 2001 M_w 7.8 Kunlun earthquake occurred in northern Tibet, and produced a 450-km-long surface rupture zone along the western segment of the strike-slip Kunlun fault. There are, however, no historic or instrumental records of large earthquakes in this fault segment. Field investigations of liquefaction structures and radiocarbon dating results reveal that at least three large earthquakes, including the 2001 earthquake, occurred in the western segment of the Kunlun fault during the past seven to nine centuries. Liquefaction structures formed in alluvial deposits composed of sand-gravel yielding ^{14}C ages of 679–901 yr BP are observed on the current stream channel which is sinistrally offset 75–82 m, including 3–6 m displacement produced by the 2001 event. On the basis of the field investigations and ^{14}C dating results, we conclude that the liquefaction structures and subsequent faulting events were caused by at least two large earthquakes of $M > 7$ prior to the 2001 earthquake and the average recurrence interval of large earthquakes is estimated to be about 400 years in the late Holocene.

The M_w 7.8 Kunlun (M_S 8.1) earthquake occurred on 14 November 2001, and produced a 450-km-long surface rupture with a large strike-slip up to *c*. 16 m along the western segment of the strike-slip Kunlun fault (Lin *et al.* 2002, 2003; Lin & Nishikawa 2007; Lin 2008). Although this earthquake displaced the Qinghai–Tibet railway which was under construction in 2001 and also caused liquefactions and avalanches of snow and glacial ice, there are no reports of casualties or great damage because the earthquake occurred in a sparsely populated region in the remote high mountains of northern Tibet. Two large earthquakes of $M \geq 7.5$ occurred prior to the 2001 Kunlun earthquake in the last century in both the eastern and western fault segments bounded by the 2001 rupture segment of the Kunlun fault (e.g. Jia *et al.* 1988; Peltzer *et al.* 1999; Guo *et al.* 2006). Recently, based on trench and field investigations, an average strike-slip rate of 16.4 mm/a and an average recurrence interval of 300–400 years for large earthquakes have been estimated in the fault segment associated with the 2001 earthquake (Lin *et al.* 2006). However, there are no historic or instrumental records of large earthquakes in the 450-km-long fault segment due to the remoteness and sparse population of the mountain region. The absence of historical and instrumental records of large earthquakes hinders further assessment of past long-term seismic behaviour of large intracontinental strike-slip faults in the Tibet plateau.

In this study, we report at least two large prehistoric earthquakes revealed by liquefaction-related structures found in the 2001 rupture segment and discuss the late Holocene activity of the strike-slip Kunlun fault by field investigations and interpretations of 1-m-resolution IKONOS images. Liquefaction of saturated sands during a large earthquake often causes great damage to buildings, earth embankments and retaining structures, and has been reported and documented in many large intracontinental earthquakes worldwide (e.g. Seed & Idriss 1967; Obermeier *et al.* 1985; Lin 1997; Lin *et al.* 2001). Archaeological and palaeoseismic studies show that liquefied sediments are believable indicators of historic and prehistoric seismicity (Rodbell & Schweig 1993; Obermeier 1995; Marco *et al.* 1996; Tuttle & Schweig 1996; Lin 2006). The study on earthquake-induced liquefaction structures would help us evaluate palaeoseismicity and seismic hazard for possible engineering damage associated with future large earthquakes on intracontinental active faults.

Seismic activity of the Kunlun fault

The study area is located along the Kunlun fault in the central Kunlun mountain range of northern Tibet, with an average elevation of >4500 m (Fig. 1).

From: REICHERTER, K., MICHETTI, A. M. & SILVA, P. G. (eds) *Palaeoseismology: Historical and Prehistorical Records of Earthquake Ground Effects for Seismic Hazard Assessment*. The Geological Society, London, Special Publications, **316**, 145–154. DOI: 10.1144/SP316.8 0305-8719/09/$15.00

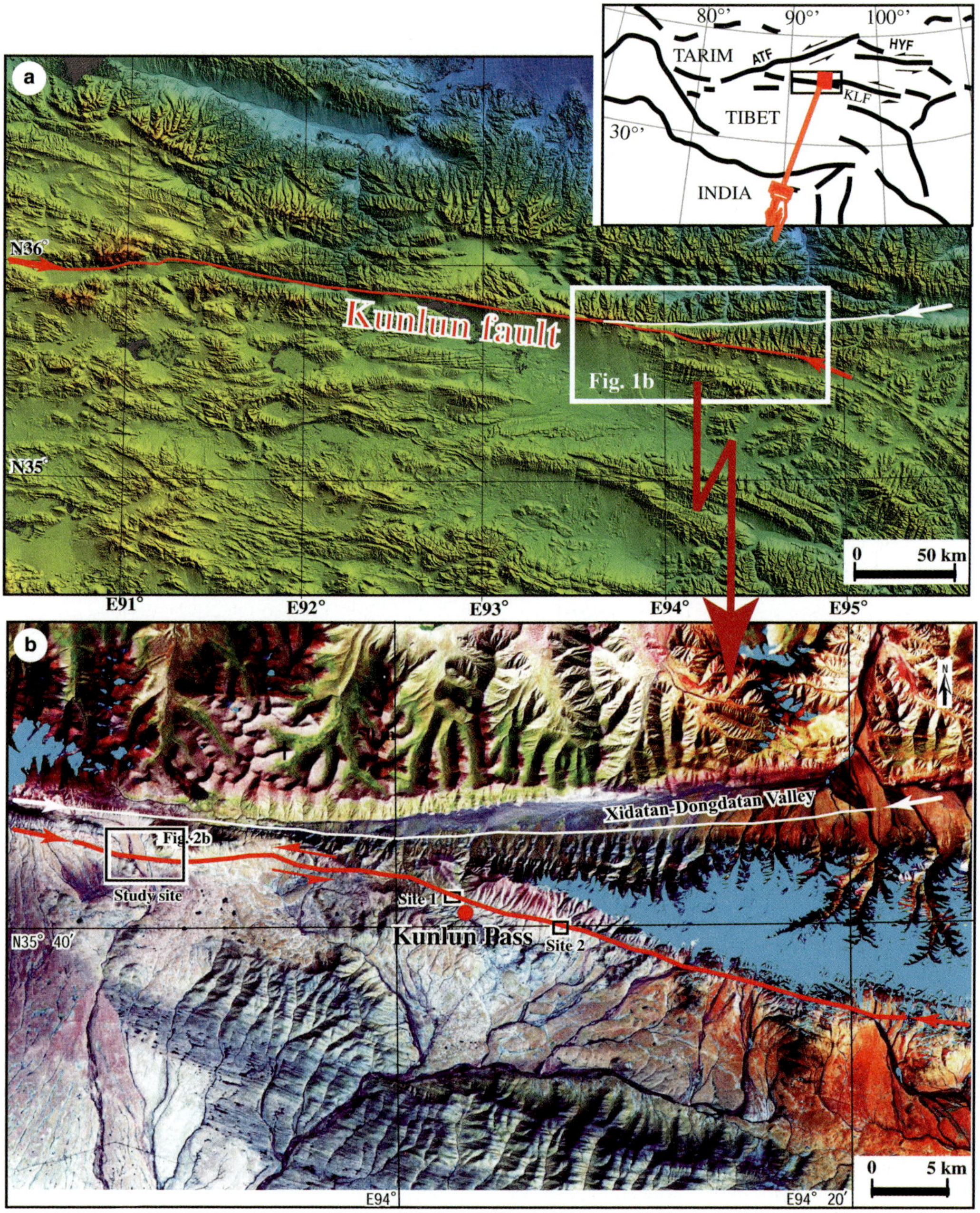

Fig. 1. (**a**) Satellite image showing the tectonic topography in the area around the Kunlun fault. (**b**) Close-up view of the east segment of the Kunlun fault in the study area. Thin red lines and arrows indicate the surface rupture zone produced by the 2001 M_w 7.8 earthquake. Thin white lines and arrow indicate the Xidatan–Dongdatan segment of the Kunlun fault which was not ruptured by the 2001 earthquake. Sites 1 and 2 indicate the trench locations of Lin *et al.* (2006). Abbreviations: ATF, Altyn Tagh fault; HYE, Haiyuan fault; KLF, Kunlun fault.

The Kunlun fault strikes east–west to WNW–ESE over *c.* 1200 km, and is considered as one of the major strike-slip faults along which strike-slip partitioning occurs in accommodating both the northeastward shortening and eastward extrusion of Tibet (e.g. Tapponnier & Molnar 1977; Meyer *et al.* 1998; Wang *et al.* 2001). Two historic earthquakes of $M \geq 7.5$ that occurred in the last century, producing left-lateral offsets of 3–7 m along the Kunlun fault, indicate that the fault is currently active as a large earthquake source fault. The first was the 1937 M 7.5 Tuosuo Lake earthquake that ruptured the 150–180 km long Tuosuo Lake segment (Jia *et al.* 1988; Guo *et al.* 2007). The other was the 1997 M_w 7.8 (Ms 7.9) Manyi earthquake that produced a 170-km-long surface rupture zone along the westernmost segment of the Kunlun fault (Peltzer *et al.* 1999). Recently, the 2001 M_w 7.8 Kunlun earthquake ruptured a 450-km-long fault segment along the Kunlun fault (Fig. 1; Lin *et al.* 2002, 2003, 2004; Xu *et al.* 2002; Van der Woerd *et al.* 2002*a*) bounded by the 1997 Manyi coseismic surface rupture zone in the west. Based on the trench and field studies (at sites 1 and 2 shown in Fig. 1b), it is inferred that: (1) at least four prehistoric earthquakes occurred in the past 6200 years; (2) the penultimate event occurred within the past 400 years; and (3) the average slip rate is 16.4 mm/a and the average recurrence interval of M 8 earthquakes is 300–400 years on the 2001 rupture segment of the Kunlun fault (Lin *et al.* 2006).

Liquefaction structures and faulting events

The study site is located in the eastern side of the 2001 coseismic surface rupture zone where the current stream channel is sinistrally displaced 75–82 m (Fig. 2b). The coseismic offsets produced by the 2001 earthquake are 3–6 m in this location (Lin *et al.* 2002, 2003; Xu *et al.* 2002, 2006; Lin & Nishikawa 2007). This indicates that large earthquakes repeatedly occurred and coseismic displacements are accumulated on the stream channel along the Kunlun fault in the study area.

The liquefaction structures produced by the 2001 M_w 7.8 Kunlun earthquake were observed in many locations along the coseismic surface rupture zone (Fig. 3; Lin *et al.* 2002, 2003). Sand boils formed by liquefaction during the 2001 earthquake occurred as a cone shape on the current stream channels, riversides and lakes along the surface rupture zone (Fig. 3). These sand-boil cones vary from 50 cm to 10 m in diameter, generally 2–5 m; they are well preserved and easily recognized after the earthquake due to freezing of boiled sands (Fig. 3a and d).

In the study area, liquefaction structures formed prior to the 2001 earthquake were found at several locations along the 2001 coseismic rupture zone (Figs 4 and 5). Boiled sand–gravel layers were typically exposed on the extensional crack walls of >20 m in length and >2 m in depth along the coseismic surface rupture on the current stream channel which was displaced 3–6 m sinistrally by the 2001 earthquake (Fig. 3; Lin *et al.* 2002, 2003; Lin & Nishikawa 2007; Lin 2008). Two typical exposures are sketched in Figures 4 and 5. Based on our field investigation carried out immediately after the 2001 earthquake, no sand boils formed during the 2001 earthquake on these exposed sections where the frozen river water was ruptured on the current stream channel (Fig. 3a and b). Therefore, it is possible to distinguish the coseismic liquefaction structures produced by the 2001 earthquake from the old liquefaction structures by the topographic deformation features and geological structures observed in the field immediately after the 2001 earthquake.

The sedimentary deposits exposed on the crack-walls consist of unconsolidated alluvial sand–gravel and silt–sand which can be divided into four major stratigraphic units (Units 1, 2, 3 and 4) from the top to the base at the exposed sidewalls of coseismic cracks (Fig. 5). Unit 1 is mainly composed of coarse-grained sand and gravel, which capped the top 20–30 cm of exposures. Unit 2 is composed of fine-grained silt–sand containing some brown soil, which is 10–30 cm in thickness. The deposits in this unit are unliquefied. Unit 3 consists of fine- to medium-grained yellowish-grey sand, which was strongly liquefied (Fig. 5). Some sand dykes are exposed at the surface on the current stream channel and can be backed to the source sand layer (Figs 4 and 5). Unit 4 consists of gravel with medium- to coarse-grained sand including some organic soil lenses, which is also liquefied (Fig. 5). The total thickness of this unit is >1 m. For collecting radiocarbon dating samples, several pits were dug at the study site (Loc. 1). Two samples of organic soil lenses including charcoals were collected from depths of 0.2–0.4 m in unit 2 and 0.6–0.8 m in unit 4, which yielded ^{14}C ages of 679–901 yr BP and 4828–5025 yr BP, respectively (Table 1).

The liquefied sand–silt–gravel layers are cut by subsequent faults (or cracks) that are sealed and were not activated during the 2001 earthquake (Figs 4 and 5). Numerous striations and grooves produced the 2001 coseismic slipping on the crack walls of liquefied sedimentary deposits; they are imposed obliquely on the fault (Fig. 4b) and indicate a left-lateral strike-slip movement (Fig. 5). These structural features suggest that the subsequent faults formed during a period between the 2001 earthquake and the liquefaction.

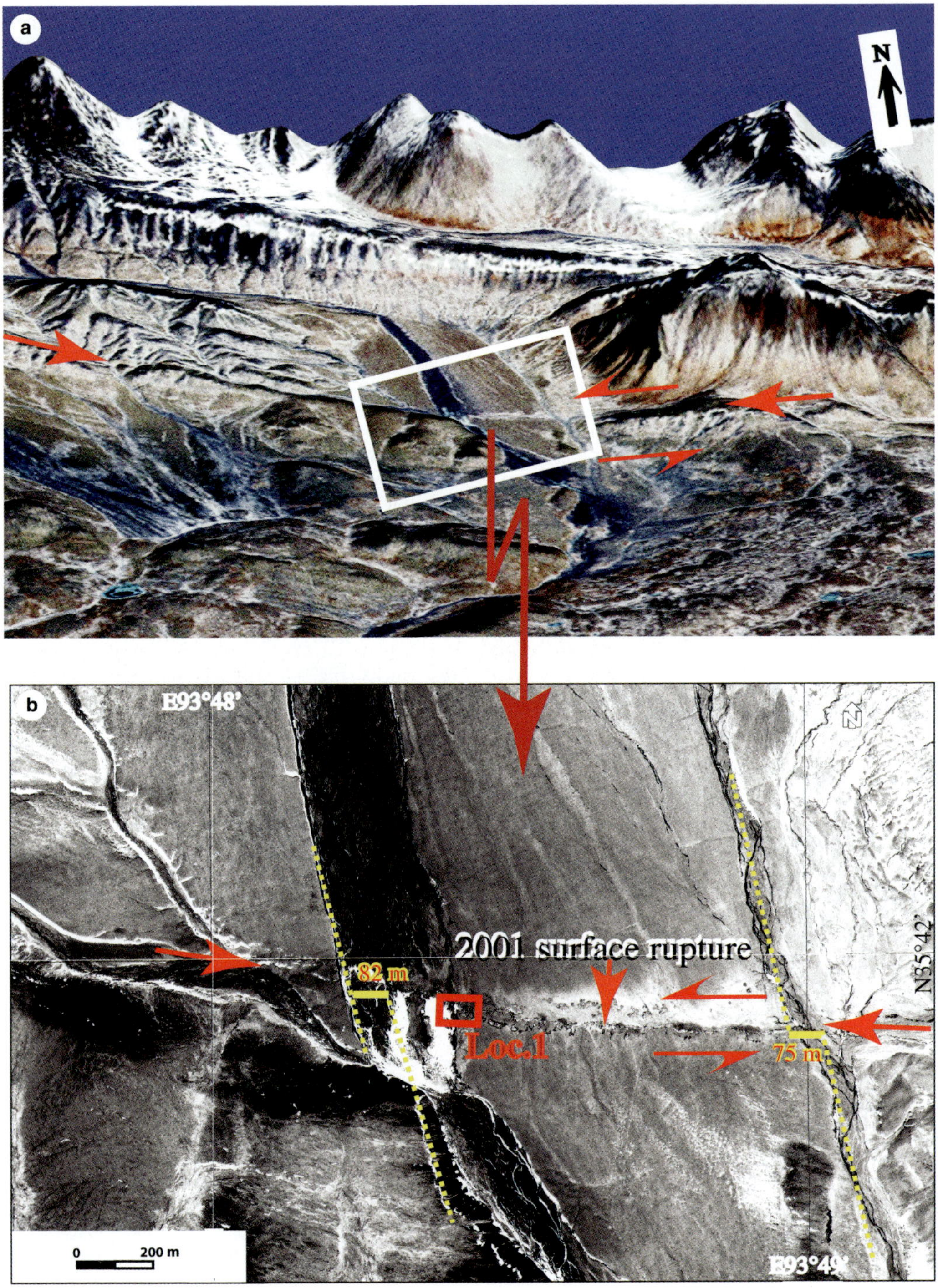

Fig. 2. Shade relief map (SRTM) (**a**) and IKONOS image (**b**) showing the tectonic topography of the Kunlun fault in the study area. (a) North-looking perspective view of the Kunlun fault at the study site (generated by draping the Landsat image over GTOPO30 DEM data). The size of the image is about 8 km wide by 7 km long. (b) IKONOS image (1 m resolution) acquired in January 2002 immediately after the 2001 earthquake showing the displaced stream channel and the 2001 coseismic surface ruptures in the study area. Loc. 1 indicates the main liquefaction site. Red arrows indicate the 2001 coseismic surface rupture along the pre-existing fault. Numbers (75 m and 82 m) indicate the displacement amounts of the current stream channel in the east and west banks.

Fig. 3. Photographs showing the 2001 coseismic surface ruptures (**a–c**) and related sand boils (**d**) in the study area. (a) Surface rupture on the iced (white colour) current stream channel. (b) Coseismic extensional cracks where liquefied sand–silts formed prior to the 2001 earthquake are observed (see Fig. 4 for details). Note that there is no liquefaction produced by the 2001 Kunlun earthquake in this location. (c) Coseismic surface ruptures with mole track (in the centre). Red arrows indicate the coseismic surface rupture. (d) Sand boils (indicated by red arrows) produced by the 2001 Kunlun earthquake. The field note (indicated by white arrow) shown for scale is *c*. 18 cm long.

Discussion and conclusions

Magnitude of earthquake-induced liquefaction

Liquefaction is a phenomenon of transformation of loose saturated sand–soil deposits from a solid to a liquid state as a result of increased pore water pressure and reduced effective stress, which can be caused by strong earthquake shaking and other geological processes. Criteria for distinguishing earthquake-induced liquefaction features from others types of unconsolidated sediment deformation structures have been reported in many studies (e.g. Tuttle & Seeber 1991; Sims & Garvin 1995; Tuttle & Schweig 1996; Lin 1997, 2006). On the basis of the liquefaction structures of the sedimentary deposits described in this study, we conclude that the vented sand veins and dykes and boiled loose sand–gravel were produced from liquefaction caused by strong earthquake shaking. Other non-earthquake origins considered for the liquefaction structures described here include floods, landslides, avalanches of snow and glacial ice; others can be rejected owing to the following three main features of the silt–sand veins: (1) upward injections of sand and gravel veins; (2) disruption of overlying unliquefied sediment layers; and (3) wide distribution and similar deformation features of liquefied sediments. The morphology of sand and gravel veins indicates that these veins were injected from the lower source layers that were liquefied. The liquefaction structures described in this study show that the liquefied sand and gravel layers are capped by an unliquefied fine-grained silt–sand layer and the top sand–gravel layer (Fig. 5). This fine-grained cap layer may play an important role in preserving high pore water pressure within underlying sand and gravel layers to produce liquefaction. The presence of liquefied sand–gravel deposits and the unliquefied overlying fine-grained silt–sand layer suggests the vented

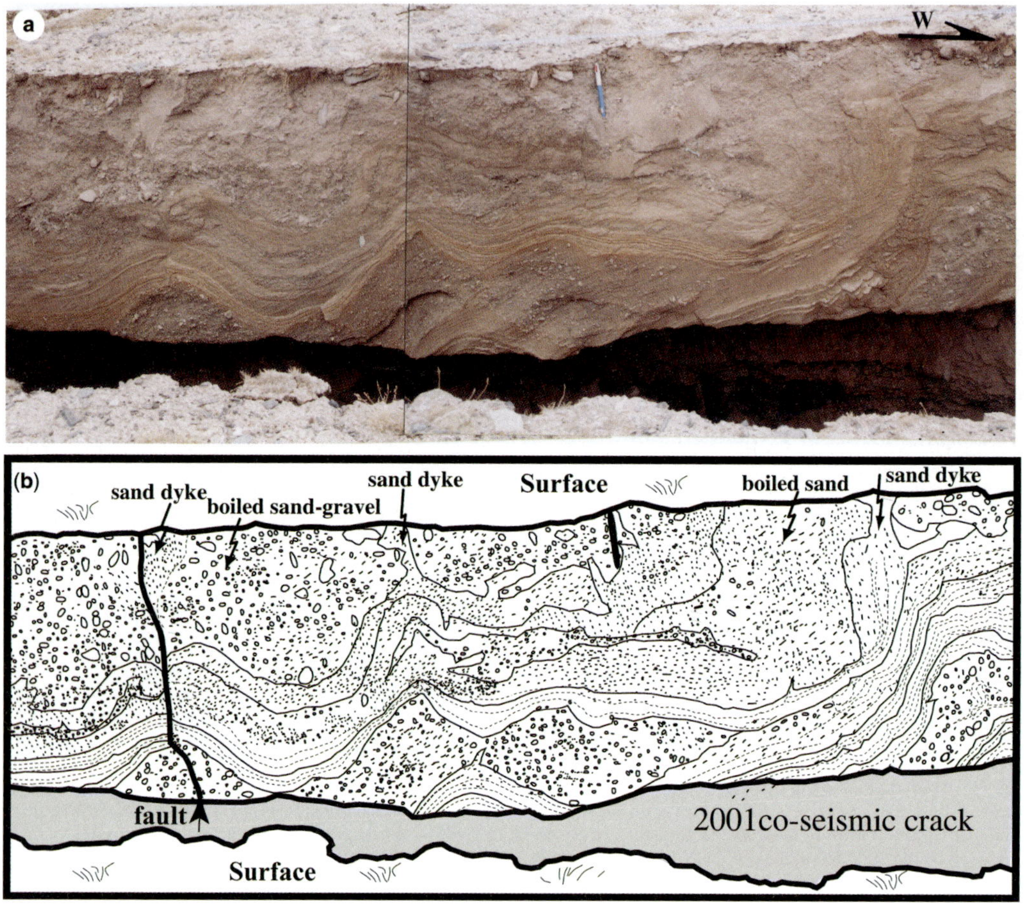

Fig. 4. Photograph (**a**) and corresponding sketch (**b**) of the exposure (crack wall shown in Fig. 2b) along the 2001 coseismic surface rupture. Note that the sand–silt layers are strongly deformed by the sand boils. Ball-point pen (on the top of the exposure) indicates the scale.

sand–gravel dykes formed by strong ground shaking and were injected rapidly and forcefully, consistent with an earthquake origin. Such structures of liquefied sand–gravels are also reported from the epicentral area of the 1995 M_w 7.2 Kobe earthquake, which were caused by strong ground shaking during the earthquake (Nirei *et al.* 2001). The similar features of liquefaction structures observed at the same site for a long continuous exposure over 200 m also suggest that the liquefaction occurred within the coseismic surface ruptures produced by at least one past large earthquake. Similar liquefaction features including sand dyke structures caused by the 1937 M7.5 Tuosho Lake earthquake and palaeoearthquakes that occurred in the eastern segment of the Kunlun fault are also observed along the coseismic rupture zone (Guo *et al.* 2007). Such earthquake-induced liquefaction features are consistent with those reported by others (e.g. Obermeier *et al.* 1985; Sims & Garvin 1995; Tuttle & Schweig 1996; Lin 1997, 2006).

The most important factors contributing to earthquake-induced liquefaction are the amplitude of cyclic shear stresses and the number of applications of these shear stresses related to earthquake magnitude (National Research Council 1985). Although earthquake-induced liquefaction can develop in the epicentral areas of earthquakes at magnitudes as low as 5.5 (Kuribayashi & Tatsuoka 1975; Audemard & de Santis 1991), the soft-sediment deformational structures formed in this level of shaking intensity are commonly small scale (Sims 1975). Similar liquefaction features including vented sand–gravel deposits observed along the fault trace on a continuous exposure of >200 m in the study site indicate that their formation requires strong shaking and may be related to coseismic surface ruptures produced by large

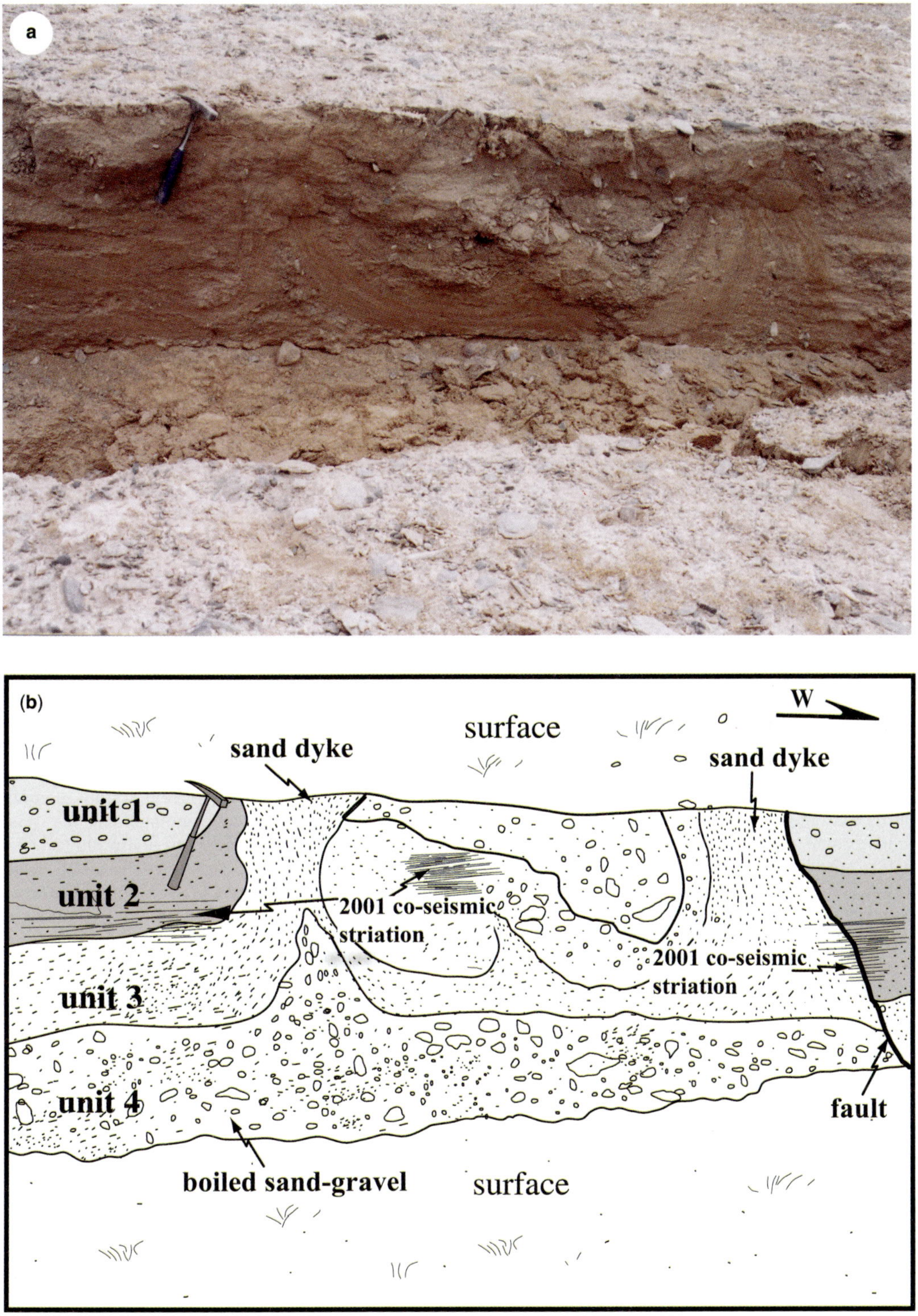

Fig. 5. Photograph (**a**) and corresponding sketch (**b**) of the exposure (crack wall) along the 2001 coseismic surface rupture. Units 1–4 are sediment units, see text for details. Hammer (on the top of the exposure) indicates the scale.

Table 1. *Radiocarbon dates from the organic soil taken from the study site*

Sample date code	Laboratory ID*	Radiocarbon age[†] (yr BP)	Calibrated age[‡] (yr BP)	Sampling depth (m)	Description
2004-C04	196974	840 ± 40	679–901	0.2–0.4	Organic soil
2004-C02	196973	4310 ± 40	4828–5025	0.6–0.8	Organic soil

Samples taken from location indicated by Loc. 1 in Figure 2.
*Samples were analysed at Beta Analytic Inc.
[†]Radiocarbon ages were measured using accelerator mass spectrometry (AMS).
[‡]Dendrochronologically calibrated calendar age by Method A from program of Stuiver *et al.* (2003) with two standard deviation uncertainty.

earthquakes. Historic records show that the coseismic surface ruptures are generally produced by large earthquakes of M > 7 in the northern Tibet–Xingjian region (Feng 1997). The coseismic surface ruptures were produced by three large historic earthquakes of M ≥ 7.5 including the 2001 earthquake along the Kunlun fault, but were not produced by the 1971 M 6.3 earthquake that occurred in the eastern segment of the Kunlun fault (Jia *et al.* 1988; Guo *et al.* 2007). Magnitude 7, therefore, is considered as a conservative lower bound for the liquefaction described here.

Timing and average recurrence interval of large earthquakes

The result of ^{14}C dating of an organic sample taken from a depth of 0.4 m in this study shows the liquefaction formed in the past 679–901 yr BP. The fact that the liquefied sediments are vented and exposed on the current river channel and cut by subsequent faults formed prior to the 2001 earthquake indicates that at least one seismic event occurred during a period between the 2001 earthquake and the identified liquefaction event. This means that at least two seismic events occurred in the past 679–901 yr BP before the 2001 earthquake with an average recurrence interval of 400 (350–450) years. This result coincides with the average recurrence interval of 300–400 years estimated from the trench and field investigations and supports that the penultimate event prior to the 2001 earthquake occurred in the past 400 years in the western segment of the Kunlun fault associated with the 2001 M_w 7.8 Kunlun earthquake (Lin *et al.* 2006).

Palaeoseismic and trench studies show that return times of surface rupturing events can be placed within relatively confined recurrence intervals, and provide the most direct measure of the past recurrence intervals of moderate to large earthquakes on active faults (Yeats *et al.* 1997). The displacement of 75–82 m observed on the current stream channel in the study site is considered to be an offset amount accumulated in the late Holocene such as those observed in the Kunlun Pass area (around sites 1 and 2 shown in Fig. 1b) 7–8 km east of the study site (Lin *et al.* 2006). The accumulation of offset on the fault indicates that seismic slip occurred repeatedly on the fault. Based on ^{14}C dates and the relationship between the height of terraces and average stream incision rate of 1.0–2.0 mm/a, it is inferred that the main terraces and alluvial fans with a height of <10 m from the current stream channel developed in the Kunlun Pass area east-bounded to the study site formed in the last 7000 years (Lin *et al.* 2006). The height of terraces from the current stream channel is generally lower than 5 m in the study site as stated above, so we estimate that the main terraces and alluvial fans formed in the late Holocene. The result of ^{14}C dating of an organic sample taken from a depth of 0.6–0.8 cm in Unit 4 shows that the main alluvial deposits formed in the past 4828–5025 yr BP. Therefore, the 75–82 m displacements of the current stream channel are considered to have cumulated in the past period of 4800–7000 years, and a slip rate of 11–16.7 mm/a is obtained. This result is comparable with the average slip rate of 16.4 mm/a inferred from the trench and field studies using radiocarbon dating results (26 ^{14}C samples) in the same fault segment related to the 2001 earthquake (Lin *et al.* 2006). This average slip rate is larger than that (10.0 ± 1.5 mm/a) estimated by Li *et al.* (2005) but coincides with the Pleistocene average slip rate of 10–20 mm/a estimated by Kidd & Molnar (1988) for the same segment of the Kunlun fault. Li *et al.* (2005) estimated the average slip rate by using the offset amount of terrace risers developed on one alluvial fan with four thermoluminescence ages of alluvial deposits but without trench and ^{14}C dating results. The difference in estimated slip rates between our studies (Lin *et al.* 2006 and this study) and Li *et al.* (2005) is probably caused by the different dating results. A recent study by Kirby *et al.* (2007) also shows that the slip rate estimated from radiocarbon dating results is quite different from those estimated mainly from cosmogenic dating ages (e.g. Van der Woerd *et al.* 2002*b*) along the easternmost segment of the Kunlun fault. Our recent study shows that the slip rate is non-uniform and diminishes from the west to the east with an average gradient of 1 mm/100 km along the strike-slip Kunlun fault (Lin & Guo 2008). One of the major reasons for over- or underestimation of slip rate for the Kunlun fault is probably caused by the uncertainty of cosmogenic and thermoluminescence dating methods for the glacial deposits in the Tibet plateau. Strike-slip offset produced by the 2001 M_w 7.8 earthquake is 3–6 m with an average offset of 4.5 m in the study site (Lin *et al.* 2002, 2003). If we use this average offset of 4.5 m caused by an individual faulting event and the total offset of 75 m, an average recurrence interval of magnitude 8 earthquakes is calculated to be *c.* 350 years for the past 4800–7000 years which is comparable with both our previous results (300–400 years) obtained from trench and field investigations (Lin *et al.* 2006) and that (400 years) estimated in this study. This recurrence interval is also comparable with that (300 years) estimated by Li *et al.* (2005) in the same segment of the Kunlun fault.

Comments and discussions with Professor J. Suppe of Princeton University are gratefully acknowledged. Thanks are also due to the reviewer Professor P. Alfaro for his critical review. This contribution was inspired by discussions with students and staff of Shizuoka University and Princeton University where the author (A. Lin) spent one year as a visiting scholar. We thank the U.S. Geological Survey Land Processes Distributed Active Archive Center for releasing GTOPO30 data, the University of Maryland Global Land Cover Facility for Landsat ETM+ images, and NASA Jet Propulsion Laboratory for ASTER images. This work was supported by the Nuclear and Industrial Safety Agency, Japan, the Active Fault Research Center, National Institute of Advanced Industrial Science and Technology, and the Science Project (Project No. 18340158 for A. Lin) of the Ministry of Education, Culture, Sports, Science and Technology of Japan.

References

AUDEMARD, F. A. & DE SANTIS, F. 1991. Survey of liquefaction structures induced by recent earthquakes. *Bulletin of International Association of Engineering Geology*, **44**, 5–16.

FENG, X. 1997. *The paleoearthquakes in Xinjiang region, China. Urumqi, China.* Xinjiang Science & Technology & Hygiene Publishing House, Urumqi [in Chinese].

GUO, J., LIN, A., MARUYAMA, T., ZHENG, J. & SUN, G. 2006. New constraints on recent large earthquakes along the Xidatan-Dongdatan segment of the Kunlun fault, western China. *Bulletin of Seismological Society of America*, **96**, 48–58.

GUO, J., LIN, A., SUN, G. & ZHENG, J. 2007. Characteristic features of the surface ruptures associated with the 1937 M 7.5 and the 1963 M 7.0 earthquakes and the paleoseismicity along the Tuosuo Lake segment of the strike-slip Kunlun Fault, China. *Bulletin of Seismological Society of America*, **97**, 474–496.

JIA, Y., DAI, H. & SU, X. 1988. Tuosuo Lake earthquake fault in Qinghai province. *In*: Xinjiang Seismological Bureau, (ed.) *Research on Earthquake Faults in China*. Xinjiang People Press, Urumqi 66–71 [in Chinese].

KIDD, W. S. F. & MOLNAR, P. 1988. Quaternary and active faulting observed on the 1985 Academia-Sinica Royal-Society Geotraverse of Tibet. *Philosophical Transactions of the Royal Society London*, **327**, 337–363.

KIRBY, E., HARKINS, N., WANG, E., SHI, X., FAN, C. & BURBANF, D. 2007. Slip rate gradients along the eastern Kunlun fault. *Tectonics*, **26**, TC2010. DOI: 10.1029/2006TC002033.

KURIBAYASHI, E. & TATSUOKA, F. 1975. Brief review of liquefaction during earthquakes in Japan. *Soil Foundation*, **15**, 81–92.

LI, H., VAN DER WOERD, J., TAPPONNIER, P., KLINGER, Y., QI, X., YAN, J. & ZHU, Y. 2005. Slip rate on the Kunlun fault at Hongshui Gou, and recurrence time of great events comparable to the 14/11/2001, Mw ~ 7.9 Kokoxili earthquake. *Earth and Planetary Science Letters*, **237**, 285–299.

LIN, A. 1997. Instantaneous-shaking liquefaction induced by the M7.2 Southern Hyogo Prefecture earthquake, Japan. *Geology*, **25**, 435–438.

LIN, A. 2006. Repeated large subduction zone earthquakes in the Nankai-Suruga trough: evidence from submarine liquefactions. *Geophysical Research Letters*, **33**, L20314. DOI: 10.1029/2006GL027952.

LIN, A. 2008. Geometry and slip distribution of co-seismic surface ruptures produced by the 2001 M_w 7.8 earthquake along the strike-slip Kunlun fault, northern Tibet. *In*: FUKUYAMA, E. (ed.) *Fault-zone Properties and Earthquake Rupture Dynamics*. Elsevier, Amsterdam, 15–36.

LIN, A. & GUO, J. 2008. Non-uniform slip rate and millennial recurrence interval of large earthquakes on the eastern segment along the Kunlun fault, northern Tibet. *Bulletin of Seismological Society of America*, **98**. DOI: 10.1785/0120070193.

LIN, A. & NISHIKAWA, M. 2007. Coseismic strike-slip offsets of surface rupture produced by the 2001 M_w 7.8 Kunlun (Tibet) earthquake, measured from high-resolution IKONOS and QuickBird imagery. *International Journal of Remote Sensing*, **27**, 2431–2445.

LIN, A., OUCHI, T., CHEN, A. & MARUYAMA, T. 2001. Co-seismic displacements, folding and shortening structures along the Chelungpu surface rupture zone occurred during the 1999 Chi-Chi (Taiwan) earthquake. *Tectonophysics*, **330**, 225–244.

LIN, A., FU, B., GUO, J., ZENG, Q., DANG, G., HE, W. & ZHAO, Y. 2002. Co-seismic strike-slip and rupture length produced by the 2001 Ms 8.1 Central Kunlun earthquake. *Science*, **296**, 2015–2017.

LIN, A., KIKUCHI, M. & FU, B. 2003. Rupture segmentation and process of the 2002 Mw 7.8 Central Kunlun earthquake, China. *Bulletin of Seismological Society of America*, **93**, 2477–2492.

LIN, A., GUO, J. & FU, B. 2004. Co-seismic mole-track structures produced by the 2001 Ms 8.1 Central Kunlun earthquake, China. *Journal of Structural Geology*, **26**, 1511–1519.

LIN, A., GUO, J., KANO, K. & AWATA, Y. 2006. Average slip-rate and recurrence interval of large magnitude earthquakes on the western segment of the strike-slip Kunlun fault, northern Tibet. *Bulletin of Seismological Society of America*, **96**, 1591–1611.

MARCO, S., STEIN, M. & AGNON, A. 1996. Long-term earthquake clustering: A 50,000-year paleoseismic record in the Dead Sea Graben. *Journal of Geophysical Research*, **10**, 6179–6191.

MEYER, B., TAPPONNIER, P. *ET AL.* 1998. Crustal thickening in Gansu-Qinghai, Lithospheric mantle subduction, and oblique, strike-slip controlled growth of the Tibet plateau. *Geophysical Journal International*, **135**, 1–47.

NATIONAL RESERACH COUNCIL. 1985. *Liquefaction of Soils during Earthquakes*. National Academy Press, Washington. D.C.

NIREI, H., KUSUDA, O. *ET AL.* 1995. J. Liquefaction phenomena of instantaneous state-collapsed type. *In*: *Proceedings of Symposium on the Great Hanshin-Awaji Earthquake and its Geo-environments*. Geological Society of Japan, Committee of Environmental Geology, 137–142 [in Japanese with English abstract].

OBERMERIER, S. F. 1995. Using, liquefaction-induced features for paleoseismic analysis. *In*: OBERMERIER, S. F. & JIBSON, R. W. (eds) *Using Ground Failure Features for Paleoseismic Analysis*. U.S. Geological Survey, Open-File Report, **94-683**, 1–98.

OBERMERIER, S. F., GOHN, G. S., WEEMS, R. E. & RUBIN, M. 1985. Geologic evidence for recurrent moderate to large earthquakes near Charleston, South Carolina. *Nature*, **227**, 408–411.

PELTZER, G., CRAMPE, F. & KING, G. 1999. Evidence of nonlinear elasticity of the crust from the Mw 7.6 Manyi earthquake. *Science*, **286**, 272–276.

RODBELL, D. T. & SCHWEIG, E. S. III 1993. The record of seismicity induced liquefaction on late Quaternary terraces in northwestern Tennessee. *Bulletin of Seismological Society of America*, **83**, 269–278.

SEED, H. B. & IDRISS, I. M. 1967. Analysis of liquefaction: Niigata earthquake. *American Society of Civil Engineering Journal of Soil Mechanics and Foundation Division*, **98**, 83–108.

SIMS, J. D. 1975. Determining earthquake recurrence intervals from deformational structural in young lacustrine sediments. *Tectonophysics*, **29**, 141–153.

SIMS, J. D. & GARVIN, C. D. 1995. Recurrent liquefaction induced by the 1989 Loma Prieta earthquake and 1990 and 1991 aftershocks: implications for paleoseismicity studies. *Bulletin of Seismological Society of America*, **85**, 51–65.

STUIVER, M., REIMER, P. J. & REIMER, R. 2003. *CALIB Radiocarbon Calibration Version 4.4*. Available at: http://radiocarbon.pa.qub.ac.uk/calib/

TAPPONNIER, P. & MOLNAR, P. 1977. Active faulting and tectonics in China. *Journal of Geophysical Research*, **82**, 2905–2930.

TUTTLE, M. & SCHWEIG, E. S. 1996. Recognizing and dating prehistoric liquefaction features: lessons learned in the New Madrid seismic zone, central United States. *Journal of Geophysical Research*, **101**, 6171–6178.

TUTTLE, M. & SEEBER, L. 1991. Historic and prehistoric earthquake-induced liquefaction in Newbury, Massachusetts. *Geology*, **19**, 594–597.

VAN DER WOERD, J., MERIAUX, A. S., KLINGER, Y., RYERSON, F. J., GAUDEMER, Y. & TAPPONNIER, P. 2002*a*. The 14 November 2001, Mw = 7.8 Kokoxili earthquake in northern Tibet (Qinghai Province, China). *Seismological Research Letters*, **73**, 125–135.

VAN DER WOERD, J., TAPPONNIER, P. *ET AL.* 2002*b*. Uniform postglacial slip-rate along the central 600 km of the Kunlun Fault (Tibet), from ^{26}Al, ^{10}Be, and ^{14}C dating of riser offsets, and climatic origin of the regional morphology. *Geophysical Journal International*, **148**, 356–388.

WANG, Q., ZHANG, P. *ET AL.* 2001. Present-day crustal deformation in China constrained by global positional system measurements. *Science*, **294**, 574–578.

XU, X., CHEN, W., MA, W., YU, G. & CHEN, G. 2002. Surface ruptures of the Kunlunshan earthquake (Ms 8.1), northern Tibetan Plateau, China. *Seismological Research Letters*, **73**, 884–892.

XU, X., MA, W., YU, G., TAPPONNIER, P., KLINGER, Y. & VAN DER WOERD, J. 2006. Re-evaluation of surface rupture parameters and faulting segmentation of the 2001 Kunlunshan earthquake (Mw 7.8), Northern Tibetan Plateau, China. *Journal of Geophysical Research*, **111**, B05316. DOI: 10.1029/2004JB003488.

YEATS, R. S., SIEH, K. & ALLEN, C. R. 1997. *The Geology of Earthquake*. Oxford University Press, Oxford.

The Muzaffarabad, Pakistan, earthquake of 8 October 2005: surface faulting, environmental effects and macroseismic intensity

ZAHID ALI[1], MUHAMMAD QAISAR[1], TARIQ MAHMOOD[1]*, MUHAMMAD ALI SHAH[1], TALAT IQBAL[1], LEONELLO SERVA[2], ALESSANDRO M. MICHETTI[3] & PAUL W. BURTON[4]

[1]*Micro Seismic Studies Programme, Ishfaq Ahmed Research Laboratories, P. O. Nilore, Islamabad, Pakistan*

[2]*ISPRA – Geological Survey of Italy, Via V. Brancati 48, 00144 Roma, Italy*

[3]*Dipartimento di Scienze Chimiche e Ambientali, Università dell'Insubria, Via Valleggio 11, 22100 Como, Italy*

[4]*School of Environmental Sciences, University of East Anglia, Norwich, NR4 7TJ, UK*

**Corresponding author (e-mail: tariqmssp@yahoo.com)*

Abstract: The Mw 7.6 Muzaffarabad earthquake of 8 October 2005, occurred on a lateral equivalent of the main ramp of the Hymalaia frontal thrust, and is the result of the collision tectonics between the Indian and Eurasian plates. The epicentre was located near the town of Basantkot (Muzaffarabad), and the focal depth was about 13 km. The Muzaffarabad earthquake provides unequivocal evidence about the localization of severe damage, intense ground shaking and secondary environmental effects near the surface expression of the source fault. We analyse its nature, and impact on man-made structures and the physical environment, on the basis of a detailed survey and macroseismic study of the affected areas conducted by the Micro Seismic Studies Programme (MSSP) Team (Ishfaq Ahmad Research Laboratories, Pakistan Atomic Energy Commission) immediately after the mainshock, assisted by a careful review of the subsequent data and literature. In the course of the field survey, the displacement and surface expression of the causative fault, and accompanying secondary environmental effects were observed at a number of places along a capable thrust fault structure. We refer to this structure as the Kashmir Thrust (KT) capable fault following the terminology of local research geologists in Pakistan; the seismological evidence of this structure is already known in the literature as the Indus–Kohistan Seismic Zone. A complex, clearly segmented, at least 112-km-long surface rupture was mapped along the KT. The maximum values of vertical displacement (on the order of 4 to 7 m) were observed mainly between Muzaffarabad and Balakot, along the central segment of the rupture (52 km) associated with maximum slip at depth and a major portion of the energy release. Both the NW Alai segment (38 km) and SE Bagh segment (22 km) are characterized by scattered minor surface ruptures with a few centimetres of displacement, accompanied by extensive surface cracking, landslides and severe damage, concentrated in a narrow belt along the fault trace. A maximum intensity of XI on the Modified Mercalli Intensity (MMI) scale and on the Environmental Seismic Intensity scale (ESI 2007) was recorded in the epicentral area between Muzaffarabad and Balakot. Extremely severe damage and very important secondary environmental effects in the hanging wall adjacent to the trace of the causative fault plane are mainly due to near-fault strong motion and rupture directivity effects. To our knowledge, this is the first study to present field observations over the whole near-field of the earthquake, and to include the intensity map of the entire meizoseismal region.

The 8 October 2005 Muzaffarabad earthquake, with an M_L magnitude of 7.0 ($M_S = 7.7$, $m_b = 6.8$ and $M_w = 7.6$ reported by USGS) occurred at 03:50:38 GMT (08:50:38 local time) near the city of Muzaffarabad (Fig. 1; Table 1). The mainshock was followed over the ensuing month by more than 6400 aftershocks, 296 of which had magnitudes (M_L) greater than 4, as recorded by the Micro Seismic Studies Programme seismic network within one month after the mainshock (MSSP 2005).

The Pakistan official death toll due to the earthquake is 74 698 fatalities, which probably minimizes the dimension of the disaster (more than 86 000 fatalities have been estimated by USGS). About 100 000 people were injured, and *c.* 4 million were left homeless. Most buildings were

From: REICHERTER, K., MICHETTI, A. M. & SILVA, P. G. (eds) *Palaeoseismology: Historical and Prehistorical Records of Earthquake Ground Effects for Seismic Hazard Assessment*. The Geological Society, London, Special Publications, **316**, 155–172. DOI: 10.1144/SP316.9 0305-8719/09/$15.00

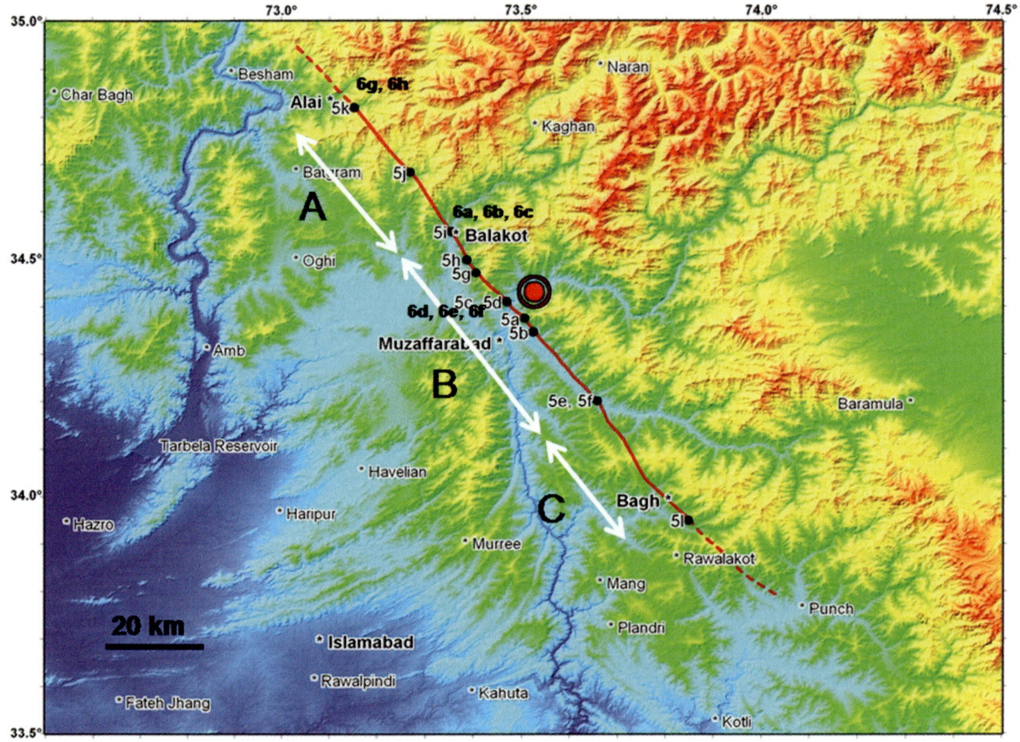

Fig. 1. Map showing the epicentre of the 8 October 2005, Muzaffarabad earthquake (red circle; MSSP 2005) and the trace of coseismic surface rupture (red line). The double-pointed white arrows show the major surface-rupture segments: A, the 38-km-long Alai segment; B, the 52-km-long Muzaffarabad segment; C, the 22-km-long Bagh segment (as mapped by MSSP 2005 immediately after the mainshock). Also shown are the possible extension of surface rupture (dashed red line), the location of environmental effects (shown in Fig. 5a–l) and of associated damage (shown in Fig. 6a–h) along the source fault, here referred to as the Kashmir Thrust (KT).

destroyed or heavily damaged in the Azad Kashmir and Hazara areas of Pakistan. The heaviest damage occurred in the cities of Muzaffarabad, Balakot, Bagh, Alai and in the valleys of Jhelum, Kaghan, Neelum and Siran rivers.

The earthquake was the consequence of the collision between the Indian and Eurasian plates along a prominent, structurally complex and segmented structure (Fig. 1), belonging to the frontal system of the Pakistan and Indian Himalaya (e.g. Nakata *et al.* 1991; Tapponnier *et al.* 2006*a*; Kumar *et al.* 2006; Rao *et al.* 2006). The earthquake rupture reactivated previously mapped active (capable) faults, such as the Muzaffarabad fault and the Tanda fault (Calkins *et al.* 1975; Nakata *et al.* 1991; Kumahara & Nakata 2006). The relations of

Table 1. *Source parameters of the 8 October 2005, Muzaffarabad earthquake*

Date/Time	8 Oct 2005/08:50:38PST
Magnitude (M_L)	7.0 (USGS $m_b = 6.8$, $M_s = 7.7$, $M_w = 7.6$)
Epicentre	Longitude 73.52°E, latitude 34.42°N
Focal depth	13 km
Fault movement	Predominantly thrust
Fault plane strike	338°
Fault plane dip	60°NE
Rake	138°
Vertical max. displacement	4.2 ± 0.5 m

the 8 October 2005 earthquake causative fault to known active faults and geological structures is fully discussed by Avouac *et al.* (2006). The geometry and kinematics of the ruptured thrust fault are consistent with the observation of a recent reversal of the sense of motion on the Main Boundary Thrust (Calkins *et al.* 1975; Figs 2 and 3) and with the seismological evidence that recent deformation cuts across the Hazara Syntaxis (Armbruster *et al.* 1978). However, it should be remarked that these authors did not recognize the NW segment of the surface rupture in the Alai Valley, as described below, which represents one of the goals of this paper (Fig. 1).

Kaneda *et al.* (2008) illustrated the details of the surface rupture between Balakot and Dhallan, essentially based on 11 days of field survey conducted about 5 months after the mainshock. To our knowledge, this is the best study conducted in the field and available to date on earthquake surface faulting. Field mapping along the source fault immediately after the mainshock was in fact very difficult. The devastation in the epicentral area was immense and the logistic conditions extremely challenging. The high mountain topography, the snow-cover, the security problems related to landslide hazards, all these factors did not allow a complete and timely survey of the coseismic ground effects. Based on these field observations, satellite imagery analysis, and modelling of seismographic data, some authors referred to the coseismic surface rupture as the Balakot–Bagh fault (e.g. Parsons *et al.* 2006; Kaneda *et al.* 2008) or the Balakot-Ghari fault (Kumahara & Nakata 2006).

The Micro Seismic Studies Programme (MSSP) Team (Ishfaq Ahmad Research Laboratories, Pakistan Atomic Energy Commission), conducted extensive field survey and macroseismic assessment immediately after the mainshock over the whole epicentral area (MSSP 2005; Table 2 lists all the visited sites, also mapped in Fig. 4). In particular,

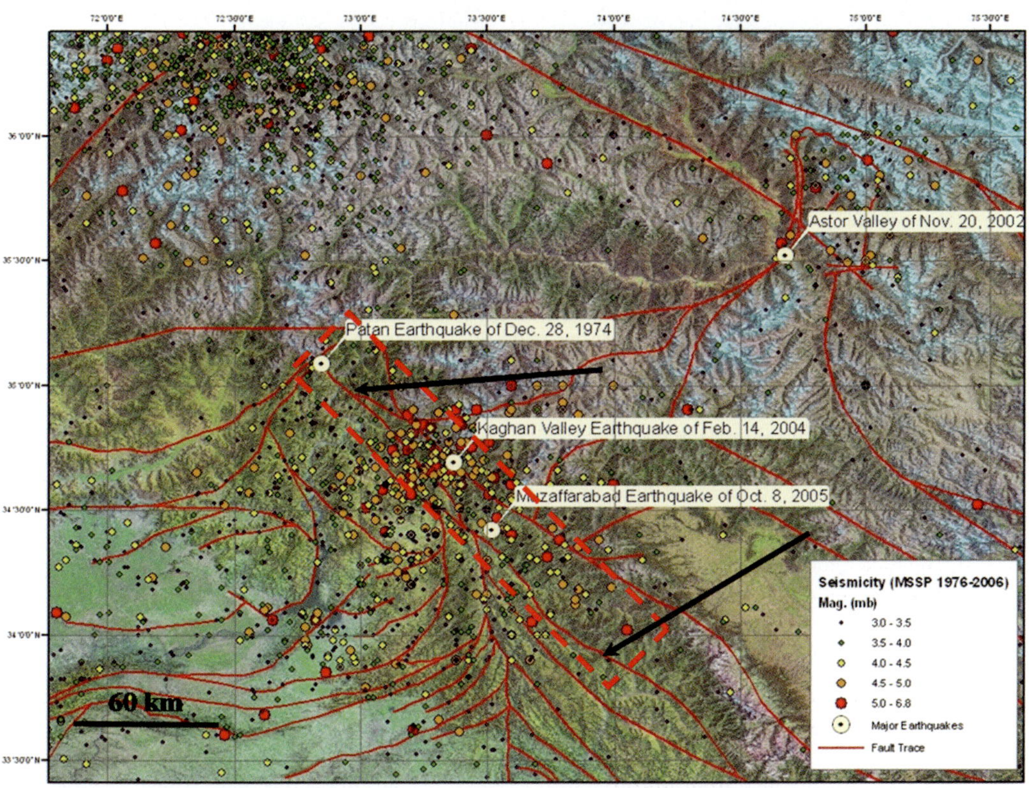

Fig. 2. Map showing the tectonic framework of north Pakistan (modified from Kazmi & Jan 1997), the recent seismicity recorded by the MSSP network in the period 1976–2006, and epicentres of significant earthquakes that occurred in the area in the past 35 years. The NW-trending belt defined by the instrumental seismicity (dashed red rectangle) is the Indus–Kohistan Seismic Zone (IKSZ) described in Armbruster *et al.* (1978). Black arrows show the trace of the Kashmir Thrust (KT), extending more than 150 km SE of Patan. Surface rupture along the KT during the 8 October 2005 earthquake, was observed by the MSSP team over a length of 112 km (MSSP 2005; see Fig. 1).

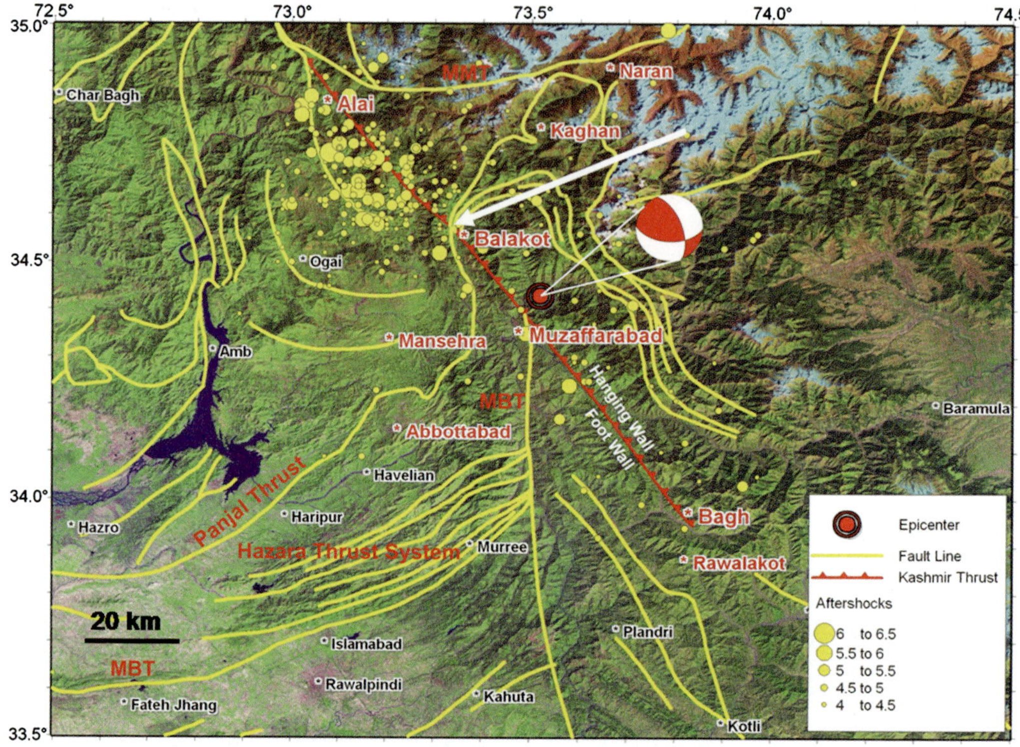

Fig. 3. Map showing the epicentre of the 8 October 2005, Muzaffarabad earthquake, the mainshock focal mechanism solution (Harvard CMT), the aftershock distribution and the Kashmir Thrust surface rupture, within the structural framework of north Pakistan (modified from Kazmi & Jan 1997). The white arrow marks the point NW of Balakot where the coseismic rupture cuts across the Hazara Syntaxis, and follows the Indus–Kohistan Seismic Zone (IKSZ; see Fig. 2 for the complete map view of the IKSZ) as defined by Armbruster *et al.* (1978). In the sector between Muzaffarabad and Balakot, the trace of the KT surface rupture is parallel and very close to the Main Boundary Thrust (MBT; e.g. Kazmi & Jan 1997), but with opposite dip (at the scale of this figure the two traces cannot be separated and are mapped as coincident).

systematic observations were made in the Alai area, NW of Balakot, in the first weeks after the mainshock; to our knowledge, no report of field survey in this area has been published yet. As already mentioned, due to the local climatic and topographic environment the field survey cannot be regarded as complete. In our opinion, the collected data set is, however, suitable for the purpose of this research. Based on these investigations, the fault surface rupture extended significantly from the Alai Valley to the NW to SE of Bagh. The coseismic surface faulting followed a well-defined NNW–SSE, NE-dipping, complex segmented structure. Part of this fault, for instance from Muzaffarabad to Chatter Jhatian (respectively, sites 1 and 11 in Fig. 4), was already known on geological maps (e.g. Kazmi & Jan 1997; Fig. 2). In the following we will refer to the overall surface rupture accompanying the 8 October 2005, Muzaffarabad earthquake and mapped in Figure 1 as the Kashmir Thrust (KT) fault, following the terminology established by local researchers in Pakistan. We regard the KT as the surface expression of the so-called Indus–Kohistan Seismic Zone (IKSZ; Armbruster *et al.* 1978; Seeber & Armbruster 1979; Avouac *et al.* 2006), a basement thrust ramp that intersects the Hazara Syntaxis (Figs 2 and 3) and should be considered a lateral equivalent of the main ramp of the Hymalaia frontal thrust (Tapponnier *et al.* 2006*a*).

The existing instrumental seismological data do not attest to any important seismic event generated by the KT in the last few decades. However, the area affected by the earthquake does constitute a major tectonically active area. Along with the KT, the Main Boundary Thrust (MBT), Main Mantle Thrust (MMT) and Panjal Thrust faults are regarded as the most prominent seismogenic structures (Nakata *et al.* 1991; Yeats *et al.* 1992; Kazmi & Jan 1997; Figs 2 and 3). The Muzaffarabad area itself had no history of large earthquakes before

Table 2. *MMI (Wood & Neumann 1931) values of the Muzaffarabad earthquake, 8 October 2005*

Site no.	Location	Longitude °E	Latitude °N	Distance from epicentre (km)	MMI
1	Muzaffarabad (Main City)	73.471	34.370	6.95	X
2	Chella Bandi	73.470	34.393	5.38	XI
3	Nisar Camp	73.476	34.394	4.91	XI
4	Domel	73.470	34.351	8.79	IX
5	Chatter Plain	73.458	34.341	10.54	VIII
6	Lower Plate	73.465	34.376	6.95	X
7	Bala Pir	73.437	34.344	11.42	VIII
8	Dhanni	73.476	34.406	4.39	XI
9	Majhoi	73.587	34.251	19.65	X
10	Ghari Dopatta	73.636	34.216	24.95	XI
11	Jhatian	73.660	34.198	27.79	XI
12	Jigal	73.755	34.167	35.43	VIII
13	Siri	73.679	34.167	31.61	XI
14	Bankot	73.882	34.121	46.97	VII
15	Muhuri	74.063	34.011	67.47	VII
16	Basantkot	73.537	34.427	2.20	X
17	Kandar	73.521	34.427	2.20	X
18	Maira	73.680	34.443	14.74	VIII
19	Shahdara	73.656	34.432	12.43	VIII
20	Chelihan	73.658	34.459	13.36	VIII
21	Bhaler	73.478	34.276	16.44	VIII
22	Nammal	73.481	34.197	24.95	VIII
23	Majuhan	73.483	34.208	23.77	VIII
24	Sangal	73.485	34.161	28.90	VII
25	Balakot City	73.352	34.550	21.19	XI
26	Shinkiari	73.269	34.471	23.77	VIII
27	Daadar	73.283	34.613	30.52	X
28	Basian	73.342	34.462	16.88	VIII
29	Bela	73.352	34.554	21.30	XI
30	Patlang	73.358	34.572	22.51	XI
31	Dheri	73.358	34.530	19.15	X
32	Kashtra	73.361	34.432	14.41	VIII
33	Ghari Habibullah	73.383	34.399	12.81	VIII
34	Ramkot	73.398	34.371	12.43	VIII
35	Bhuraj	73.354	34.385	15.54	VIII
36	Mang	73.349	34.594	24.86	XI
37	Baso	73.275	34.672	35.83	XI
38	Manda Ghucha	73.266	34.663	35.63	X
39	Deoli	73.233	34.683	39.30	X
40	Patti	73.305	34.683	35.22	X
41	Kalas	73.236	34.629	34.88	VIII
42	Baleja	73.194	34.724	45.08	IX
43	Alai Town	73.094	34.821	59.16	IX
44	Gangwal	73.151	34.804	54.40	X
45	Shahidpatti	73.127	34.740	50.58	VIII
46	Battamori	73.086	34.687	49.57	VII
47	Ganda	73.086	34.677	48.93	VII
48	Bela	73.112	34.698	48.49	VIII
49	Murad Banda	73.125	34.709	48.34	VIII
50	Shumlai	73.116	34.705	48.69	VIII
51	Gidar	73.174	34.713	45.40	IX
52	Baleja	73.194	34.724	45.08	IX
53	Rashang	73.124	34.819	57.21	IX
54	Pokal	73.080	34.822	60.13	VIII
55	Karag	73.063	34.834	62.14	VIII
56	Rupkani	73.105	34.838	59.97	IX
57	Palaag	73.133	34.855	59.89	X
58	Jabbar	73.115	34.865	61.75	IX

(Continued)

Table 2. *Continued*

Site no.	Location	Longitude °E	Latitude °N	Distance from epicentre (km)	MMI
59	Bagh Bridge	73.773	33.980	54.13	IX
60	Bagh City	73.788	33.973	55.41	IX
61	Mohri	73.783	33.980	54.53	IX
62	Chattar	73.830	33.951	59.36	X
63	Serimang	73.902	33.927	65.07	IX
64	Dharian	73.833	33.934	61.17	IX
65	Dhuli	73.925	33.930	65.95	VIII
66	Rakot	73.966	33.950	66.39	VIII
67	Dhand	73.759	34.023	49.28	X
68	Surul	73.772	34.040	48.19	X
69	Mastan	73.833	34.077	47.73	VIII
70	Chakothi	73.883	34.116	47.48	VII
71	Mansehra	73.196	34.336	31.15	VII
72	Mang	73.646	33.809	68.85	VII
73	Batgram	73.011	34.681	54.84	VII
74	Abbottabad	73.210	34.148	41.51	VII
75	Murree	73.367	33.895	60.01	VII
76	Islamabad	73.059	33.692	91.33	VI
77	Kaghan	73.509	34.776	39.55	VII
78	Havelian	73.148	34.049	53.55	VI
79	Rawalakot	73.806	33.866	66.93	VIII
80	Rawalpindi	73.054	33.604	100.30	VI
81	Amb	72.823	34.305	65.18	VI
82	Fatehjang	72.640	33.564	124.97	V
83	Kahuta	73.381	33.584	93.76	VI
84	Besham	72.880	34.886	78.11	VII
85	Paras	73.446	34.666	28.14	VIII
86	Shogran	73.461	34.641	25.15	VIII
87	Mahandri	73.566	34.683	29.56	VII
88	Khannian	73.502	34.730	34.46	VII

See Figure 4 for site locations.

this event. The analysis of recorded seismic data shows that the area is dominated by frequent low to moderate seismicity nucleating at relatively shallow crustal depth (10–30 km). Significant earthquakes occurring in the IKSZ and surrounding region in the near past are the Pattan earthquake ($m_b = 6.0$) of 28 December 1974 (Wayne 1979), the Astor Valley earthquake ($m_b = 6.2$) of 1 November 2002 (Mahmood *et al.* 2002) and the Kaghan Valley earthquake ($m_b = 5.6$) of 14 February 2004 (Mahmood *et al.* 2004) (Fig. 2).

In the following sections we review the available seismological and geological information based on the results of the MSSP field survey (MSSP 2005). The purpose is to compare macroseismic intensity data and distribution of earthquake environmental effects over the whole epicentral area. To reach this goal, we describe several examples of coseismic ground effects and earthquake damage at relevant sites along the KT fault. We show that the KT surface rupture can be mapped in the field for at least 112 km from Alai Valley to SE of Bagh. Our findings suggest that ground deformation resulting from fault displacement caused particularly severe damage. Ground shaking was especially violent adjacent to the source fault, where secondary environmental effects, such as landslides and ground fracturing, were of extremely great magnitude. High macroseismic intensity appears to closely follow the trace of surface faulting, with highest values measured where the maximum surface displacement was observed.

Seismological observations

As indicated in Table 1, the focal mechanism solution obtained for the Muzaffarabad earthquake was predominantly thrust, striking NNW and steeply dipping NE, with a slight strike-slip component (Fig. 3; Parsons *et al.* 2006; Pathier *et al.* 2006; Avouac *et al.* 2006; Mandal *et al.* 2007). This solution coincides well with the slip nature of KT and is also supported by surface evidence observed from a thorough geological survey of the area. The rupture was initiated close to the northern

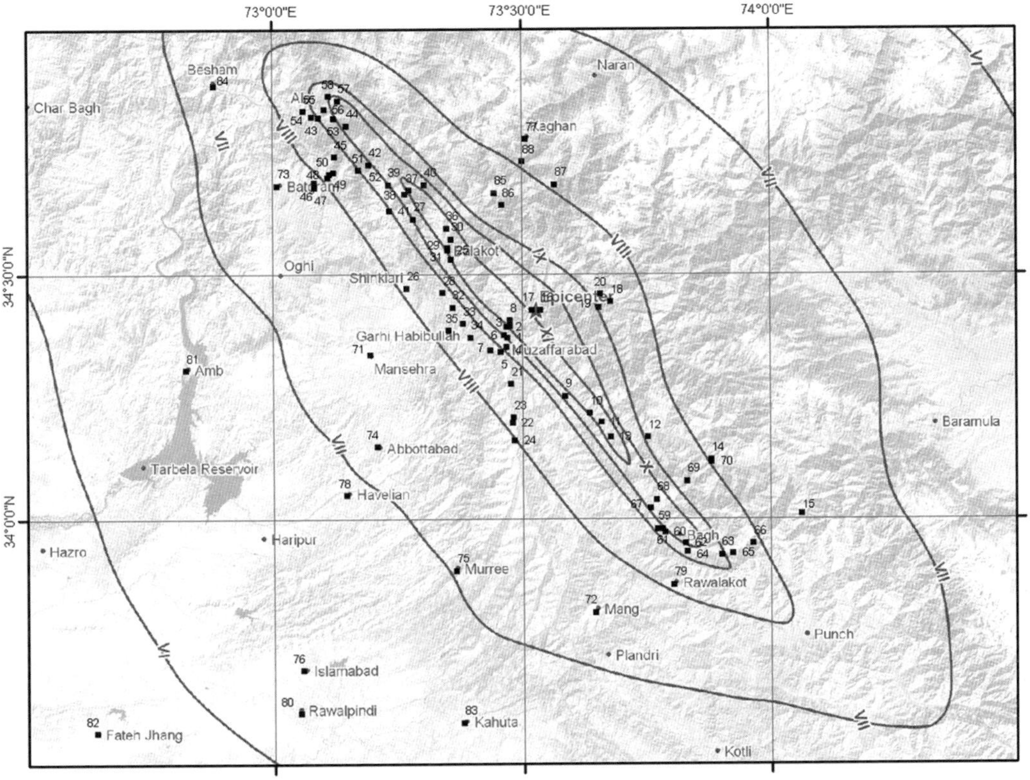

Fig. 4. Map showing the distribution of MMI intensity observations in the near region of the 8 October 2005, Muzaffarabad earthquake, along with the intensity isoseismals of the mainshock; it should be noted that in the survey region some areas are not accessible due to high/sharp relief and/or snow covering; in sparsely populated areas without man-made structures, intensity values were assessed through environmental features such as ground fractures, landslides, rock falls, and slope failures (MSSP 2005; see also Table 2). Locations of sites listed in Table 2 are shown.

margin of the Indian Plate by thrust along the alignment of the Bagh, Muzaffarabad, Balakot and Alai areas, where the southwestern fault block of KT acted as a foot wall (Figs 2 and 3). The instrumental epicentral location, near the town of Kandar (Neelum Valley) in the Muzaffarabad district, is also supported by the geological field survey and macroseismic observations showing that some epicentral features typical of high intensity, like sharp jerk, an explosion-like sound, and damage to structures at the first impetus were reported in the Kandar and Basantkot areas. Seismological observations and modelling of satellite imagery data for the mainshock and aftershocks indicated that coseismic displacement concentrated in the northwestern portion along strike of the KT at shallow depths (Avouac *et al.* 2006; Parsons *et al.* 2006; Pathier *et al.* 2006).

No strong-motion stations are available in the near-source region. Ground motion estimates based both on empirical analytical source mechanism models and stochastic finite fault seismological models (Singh *et al.* 2006; Raghukanth 2008) predict PGA (peak ground acceleration) values exceeding 1g at hard sites in the epicentral region, and a high stress drop (greater than 100 bars).

Field evidence relative to the earthquake

Landsat ETM (0.5 arc second) imagery, SRTM (3 arc second) digital elevation model (DEM), remote sensing modelling made available immediately after the mainshock (e.g. COMET 2005), GPS, GIS, along with preliminary information on source parameters, were utilized in the field survey conducted after 10 October 2005, in order to detect and understand the surface features of the causative fault, and to map coseismic environmental effects. In the following section, we describe the main features of the observed surface ruptures. We identify three major geometric segments of the KT separated by small gaps, as mapped by MSSP (2005) immediately after the mainshock (Fig. 1). We refer to the 38-km-long northern segment as the Alai segment ('A' in Fig. 1; from Sagwal to the NW to Daadar to the SE; respectively, close to sites 58 and 27 in Fig. 4, Table 2); the 52-km-long central segment as the Muzaffarabad segment ('B' in Fig. 1; from near Balakot to the NW to Siri to the SE; respectively, sites 36 and 13 in Fig. 4, Table 2), and the 22-km-long southern segment as the Bagh segment ('C' in Fig. 1; from near Surul to the NW to near Dhuli to the SE; respectively, sites 13 and 65 in Fig. 4, Table 2). The detailed mapping of the fault slip distribution is beyond the scope of this research. For this kind of information the reader is referred to Kaneda *et al.* (2008), for the section between Balakot and Bagh (segments 'B' and 'C' in Fig. 1). The detailed mapping of surface ruptures in the Alai segment ('C' in Fig. 1), which to our knowledge was only surveyed by the MSSP Team, is the object of a companion paper that will be published elsewhere.

Surface faulting

The overall mapped length of the KT is estimated at about 154 km in the North Pakistan region (Kazmi & Jan 1997; Fig. 2). The 8 October 2005, Muzaffarabad earthquake ruptured the KT for an overall length of at least 112 km (Figs 1 and 3). The subsurface dimension of the KT coseismic rupture may extend beyond the diffuse ends of surface faulting, that is Dhuli (Bagh) to the SE and Sagwal (Alai Valley) to the NW (Figs 1 and 3). It should be noted that rupture length from aftershock distribution and teleseismic body waveforms is on the order of 120 km (e.g. Parsons *et al.* 2006), while estimated surface rupture based on remote sensing ranges between 75 and 90 km (Fujiwara *et al.* 2006; Pathier *et al.* 2006; Avouac *et al.* 2006).

Our observations show a greater surface rupture length than that assessed by other authors (e.g. AIST 2006; Nakata & Kumahara 2006; Kaneda *et al.* 2006; Tapponnier *et al.* 2006*b*; Avouac *et al.* 2006; Bendick *et al.* 2007; Kaneda *et al.* 2008). As already mentioned, this is essentially due to the lack of field survey in the Alai Valley area, which was visited only by the MSSP Team. Available estimates of coseismic surface rupture length are based in fact on very limited fieldwork, due to the very demanding environment existing in the epicentral area immediately after the mainshock.

To this end, it is important to remark that the epicentral area is mostly characterized by sharp relief, with steep mountain slopes and narrow valleys. Therefore, assessment of surface rupture requires careful and detailed field mapping, which is only seldom possible. As also discussed by Kaneda *et al.* (2008), in the local structural and geomorphic setting the evidence for rupture along the causative fault might not be crystal-clear, and might also exhibit some misleading features. The tectonic offset at a number of places occurs at the base of range fronts and is accompanied by landsliding; the resulting surface deformation is not the true representation of fault displacement. We checked in the field and took into account these instances, and also carefully considered the local geological conditions, including the occurrence of ground subsidence, gravity slope deformation, lithological features, bedding orientation in sedimentary formations, fracture patterns and association of these features with the surrounding geomorphic features along the surface fault rupture (MSSP 2005). Precise fault displacement measurements are justified at only a few sites, because of the abrupt and high relief of the area, the presence along the KT fault of soft sedimentary rocks that may produce distributed displacement (leading to surface offset estimates lower than the true tectonic displacement) and significant landsliding (showing exaggerated rupture displacement). The thrust fault nature of the earthquake rupture rarely produces the appropriate surface evidence required to precisely estimate fault slip. At several locations, the best evidence of the earthquake rupture was the occurrence of very extensive surface fracturing, caused by the morphology of the thrust scarps. Very often, however, surface fracturing was only visible in the first weeks following the earthquake.

However, the field survey along the fault zone provided enough resolution to clearly differentiate the three major fault segments described in Figure 1 based on the average amount of surface displacement. The Muzaffarabad segment (Fig. 1) was characterized by almost continuous earthquake scarps, with average vertical offset on the order of 2–4 m. The Alai and Bagh segments (Fig. 1) are characterized by irregular surface faulting with small vertical displacement.

On the whole, we based our assessment of surface faulting on (A) mapping the rupture traces with solid geological and survey features along the fault zone, (B) comparison with macroseismic observations, (C) detailed intensity distribution in the epicentral area, and (D) field inspection of the local relations with the source fault. The occurrence of NW-trending surface cracks along the KT trace, accompanied by extremely severe intensity along narrow belts aligned on the fault hanging wall, and extensive landslides on the adjacent mountain slopes, as well illustrated at the Nisar Camp near Muzaffarabad (Figs 5c, d, 6, 7; see also Kaneda *et al.* 2008; and fig. 3 of Sato *et al.* 2007) has been considered as diagnostic evidence of tectonic surface rupture. It was very clear in the field that severe damage (I = X and XI in the Modified Mercalli Intensity (MMI) scale of Wood & Neumann 1931) and major primary and secondary environmental effects (essentially surface faulting, surface cracking and landslides) were occurring along a narrow belt, a few hundred metres wide, along the KT trace, and mostly concentrated in the fault hanging wall.

Typical geological surface features accompanying the coseismically reactivated segments of the KT (i.e. reverse fault scarps, pressure ridges and ground warps with extensional cracks on the fold crest, thrust-induced landslides, and intense surface fracturing) were observed at Bandi (Neelum Valley), Chatter Jhatian (Upper Jhelum Valley), Jabori (Siran Valley, NW of Balakot), Balakot City (Kunhar Valley), and Sagwal (Alai Valley) (Fig. 5).

The central segment (Fig. 1) includes the earthquake epicentre, and is characterized by the most important earthquake ground effects. Vertical displacement exceeding 4 m was observed at several sites between the Muzaffarabad and Balakot areas (Fig. 5). The maximum vertical displacement of more than 7 m was reported by Kaneda *et al.* (2008) at a site a few kilometres south of Balakot.

The areas of Bagh (Dhuli, Sudhan Gali; Fig. 1) and Alai Valley (Karg; Fig. 1) are identified as diffuse ends of the KT rupture. Along these segments of rupture termination the observed surface displacement is discontinuous, and limited to no more than few tens of centimetres. Our observations consistently show that surface faulting extended significantly NW of Balakot, where we observed surface cracking and severe damage (I = X MMI scale; Figs 5j, k and 6g, h).

Other ground effects

As pointed out by all early reports (MSSP 2005; EERI 2005, 2006; Peiris *et al.* 2006; Durrani *et al.* 2006), extensive landsliding was a particular feature of this event, and played a major role in the large economic losses and number of fatalities (Sato *et al.* 2007; Dunning *et al.* 2007; Owen *et al.* 2008). A white belt appeared during the mainshock along the slopes located on the KT surface rupture (Figs 5a, b, d, g, h, j, 6d). These barren slopes were the result of widespread, shallow disaggregated slides and rockfalls, affecting virtually all the lithologies along the KT rupture, and mostly within a few hundred metres from the trace of the surface rupture. Deep-seated landslides were much less common. Among them, however, two very significant ones were noted in Muzaffarabad (the Chela Bandi landslide in dolomitic limestone, 500 m high and 2 km long; Figs 5d, 6d) and in the Jhelum Valley (the 68×10^6 m^3 Hattian Bala rock avalanche; see EERI (2005) and Dunning *et al.* (2007) for a review of the data on this extremely large landslide).

As already mentioned, ground fracturing was also extremely common near the KT fault trace. Some of the fractures were clearly of tectonic nature, due to the extensional jointing on the top of fault-related fold scarps (such as at Nisar Camp; Fig. 5c, d). Others were clearly related to landslides and gravity slope deformation triggered by the surface rupture (such as those at Chatter Jhatian; Fig. 5e, f, g). However, in many cases there was no obvious, direct relation with the tectonic surface rupture or with landslides; in these instances ground fracturing must be related to violent ground shaking along the surface expression of the KT (Figs 5k, l, 6c).

The only observation of coseismic liquefaction available to our knowledge is described in Jayangondaperumal & Thakur (2008) near Jammu, at a site located *c.* 230 km SE of the epicentre; based on accounts from local witnesses, the occurrence of liquefaction should be related to the mainshock. This evidence would confirm the large area affected by secondary environmental effects.

Intensity distribution

The meizoseismal area of the Kashmir 2005 earthquake was assigned an intensity X on the MSK/EMS scales by Ahmed *et al.* (2006), EERI (2006) and Burton & Cole (2006*a*), and an intensity XI on the MMI/EMS scale by MSSP (2005) and Mahajan *et al.* (2006). According to GSP (2005), MMI XII was observed at the epicentre in Gori, 20 km from Muzaffarabad, X in Mansehra, Bagh and Rawalakot, IX in Batgram. Durrani *et al.* (2006) describe a tendency for intensity XI on the MMI scale at Muzaffarabad and Balakot, stating that 'absence of evidence' (for some indicators of MMI XI in the epicentral area, such as 'railroad tracks are badly bent') 'should not be construed as evidence of absence'. Therefore, the absence of railroads and collapsed bridges should not lead to underestimating the intensity in the near-field of this event. We agree with this statement. Taking into account the specific built environment in the area between Muzaffarabad and Balakot and the devastating scene observed at selected sites along the trace of surface rupture, we conclude that maximum intensity in the epicentral area should be assessed as XI MMI; the use of the MSK or EMS scales yields the same value, as discussed below in detail. The results of the extensive macroseismic survey conducted by the MSSP Team in a number of cities and towns around the epicentral area are presented in Table 2 and Figure 4. To our knowledge, this is the first intensity map that covers all the meizoseismal area and is based on direct field observation of a number of localities in the near-field. Mahajan *et al.* (2006) presented a regional intensity map showing results very similar to our assessment, but largely confined to a narrow conical sector extending from Balakot and

Fig. 5. (**a**) The geomorphic setting of the KT coseismic surface rupture near the town of Muzaffarabad (see location in Fig. 1). Note the barren white slopes marking the area of extensive shallow landslides along the young, fault-generated mountain slope in the KT hanging wall. Photo taken by MSSP on 15 October 2005. (**b**) Geomorphic setting of the KT coseismic surface rupture near the town of Muzaffarabad. Shallow and deep-seated landslides are typically associated with the surface rupture along the base of the mountain front. Photo taken by MSSP on 15 October 2005 (see Fig. 1 for location). (**c**) Nisar Camp, near Muzaffarabad (see location in Figs 1 and 6d); cracks in the road pavement along the trace of the KT surface rupture. Similar tension cracks have been typically observed close to the top of earthquake fold scarps such as the one at the lateral ramp shown in (d), where a coseismic surface displacement of 4.2 ± 0.5 m was observed. Damage at this site was extremely severe, as described by others (e.g. AIST 2006; Durrani *et al.* 2006; Peiris *et al.* 2006). Photo taken by MSSP on 15 October 2005. (**d**) Nisar Camp, north of Muzaffarabad, is the site where a surface displacement of 4.2 ± 0.5 m was measured; the double-pointed white arrow shows the vertical component of cumulative long-term surface offset. The WSW-trending tension cracks on the road (see detail in Fig. 5c) are clearly associated with the coseismic deformation of the fold ridge in the background; note the tilted trees on the fold scarp; this scarp formed along a lateral ramp of the main KT (e.g. AIST 2006; Pathier *et al.* 2006; Kaneda *et al.* 2008; see map view in Fig. 6d). Intensity XI in the MMI and ESI 2007 scale has been assessed at this site (see also Fig. 4). Note in the background the Chela Bandi landslide along the mountain front bounded by the KT (see also Fig. 6d), which partially dammed the Neelum River flow southward. Similar relations between surface cracks, tectonic surface deformation and displacement, landslides and extremely high intensity, were consistently observed along the whole trace of the

Fig. 5. (*Continued*) KT rupture shown in Figure 1. Photo taken by MSSP on 15 October 2005 (see Figs 1 and 6d for location). (**e**) Thrust-generated surface cracks parallel to the strike of Kashmir Thrust in Chatter Jhatian, Muzaffarabad region, most likely related to deep-seated slope gravity deformation triggered by the coseismic surface rupture. Photo taken by MSSP on 20 October 2005 (see Fig. 1 for location). (**f**) Thrust-generated surface cracks and ground displacement along the strike of Kashmir Thrust in Chatter Jhatian, Muzaffarabad region, most likely related to slope gravity deformation triggered by the coseismic surface rupture. Photo taken by MSSP on 20 October 2005 (see Fig. 1 for location). (**g**) Thrust-generated sliding along the Kashmir Thrust coseismic rupture between Muzaffarabad and the Balakot region. Photo taken by MSSP on 14 October 2005 (see Fig. 1 for location). (**h**) Trace of the coseismic thrust scarp along the KT between Muzaffarabad and the Balakot region. Photo taken by MSSP on 14 October 2005 (see Fig. 1 for location). (**i**) Vertical offset of *c*. 2.0 m across the KT surface rupture in the Balakot City area. Photo taken by MSSP on 19 November 2005 (see Figs 1 and 6a for location). (**j**) Thrust-generated sliding along the strike of Kashmir Thrust in Baso Village, about 23 km NW of Balakot. Photo taken by MSSP on 28 November 2005 (see Fig. 1 for location). (**k**) NW-trending, thrust-related surface cracks parallel to the strike of the KT in Alai City, in the area of the NW termination of the surface rupture. Photo taken by MSSP on 27 October 2005 (see Fig. 1 for location). (**l**) Thrust-generated surface cracks parallel to the strike of Kashmir Thrust in Dhuli (Bagh region). Photo taken by MSSP on 19 October 2005 (see Fig. 1 for location).

Fig. 6. (**a**) Satellite image of Balakot after the earthquake (image from http://www.digitalglobe.com; modified after EEFIT 2006), and damage and coseismic surface rupture observations along the main trace of the KT (dashed yellow line); collapsed or severely damaged commercial and residential areas are concentrated along a narrow belt in the KT hanging wall; vertical displacement along the KT surface rupture ranges from 1.8 to *c*. 4.0 m in this area (see Fig. 1 for location). (**b**) Collapsed and heavily damaged masonry residential units on a hill, NW of Balakot city centre (see location in Fig. 6a), that is in the hanging wall of the KT; surface faulting (Fig. 5i) and tension cracks (Fig. 6c) run along the base of the hill; intensity XI on the MSK scale has been assessed at this site. Comparatively minor damage occurred in the footwall of the fault, a few hundreds metres from the fault trace, as indicated in (a). Photo taken by the EEFIT Team on 23 November 2005 (Peiris *et al.* 2006). (**c**) Ground cracks associated with the surface rupture of the KT in Balakot, extending from the road toward the hill west of Kunhar river (see location in a). Photo taken by the EEFIT Team on 23 November 2005 (Peiris *et al.* 2006). (**d**) Satellite image of Muzaffarabad city (image from http://www.digitalglobe.com) taken after the earthquake showing the location of damage observations in (e) and (f); note also the continuous belt of landslides (white slopes) triggered by coseismic surface faulting (yellow arrows mark some detail of the mapped surface ruptures)

Fig. 6. (*Continued*) along the Kashmir Thrust (the dashed yellow line shows the main NW trend of the KT), and the WSW-trending lateral ramp (red arrows) in Nisar Camp, Muzaffarabad region, where maximum displacement was measured (Fig. 5d; see also Pathier *et al.* 2006; Kaneda *et al.* 2008). (**e**) Damage related to fault displacement in the KT footwall, a few kilometres north of Muzaffarabad (see location in d). The KT rupture is near the newly exposed white area of limestone landslide; the reinforced concrete columns lean out and right from the photograph as the foundation moved inwards and left towards the fault. Photo taken by EEFIT Team on 24 November 2005 (see also Burton & Cole 2006b). (**f**) Damage related to landslide in the alluvial deposits of Muzaffarabad City (see location in d); the house was buried by the gravel and soil that slid down from the slope. The sliding also created instability in the residential building at the top of the slope and a risk of damage to the commercial building at road level. This is an example of a small landslide in the KT fault footwall; much larger landslides developed in the hanging wall, as shown in (d). Landslides were a major secondary hazard clearly related to the KT rupture, as discussed by Sato *et al.* (2007); about one-quarter of the 8 October 2005 casualties resulted from coseismic landslides (e.g. Dunning *et al.* 2007). Photo taken by EEFIT

Fig. 6. (*Continued*) Team on 22 November 2005 (Peiris *et al.* 2006). (**g**) Earthquake–induced landslide in the Alai–Batgram area. Photo taken by MSSP on 27 October 2005 (see Fig. 1 for location). (**h**) Detail of the damaged buildings at the site shown in (g). Photo taken by MSSP on 27 October 2005 (see Fig. 1 for location).

Muzaffarabad to epicentral distances over 600 km in the SE direction, that is towards India.

As clearly shown in Figures 4 and 7, intensity levels were observed to be remarkably high (up to XI on MMI) along the trace of the KT, and decreased rapidly perpendicular to the strike of the causative fault. This may be due to near-rupture and/or rupture directivity effects (Archuleta & Hartzell 1981; Somerville *et al.* 1997; Mahmood *et al.* 2004; EERI 2006; Fig. 6e). One main factor that

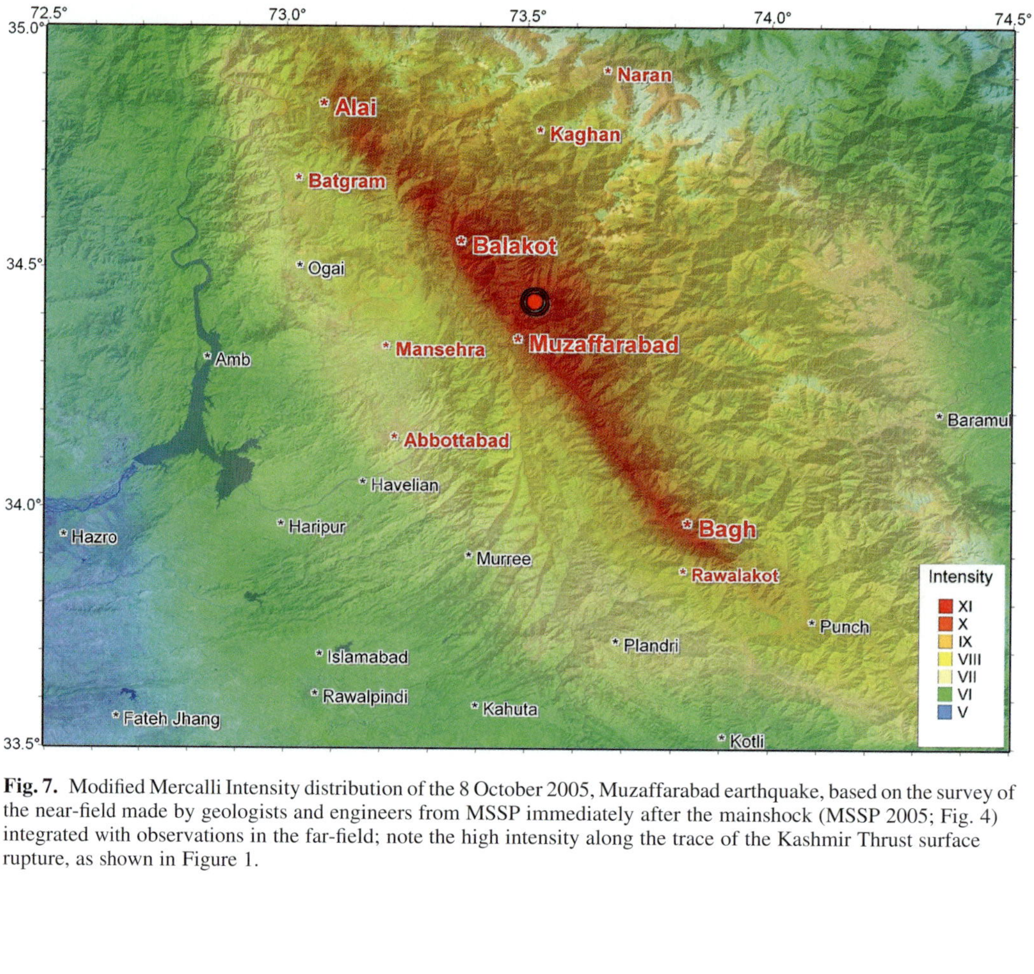

Fig. 7. Modified Mercalli Intensity distribution of the 8 October 2005, Muzaffarabad earthquake, based on the survey of the near-field made by geologists and engineers from MSSP immediately after the mainshock (MSSP 2005; Fig. 4) integrated with observations in the far-field; note the high intensity along the trace of the Kashmir Thrust surface rupture, as shown in Figure 1.

certainly increased the level of damage along the fault rupture zone was the occurrence of the very dense, high-frequency band of landslides and ground failure, already mentioned (Fig. 6f, g). Also, the sites on the hanging wall very close to the KT plane retain relatively high intensities as compared to those at comparable distances on the footwall (Figs 5d, 6a, d). The tectonic mechanism of the earthquake is consistent with the considerable differences in intensity distribution and corresponding strong ground motion along the fault trace, that is greater at the sites on the hanging wall of KT than at those on the footwall, as observed for instance by Abrahamson & Somerville (1996) and Yu & Gao (2001) during similar thrust faulting earthquakes in Northridge, California, and Chi-Chi, Taiwan, respectively.

In particular, intensity XI MMI was assigned, after detailed consideration, at the site of remarkable surface faulting in Nisar Camp, in the Muzaffarabad neighbourhood a few kilometres north of the city centre (Fig. 5d). Here, the total collapse of buildings in the fault hanging wall (fault-generated ridge to the left of Fig. 5d) may be indicative of XI in the MMI, MSK and EMS scales (e.g. Mahajan *et al.* 2006), depending on the state of vulnerability of the damaged and collapsed buildings. These buildings, although nominally reinforced concrete, were not adequately engineered, and their vulnerability class was probably more like type B than type C; general collapse resulting in damage grade 5 is obvious. Damage grade 5 for class B buildings is indicative of intensity X on the MSK scale if there are 'many' examples of grade 5, 'many' being defined as 20–50%. The scale is not explicit with respect to 'most' or $\geq$60% examples of grade 5 damage to B types, but it is implicit of XI MSK. However, if these buildings were as weakly constructed as class A buildings then collapse of most is indicative of X MSK. If the buildings were of class B, or better, e.g. a reasonable standard reinforced concrete, then there can be no doubt of XI MSK based on local building damage evidence. Damage grade 5 for class B buildings is indicative of intensity XI in the EMS scale (Grünthal 1998). Nisar Camp is on the alluvial terrace of the west bank of the Neelum River, mainly composed of rounded fluvial deposits. Heavy roofs collapsed to ground level can be seen on the sloping edge to this ridge; the ridge crest was affected by extensive tensional cracking. To the right of this ridge, buildings were left standing although heavily damaged (intensity VIII or IX MSK/EMS). To Burton & Cole (2006), this suggested amplification due to ridge effects and landsliding on the ridge slope; they assessed intensity X MSK/EMS at this site on evidence of damage to weak buildings alone. However, subsequent investigation clearly demonstrated that the ridge itself was heavily affected by thrust faulting at the ground surface, without occurrence of landslides. Therefore, the total collapse of buildings along the Nisar Camp ridge should be related to extremely severe shaking in the fault hanging wall, as also noted by others (EERI 2006; Peiris *et al.* 2006; Kaneda *et al.* 2006; AIST 2006). For these reasons, we confirm the MSSP (2005) assessment of intensity XI MMI at this site. Likewise, more than 4 m of vertical displacement accompanied by strong fracturing is consistent with the description of effects on nature given in the MSK (1980) scale for intensity XI: 'Ground considerably distorted by broad cracks and fissures, as well as by movement in horizontal and vertical directions; numerous land slips and falls of rock'. This feature was not diagnosed in this way at the time; the role of coseismic surface faulting was not commonly recognized immediately after the earthquake (e.g. Durrani *et al.* 2006; EERI 2006; see Kaneda *et al.* (2008) for the complete story of the recognition of surface faulting as a very important feature of the 8 October 2005, Muzaffarabad earthquake).

The overall weight of combined evidence, damage to buildings, building vulnerability, and emplacement of buildings on a thrust faulted surface, together are consistent with intensity XI on the Nisar Camp ridge.

The same relations have been observed in Balakot, where the total collapse of buildings, fault displacement and fracturing occurred on the ridge in the NW part of the town (Figs 5i, 6a–c). Therefore, we assessed intensity XI MMI also at this site. Intensity X in the MMI scale was recorded in Bagh and in Alai, which represent the two closest localities to the NW and SE surface rupture terminations, respectively. Bagh is located about 45 km from the epicentre (Figs 1, 3). The KT rupture passes about 2 km north of the city centre. Severe damage to concrete structures and total collapse of loose masonry construction was observed (MSSP 2005; Mahajan *et al.* 2006). Alai is a sparsely populated township situated in the Alai Valley, about 55 km NW of the epicentre. Both the directivity effects due to the northwestward rupture propagation and the large number of aftershocks contributed to the heavy damage recorded in this area (Fig. 6g, h).

Figure 4 and Table 2 summarize our observations on a number of other sites in the near-field. The detailed description of intensity data at each site is beyond the scope of this paper; likewise, the scale of the figure does not allow us to represent the large variations of intensity at each site, due to fault hanging wall effects and other local factors. However, the objective of mapping the intensity distribution at the scale of the whole epicentral area is to allow an easy comparison between Figure 1 and Figure 7. This makes one of the main points of this paper, that is the close relations between intensity and surface faulting during the Muzaffarabad earthquake, a point that was very clear to all the authors working in the field immediately after the mainshock, but has not been properly documented in the literature until now.

Epicentral intensity has also been evaluated using the ESI 2007 scale, based only on earthquake environmental effects (Michetti *et al.* 2004, 2007). Rupture length on the order of 80 to 100 km, and maximum surface displacement on the order of 4 to 7 m are indicative of ESI 2007 epicentral intensity XI. The total area affected by the earthquake and displaying significant primary and secondary effects is estimated to be about 9400 km^2 (MSSP 2005; Vinod Kumar *et al.* 2006; Sato *et al.* 2007) which is in agreement with ESI 2007 epicentral intensity XI. A detailed survey for assessing ESI 2007 intensity at each relevant site in the region affected by the earthquake has not been conducted. However, the available observations consistently confirm that ESI 2007 intensity XI can be assigned at several locations in the epicentral area, based on both maximum surface displacement and frequency and size of large landslides.

Discussion and conclusions

The magnitude Ms = 7.7 assigned to this event together with the rupture on the central segment of the fault (52 km in length, with *c.* 4 m average displacement) that can be supposed to account for a major portion of the total energy release and the thrust fault mechanism, all imply a high stress drop (>100 bars) that would generate very high accelerations in the meizoseismal area (cf. Mohammadioun & Serva 2001; Singh *et al.* 2006; Raghukanth 2008).

The intensity distribution and respective strong ground motions generated by the mainshock were a function of the three-dimensional spatial geometry of its causative fault, that is the Kashmir Thrust. It was found that the ground shaking was not distributed symmetrically around the epicentre, but rather was intense along the strike of the KT and in an area on the hanging wall, mainly due to fault rupture directivity and near-fault effects, respectively. Therefore, in addition to the seismic potential and tectonic structure of an area, understanding the spatial spreading and orientation of seismogenic structures constitutes an essential consideration in seismic hazard assessment.

One of the most important lessons learned from the 8 October 2005, Muzaffarabad earthquake is the very close relationship between macroseismic intensity, location and amount of surface faulting, distribution and magnitude of landslides, and density of ground fracturing. Intensity XI can be assessed at several sites near Muzaffarabad and Balakot where the maximum vertical displacement along the Kashmir Thrust surface rupture has been measured. Once a detailed field survey is available, this assessment of epicentral intensity is independent of the adopted macroseismic scale. Degree XI can be measured at specific sites between Muzaffarabad and Balakot using either the MSK, MMI or EMS scale.

Zones where intensity was remarkably severe, such as in the town of Balakot and at Nisar Camp north of Muzaffarabad city centre, were either right on the KT surface rupture or very close to it. Clearly a relevant component of damage along the belt affected by surface faulting is due to primary ground displacement or deformation, and secondary ground failure and landsliding. However, most of the damage was obviously related to extremely intense ground shaking. In the major urban areas, the percentage of house collapse is systematically greater near the fault than in the surrounding areas. In poorly inhabited areas, the density of shallow disrupted rock slides can be used instead as a valuable indicator of the ground shaking level. Of the 2424 landslides identified by Sato *et al.* (2006) along the central segment of the KT (Fig. 1), 75% were small slides, mostly located on the KT hanging wall; and more than one-third of the landslides occurred within 1 km from the KT fault trace. Small, shallow rock slides are primarily triggered by the violent ground shaking, and not by the fault displacement. A similar reasoning holds for the degree of ground fracturing along the KT rupture. These findings strongly suggest that in the near-field of strong earthquakes the environmental effects can be effectively used as intensity diagnostics. This is consistent with the original definition of the MCS (Mercalli–Cancani–Sieberg; Sieberg 1912), MM (Modified Mercalli; Wood & Neumann 1931; Richter 1958) and MSK (Medvedev–Sponheuer–Karnik; Sponheuer & Karnik 1964) scales, and represents the foundation for the new ESI 2007 scale (Michetti *et al.* 2004, 2007).

We are grateful to Bagher Mohammadioun and Koji Okumura for constructive reviews of the manuscript, that greatly improved the quality of this paper. The fieldwork of P.W.B. with the EEFIT team was funded by the Engineering and Physical Sciences Research Council, UK.

References

ABRAHAMSON, N. A. & SOMERVILLE, P. G. 1996. Effects of the hanging wall and footwall on ground motion recording during the Northridge Earthquake. *Bulletin of the Seismological Society of America*, **86**, S93–S99.

AHMED, M. A., ULLAH, A. & NAZIRULLAH, R. 2006. Seismic intensities and earthquake hazard zones of

Pakistan. *In*: KAUSAR, A. B., KARIM, T. & KHAN, T. (eds) *International Conference on 8 October 2005 Earthquake in Pakistan: Its Implications & Hazard Mitigation*, 18–19 January 2006, Islamabad, 121–124.

AIST 2006. *On-Site Survey Identifies Surface Fault associated with the 2005 Pakistan Earthquake*. Translation of AIST press release on 3 February 2006, National Institute of Advanced Industrial Science and Technology, Japan (available at: http://www.aist.go.jp/aist_e/latest_research/2006/20060222/20060222.html).

ARCHULETA, R. J. & HARTZELL, S. H. 1981. Effects of fault finiteness on near ground motion. *Bulletin of the Seismological Society of America*, **71**, 939–957.

ARMBRUSTER, J., SEEBER, L. & JACOB, K. H. 1978. Northwestern termination of Himalayan mountain front—active tectonics from microearthquakes. *Journal of Geophysical Research*, **83**, 269–282.

AVOUAC, J. P., AYOUB, F., LEPRINCE, S., KONCA, O. & HELMBERGER, D. V. 2006. The 2005, Mw 7.6 Kashmir earthquake: Sub-pixel correlation of ASTER images and seismic waveforms analysis. *Earth and Planetary Science Letters*, **249**, 514–528.

BENDICK, R., BILHAM, R., KHAN, M. A. & KHAN, S. F. 2007. Slip on an active wedge thrust from geodetic observations of the 8 October 2005 Kashmir earthquake. *Geology*, **35**(3), 267–270.

BURTON, P. W. & COLE, S. 2006*a*. Pakistan earthquake 2005 October 8: Earthquake expectation and seismic hazard through to impact. *Proceedings of the First European Conference on Earthquake Engineering and Seismology*, Geneva, Switzerland, 3–8 September 2006, Paper Number 811.

BURTON, P. W. & COLE, S. 2006*b*. Two contrasting earthquakes and development towards seismic hazard mapping: 2005 October 8, Kashmir and 2006 May 26, Java. *Proceedings of the 3rd International Symposium and Exhibition on Earth Resources and Geological Engineering Education*, 72–80.

CALKINS, J. A., OFFIELD, T. W., ABDULLAH, S. K. M. & ALI, S. T. 1975. *Geology of the southern Himalaya in Hazara, Pakistan and adjacent areas*. US Geological Survey Professional Paper, **716-C**, U.S. Government Printing Office, Washington, D.C.

COMET 2005. *Locating the Kashmir Fault*. Centre for the Observation and Modelling of Earthquakes and Tectonics (available at: http://comet.nerc.ac.uk/news_kashmir.html).

DUNNING, A., MITCHELL, W. A., ROSSER, N. J. & PETLEY, D. N. 2007. The Hattian Bala rock avalanche and associated landslides triggered by the Kashmir Earthquake of 8 October 2005. *Engineering Geology*, **93**(3-4), 130–144.

DURRANI, A. J., ELNASHAI, A. S., HASHASH, Y. M. A., KIM, S. J. & MASUD, A. 2006. The Kashmir earthquake of October 8, 2005: A quick look report. Mid-America Earthquake Center, Report **05-04**.

EERI 2005. The Kashmir Earthquake of October 8, 2005: Impacts in Pakistan. *EERI Newsletter*, December, **39**(12) (available at: http://www.eeri.org/lfe/pdf/kashmir_eeri_1st_report.pdf).

EERI 2006. The Kashmir Earthquake of October 8, 2005: Impacts in Pakistan. *EERI Newsletter*, February, **40**(2) (available at: http://www.eeri.org/lfe/pdf/kashmir_-eeri_2nd_report.pdf).

FUJIWARA, S., TOBITA, M. *ET AL*. 2006. Satellite data gives snapshot of the 2005 Pakistan earthquake. *EOS Transactions (AGU)*, **87**, 73–77.

GRÜNTHAL, G. (ed.) 1998. *European Macroseismic Scale 1998 EMS-98*. Conseil de l'Europe, Cahiers du Centre Europeen de Geodynamique et de Seismologie, **15**, Luxembourg.

GSP 2005. *October 08 2005 Earthquake*. Geological Survey of Pakistan, GSP Online (available at: http://www.gsp.gov.pk/gsps_role.html).

JAYANGONDAPERUMAL, R. & THAKUR, V. C. 2008. Co-seismic secondary surface fractures on southeastward extension of the rupture zone of the 2005 Kashmir earthquake. *Tectonophysics*, **446**, 61–76.

KANEDA, H., AWATA, Y. *ET AL*. 2006. Extensive surface fault rupture associated with the 2005 Mw 7.6 Pakistan earthquake. *AOGS 3rd Annual General Meeting*, 10–14 July 2006, Asia Oceania Geosciences Society, Singapore, Abstracts, 922–923, (available at: http://www.asiaoceania.org/pdf2006/se/59-SE-A1046.pdf).

KANEDA, H., NAKATA, T. *ET AL*. 2008. Surface rupture of the 2005 Kashmir, Pakistan, earthquake and its active tectonic implications. *Bulletin of the Seismological Society of America*, **98**(2), 521–557. DOI: 10.1785/0120070073.

KAZMI, A. H. & JAN, M. Q. 1997. *Geology and Tectonics of Pakistan*. Graphics Publishers, Karachi.

KUMAHARA, Y. & NAKATA, T. 2006. *Active faults in the epicentral area of the 2005 Pakistan earthquake*. Research Center for Regional Geography, Hiroshima University, Japan.

KUMAR, S., WESNOUSKY, S. G., ROCKWELL, T. K., BRIGGS, R., THAKUR, V. C. & JAYANGONDAPERUMAL, R. 2006. Paleoseismic evidence of great surface-rupture earthquakes along the Indian Himalaya. *Journal of Geophysical Research*, **111**. DOI: 10.1029/2004JB003309.

MAHAJAN, A. K., KUMAR, N. & ARORA, B. R. 2006. Quick look isoseismal map of 8 October 2005 Kashmir earthquake. *Current Science*, **91**(3), 356–361.

MAHMOOD, T., QAISAR, M. & ALI, Z. 2002. Source mechanism of Astor Valley Earthquake of November 20, 2002 inferred from teleseismic body waves. *Geological Bulletin of the University of Peshawar*, **35**, 151–161.

MAHMOOD, T., QAISAR, M. & ALI, Z. 2004. *Intensity distribution and impact of Kaghan Valley (Pakistan) Earthquake February 14, 2004*. Micro Seismic Studies Programme, MSSP Internal Report, **MSSP-73/2004**, P.O. Nilore, Islamabad.

MANDAL, P., CHADHA, R. K., KUMAR, N., RAJU, I. P. & SATYAMURTY, C. 2007. Source parameters of the deadly M_w 7.6 Kashmir earthquake of 8 October, 2005. *Pure and Applied Geophysics*, DOI: 10.1007/s00024-007-0258-8.

MICHETTI, A. M., ESPOSITO, E. *ET AL*. 2004. *The INQUA scale: An innovative approach for assessing earthquake intensities based on seismically-induced ground effects in the environment*. Servizio Geologico d'Italia – Dipartimento Difesa del Suolo, APAT,

Roma, Italy, Special Paper, Memorie Descrittive della Carta Geologica d'Italia, **67**.

MICHETTI, A. M., ESPOSITO, E. *ET AL.* 2007. *Environmental Seismic Intensity Scale 2007–ESI 2007.* Servizio Geologico d'Italia – Dipartimento Difesa del Suolo, APAT, Roma, Italy, Memorie Descrittive della Carta Geologica d'Italia, **74**, 7–54.

MOHAMMADIOUN, B. & SERVA, L. 2001. Stress drop, slip type, earthquake magnitude- and seismic hazard. *Bulletin of the Seismological Society of America*, **91**(4), 694–707.

MSK 1980. After: Report on the Ad-hoc Panel Meeting of Experts on Up-dating of the MSK-64 Seismic Intensity Scale, Jena (GDR), 10–14 March 1980. *Gerlands Beitrage Geophysit*, **90**(3), 261–268.

MSSP 2005. *Muzaffarabad earthquake of October 8, 2005.* Micro Seismic Studies Programme, Ishfaq Ahmad Research Laboratories, Pakistan Atomic Energy Commission, MSSP Internal Report, **MSSP-76/2005**, P.O. Nilore, Islamabad.

NAKATA, T. & KUMAHARA, Y. 2006. *Active Faults of Pakistan with Reference to the Active Faults in the Source Area of the 2005 North Pakistan Earthquake. In*: KAUSAR, A. B. & KHAN, T. K. T. (eds) *International Conference on 8 October 2005 Earthquake in Pakistan: Its Implications & Hazard Mitigation*, 18–19 January 2006, Geological Survey of Pakistan, Islamabad, Extended Abstracts, 18–22.

NAKATA, T., TSUTSUMI, H., KHAN, S. H. & LAWRENCE, R. D. 1991. *Active faults of Pakistan.* Map sheets and inventories. Hiroshima University Research Centre for Regional Geography, Special Publication **21**.

OWEN, L. A., KAMP, U., KHATTAK, G. A., HARP, E. L., KEEFER, D. K. & BAUER, M. A. 2008. Landslides triggered by the 8 October 2005 Kashmir earthquake. *Geomorphology*, **94**, 1–9.

PARSONS, T., YEATS, R. S., YAGI, Y. & HUSSAIN, A. 2006. Static stress change from the 8 October, 2005 M = 7.6 Kashmir earthquake. *Geophysical Research Letters*, **33**, L06304. DOI: 10.1029/2005GL025429, 2006.

PATHIER, E., FIELDING, E. J., WRIGHT, T. J., WALKER, R. & PARSONS, B. E. 2006. Displacement field and slip distribution of the 2005 Kashmir earthquake from SAR imagery. *Geophysical Research Letters*, **33**, L20310. DOI: 10.1029/2006GL027193, 2006.

PEIRIS, N., ROSSETTO, T., BURTON, P. & MAHMOOD, S. 2006. *EEFIT Mission: October 8, 2005 Kashmir Earthquake, Preliminary Report.* Earthquake Engineering Field Investigation Team (EEFIT) (available at: http://www.istructe.org.uk/eefit/files/EEFIT%20Mission%20Pakistan%20-%20prelim%20report.pdf).

RAGHUKANTH, S. T. G. 2008. Ground motion estimation during the Kashmir earthquake of 8th October 2005. *Natural Hazards*, **46**, 1–13. DOI: 10.1007/s11069-007-9178-2.

RAO, N. P., KUMAR, P., KALPNA, T., TSUKUDA, E. & RAMESH, D. S. 2006. The devastating Muzaffarabad earthquake of 8 October 2005: New insights into Himalayan seismicity and tectonics. *Gondwana Research*, **9**, 365–378.

RICHTER, C. F. 1958. *Elementary Seismology.* Freeman, San Francisco.

SATO, H. P., HASEGAWA, H., FUJIWARA, S., TOBITA, M., KOARAI, M., UNE, H. & IWAHASHI, J. 2007. Interpretation of landslide distribution triggered by the 2005 Northern Pakistan earthquake using SPOT 5 imagery. *Landslides*, **4**, 113–122. DOI: 10.1007/s10346-006-0069-5, 2007.

SEEBER, L. & ARMBRUSTER, J. G. 1979. Seismicity of the Hazra arc in northern Pakistan: Decollement vs. basement faulting. *In*: FARAH, A. & DEJONG, K. A. (eds) Geodynamics of Pakistan. *Geological Survey of Pakistan, Quetta*, 131–142.

SIEBERG, A. 1912. Über die makroseismische Bestimmung der Erdbebenstärke. *Gerlands Beitrage Geophysik*, **11**, 227–239.

SINGH, S. K., IGLESIAS, A. *ET AL.* 2006. Muzaffarabad earthquake of 8 October 2005 (Mw 7.6): A preliminary report on source characteristics and recorded ground motions. *Current Science*, **91**, 689–695.

SOMERVILLE, P. G., SMITH, N. F., GRAVES, R. W. & ABRAHAMSON, N. A. 1997. Modification of empirical strong ground motion attenuation relation to include the amplitude and duration effects of rupture directivity. *Seismological Research Letters*, **68**, 199–222.

SPONHEUER, W. & KARNIK, V. 1964. Neue seismische Skala. *In*: SPONHEUER, W. (ed.) *Proceedings of 7th Symposium of the ESC*, Jena, 24–30 September 1962. Veröffentlichungen des Institutes für Bodendynamik und Erdbebenforschung in Jena, Deutsche Akademie der Wissenschaften zu Berlin, **77**, 69–76.

TAPPONNIER, P., KING, G. & BOLLINGER, 2006*a*. *Active Faulting and Seismic Hazard in the Western Himalayan Syntaxis, Pakistan. In*: KAUSAR, A. B., KARIM, T. & KHAN, T. (eds) *International Conference on 8 October 2005 Earthquake in Pakistan: Its Implications & Hazard Mitigation*, 18–19 January 2006, Islamabad, Extended Abstracts, 6–7.

TAPPONNIER, P., KING, G., BOLLINGER, L. & GRASSO, J. 2006*b*. Surface faulting during the October 8th, 2005, Muzaffarabad earthquake and Coulomb stress increase on the Jhelum Fault. *BSSA 2006 Annual Meeting*, San Francisco, California, 18–22 April 2006, Abstract (available at http://www3.seismosoc.org/abstracts/).

VINOD KUMAR, K., MARTHA, T. R. & ROY, P. S. 2006. Mapping damage in the Jammu and Kashmir caused by 8 October 2005 Mw 7.3 earthquake from the Cartosat–1 and Resourcesat–1 imagery. *International Journal of Remote Sensing*, **27**(20), 4449–4459.

WAYNE, P. D. 1979. *A summary of field and seismic observations of the Pattan Earthquake – 28 December 1974.* Department of Geology and Geophysics, University of Wisconsin.

WOOD, H. O. & NEUMANN, F. 1931. Modified Mercalli intensity scale of 1931. *Bulletin of the Seismological Society of America*, **21**, 277–283.

YEATS, R. S., NAKATA, T., FARAH, A., MIZRA, M. A., PANDEY, M. R. & STEIN, R. S. 1992. The Himalayan frontal fault system. *Annals of Tectonics*, **VI**, 85–98.

YU, Y. X. & GAO, M. T. 2001. Effects of the hanging wall and footwall on peak acceleration during the Jiji (Chi-Chi), Taiwan Province, earthquake. *Acta Seismologica Sinica*, **14**(6), 654–659.

Stress change over short geological time: the case of Scandinavia over 9000 years since the Ice Age

SOREN GREGERSEN* & PETER VOSS

Geological Survey of Denmark and Greenland (GEUS), Oster Voldgade 10, DK-1350 Copenhagen K, Denmark

**Corresponding author (e-mail: sg@geus.dk)*

Abstract: Palaeoseismological investigations are used in many regions of the world to extend back in time the earthquake statistics of historical written or oral records as well as instrumental information. This is very valuable for discussions of earthquake hazard, but it only applies to areas of stable stress regime. Although the intraplate areas of Scandinavia and Greenland have experienced only rather small earthquakes within the human timescale, they serve as a clear warning on the application of palaeoseismology for hazard studies in regions where the stresses have changed.

In a small part of Scandinavia, where recent earthquake activity is not significantly different from that of its surroundings, large faults have been discovered and several have been investigated via palaeoseismology. They are interpreted to show the occurrence of large earthquakes about 9000 years ago. Signs of this are coincident landslides as well as liquefaction in loose sediments, which are well dated through varve-counting.

In contrast to this the present-day stress release in earthquakes and in surface rock deformations is mainly caused by plate motion. Regional investigations in Scandinavia and Greenland/North America, as well as those included in the World Stress Map Project of the 1990s, have shown compression within the plate, mainly in the direction of absolute plate motion. The ice cap influence has disappeared. So stress reorganization is clearly indicated over the short geological timespan of 9000 years. Into this argument goes the observation from Greenland and Antarctica, that no earthquakes occur under the ice caps. For Scandinavia the argument is that no earthquakes occurred under the ice sheet during the Ice Age, and that the stored stresses were released when the ice sheet melted 9000 years ago. This does emphasize a warning. There are regions of the globe where palaeoseismological investigations can give a fantastic extension of the short-term historical earthquake records. But in some regions stress reorganization has changed this condition.

Intraplate earthquake regime is the general term for areas such as Scandinavia and Greenland discussed in the present paper. There is good reason for treating these areas together, apart from the fact that the Geological Survey of Denmark and Greenland (GEUS) has seismograph networks in both areas. Ice cap loads and the resulting lithospheric stresses are or have been important for both areas. For both areas it has been discussed whether the causes of the earthquakes are plate tectonics motions, or changes or disappearance of ice cap load (Gregersen 1992, 2002; Ekman 1993; Arvidsson 1996; Chung & Gao 1997; Chung 2002). For Scandinavia the ice cap of the Ice Age left the area about 9000 years ago. For Greenland the shrinking of the ice cap and the ensuing uplift ended about 5000–6000 years ago, except on the small scale. The recent phenomenon of the shrinking ice cap in Greenland has only local influence on the lithosphere.

Discussion on Scandinavian stresses from plate tectonics versus post-glacial uplift

The earthquake geography in Scandinavia is displayed in Figure 1. Like displays of earthquake activity in other parts of the world it does not matter much which is the time period displayed; the general pattern is the same in any time period, although small details can be different. The figure shows that the earthquake activity is most concentrated along the Norwegian coast and continental margin, along the Swedish east coast, and in and around the Oslo Fjord. In more detailed compilations it is well established that the earthquake activity in Denmark is the southern limit of Scandinavian seismicity (Gregersen *et al.* 1998). Figure 1 shows lower earthquake activity in northern Germany, in Poland and in Estonia, Latvia and Lithuania. In the latter areas the seismograph coverage has until recently been significantly poorer, so

From: REICHERTER, K., MICHETTI, A. M. & SILVA, P. G. (eds) *Palaeoseismology: Historical and Prehistorical Records of Earthquake Ground Effects for Seismic Hazard Assessment*. The Geological Society, London, Special Publications, **316**, 173–178. DOI: 10.1144/SP316.10 0305-8719/09/$15.00

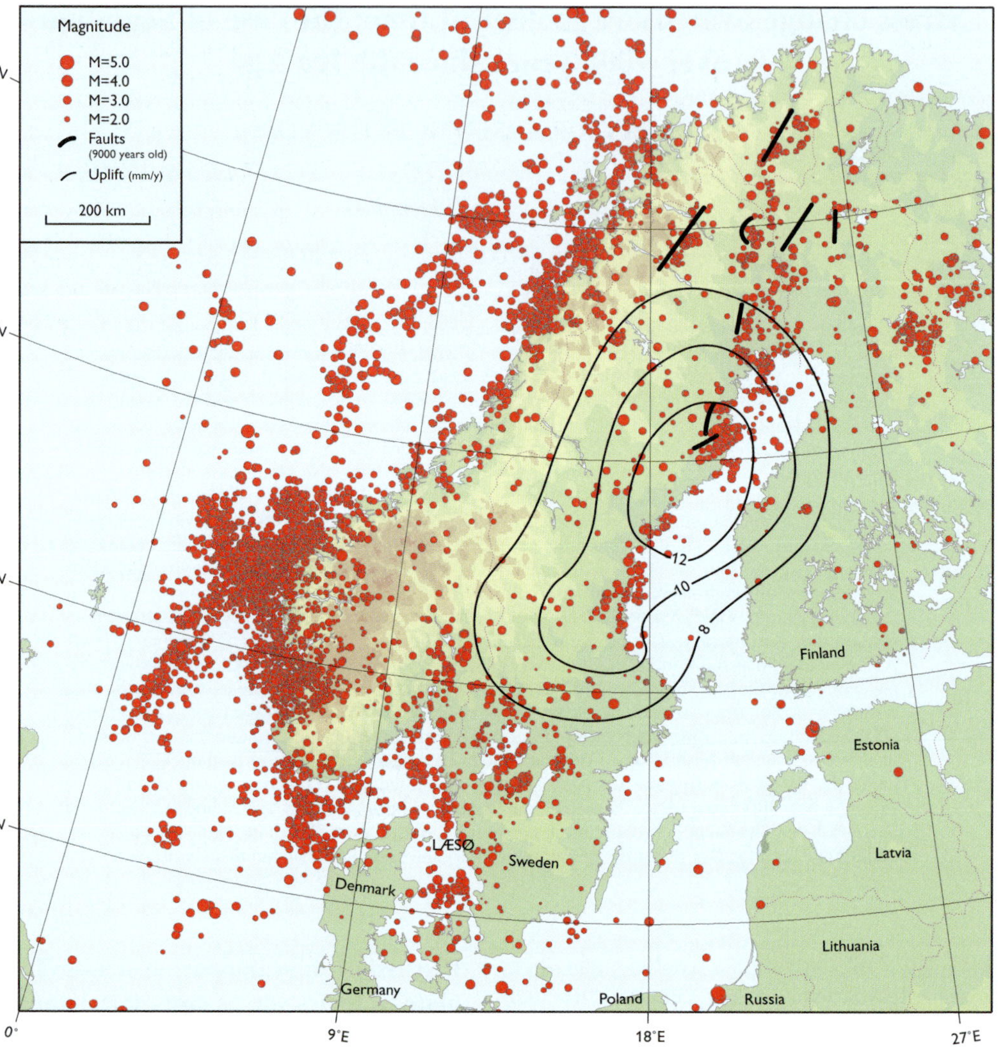

Fig. 1. Earthquake map of Scandinavia for the years from January 1970 to December 2004. Earthquakes in the Danish area are extracted from the GEUS earthquake catalogue and earthquakes located outside this area are extracted from the Scandinavian catalogue from Helsinki University. The earthquake magnitude scale is given in the upper left corner. The contours show the area of maximum post-glacial uplift (Scherneck *et al.* 2001). Very thick black lines show large faults with an age close to 9000 years (Lagerbäck 1991).

the information in the map is influenced by less sensitivity to small earthquakes.

Also shown in Figure 1 are short thick lines in northern Norway, Sweden and Finland. These show the locations of large post-glacial faults, which are convincingly argued to have developed in large earthquakes (e.g. Lagerbäck 1991; Olesen 1991). The dating of these earthquakes is very accurate via disturbances of sediments in liquefaction, and counting of varve layers. Coincident with the liquefaction phenomena are large landslides, which support the interpretation of these large faults as signs of earthquakes 9000 years ago.

Figure 1 also shows present-day uplift contours determined via GPS (Scherneck *et al.* 2001). Neither the specific area of largest uplift, nor the zone around it of the shape of the uplift area stand out with special earthquake activity. This has been recently elaborated in a paper by Gregersen (2002). Also the updated illustration in Figure 1 shows significant differences in earthquake geography 9000 years ago and now. We argue that this shows a dramatic change of the stress pattern: 9000 years ago the stresses were determined by the latest ice locations in the area of the large post-glacial faults of Figure 1. The influence of the ice cap is discussed in the next section of this paper.

It was shown by Slunga (1989) for the Baltic Shield, and emphasized by Gregersen (1992) for Scandinavia, that the dominant stress field nowadays is compression in the orientation NW–SE. This is in agreement with the results of the World Stress Map Project (Zoback *et al.* 1989) showing a worldwide pattern of compressional stress in the interiors of the lithospheric plates along the direction of the absolute plate motion. For Scandinavia this is the abovementioned NW–SE direction. This pattern is shown in Figure 2, which is a recent printout of the file of the continuously updated World Stress Map (Reinecker *et al.* 2005). Orientations of various geological/geophysical stress indicators are shown. For Scandinavia the main sources of stress information are the small earthquakes of Figure 1. The picture is not indisputable. Even if the dominant NW–SE compression is distinguished, there is much scatter. The present

Fig. 2. Scandinavian map of the World Stress Map Project (Reinecker *et al.* 2005). Orientations are shown of maximum horizontal compressional stress via many different geophysical/geological measurement methods.

authors claim that the form of the uplift pattern, displayed in Figure 1 as the uplift contours, cannot be distinguished in either the seismicity of Figure 1 or in the stress orientations of Figure 2. The same conclusion on the dominance of plate motion stresses nowadays has been reached by Pascal *et al.* (2005) based on geological stress indicators. Discussions of data displays like Figures 1 and 2 have led Arvidsson (1996) to the opposite conclusion, namely that uplift since the Ice Age is the dominant cause of earthquakes in the intraplate region of Scandinavia. Recently it has been possible to establish horizontal motions caused by the post-glacial uplift through GPS measurements on permanent stations. They are 1–2 mm/year (Milne *et al.* 2001) in outward directions, from the contours in Figure 1, but more elongated along the Swedish geographic length axis. This outward motion cannot be distinguished in Figures 1 and 2.

The stress change as presented here, built on arguments collected over many years (e.g. Gregersen & Basham 1989; Stewart *et al.* 2000; Gregersen 2002, 2006), is in agreement with a basic conclusion of Mörner (2003). Mörner summarizes his research of the high post-glacial seismic activity in Sweden into this quote: 'the seismic mode – in intensity as well as in driving forces – was simply not the same as today'. This overall agreement with Mörner's claim is independent of particulars in interpretation of post-glacial movements of the island of Læsø (Figs 1 and 2) in Kattegat (Hansen 1994) or of the Stockholm area (Mörner 2003). It depends only on the acceptance of large earthquakes caused by post-glacial geodynamics in one or another area of Scandinavia.

Supplementary stress information from Greenland

The stresses in North America and Greenland, which also supported a large ice cap during the Ice Age, were treated in parallel to Scandinavia by Gregersen & Basham (1989). Similarly the lithospheric plate motions were found to have been more important than the uplift stresses. This conclusion is in line with the results of the World Stress Map Project (Zoback 1992). The earthquake map for Greenland (Gregersen 1989) showed, without any doubt, that earthquakes occur only in the coastal regions, and not under the inland ice. This is also convincing when historical reports are taken into account (Gregersen 1982), and it is supported in the new map of the intraplate earthquake activity, which has just been compiled in Figure 3. This present map contains much more data than previous maps. It is updated to 2004, and for many of the recent years the seismograph network has an expanded number of 10–20 instruments in Greenland.

The same observation of no earthquakes under the ice cap has been put forward for Antarctica with one exception (Adams *et al.* 1985). We apparently have an observational rule with an exception emphasizing the rule. This rule has been used by Johnston (1987, 1989) in arguments for the appearance of the special suite of Scandinavian earthquakes 9000 years ago. Johnston (1987, 1989) argues that part of the cause is that the glacial load disappeared. But a supplementary cause of the earthquake activity pulse is release of the stresses built up by plate motion under the ice cap, while it existed in Scandinavia. Those were not released, because the ice cap prohibited the stress release in earthquakes just like the present-day situation in Greenland and Antarctica. A recent report by

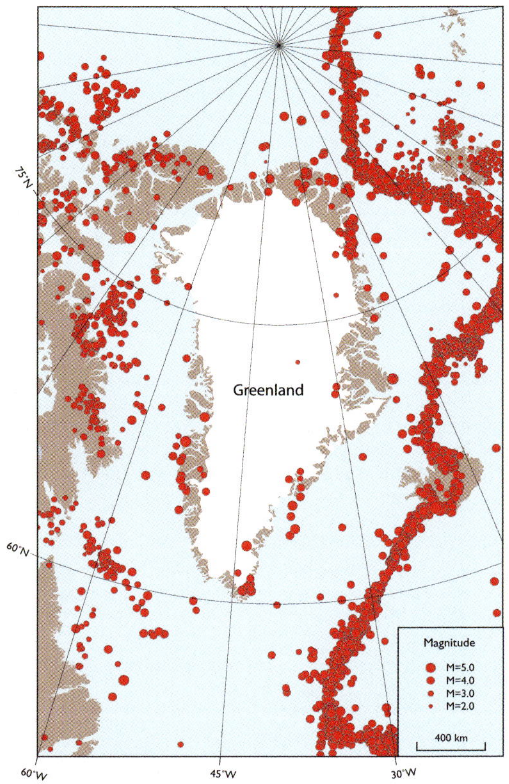

Fig. 3. Earthquake map of Greenland for the years from January 1970 to June 2004. The earthquake locations and magnitudes are extracted from the GEUS earthquake catalogue. Earthquakes located along the mid-Atlantic ridge and in Canada are extracted from the ISC earthquake catalogue (ISC 2001). The earthquake magnitude scale is given in the lower right corner.

Kaminuma (2006) on earthquake activity in Antarctica has confirmed the general lack of earthquakes in the interior of Antarctica. The earthquake activity in Greenland is slightly better studied than that in Antarctica. A few focal mechanisms are interpreted to show dominance of glacial off-loading (Chung & Gao 1997; Chung 2002). On the other hand Gregersen (1989, 2006) has argued, based on interpretations of the same data, that the main stress field nowadays is related to lithospheric motion.

Scandinavia again

Slowly a more certain and more detailed interpretation is within reach after the main conclusion concerning the dramatic change of stress in Scandinavia. This change of stress is very significant in evaluations of the earthquake hazard. The earthquake regime is significantly changed since the large post-glacial earthquakes occurred in northern Scandinavia.

Many students have over the years contributed to the computation of the GEUS earthquake catalogue of Denmark and Greenland. Recently the locations were made by Martin Glendrup, Signe K. Poulsen and Sebastian Simonsen. Our colleague H. P. Rasmussen has made many of the P and S arrival readings. Tine B. Larsen commented on the text. Illustrations were created with GMT (Wessel & Smith 1991) and improved by Eva Melskens. We thank all of these colleagues.

References

ADAMS, R. D., HUGHES, A. A. & ZHANG, B. M. 1985. A confirmed earthquake in continental Antarctica. *Geophysical Journal of the Royal Astronomical Society*, **81**, 489–492.

ARVIDSSON, R. 1996. Fennoscandian earthquakes: Whole crustal rupturing related to postglacial rebound. *Science*, **274**, 744–746.

CHUNG, W.-Y. 2002. Earthquakes along the passive margin of Greenland: Evidence for postglacial rebound control. *Pure and Applied Geophysics*, **159**, 2567–2584.

CHUNG, W.-Y. & GAO, H. 1997. The Greenland earthquake of July 11 1987 and postglacial fault reactivation along a passive margin. *Bulletin of the Seismological Society of America*, **87**, 1058–1068.

EKMAN, M. 1993. Postglacial rebound and sea level phenomena, with special reference to Fennoscandia and the Baltic Sea. *In*: KAKKURI, J. (ed.) *Geodesy and Geophysics*. Finnish Geodetic Institute, Publication **115**, 7–70.

GREGERSEN, S. 1982. Earthquakes in Greenland. *Bulletin of the Geological Society of Denmark*, **31**, 11–27.

GREGERSEN, S. 1989. The seismicity of Greenland. *In*: GREGERSEN, S. & BASHAM, P. (eds) *Earthquakes at North-Atlantic Passive Margins: Neotectonics and Postglacial Rebound*. Kluwer, Dordrecht, 345–353.

GREGERSEN, S. 1992. Crustal stress regime in Fennoscandia from focal mechanisms. *Journal of Geophysical Research*, **97**, 11821–11827.

GREGERSEN, S. 2002. Earthquakes and change of stress since the ice age in Scandinavia. *Bulletin of the Geological Society of Denmark*, **49**, 73–78.

GREGERSEN, S. 2006. Intraplate earthquakes in Scandinavia and Greenland. Neotectonics or postglacial uplift. *Journal of the Indian Geophysical Union*, **10**, 25–30.

GREGERSEN, S. & BASHAM, P.W. (eds) 1989. *Earthquakes at North-Atlantic Passive Margins: Neotectonics and Postglacial Rebound*. Kluwer Academic Press, Dordrecht.

GREGERSEN, S., HJELME, J. & HJORTENBERG, E. 1998. Earthquakes in Denmark. *Bulletin of the Geological Society of Denmark*, **44**, 115–127.

HANSEN, J. M. 1994. *Læsø's Evolution and Landscapes – about the Island That Rocks and Jumps*. Danish Geological Survey, Geografforlaget, Brenderup, Denmark [in Danish].

ISC. 2001. *On-line Bulletin*. Available at: http://www.isc.ac.uk/Bull. International Seismological Centre, Thatcham, UK.

JOHNSTON, A. 1987. Suppression of earthquakes by large continental ice sheets. *Nature*, **330**, 467–469.

JOHNSTON, A. 1989. The effect of large ice sheets on earthquake genesis. *In*: GREGERSEN, S. & BASHAM, P. (eds) *Earthquakes at North-Atlantic Passive Margins: Neotectonics and Postglacial Rebound*. Kluwer, Dordrecht, 581–599.

KAMINUMA, K. 2006. Seismicity in the Antarctic and surrounding ocean. *Journal of the Indian Geophysical Union*, **10**, 15–24.

LAGERBÄCK, K. 1991. *Seismically deformed sediments in the Lansjärv area, northern Sweden*. Svensk Kärnbränslehandtering AB, Stockholm, SKB Technical Report 91–17.

MILNE, G. A., DAVIS, J. L., MITROVICA, J. X., SCHERNECK, H.-G., JOHANSSON, J. M., VERMEER, M. & KOIVULA, H. 2001. Space-geodetic constraints on the glacial isostatic adjustment in Fennoscandia. *Science*, **291**, 2381–2385.

MÖRNER, N. A. 2003. *Paleoseismicity of Sweden. A Novel Paradigm*. JOFO Grafiska AB, Stockholm.

OLESEN, O. 1991. *A geophysical investigation of the relationship between old fault structures and postglacial faults in Finnmark, northern Norway*. PhD thesis, Technical University of Trondheim, Norway.

PASCAL, C., ROBERTS, D. & GABRIELSEN, R. H. 2005. Quantification of neotectonic stress orientations and magnitudes from field observations in Finnmark, northern Norway. *Journal of Structural Geology*, **27**, 859–870.

REINECKER, J., HEIDBACH, O., TINGAY, M., SPERNER, B. & MÜLLER, B. 2005. *The release 2005 of the World Stress Map*. Available at: www.world-stress-map.org

SCHERNECK, H.-G., JOHANSSON, J. M., VERMEER, M., DAVIS, J. L., MILNE, G. A. & MITROVICA, J. X.

2001. BIFROST project: 3-D crustal deformation rates derived from GPS confirm postglacial rebound in Fennoscandia. *Earth Planets Space*, **53**, 703–708.

SLUNGA, R. S. 1989. Focal mechanisms and crustal stresses in the Baltic Shield. *In*: GREGERSEN, S. & BASHAM, P. W. (eds) *Earthquakes at North-Atlantic Passive Margins: Neotectonics and Postglacial Rebound.* Kluwer, Dordrecht, 261–276.

STEWART, I. S., SAUBER, J. & ROSE, J. 2000. Glacioseismotectonics: ice sheets, crustal deformation and seismicity. *Quaternary Science Reviews*, **19**, 1367–1389.

WESSEL, P. & SMITH, W. H. F. 1991. Free software helps map and display data. *EOS Transactions AGU*, **72**, 441.

ZOBACK, M. L. 1992. First and second-order patterns of stress in the lithosphere: The World Stress Map Project. *Journal of Geophysical Research*, **97**, 11,703–11,728.

ZOBACK, M. L., ZOBACK, M. D. *ET AL.* 1989. Global patterns of tectonic stress. *Nature*, **341**, 291–298.

Late Holocene earthquake geology in Sweden

NILS-AXEL MÖRNER

Palaeogeophyics & Geodynamics, Rösundavägen 17, S-13336 Saltsjöbaden, Sweden (e-mail: morner@pog.nu)

Abstract: As a function of the rapid rate of glacial isostatic uplift, deglacial palaeoseismicity in Sweden was exceptionally high, in magnitude as well as frequency. Today, seismic activity is low to moderately low with occasional events reaching M 4–5. In the Late Holocene, 11 events in the order of M 6–7 are recorded. These palaeoseismic events seem also to be recorded in several old place names, as in the tale of the Fenris Wolf. Some of the events generated local to regional tsunamis. The palaeoseismic activity recorded in Late Holocene time implies that our short-term seismic hazard assessment must include the possibility of future events in the order of up to M 7. For long-term hazard assessment, repeating glacial/deglacial phases, we must work with magnitudes of M 8 to 9.

Geology is the key to a meaningful registration of past seismic activity, and through that to long-term seismic hazard assessment. In Sweden, this is especially clear because the mode of seismic activity has changed dramatically over the last 13 000 years (i.e. the time since deglaciation). Today, there is low to moderately low seismic activity with maximum magnitudes around M 4, or just above. The historical data include only three major events, namely M 5.4 in 1904, M 5.3 in 1759 and M 4.8 in 1497.

During recent decades, we have been able to record and date numerous palaeoseismic events (Mörner 2003, 2004, 2005). The fault scarps of the events in the north tell about mega-events of M > 8. In the middle and southern parts of Sweden, the secondary effects provide quite clear information of mega-events, too. Liquefaction events have been recorded and dated to single years in the Swedish Varve Chronology (in a few cases even to the season of a year) allowing us to record the spatial distribution of liquefaction at separate events. This indicates the occurrence of mega-events practically all over Sweden at around the time of deglaciation.

Our Swedish Palaeoseismic Catalogue now includes 56 events, 50% of which occurred 9000–11 000 BP during the phase of maximum rates of glacial isostatic uplift. Even during the Middle and Late Holocene, there were high-magnitude events, however. At 6100 BP, there was a major event with venting of coarse gravel and a tsunami that broke into lakes, at least 20 m above the sea level of that time. Obviously, we seem to be dealing with a M > 8 event. In the last 5000 years, eleven (or nine) M 6–7 events have occurred. This seismicity also seems to be recorded in old place names and in legends (Mörner 2007).

Late Holocene records

Only rarely and often by chance has it been possible to find and record palaeoseismic events in Late Holocene time (Mörner 2003). Figure 1 gives a histogram of the entire palaeoseismic catalogue as it stands in 2006. In the last 5000 years – the Late Holocene – eleven events are recorded. 50% of the events refer to the period 9000–11 000 varve years BP when the glacial isostatic rate of uplift peaked. Two new events come from the interstadial period 28–32 C14 ka BP. At around 8000 C14 years BP, there is a second maximum which coincides with a new uplift pulse (Mörner 1980). A third small maximum occurred in Late Holocene time within the period 5000–2000 BP. This period falls around the time when the centre of uplift moved from central north Sweden to the innermost part of Bothnian Bay (Mörner 2003). In this paper I will confine the discussion to the records from the last 5000 years (Fig. 2).

In our studies of the palaeoseismicity of Sweden (Mörner 2003), we introduced 'a multi-parameter characterization', where no event entered the Palaeoseismic Catalogue unless it was recorded by multiple factors and firmly dated. I have tried to apply the same methodology to the Late Holocene events (Table 1). This has hitherto only been possible in a few cases (events 1 and 6, and maybe 11 too).

Faults in sedimentary beds have been treated with care because they may represent slides of non-seismic origin. Figure 3 shows a sedimentary sequence that is vertically dislocated (faulted) by 20 cm. The age of faulting is <7500–7000 C14 years BP. Because this structure represents a local sedimentary fault, which might have been caused by a slide, and because it has not yet been correlated

From: REICHERTER, K., MICHETTI, A. M. & SILVA, P. G. (eds) *Palaeoseismology: Historical and Prehistorical Records of Earthquake Ground Effects for Seismic Hazard Assessment*. The Geological Society, London, Special Publications, **316**, 179–188. DOI: 10.1144/SP316.11 0305-8719/09/$15.00

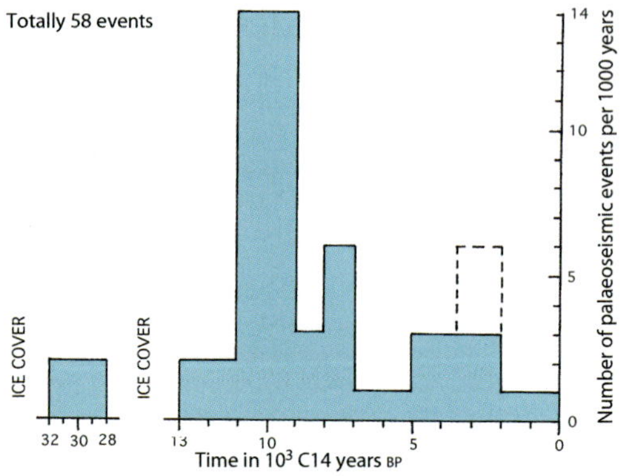

Fig. 1. Time distribution of the number of palaeoseismic events per 1000 years over the last 32 ka. There is a total of 56 events recorded; two in the interstadial beds at 32–28 ka and 54 during the last 13 ka with 50% (28 events) occurring during the phase of deglaciation and maximum rates of uplift 11–9 ka BP. A second peak (of six events) occurred at 8–7 ka, when a second uplift factor commenced. A third peak may have occurred in the interval 5–2 ka BP. If grouped in 1500 years intervals, the third peak increases to six events in the period 3.5–2.0 ka BP (dashed line).

to any other structure or bed in the vicinity that could be assigned a seismic origin, this structure is left outside our catalogue.

Table 1 lists the 11 events hitherto recorded in the last 5000 C14 years with respect to multiple criteria and assessments of magnitudes and intensities. Some of the events remain preliminary and call for additional investigations. A short description is given of each of these 11 events.

1. The Båstad–Torekov event, 4800 C14 years BP

This was a strong palaeoseismic event, estimated at M *c*. 7 (Mörner 2003). The Råle Fault was displaced by 1.0–1.4 m (Fig. 4). The beaches were extensively covered by talus. The bedrock cliffs were severely fractured. Intensive liquefaction was recorded in the River Stensån deposits (Fig. 5). Large earth slides occurred along the northern slope of Mt. Hallandsåsen spreading alluvial cones that are also recorded on top of the liquefied beds at River Stensån (Fig. 5). The main structure displaced must have been the important trans-Kattegatt fault passing along the northern slope of the Mt. Hallandsåsen horst. In terms of the INQUA intensity scale (Guerrieri *et al.* 2006), we are dealing with a degree X, maybe even XI, event. No doubt this was a very strong event and it is, indeed, recorded by multiple parameters (Table 1).

2. The Umeå event, about 4000 C14 years BP

A violent liquefaction event was recorded in a gravel pit at Umeå (Mörner 2003). When this occurred sea level was in the order of +35 m. The event seems to have generated a tsunami. Our research programme is still in progress, and our lake coring programme has not yet started. The liquefaction structures observed are impressive,

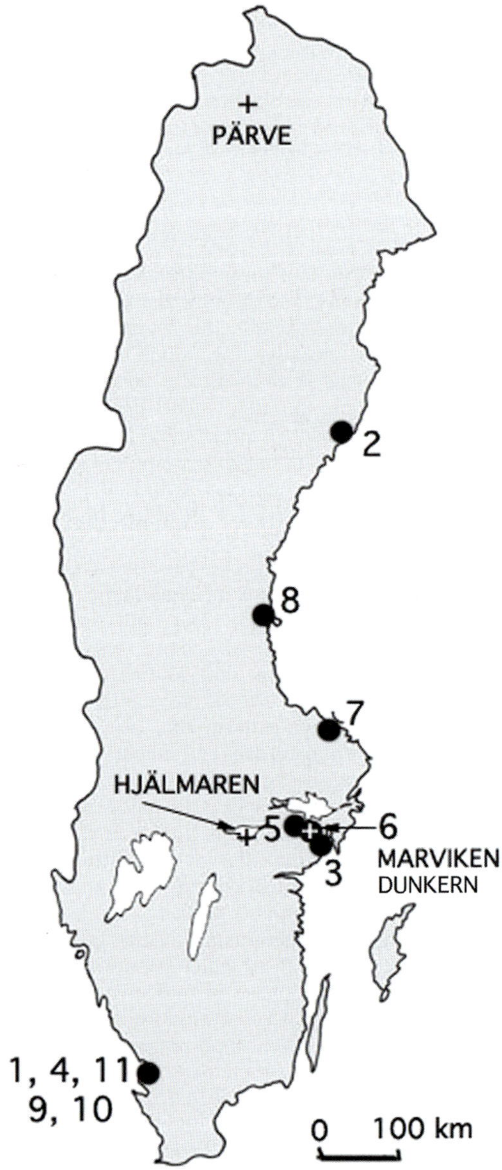

Fig. 2. Location of the 11 Late Holocene palaeoseismic events (Table 1) and some place names (+) discussed in the text.

Table 1. *Palaeoseismic events in the last 5000 years*

Event	Age	Fault	Talus	Slides	Liquefaction	Tsunami	Intensity	Magnitude
1	4800	+	+	+	+		X	~7
2	4000				+	+	XI	6–7
3	4000					+	(XII)	(>6)
4	3500	+	+				X	6–7
5	3250			+			–	–
6	3250	+	+	+	+	+	X	6–7
7	2900	(+)				+	(XI–XII)	(~7)
8	2000	+		(+)		+	(XII)	~7
9	2000	+	+				VII	>5
10	1400					+	(X)	>5
11	900	+	+			+	X	~7

Of the 11 events recorded, events 3, 5 and 10 are still preliminary. Various field expressions and interpretation with respect to intensity (Vittori & Comerci 2004; Guerrieri *et al.* 2006) and magnitude (Mörner 2003) are given. Intensities of X, XI and maybe even XII are recorded. The magnitudes reached 6–7 and even *c.* 7. This calls for a revised seismic hazard assessment. Site locations are given in Figure 2.

and seem to represent a degree XI event on the INQUA intensity scale (Guerrieri *et al.* 2006). There are some archaeological sites in the +40 m level that are covered by littoral sand, which seems to signify a tsunami event at about 4000 BP.

3. The Tystberga event, some 4000 C14 years BP

This event refers to an old section, which seems to record a significant tsunami wave flushing up over land from +25 m to at least +35 m (Mörner 2003). Further studies are necessary. We may note, however, that a 10 m tsunami run-up would represent a degree XII event on the INQUA intensity scale (Guerrieri *et al.* 2006).

4. The Båstad–Torekov event, 3500 ± 600 C14 years BP

This was another strong event recorded along the main fault passing along the northern slope of the Mt. Hallandsåsen horst (Mörner 2003). The Råle Fault moved by about 0.9 m (Fig. 4). The beaches were extensively covered by talus. In terms of the INQUA intensity scale (Guerrieri *et al.* 2006), we seem to be dealing with a degree X event.

5. The Katrineholm event, some 3000–3500 C14 years BP

At Skirtorp, a very large earth slide covers a brushwood peat of late Sub-Boreal age, suggesting an origin in earthquake shaking (Mörner 2003). This event remains preliminary, however, until additional data are found. No meaningful intensity can be assigned.

6. The Lake Marviken event, 3000–3500 C14 years BP

This is a very well documented palaeoseismic event (Mörner 2003, 2004). It occurred along an old fracture valley named Lake Marviken. A fault lineament is traced for 3.5 km. Nine earth and rock slides are mapped along the western side of the zone over a length of 6 km. One slide went into the lake setting up a local inner-fjord tsunami, which was recorded by coring over a distance of 5.2 km. Along the side of one slide, an erosional gauge was cut by water thought to originate from liquefaction. The erosional masses were deposited in an alluvial cone on land, indicating a water level in the narrow Baltic fjord of an age of about 3000 C14 years BP. This was also the C14 age obtained for the lowermost sediment covering the slide masses in the middle of the lake. At another slide going into an overgrown land surface (recorded by tree trunks and plant remains, including several well-preserved leaves), the slide sediment is bracketed by a lower date of 4000 and an upper date of 3000 C14 years BP. The slides include slides of till, sediments, mixed slope material and rock fragment. A grave from the Bronze Age has slid downhill (on the eastern side a similar grave remains in place). On the whole, it is the western side of the lineament that suffered sliding and liquefaction.

About 5 km to the west, a major slide dated around 3000 C14 years BP dammed a lake so that a bog was flooded. In cores, this is seen as a lower peat covered by 2 m of lacustrine gyttja with a thin cap of the present peat bog (the sequence is covered by ten C14 dates).

A magnitude 'M 6–7 or even M ~7' was estimated (Mörner 2003). On the INQUA intensity scale, we seem to be dealing with an event of at least degree X.

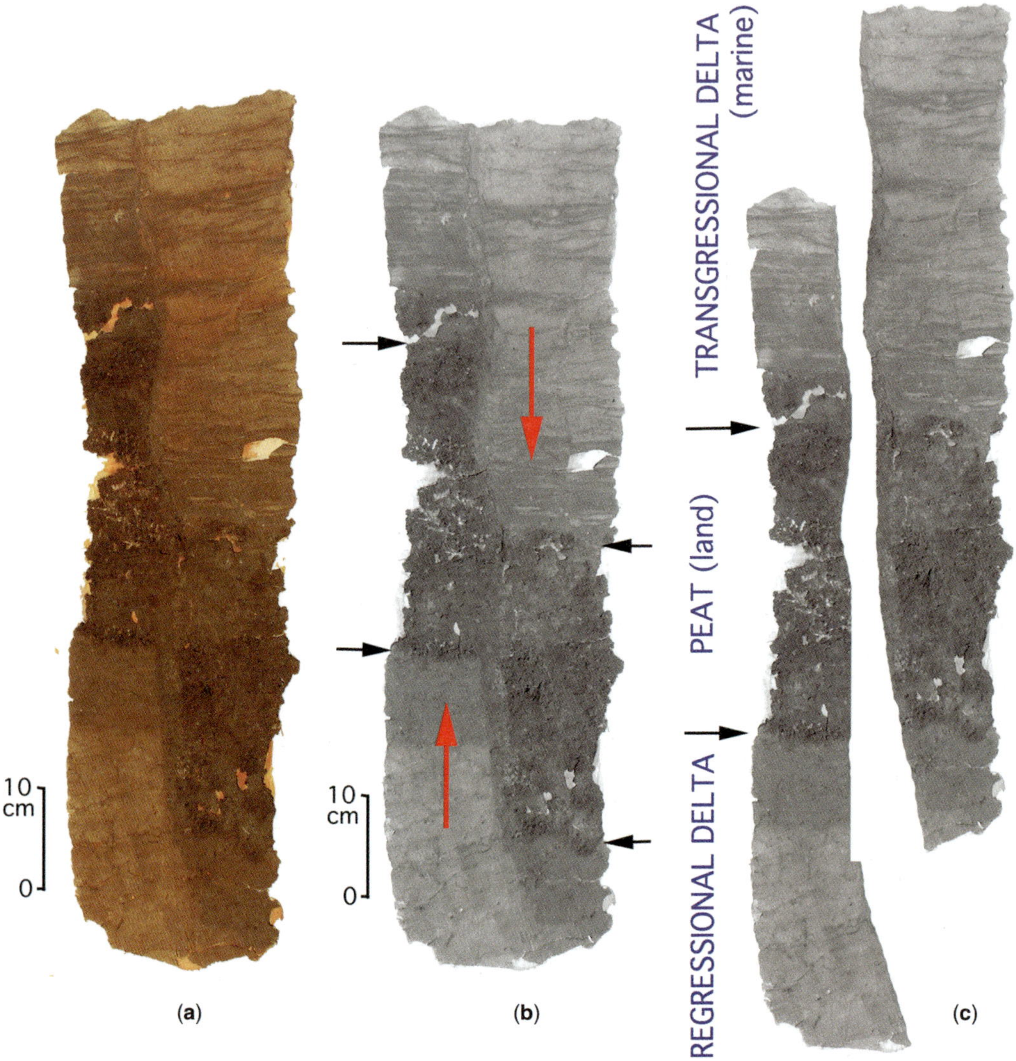

Fig. 3. Lack peal of faulted Holocene deposits from the Viskan Valley (Mörner 1969). An intra-marine peat bed of about 8000 C14 years BP is dislocated by about 20 cm. (**a**) View of the deposits as recorded in the field. (**b**) Interpretation of marker surfaces and post-depositional dislocation. (**c**) Back-faulted sequences showing the original three-part division of the sedimentary sequence of regressional delta sediments with a humus soil at its top, a terrestrial peat and a transgressional delta sequence (further discussed in Mörner 1969). This fault is interpreted as originating from a local earth slide and not of seismic origin.

7. The North Uppland (Forsmark) event, 2900 C14 years BP

In several lakes in northern Uppland (the Forsmark region), we recorded a major tsunami event (Mörner 2008). A coring and dating programme was conducted in 2004 (N.-A. Mörner, unpublished work). A tsunami bed was recorded in offshore sediments, in shore-zone sediments, and in lake and bog sediments at elevations up to 20 m (or at least 6 m) above the corresponding sea level. A tsunami with a run-up of 20 m implies a significant event.

We followed the tsunami bed from offshore basins (15 to 35 cm sand and gravel in graded bedding), via lagoonal basins (with 70 cm sandy beds at the clay/gyttja interface) up into lake basins above the corresponding shore (40–50 cm sandy-gravelly beds erosively deposited between the marine clay and lacustrine gyttja). Six C14 dates provide a close age for the offshore and

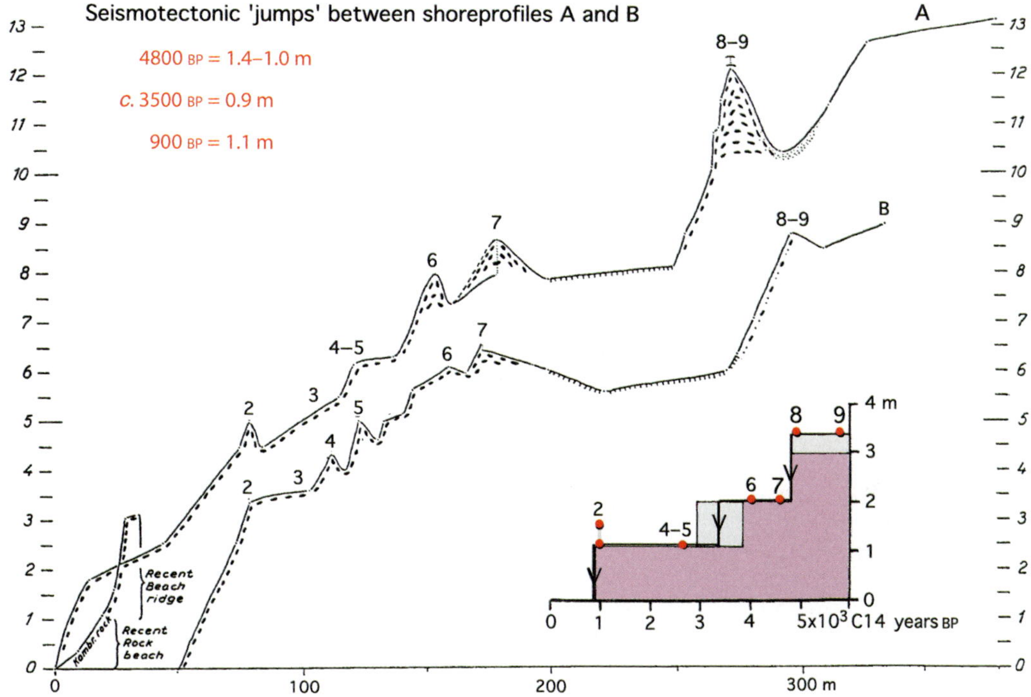

Fig. 4. Levelled shoreline profiles at Norra Ängalag (**a**) and Perstorp (**b**) north of Torekov (Mörner 1969, fig. 118). The elevation of the palaeoshorelines (2–9) in the two profiles differs significantly although the spacing between the two profiles is only 400 m. In the lower-right diagram, the differences are plotted against time indicating three major 'jumps' each of about 1.0 m off-set, at 4800, *c.* 3500 and at 900 C14 years BP (Mörner 2003). The 'jumps' are interpreted in terms of repeated seismotectonic movements along the Råle Fault which passes between the two profiles.

lagoonal sites and a strong erosive effect in the lake basins at least up to a level 5 m above the corresponding shore as illustrated in Figure 6. The data record a vertical spread of the tsunami beds from −20 m to + 6 m. The lake and bog coring suggests that the tsunami may have had a run-up of 20 m. This is not yet supported by dates, which suggest only a 6 m run-up (Fig. 6).

The Singö Fault zone crosses the area. This zone seems to have been reactivated during the deglacial phase some 10 000 years ago (Mörner 2003, 2004). Therefore, it seems likely that even this 2900 BP event represents a reactivation of this zone. We have recently investigated the tsunami signals in the lake and bog records. Judging from the tsunami run-up, we seem to be dealing with an intensity XII (20 m) or XI (6 m) event with respect to the INQUA intensity scale.

8. The Hudiksvall (Skålbo) event, 2000 C14 years BP

An event of violent methane venting was recorded at Skålbo, north of Hudiksvall (Mörner 2003), when sea level was at about +18 m about 2000 C14 years BP. Deposits interpreted as tsunami beds were recorded in five lakes ranging from +8 m up to +38 m. The +38 m site is a bog which, above the isolation level and between freshwater gyttja and covering peat, has a 2.65 m thick bed of gravel with numerous shells of Baltic brackish-water origin. This implies a tsunami wave that reached at least from +18 m to +38 m, i.e. having a run-up of 20 m or more. Consequently, this was a very strong event.

Lake Dellen, now at +37 m, previously had a 3 m lower water level dated at about 2000 C14 years BP. This implies that the rise in level of Lake Dellen by about 3 m occurred at the same time as the 20 m tsunami wave at Skålbo. Therefore, it seems likely that the Dellen rise was an effect of the tsunami (Mörner 2008).

In terms of the INQUA intensity scale (Guerrieri *et al.* 2006), the tsunami height suggests a degree XII event. At the same time, methane-venting tectonics is a new concept that is hard to convert into any intensity degree or magnitude. The blocks thrown up in a huge pile (over Late Holocene

Fig. 5. Strongly liquefied beds in the banks of River Stensån (Mörner 2003, fig. 3, p. 272). Between the liquefied beds and the covering lagoonal and lacustrine deposits, there is a thin gravel bed with stones representing the distal part of the earth slides of the palaeoseismic event at 4800 C14 years BP.

beach material) are often of immense size calling for the action of very strong forces.

9. The Båstad–Torekov event, about 2000 C14 years BP

Liquefaction is recorded in the River Stensån riverbank (Mörner 2003). The age was estimated at <2700 C14 years BP. Talus shattering is recorded on the beach ridges at a position dated around 2000 C14 years BP (Mörner 2003). We seem to be dealing with a degree VII intensity event (Guerrieri *et al.* 2006) and >5 magnitude event.

10. The Torekov–Båstad event, about 1400 C14 years BP

A phase of talus shattering on to the beach ridges along the coast is recorded at around 1300–1400 C14 years BP (Mörner 2003). At about the same time there are records of an extreme storm or a tsunami throwing shells far above the corresponding shore. The shells were dated at 1360 ± 100 C14 years BP (Mörner 1969). An intra-peat layer of gyttja dated at 1535 ± 100 C14 years BP (Mörner 1969) might perhaps now be understood in terms of a tsunami wave washing at least 5 m up above the corresponding sea level and creating an open water-pool in the back-shore swamp. Complementary studies are necessary in order to verify the event.

From the tsunami run-up, we may be dealing with a degree X intensity event (Guerrieri *et al.* 2006).

11. The Torekov–Båstad event, 900 C14 years BP

The Råle Fault (north of Torekov) moved by 1.1 m shortly after the formation of the Viking shoreline some 980–970 C14 years BP (Fig. 4; Mörner 2003). At the same time, a fault south of Torekov fractured the Viking shoreline and displaced it by about 1 m. The Viking shoreline is fairly strongly affected by talus shattering of this event (Fig. 7). Some 65 km to the north, two ships from the Viking era were suddenly silted over in the old harbour of Galtabäck. This may now be understood in terms of a tsunami wave. The ships are dated at 825 ± 65 C14 years BP (Mörner 1969, p. 310), fitting well with the expected age of the faults.

The corresponding intensity degree seems to be in the order of X (Guerrieri *et al.* 2006). The magnitude was set at M *c.* 7 (Mörner 2003).

Place names and tales

The oldest Swedish place names are held to originate in the Bronze Age. There are several Swedish place names that refer to sound and fractured bedrock (Fig. 2; Mörner 2003, 2007). The Pärve Fault in northern Sweden moved some 9000 years ago. The Lappish word 'pärve' means 'sound from the underground', suggesting that the fault was also active in Late Holocene time. Lake Hjälmaren, crossed by extensive faults, means 'the sounding', which we now interpret in terms of palaeoseismics (Mörner & Strandberg 2003). The name 'marviken' in Lake Marviken (where an event occurred some 3000 years ago) seems to refer to 'fractured bedrock' (N.-A. Mörner & S. Strandberg, unpublished work). Lake Dunkern (where a palaeoseismic event is dated at 8000 BP; Mörner 2003) refers to 'deep sound'.

Most remarkable is the tale of the Fenris Wolf (Mörner 2007) – a giant wolf, chained deep in the bedrock, which, when howling made the ground fracture and tremble violently; this is indeed a perfect description of an earthquake.

Therefore, it now seems reasonable to assume that the inhabitants of Sweden, in Late Holocene

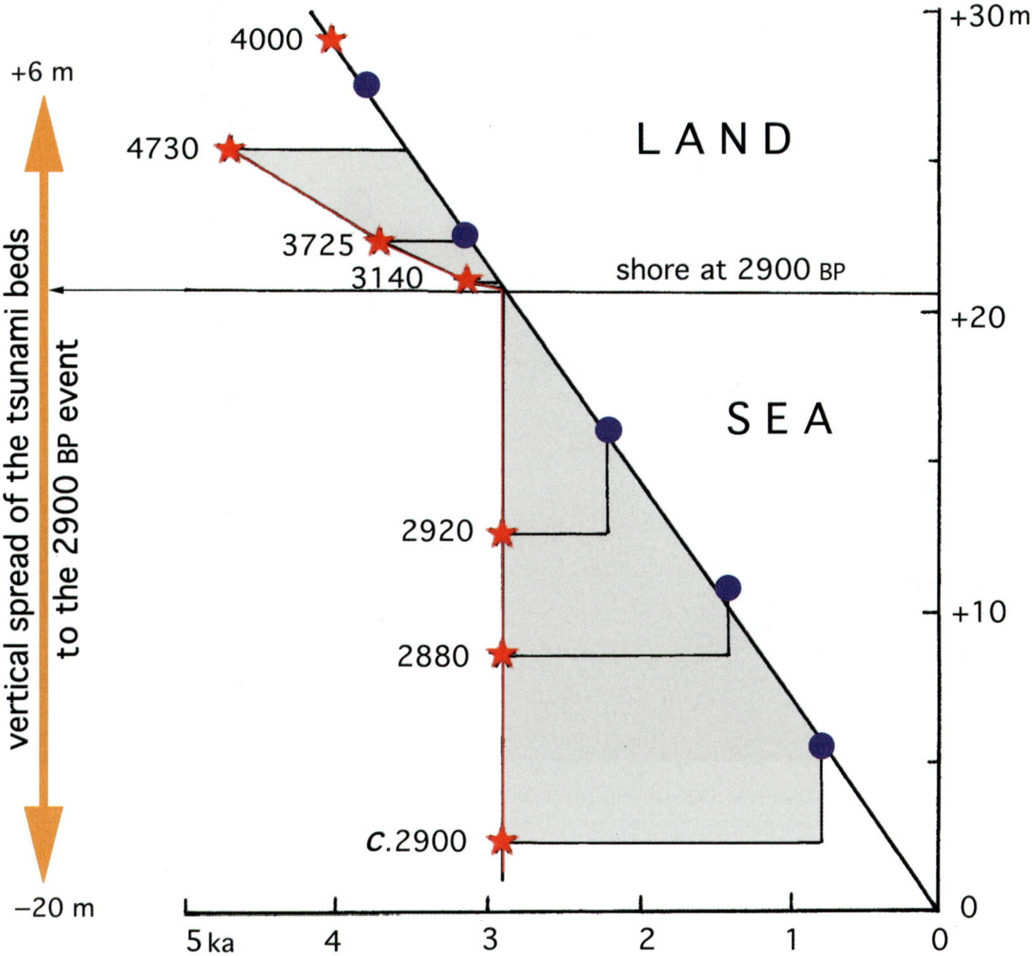

Fig. 6. The 2900 C14 years BP tsunami event in northern Uppland with respect to the rate of land uplift and shore displacement over the last 4000 years (the oblique line of *c.* 7 m uplift per millennium passing through dated anchor points marked by black dots). At 2900 C14 years BP the shore was at +20.7 m (with land above and sea below as marked on the right side of the diagram). The stars mark tsunami beds recorded and dated in off-shore sediments (all falling sharply at the 2900 BP level), in coastal deposits and in lakes and bogs on land where the tsunami beds have eroded down into the sediments. The supposed tsunami bed in the +29 m basin has an age that coincides with the time of isolation. Therefore, we can in this case not discriminate between a normal shore sand from the isolation and a tsunami bed. Consequently, the graph gives evidence of a tsunami event that deposited typical tsunami beds over a vertical range from −20 m to +6 m. The tsunami run-up might have reached 20 m above the shore level, judging from lake and bog coring at higher altitudes, however.

time, experienced significant earthquake events. Consequently, our cultural heritage and our palaeoseismic data give the same message: high intensity/high magnitude events occurred in Sweden in Late Holocene time.

Discussion

Geological data from the Late Holocene interval 4800 to 900 C14 years BP provide records, or at least traces, of 11 palaeoseismic events (Table 1, Fig. 2). Three of the events remain preliminary (events 3, 5 and 10). Especially clear and well documented are events 1, 6, 7 and 8.

Faults are recorded in six cases, with three successive events along the local Råle Fault (Fig. 4). Supposedly seismically induced talus shattering is recorded in five cases (very clear for event 1 but also significant at event 11). Earth slides are recorded at three events (1, 5 and 6). The event

Fig. 7. The Viking shore is here covered by a quite significant talus cone, including loose blocks that have rolled out on the shore surface. A seismic origin is advocated (event 11).

1 slides are large and well-expressed with lateral extension into the stratal sequence of liquefied sediments (Fig. 5). At event 6 in the Lake Marviken area, nine slides are recorded, of sand, of gravel, of till, of rock debris and even of a Bronze Age mound (Mörner 2003, p. 251, fig. 26). Liquefaction was recorded at three events (1, 2 and 6). Figure 5 shows the strongly liquefied beds of event 1. It also contains a thin gravel horizon of the associated slide. Tsunami events have been recorded at seven events. Some of them are very well documented (events 2, 6, 7, 8 and possibly 11, too). This is the field observation at hand (Table 1, Fig. 2). With this material as a base, we now turn to the seismic hazard assessment.

The 11 events form a peak of three events per millennium from 5000 to 2000 BP (Fig. 1). Six of the events, however, are confined to the period 3500–2000 C14 years BP. This seems to imply that there was a third peak in seismic activity in the Late Bonze Age and Early Iron Age. This seems significant for two different reasons: one cultural and one natural historic. From a cultural point of view, this period coincides with the origin of old place names and tales in the Asa folklore. From a natural historic point of view, this period corresponds with the displacement of the centre of uplift from the inland of central north Sweden to the northern part of Bothnian Bay. Therefore, this third peak in seismic activity may be real and not just an artefact due to the incomplete data set available at present (with our recording of Late Holocene palaeoseismic events still is in its initial phase).

In the interpretation of the observational records, I have tried to apply the INQUA earthquake environmental effects (ESI) intensity scale (Vittori & Comerci 2004; Guerrieri *et al.* 2006). In most cases, the intensity degrees arrived at seem reasonable. In the case of tsunami events, however, the degrees assigned seem to be too high. This applies for event 3 with a run-up of 25 m, event 7 with a run-up of 20 m (or 6 m) and event 8 with a run-up of 20 m, with the ESI scale calling for degree XII for all three events (degree XI if event 7 only had a run-up of 6 m), which is likely to be an overestimate (therefore these values are in parentheses in Table 1). Corresponding magnitudes are assigned from fault heights, liquefaction characteristics, and tsunami heights in a more personal comparative way among all 56 events documented in the

Table 2. *Comparison of maximum magnitudes as established by four different data sets*

Data set	Time period	Magnitude
Seismology	<100 years	<4.5
Historical data	Last 600 years	<5.5
Late Holocene	Last 5000 years	>6 to ~7
Deglacial phase	9–11 ka BP	>8

Swedish Palaeoseismic catalogue (Mörner 2003, 2005). In an objective analysis of the catalogue, Adams (2005) found a good and logical distribution of magnitudes with respect to the Wells & Coppersmith (1994) distribution.

It is now a well-established fact that the deglacial phase some 9000–11 000 BP was a period of high to super-high seismic activity both in frequency and magnitude (Mörner 2003, 2004, 2005). At the same time, it is a well-established fact that the seismic activity in present and historical (last 600 years) time was low to moderately low. The data set here presented fills the gap between the late glacial and present periods. This is illustrated in Table 2 where maximum magnitudes are given for the deglacial phase, the Late Holocene, the last 600 years and the instrumental records.

In principle, one might have expected a tailing-off of the high late glacial activity. Therefore, the relatively high seismic activity now recorded in the Late Holocene might be surprising. The 11 events recorded are in magnitudes peaking at M > 6 and probably at M *c*. 7, and in intensity at X, probably in XI and possibly even in XII. There are two different ways of interpreting the data: the stochastic or the special case interpretation.

By extending our decadal to century-based records to millennia, we stochastically come to include peak events missed in our short-term record (Table 2). Therefore, our seismic hazard assessment must, with the Late Holocene records at hand, be revised to include even events of M *c*. 7 (or at least well above M 6) and intensity degree XI (maybe even XII).

The peak of six events within the period 3500–2000 C14 years BP opens another interpretation, however. At around 3500 C14 years BP, the truly glacial isostatic dome-like uplift factor had ceased (Mörner 1980) and the centre of uplift moved from the inland of central north Sweden to the inner part of Bothnian Bay (Mörner 1980, 2003). This shift implied a new direction of tilting, which must have affected both the stress and the strain in the bedrock, which in turn are likely to have affected seismic activity. If this is correct, we see an even closer correlation between uplift and seismic activity than previously recorded (Mörner 2003, 2004). Furthermore, it would suggest that the strong seismic peak in the Late Bronze Age and Early Iron Age represented a special case in the past not to be repeated in the near future. If so, it might not call for an urgent revision of the seismic hazard assessment (just a warning).

A third way of analysing the situation is to consider the site-specific recurrence intervals (Mörner 2003). The local 'Råle Fault' experienced three step-wise jumps in the last 5000 years (Fig. 4). For the Kattegatt region (i.e. the Båstad–Torekov area and the extension of the fault zone into the Kattegatt Sea), 12 palaeoseismic events are recorded in the last 12.5 ka with a mean recurrence interval in the order of 1000 years. For the last 5000 years, five events are recorded and for the last 2000 year three events. This means a reasonably consistent seismic activity. Consequently, the recorded maximum magnitude value of M *c*. 7 is valid also for our present to future seismic hazard assessment. For the Stockholm–Mälardalen area, 14 events are recorded in the last 10.5 ka. After eight closely spaced events during the deglacial phase, six events are recorded in the last 9000 years giving a mean recurrence interval of about 1500 years. Even here, the recorded maximum magnitude value of M 6–7 should apply in our present to future seismic hazard assessment.

Further investigations, recording and documentation are, of course, necessary. Still, even the data at hand today call for the introduction of the possibility of M *c*. 7 and intensity XI events in our hazard assessment for the future decades to centuries.

Conclusions

Eleven palaeoseismic events are recorded in the Late Holocene dating from 4800 to 900 C14 years BP. The intensities are estimated at X to XI, maybe even XII (judging from the tsunami heights) on the ESI intensity scale (Vittori & Comerci 2004; Guerrieri *et al.* 2006). The corresponding magnitudes on the Richter scale are estimated at M 6–7 and even *c*. 7.

The tale of the Fenris Wolf and a number of place names referring to sound and fractured bedrock seem to confirm that the inhabitants of Sweden, indeed, experienced violent seismic events (Mörner 2007).

The palaeoseismic records described fill the gap between high seismicity during the deglacial phase some 9000–11 000 years BP (Mörner 2003) and the low to moderately low seismicity recorded by seismology and historical records.

The palaeoseismic events recorded in the Late Holocene add a new dimension to the seismic records obtained from instrumental measurements and from historical records since the late fifteenth century. It implies, from a stochastic point of view, that previous seismic hazard assessments must be updated to include the possibility of future events in the order of M 6 and 7. Admittedly, however, the activity peak in the time range 3500–2000 C14 years BP might be driven by the shift in tilt direction (stress and strain) due to the relocation of the centre of uplift, and, if so, it might be a past event not to be repeated in the near future. Site-specific recurrence diagrams for the Kattegatt region and the Stockholm–Mälardalen region indicate a more or less continuous process, however, which implies that the recorded maximum values may recur in the near future. Consequently, our seismic hazard assessment must, from now on, also include the Late Holocene data set (Tables 1 and 2).

Finally, it should be pointed out that the recording of Late Holocene palaeoseismic events is a complicated issue that often includes pure luck and always includes a painstaking recording process. Therefore, we can be sure that many more events will be found by further investigations in the near future.

References

ADAMS, J. 2005. Appendix 5. On the probable rate of magnitude ≥6 earthquakes close to a Swedish site during a glacial cycle. *In*: HORA, S. & JENSEN, M. (eds) *Expert panel elicitation of seismicity following glaciation in Sweden*. SSI Report **2005:20**, 33–59 (available in full at: http://www.ssi.se/ssi_rapporter/ssirapport.html).

GUERRIERI, L. *ET AL.* (eds) 2006. *The INQUA EEE intensity scale*. Agency for the Protection of Environment and for Technical Service (APAT), Italy (available at: www.apat.gov.it/site/en-GB/Projects/INQUA_ Scale/default.html).

MÖRNER, N.-A. 1969. The Late Quaternary history of the Kattegatt Sea and the Swedish West Coast; deglaciation, shorelevel displacement, chronology, isostasy and eustasy. *Sveriges Geologiska Undersökning*, **C-640**, 1–487.

MÖRNER, N.-A. 1980. The Fennoscandian uplift: geological data and geodynamic implications. *In*: MÖRNER, N.-A. (ed.) *Earth Rheology, Isostasy and Eustasy*. John Wiley, Chichester, 251–284.

MÖRNER, N.-A. 2003. *Paleoseismicity of Sweden – A Novel Paradigm*. A contribution to INQUA from its sub-commission on Paloseismology. P & G Unit, Stockholm University.

MÖRNER, N.-A. 2004. Active faults and paleoseismicity in Fennoscandia, especially Sweden. Primary structures and secondary effects. *Tectonophysics*, **380**, 139–157.

MÖRNER, N.-A. 2005. An interpretation and catalogue of paleoseismicity in Sweden. *Tectonophysics*, **408**, 265–307.

MÖRNER, N.-A. 2007. The Fenris Wolf in the Nordic Asa creed in the light of paleoseismics. *In*: PICCARDI, L. & MASSE, B. (eds) *Myth and Geology*. Geological Society, London, Special Publications, **273**, 117–119.

MÖRNER, N.-A. 2008. Tsunami Events within the Baltic. *Proceedings of the workshop 'Relative sea level changes-from subsiding to uplifting coasts'*. Polish Geological Institute, Special Papers, **23**, 71–76.

MÖRNER, N.-A. & STRANDBERG, S. 2003. Sjönamnet Hjälmaren i geologisk beslysning. *Ortnamssällskapets i Uppsala Årsskrift*, **2003**, 79–82.

VITTORI, E. & COMERCI, V. (eds) 2004. *The INQUA Scale. An innovative approach for assessing earthquake intensities based on seismically-induced ground effects in natural environment*. Agency for the Protection of Environment and for Technical Service (APAT), Memoire Carta Geologica d'Italy, **67**, Special Paper, 1–116.

WELLS, D. L. & COPPERSMITH, K. J. 1994. New empirical relationships among magnitude, rupture length, rupture width, rupture area and surface displacement. *Bulletin of the Seismological Society of America*, **84**, 974–1002.

Testing a seismic scenario for the damage of the Neolithic wooden well of Erkelenz-Kückhoven, Germany

KLAUS-G. HINZEN[1]* & JÜRGEN WEINER[2]

[1]*Earthquake Geology Group, Institute of Geology and Mineralogy, Cologne University, Vinzenz-Pallotti-Str. 26, 51529 Bergisch Gladbach, Germany*

[2]*Landschaftsverband Rheinland, Rheinisches Amt für Bodendenkmalpflege Bonn, Aussenstelle Nideggen, Zehnthofstr. 45, 52385 Nideggen-Wollersheim, Germany*

**Corresponding author (e-mail: hinzen@uni-koeln.de)*

Abstract: A Neolithic wooden well was discovered and excavated between 1989 and 1992 near Erkelenz in the Lower Rhine Embayment. The construction, 3 × 3 m in size and 13 m deep, was exceptionally large for its time. The larger outer box-frame contained two smaller frames whose construction could be interpreted as an attempt to repair the damaged original well. The outer box was made from 160 oak elements of about 3 m length built in the blockhouse method. The large box is dated to 5090 BC and the two smaller ones to 5057 ± 5 BC by dendrochronological analysis. At *c.* 8 m depth several elements of the large box are vertically sheared off and the broken parts moved inward and downward. The cause of this damage has not yet been determined. As the well is located only 3 km from one of the active tectonic faults in the Lower Rhine Embayment, a seismogenic origin of the damage is considered and tested in this paper. This question has relevance for determination of seismic hazard in an area with present-day moderate seismicity but documented occurrence of strong surface-rupturing earthquakes from the palaeoseismic record. First, a geotechnical model for the construction pit with a total volume of *c.* 540–550 m^3 is used to prove the stability of the open pit during well construction and to help explain how the well was built. The seismogenic hypothesis is tested in a deterministic approach using theoretically derived ground motion at the site of the well for two simulated earthquakes with magnitudes 6.2 and 6.8. Ground deformation and relative displacement calculated with a finite element model of the casing are found to be too small to account for the documented damage. Among other potential sources of damage, swelling, shrinking or rotting of the wood elements are possible explanations; however, a conclusive answer to this question remains to be found.

This paper attempts to answer the question of whether seismic ground movements damaged the Neolithic wooden well of Erkelenz-Kückhoven. Archaeoseismic case studies in recent years have gained importance in supplementing the earthquake record of certain regions, especially those with low or moderate seismicity (Galadini *et al.* 2006). Archaeoseismology has relevance for seismic hazard analysis only when quantifiable descriptions of archaeologically excavated and documented building damage can be made. In addition to ruling out other possible causes of observed damage (e.g. Nikonov 1988), some conclusions about the causal earthquake are necessary. At the very least, seismic hazard studies require a rough estimate of source location, strength, and, if possible, a window of occurrence. Even if these earthquake parameters have a large range of uncertainty, they can be useful for probabilistic studies, in which the uncertainties can be quantified (Stewart *et al.* 2007). A final goal of archaeoseismic studies should be a territorial approach (Guidoboni *et al.* 2000; Galadini *et al.* 2006), in which archaeoseismic effects throughout a region are correlated in order to define the extent of the mesoseismal zone. If such a territorial approach is possible and successful, uncertainties in the earthquake parameters lessen as the number of sites increases. However, in regions with moderate or low seismicity, typical for intraplate earthquake zones, a successful territorial approach will be a bonus. Additionally, any information gained about damaging earthquakes from pre-instrumental and prehistoric times, especially in intraplate seismic zones, from palaeoseismic and/or archaeoseismic studies are valuable for the hazard analysis (Camelbeeck *et al.* 2007).

Case studies in low to moderate seismicity areas in Europe north of the Alps have shown that valuable information on certain events, whether previously known (Fäh *et al.* 2006) or unknown (e.g. Hinzen 2005*a*; Decker *et al.* 2006), can be obtained by the application of archaeoseismological techniques. In particular, the application of engineering seismological techniques helps quantify earthquake

From: REICHERTER, K., MICHETTI, A. M. & SILVA, P. G. (eds) *Palaeoseismology: Historical and Prehistorical Records of Earthquake Ground Effects for Seismic Hazard Assessment*. The Geological Society, London, Special Publications, **316**, 189–205. DOI: 10.1144/SP316.12 0305-8719/09/$15.00

parameters (e.g. Hinzen & Schütte 2003; Hinzen 2005*a*; Fäh *et al.* 2006). Most archaeoseismological studies target a time slice within the past 2000 years and stone artefacts (Stiros & Jones 1996; Galadini *et al.* 2006). The wooden structure of this case study is dated to the Neolithic period. A wooden well from the Linear Bandkeramik Culture (LBK), which lasted from *c.* 5500 to 5000 BC, was excavated in the early 1990s at a place called Kückhoven, located in the Lower Rhine Embayment (LRE). Several of the well-worked massive wooden beams of the almost square well casing were found broken at one corner (Weiner 1991, 1992, 1994, 1997). The location of the wooden well in close proximity (*c.* 3 km) to one of the major active normal faults in the LRE was an inducement to test the possibility of an earthquake as a possible cause for the damage. This scenario includes calculation of synthetic site-specific seismograms and a finite element analysis of the dynamic behaviour of the well casing. Other potential sources of damage include acts of war, decomposition or swelling effects, and sediment pressure. The latter was tested with a simplified geotechnical model. Additionally, a static stability analysis of the construction pit constrained the quality of well construction.

History and archaeology

Starting in August 1989, the Archaeological Monuments Management Service Bonn excavated an early Neolithic (LBK) large and fortified site in the forefield of a gravel pit near the village of Erkelenz-Kückhoven, *c.* 45 km NW of Cologne (Fig. 1). As the nearest watercourses to the settlement are situated *c.* 3 km north and south, the question arose of how the settlers procured their vital water. In December 1990, during further gravel extraction, a power shovel unearthed pieces of worked wooden planks from a depth of *c.* 6 m below the actual surface (Fig. 2). Preliminary 14C

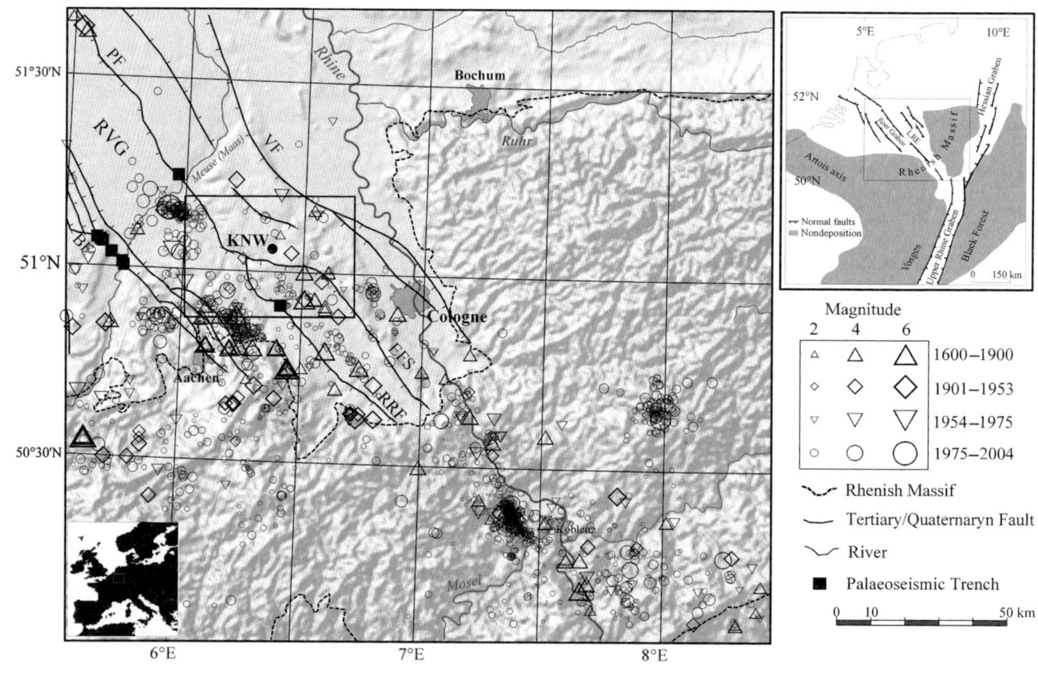

Fig. 1. Seismotectonic situation in the northern Rhine area. The inset in the lower left corner shows the map location in Europe. The small map in the upper right corner indicates the main tectonic elements of the northern part of the Rhine–Rhone rift system. Prominent Tertiary and Quaternary faults after models by Ahorner (1962), Spelter (1998), and Weber & Hinzen (2006). Open circles, triangles and diamonds show the historic and instrumental seismicity from 1600 to 2004 (Hinzen & Reamer 2007) as indicated in the ledgend. The Kückhoven Neolithic well (KNW, filled circle) is located in the centre of the Lower Rhine Embayment north of the Lövenicher Sprung, which is shown in Figure 5 in detail. The outline of the map in Figure 5 is indicated by the rectangle. Major faults are: Viersen Fault (VF), Erft Fault system (EFS), Rurrand Fault (RRF), Peel Fault (PF). The location of eight palaeoseismic trenches (filled squares) follows Camelbeeck *et al.* (2007) including those at the Bree Fault (BF). RVG indicates the central Roer Valley Graben. The digital elevation model of the Lower Rhine Embayment and its vicinity is based on Radar Topography Mission, NASA.

Fig. 2. Excavation and damage of the Neolithic wooden well at Erkelenz, Kückhoven. (**a**) An oval dark grey spot, measuring *c.* 7 × 6 m contrasts with the colour of the surrounding sandy gravel of the main terrace of the River Rhine. The spot marks the cross-section of the construction pit of the Neolithic well at Kückhoven. (**b**) Box-frames 1 and 2 of the well during the ongoing excavation process. (**c**) The oblique *in situ* position of two elements from the large box 1. Theses beams were vertically sheared off at their necks and the broken part moved inward and downward as illustrated in (**d**). The scale has a length of 1 m. (**e**) The situation in the gravel pit near Kückhoven during excavation of the well in July 1991. The excavation site is located under the plastic tent (arrow). The gravel and sand removal has created two levels at depths of *c.* 18 m and 8 m below the current surface visible in the foreground and the middle of the photo, respectively. The long steep section of the free face is in parts almost vertical. This free face was not actively mined for a period of at least three months and was stable during the whole period (photos by J. Weiner, drawing by K. Drechsel).

dating determined an LBK age, which was corroborated by dendrochronological dating (Schmidt 1992; Schmidt *et al.* 1998). The types and dimensions of the wooden fragments and, in particular, the depth of their location served to identify them as remains of a wooden well (Weiner 1991, 1992, 1994, 1997). Due to continuing gravel extraction, the upper part of the place of discovery had been completely removed by a power shovel. As a result, only part of the section's very top could be documented archaeologically.

Excavation of the well started in January 1991 at a depth of 6 m below the current surface. After initially clearing the site, an oval dark grey spot, measuring *c.* 7 × 6 m contrasted nicely with the surrounding *in situ* sandy gravel of the main terrace of the Rhine River (Fig. 2a). This feature was resolved as the horizontal cross-section of the well's former construction pit, the filling of which had subsequently changed colour due to infiltration of organic material. Test drilling revealed an additional depth of 7 m, indicating a total depth for the original construction of 13 m. In its centre, two separate horizontal, square-shaped wooden constructions were revealed, i.e. two box-frames, one built into the other. The larger, outer box-frame 1 measured *c.* 3 × 3 m, whereas the smaller box-frame 2 measured *c.* 1.6 × 1.6 m (Fig. 2b). The wood is exclusively first quality oak (*Quercus* sp.), and all wooden planks forming the four sides of each box-frame are elements of radially split oak logs, each roughly triangular in cross-section. In October 1991 a third smaller box-frame 3 was discovered, built flush into the lower section of box-frame 2, forming a telescopic extension of the latter. This box-frame measured *c.* 1.1 × 1.1 m. The relative positions and different sizes of the box-frames strongly suggest that box-frame 1 is the oldest construction and formed the first well I, and box-frames 2 and 3 are one unit, forming a younger well II, a repair measure of well I (Weiner 1998).

The box-frames have been dated by radiocarbon and dendrochronological methods. The latter delivered a date for box-frame 1 of 5090 denBC. This date is very precise because the samples had a fully developed waney edge. Box-frames 2 and 3 (well II) gave dates of 5057 ± 5 denBC (Schmidt *et al.* 1998). At the time, the date for box-frame 1 made the Kückhoven Neolithic well the world's oldest wooden construction known. Since then, several more LBK wooden wells have been discovered which are even older (Stäuble & Campen 1998; Campen 2000; Schmidt & Gruhle 2003).

Two building techniques are proposed for the construction of the box-frames. (1) A prefabricated section of a box-frame was lowered from the surface by extracting sediment from underneath its bottom, thus lowering the whole construction while simultaneously adding further construction elements on top until the final depth is reached. (2) An open construction pit was excavated from the surface down to the potential water-bearing level, subsequently building a complete box-frame from the pit's bottom upwards, while simultaneously backfilling the open space between the box-frame and the pit's wall.

Because the dynamic analysis is dependent to a large extent on the method of construction (stability of the open pit, material parameters of the sediments, possibility of shearing off necks from the beams during the construction etc.), this question is briefly discussed in the following.

Well building using method 1, the gradual lowering of a well frame, is well known from Roman times and later. This type of construction requires wooden (or even stone) constructions, free of any protruding parts on their sides or corners. Since historically known lowered box-frames, without exception, show flush sides and straight corners it is highly unlikely that method 1 would have worked at Kückhoven. Lowering the box-frame would have been extremely difficult; the worst obstacle would have been the protruding edges. In fact, lowering even the first frame would have required additionally removing a considerable amount of sandy gravel over a distance of more than half a metre beyond the outer edge of the box-frame in all four corners. Even assuming an initial positive result, the mass of the box-frame would have gradually increased to more than 10 t. It is unlikely that a construction of that weight could have been lowered without tilting, and, once tilted, the construction would invariably have been stuck. As backfilling of the gap between the casing and the pit walls would have been delayed until the complete casing was in position, total stability of the pit walls was an additional requirement for this technique. Material that caved in from the pit walls would have been extremely difficult to remove and would have endangered the construction process.

The second method also depends on the stability of the pit's walls during the construction period, estimated to be on the order of less than two months (Weiner & Lehmann 1998). However, in this case collapsing sediment would have posed a serious threat for the working crew down the pit, even though it would not have blocked the entire construction. To estimate the feasibility of method 2, a limit equilibrium stability analysis was performed for a total of 700 potential sliding surfaces for each of the two sidewalls with four methods: Bishop, Janbu, Morgenstern–Price (force and moment methods) (e.g. Coduto 1998). Figure 3 shows the geometry of the slopes of the excavation

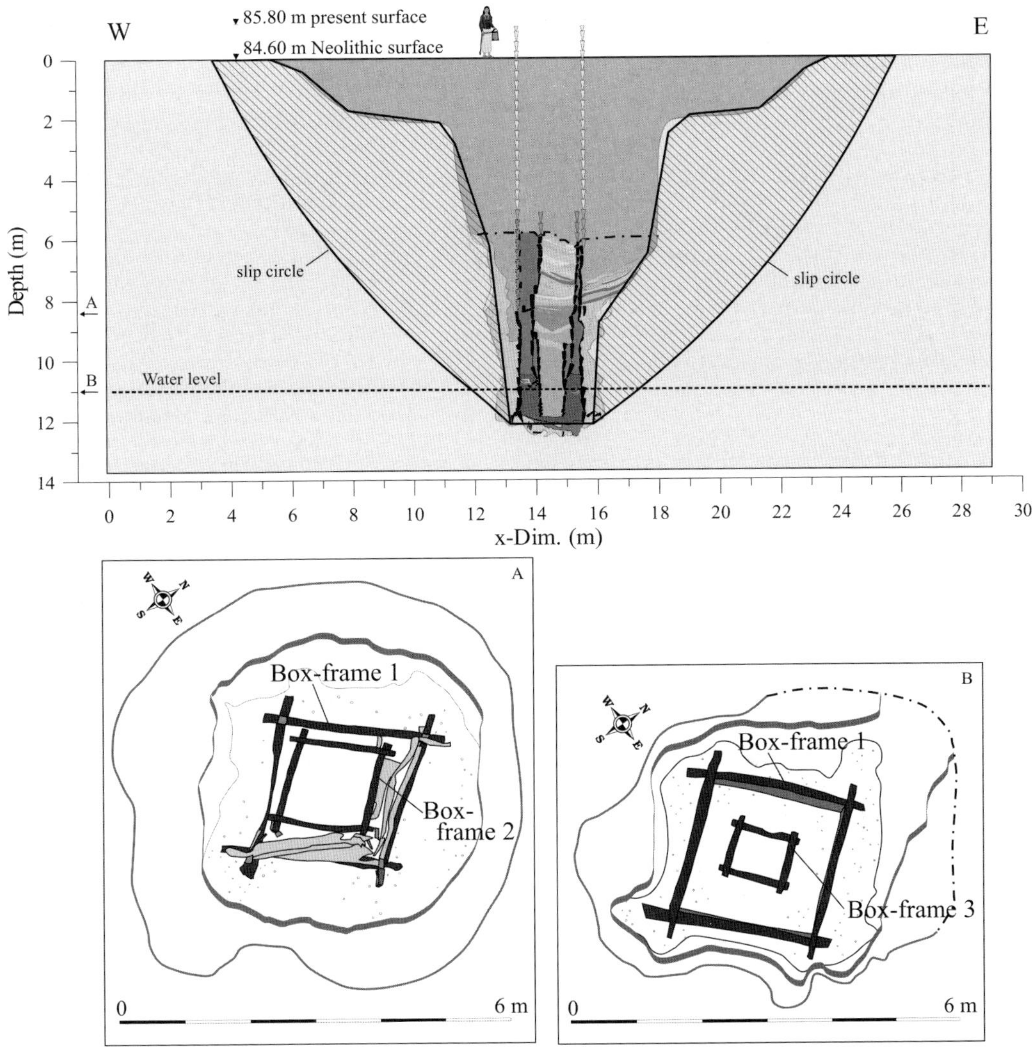

Fig. 3. (**Top**) Cross-cut through the Kückhoven Neolithic well. The archaeological findings are documented below the dash-dotted line as observed *in situ*. Above this line the situation is reconstructed. However, the size of the pit is constrained by observations of J. Weiner (drawing by K. Drechsel). Overlaid on the archaeological record is the simplified shape of the open excavation pit (heavy black lines), which was used for the static stability test. If a cohesion of less than 18–20 kN/m^2 is used for the sediments in which the pit was constructed, the walls of the pit become unstable. The resulting slip circles are indicated and the hatched areas show the sliding bodies. (**Bottom**) Two horizontal plans of the well at depth of 8.3 m (A) and 11 m (B). The shape of the pit is simplified after a drawing by K. Drechsel. The dark coloured beams in section (A) are in the original position. The light coloured beams moved into this level from above after the necks sheared off. The inner box 2 is located in the NW corner of box 1, in contrast to the cantered position at 11 m depth (B).

pit. Unit weight and angle of internal friction of 20 kN/m^3 and 37.5°, respectively, were used. Cohesion was varied between 0 and 30 kN/m^2. Results of the slope stability analysis are summarized in Figure 4. Cohesion values of 18–20 kN/m^2 are required for stable slopes of the excavation pit. For smaller values of cohesion, the slip circles for both sides are indicated in Figure 3. In both cases, it is the entire wall that shows the smallest factors of safety.

During the excavation of the well in 1991 a large section of the free face, measuring 40–60 m in width, was left untouched by the gravel production (Fig. 2e). The face had a height of at least 8 m and

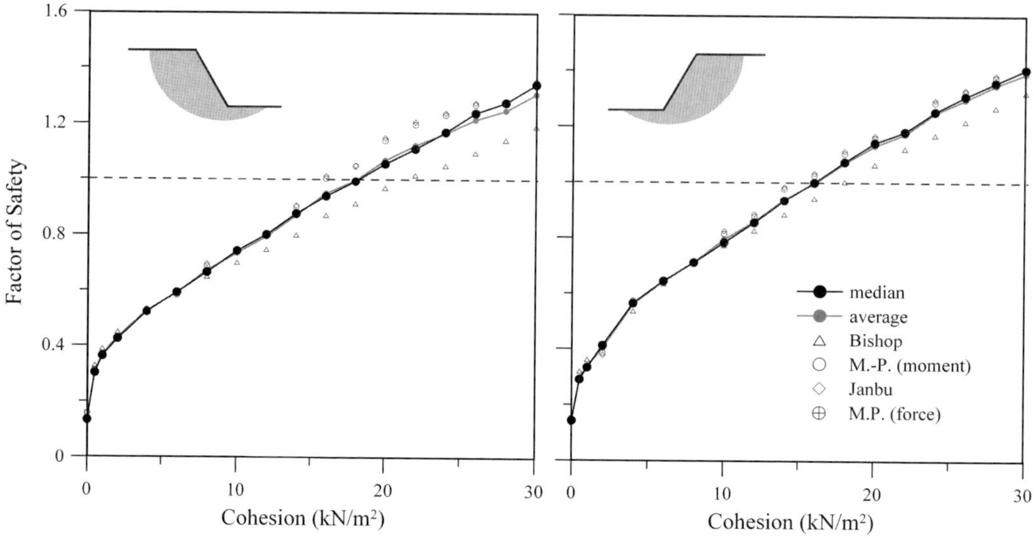

Fig. 4. Factor of safety as a function of cohesion calculated for the slopes of the left and right walls of the excavation pit shown in Figure 3. Results from four methods (see legend) are shown as well as the average and median value of the factor of safety. The dashed lines indicates a factor of safety of 1.0.

was in long, nearly vertical sections indicating a relatively high value for the effective cohesion, minimally 20 kN/m^2. This result makes construction method 2 more probable than method 1. The builders were probably experienced, and a balance was reached between stability of the slopes and minimization of the amount of excavated material. This hypothesis is supported by the varying inclination of the pit (Fig. 3). Also, provisional, temporal support of the walls by a wooden retaining system is likely but has not been proven from the archaeological excavation. The 8 m of significantly steep slopes may well indicate the maximum height of a subvertical stable wall that could be achieved under the given geotechnical conditions. With the gradual increase of the box-frame's height, it can be assumed that the space between the box-frame and the construction pit wall was immediately and continuously refilled in order to avoid collapsing of the pit wall. This system would have also stabilized the construction.

Seismotectonic setting

The Kückhoven Neolithic well is located in the Lower Rhine Embayment in the northern part of the Rhine–Rhone Rift system (Ziegler 1992, 1994; Van den Berg 1994; Ziegler & Cloetingh 2004) (Fig. 1). The current period of tectonic movements in the Lower Rhine Embayment can be closely correlated with late Tertiary graben structures and was initiated by small but widely distributed fault displacements in the late Miocene. During the Pliocene, faulting was more intense and cumulated in the late Pliocene and early Pleistocene. Considerable synsedimentary and intersedimentary crustal displacements occurred in the Quaternary when the Older and Younger Main Terraces of the Rhine and Meuse (Maas) Rivers were accumulated (Ahorner 1962).

Regional flexures, tilted blocks, basin-like subsidence, and frequent normal faulting occurring subvertically, at least at shallow depth, characterize the present geological structural activity in the Lower Rhine Embayment. The main Lower Rhine Embayment faults, with a cumulative length of more than 400 km, generally strike NW–SE (Fig. 1). Some active faults express a maximum vertical displacement of 174 m during the Quaternary. The structural features are expressions of a regional crustal tension in a NE–SW direction, confirmed by stress inversions of fault plane solutions (Hinzen 2003).

The Lower Rhine Embayment shows moderate seismicity in present and historic times (Fig. 1). Most of the stronger historic and instrumental earthquakes with magnitudes above 4 are connected to recent movements on the border faults of the Roer Valley Graben (RVG). A contemporary catalogue of instrumental seismicity since introduction of seismic networks in the northern Rhine area (Reamer & Hinzen 2004) combined with historic and early instrumental events leads to a Gutenberg-Richter model for the occurrence rate of earthquakes with magnitudes $1.8 \leq M_W \leq 5.2$ for the northern Rhine area (Hinzen & Reamer 2007):

$$\log(Nm/a) = -1.083M_W + 3.434 \quad (1)$$

Based on this empirical relation, average recurrence of an M5, M6 and M7 earthquake in the area is *c*. 100, *c*. 1200 and *c*. 14 000 years, respectively. Palaeoseismic studies at or near the border faults of the RVG show that the instrumentally and historically recorded earthquakes from the past 300 years do not include the largest events possible. Camelbeeck & Meghraoui (1998) found at least three surface-rupturing palaeoearthquakes on the Bree section of the Feldbiss Fault (a major western border fault of the RVG, Fig. 1) during the Holocene with an estimated average recurrence interval of 3500–5000 years. These earthquakes most probably attained M_W 6.3. Studies along other sections of the Bree fault (Vanneste *et al.* 2001) show evidence for six surface-rupturing events since the late Pleistocene. Several authors estimate a maximum magnitude for earthquakes in the Lower Rhine Embayment of M_W 7.0 (e.g. Ahorner 2001; Pelzing 2002; Hinzen 2005*a*). Camelbeeck *et al.* (2007) and Hinzen & Reamer (2007) provide an inventory of instrumentally recorded, historic, and palaeoseismological earthquakes of the region.

The Lövenicher Sprung (Fig. 5), the closest major fault to the Kückhoven Neolithic Well shows a vertical displacement of the bases of the Lower Pleistocene main terrace of 7 to 25 m (Ahorner 1962). With an age of *c*. 350 000 to 700 000 years (Klostermann 1988; Camelbeeck *et al.* 2007) this indicates an average displacement rate of 0.01 to 0.07 mm/year. Application of the empirical relation between average displacement on the fault plane of a normal faulting earthquake given by Wells & Coppersmith (1994) results in an average time span between two M6 earthquakes at this fault of 3000–20 000 years. Without a doubt, this estimate of the frequency of damaging earthquakes on a specific fault section contains a

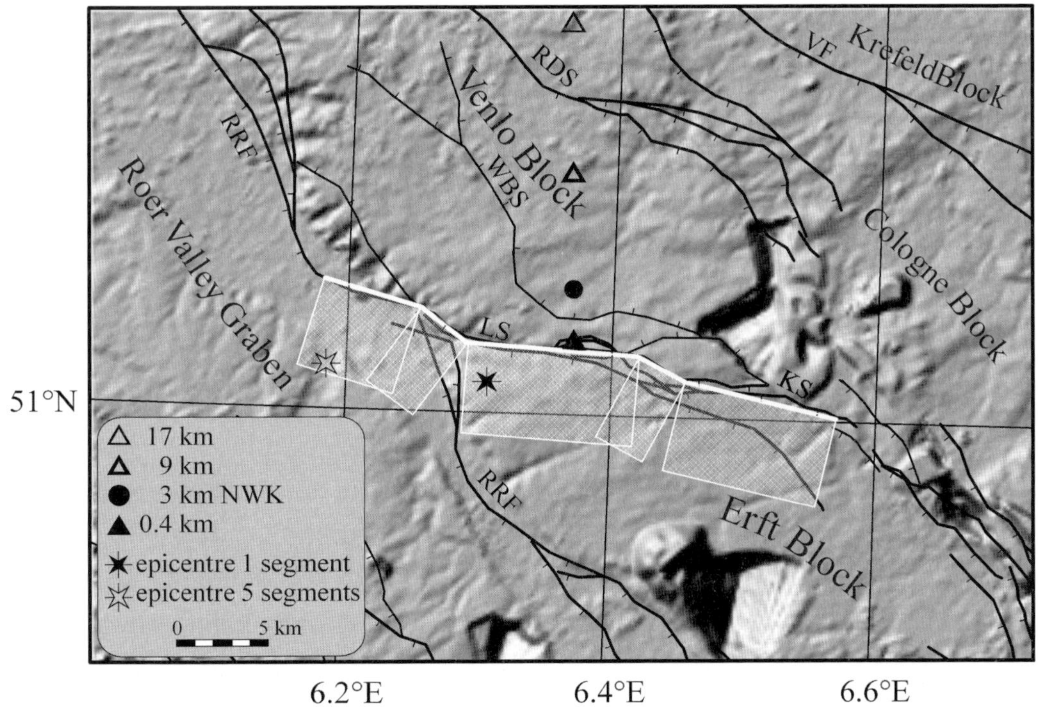

Fig. 5. The borders of this map are shown in Figure 1. The main tectonic blocks and active faults are shown. The fault names are: Viersen Fault (VF), Rheindahlener Störung (RDS), Wegberg Sprung (WBS), Rurrand Fault (RRF), Lövenicher Sprung (LS), and Kaster Sprung (KS). The geometry of the fault planes of two earthquake scenarios is indicated by the surface projection of the fault planes (white hatched rectangles). Stars indicate the epicentres for the two scenarios. The position of four sites (triangles and circle) for which the ground motion was calculated is shown; the circle is the location of the Kückhoven well (NWK). The striking topographic features in the middle part of the map are large open-pit lignite mines.

large range of uncertainty. However, ongoing movements on this fault system (Ahorner 1962) support the possibility of a damaging earthquake 7000 years BP and lend credence to the idea of an archaeoseismological basis for well damage at Kückhoven.

The Neolithic wooden well of Kückhoven

Site and stratigraphy

The Neolithic well of Kückhoven is located at 51.06°N and 6.37°E, 85.8 m a.s.l. (current surface) on the Venlo Block in the Lower Rhine Embayment. Geotechnical studies of the subsurface material in the vicinity of the well have been made in connection with the gravel-pit in which the well was discovered. The gravel-pit is located in an area where material of Tertiary age is covered by late Oligocene fine sand, middle to coarse sand and gravel of the Lower Rhine Terrace with a thickness of more than 30 m. A typical profile mapped from a borehole in the well's vicinity is shown in Figure 6. Holocene loam covers these layers with *c.* 2.5 m thickness.

Water-bearing strata are Pleistocene gravel sands of the Lower Terrace above the Tertiary clays at 35 m a.s.l. Layers at the gravel-pit site are 2 m of cohesive silt, fine sand and medium sand. The uppermost layer at present is a 0.4-m-thick humus topsoil horizon. Underneath are clastic Quaternary sediments, mainly medium to coarse sand and gravel. The sand–gravel layers are densely packed with an intermittent grain size distribution. The coarse grain is square-edged, and the angle of internal friction was determined to be 37.5°. Blow counts for 10 cm penetration in the SRS German standard test for the Quaternary sediments are well above 20, sometimes above 50. Unit weight is 20 kN/m^3. Radbruch (1969) rated stability for slopes in soil for the Oakland, California, area and introduced the four stability ratings of poor, fair, varied and good. Considering the penetration test results and the observation of the stable free face, the slopes in the gravel-pit are rated as good. For this category, Radbruch (1969) assumes an angle of internal friction of 35° and an effective cohesion of 70 kN/m^2.

The contemporary surface at the well site was 85.80 m a.s.l. before the extraction of gravel started. The walking horizon during Neolithic was at *c.* 84.60 m a.s.l. (Weiner 1998) The bottom of the construction pit is at 72 m and the bottom of the well casing at 72.5 m a.s.l. The current groundwater distribution at the location of Kückhoven Neolithic well is well known due to numerous exploration and measuring boreholes in the forefront of an opencast lignite mine in the vicinity. Spelter (1998) estimates that the leachate reaches the former groundwater horizon at a depth of 12 m within a few days. The hydrograph curve therefore shows a trend similar to the natural precipitation. With a groundwater level at +74 to +75 m a.s.l. and the well bottom at +72 m a.s.l., the latter was *c.* 2.5 m below the water table.

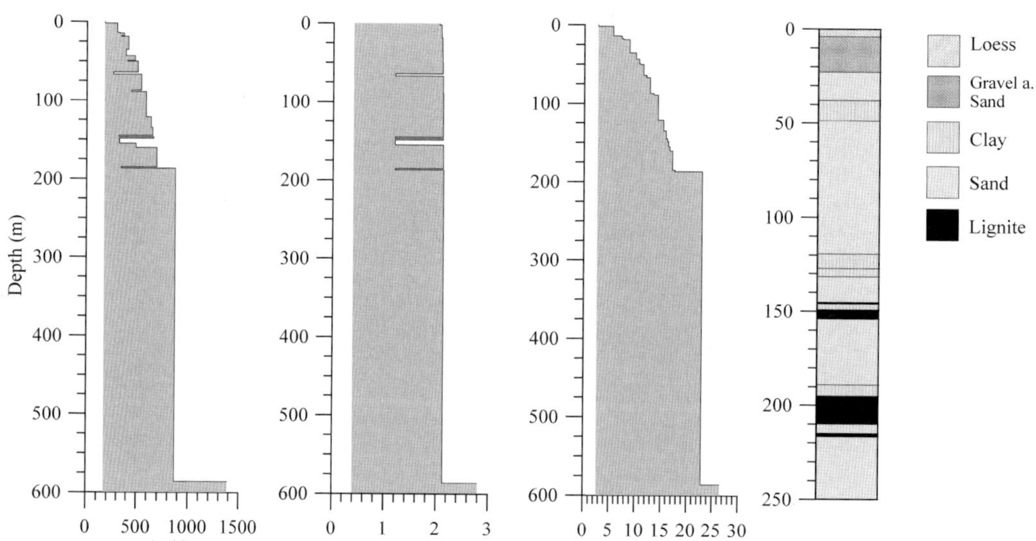

Fig. 6. Depth distribution of the shear wave velocity, v_s, density ρ and Q-factor, Q for the upper 600 m and litholog for the upper 250 m from a borehole at the location of the Kückhoven Neolithic well.

Construction

Since the well construction method and static stability of the open pit have to be determined prior to dynamic analysis, these questions are addressed in the following. Figure 3 shows a drawing with a section through the site in west–east direction. The *in situ* positions of the construction elements of well casing box-frames 1, 2 and 3 are shown below the dash-dotted line. Above this line, the positions of the box-frame of wells II and I have been reconstructed. The same applies to the shape of the construction pit. However, the excavation confirms the lateral dimension of the pit at a level of 6 m below the surface. In Figure 3, the corresponding simplified model of the construction pit is shown, which was used for further analysis. The dimension of the construction pit is estimated as follows from the archaeological documentation. Total depth of the pit, measured from the estimated walking horizon in Neolithic time, is 13 m. The cross-cut surface of the pit (Figs 2 and 3) measures 77.3 m^2. To estimate the volume of the entire excavation, the cross-cut was vertically sectioned in seven frustums. This results in a volume of 540–550 m^3, requiring the removal of over 1000 tonnes of material.

The archaeological reconstruction counts a total of 37 construction elements on each of the four sides from bottom to surface. For security and hygienic reasons a parapet grab above the surface has to be assumed, consisting of three planks measuring together roughly 1 m in height. This results in a total number of 160 (4 $\times$ 40) elements. A schematic drawing of a worked plank is shown in Figure 7 together with box-plots of the dimensions measured from 68 recovered construction elements, of which 60 were complete. Average length (L), height (H), width at the pith-channel (W_i), and at the outer edge (W_o) are 273 cm, 33 cm, 14 cm and 5 cm, respectively. With a density of 1.07 Mg/m^3 the weight of a single standard plank is more than 60 kg and the total weight of box-frame 1, forming well I is more than 10 tonnes. The box-plots show that the elements were very precisely worked, with small deviation from the standard measure. This is a strong argument against deficits in the quality of the construction as a possible cause of the damage.

Damage

Already on the visible first planum when both box-frames 1 and 2 could initially be seen *in situ*, an important difference between the two constructions was noted. Frame 2 displayed four complete sides with perfectly, and horizontally positioned wooden planks, devoid of any damage. In contrast, frame 1 presented only two sides (NW, SW sides) in their original, horizontal position in an undamaged state while the remaining two sides (SE, NE sides) displayed unusual traces of damage. On those sides, the planks' central sections between their corners

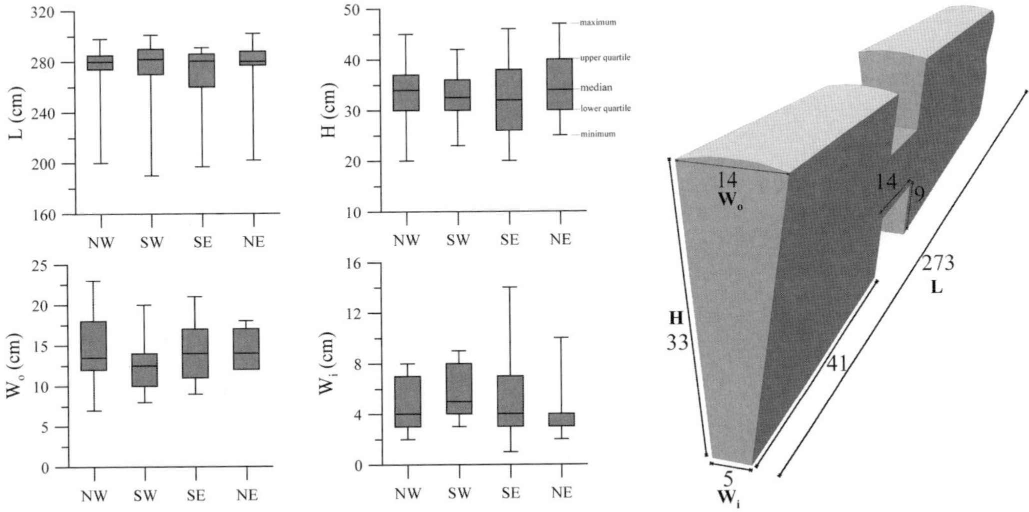

Fig. 7. The drawing on the right gives a perspective view of the worked end of a single standard oak beam of box-frame 1 of the Neolithic well of Kückhoven. Measurements given in centimetres are the average values over all recovered beams. The box plot diagrams on the left show the distribution of the free length between the notches (L), the height of the beams (H), and the width at the outer and inner perimeter, W_o and W_i, respectively. The four sides of the well casing are labelled with points of the compass.

(i.e. northern and eastern, and southern and eastern corners, respectively) were moved from their original horizontal position into a pronounced oblique one. All planks from the southeastern and the northeastern wall were dipping to the eastern corner (Fig. 2b, c, d). Additionally, all displaced planks were moved away from their former right-angle position, as if they were pushed into the box-frame's inner, open space (Fig. 3). This movement also led to a distinct, and obviously simultaneous, dislocation of the eastern corner, which had been moved away from its original vertical position into the box-frame's inner space, as well. Consequently this caused a deformation of the box-frame's original square cross-section, resulting in an irregularly lozenge-shaped one (Fig. 3).

Highly noticeable is the observation that, despite the damage-causing movement, the ends of all affected planks protruding into the refilled construction pit's open space had retained their original horizontal positions; however, they were all vertically sheared off at their weakest points, i.e. their 'necks' (Fig. 2d). Taking those specific damage features into account, it is possible to infer the direction of the movement producing the damage. It must have been directed simultaneously downwards and inwards, but the cause of the damage still remains unclear.

Damage scenarios

Since the discovery of box-frame 1 the question arose of what may have caused the particular damage. Hypotheses have been developed, and subsequently discarded (e.g. Weiner 1998). As it became clear that the damage could not have been caused by human activity, only natural causes remained plausible, the most feasible ones being static sediment pressure and/or an earthquake. The aim of this study is to test for possible damage scenarios using geotechnical and engineering geophysical models.

As described by Galadini *et al.* (2006), the development and study of damage scenarios for archaeological findings is a multifaceted problem. In the following, five damage scenarios are explored.

(1) Initial interpretations, some of which were already developed during the archaeological excavation's first stages (Weiner 1998), explained the damage as man-made. It was assumed that during a raid on the settlement, one or several persons climbed into the well to a depth of about 6 m below the LBK surface and, using adzes, chopped at some of the wooden elements, thus sabotaging the box-frame in order to make the well unusable. This interpretation is no longer feasible, as it would have been a very hazardous life-threatening enterprise. People back then were most probably aware that the well could have been rendered unusable by poisoning it, e.g. by simply dropping down a carcass. Additionally, no traces of copping were found at the sheared beam necks.

(2) In the second damage scenario, pressure from the surrounding sediment cracks the box-frame at the southeastern corner, simultaneously pushing a large section of the box-frame's southeastern and northeastern walls into the well shaft's open space. Earth pressure is the lateral force exerted by the soil on a shoring system. It is dependent on the soil structure and the interaction with the retaining system. Active earth pressure is the condition in which the earth exerts a force on a retaining system, the members of which tend to move towards the excavation. The well-known theory of Coulomb (1776) provides expressions for earth pressure for a soil mass at a stage of failure. Figure 8 shows the active earth pressure on the vertically assumed walls of box-frame 1. The grey zone indicates the depth of the observed damage to box-frame 1. The weakest parts of the wooden elements are the sections with reduced size of the cross-sectional area due to the notches at both ends of the elements. With the average dimensions of the elements this section is *c.* 0.014 m^2. Assuming shear strength of the oak wood to be 11.6 MPa (MatWeb 2005; P. Beiss & E. El-Magd pers. comm.), shear failure would occur if the cohesion were less than

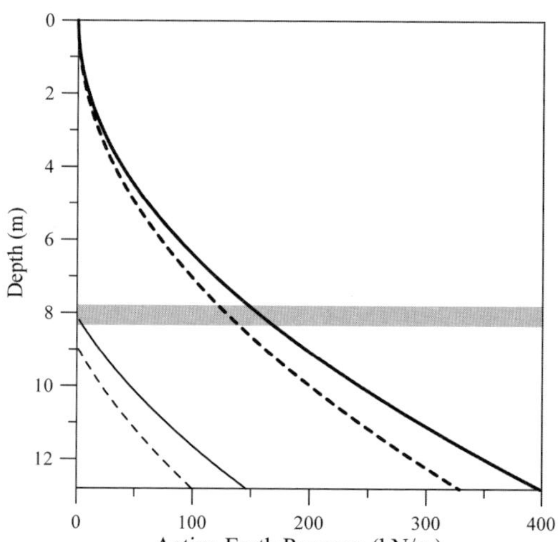

Fig. 8. Active earth pressure at the vertical walls of the well casing as a function of depth. Heavy and thin lines represent the results for the calculation without and with cohesion (20 kN/m^2), respectively; continuous and dashed lines show the cases without and with wall friction. The grey bay indicates the depth at which the damage occurred in box 1.

3.2 kN/m^2. With cohesion of at least 20 kN/m^2 the active earth pressure could not overcome the shear strength of the construction elements in the zone of observed damage. Consequently, the scenario of earth pressure acting on an otherwise intact well casing can be discarded.

(3) Decomposition of wooden construction elements poses the third damage scenario. While completely dry sections as well as saturated sections of the box-frame are sturdy, frequent changes between wet and dry conditions due to fluctuation of the water table and/or waves and splatter induced by pulling up water with bag-shaped pails might have had a negative influence on the wood's structure. However, as it could be established that this box-frame had been built using nearly exclusively first class 'straight grain' oak, severe damage by decomposition should not have occurred after around 30 years as indicated by the dating results (B. Schmidt pers. comm. 1996).

(4) The fourth proposed damage scenario is that of dynamic excitation by earthquake forces. One hypothesis is that the wider section of unconsolidated, comparatively loose backfill at the upper two-thirds of the well (Fig. 3) might have produced a bending effect and resulted in the very characteristic damage pattern, vertically shearing off the wooden element's protruding ends in the damaged section. An engineering seismological model is developed in a separate section.

(5) Planks above the water table are exposed to the atmosphere at the well side and to ground moisture at the back, while those beneath the water table are saturated. Changes of the water table in the well determined which planks became dry or eventually water-saturated. Oak wood shows quite different swelling behaviour in different directions. The German standard DIN 52184 (1979) gives the maximum swelling of oak wood as 0.3–0.6% parallel to the fibres, 4.6% in a radial direction and 10.9% tangential to the fibres. In a worst-case scenario, we assume that the well was dry for a certain period, i.e. a dry summer. With the start of a rainy period, a quick rise of the water table saturates the well. As mentioned before, the hydrograph curve shows a similar trend as the precipitation and the leachate reaches the ground water table fast at the Kückhoven Neolithic well (Spelter 1998). In this case differential swell between the uppermost saturated and the lowermost 'dry' plank might have occurred. Due to the large swelling coefficient tangential to the fibres, the 14 cm wide top of the saturated plank might have experienced a swelling of up to 1.5 cm. This swelling would result in a growing stress in the neck of the upper 'dry' plank and could have eventually sheared off the end of a plank. After such initial damage, earth pressure could have driven the sheared plank inward causing more planks to break at their necks.

Testing a seismic scenario

The Kückhoven Neolithic well is located at a lateral distance of *c*. 3 km from the Loevenicher Fault, part of the system of major tectonic faults in the Lower Rhine Embayment (Figs 1 and 5). The fault strikes 100–110°, and dips towards the SE. In a deterministic approach it is assumed that this fault ruptured in close vicinity to the well site. Two earthquake scenarios are assumed: (1) a 9.1-km-long section of the Loevenicher Fault ruptures resulting in an earthquake with magnitude 6.2; and (2) five segments, three of the Lövenicher Fault and two on the Kaster Fault (Fig. 5), rupture consecutively from west to east with a total rupture length of 28.5 km resulting in a magnitude 6.8 earthquake. As the seismotectonic potential of other faults in the LRE does not allow earthquakes significantly stronger than assumed here, we limited the test to these two scenarios. If the earthquake frequency model from Equation 1 is extrapolated to stronger events, the recurrence rates for the two earthquake scenarios for the northern Rhine area are 1860 years and 7900 years, respectively. Figure 5 shows the source geometry and position of the well.

Composite source model

A composite source model (CSM) (Zeng *et al.* 1994; Keaton 1999) was used to calculate synthetic seismograms of the assumed earthquakes specifically for the site of the well. The modelling is split into two steps. First the elastodynamic Green's function is calculated for a specified location and distance from a planar (earthquake scenario 1) or multiplanar (earthquake scenario 2) fault for seismic waves passing through a flat-layered medium. In a second step, three-component acceleration, velocity and displacement seismograms are calculated by convolving the pulse of energy from generally thousands of subevents, which are distributed as circular asperities on the fault plane(s). Rupture progress over the source area is simulated by time-delayed activation of the subevents, starting at the hypocentre and progressing with a constant rupture velocity (Zeng *et al.* 1994). Parameters of the one-dimensional velocity model are given in Table 1 and Figure 6, and dimension and properties of the sources in Table 2. Figure 9 shows the calculated strong motion seismograms at the well site.

Site effects

Overall thickness and structure of the sediments in a basin like the Lower Rhine Embayment have a

Table 1. *Parameters of the flat layer earth model used for the calculation of synthetic seismograms. Q_P and Q_S indicate the quality factor for P- and S-waves, respectively*

Layer thickness (km)	P-wave velocity (km/s)	S-wave velocity (km/s)	Q_P	Q_S	Density (Mg/m^3)
0.3	1.00	0.40	90	70	1.7
0.5	1.20	0.60	100	80	1.8
1.0	5.00	2.89	150	100	1.9
1.0	5.50	3.18	160	107	2.1
1.0	5.80	3.35	300	200	2.4
7.0	6.00	3.46	500	333	2.5
8.0	6.25	3.61	800	533	2.8
12.0	6.90	3.98	800	533	3.1
10.0	8.10	4.68	900	600	3.3
Halfspace	8.11	4.68	1000	667	3.3

Table 2. *Source parameters used in the calculation of synthetic seismograms for the earthquake scenarios 1 and 2*

	Scenario 1	Scenario 2
Moment magnitude	6.19	6.65
Rupture length (km)	9.1	28.5
Rupture width (km)	9.3	13.1
Seismic moment (Nm)	2.40E+18	1.17E+19
Average displacement (m)	0.78	1.09
Dip of fault plane	70°	70°
Rake	−85°	−85°
Number of subsources	1633	2586
Min. subsource radius (km)	0.03	0.06
Max. subsource radius (km)	3.93	9.85

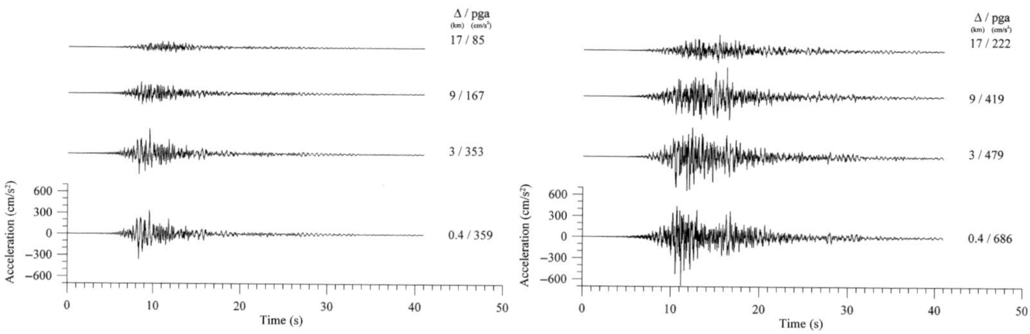

Fig. 9. Synthetic seismograms calculated with the composite source model for the two earthquake scenarios described in the text. The seismograms on the left and right sides show the horizontal (EW) acceleration at the four sites shown in Figure 5 for earthquake scenarios 1 and 2, respectively. Distance from the surface trace of the fault and the maximum acceleration are given next to each seismogram. The trace at 3 km distance represents the situation at the Kückhoven Neolithic well site.

significant influence on amplitudes, frequency and duration of seismic ground movements. The model and method described by Weber & Hinzen (2006) are used to compute site effects and strain in the sediment layers at the Kückhoven Neolithic well site. Calculations are based on the one-dimensional model from Figure 6 and interaction with the well is not accounted for. The almost 600 m of sediment have a damping effect at frequencies above 1.45 Hz (Fig. 10). At lower frequencies the maximum amplification of 2.9 is reached at 0.32 Hz and 0.29 Hz for earthquake scenarios 1 and 2, respectively. The slightly lower frequency of the maximum amplification and the overall smaller amplification is due to the non-linear behaviour of the sediments, which show an increase in damping with increasing shear strain.

Though the soil structure interaction is not accounted for in the simple 1D calculation of the soil movements, the depth distribution of strain in the top layers gives an upper bound for the horizontal deformation of the wooden well casing during seismic excitation from the two earthquake scenarios. The maximum strain of the sediments at the Kückhoven Neolithic well site following the model from Figure 6 was used to estimate the corresponding horizontal deformation of box-frame 1. The depth-dependent deformation is shown in Figure 11. At 12 m depth the maximum strain comes close to 0.03%, equivalent to a deformation of the box-frame of less than 1 mm.

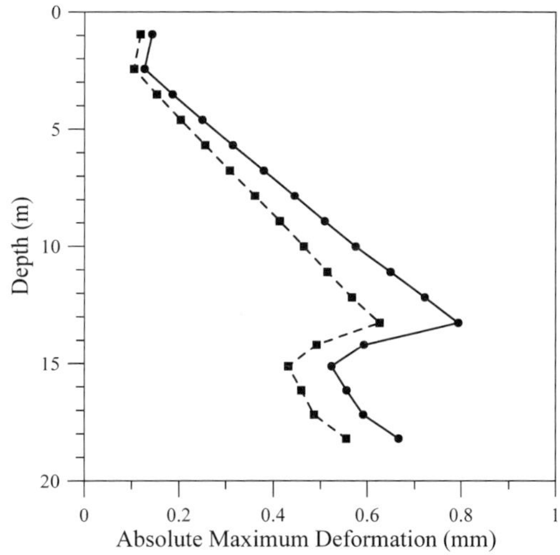

Fig. 11. Maximum deformation on 2.8 m length as a function of depth within the sediments. The continuous and dashed curves were derived for earthquake scenarios 1 and 2, respectively.

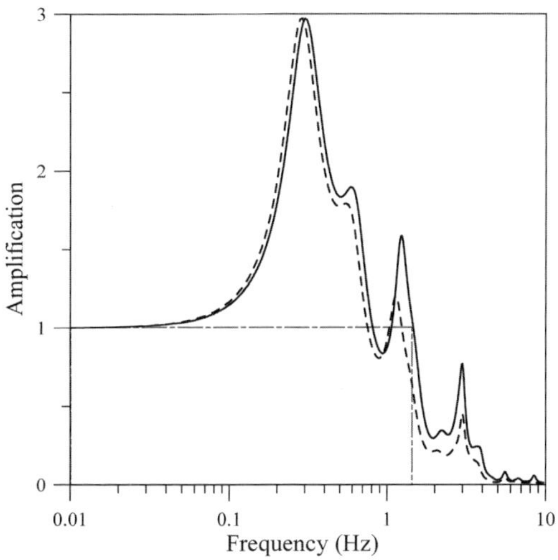

Fig. 10. Frequency-dependent ground amplification at the Kückhoven Neolithic well site for earthquake scenario 1 (continuous line) and scenario 2 dashed line. The dash-dotted line indicates the frequency of 1.45 Hz above which the sediments reduce the ground motion amplitudes.

Dynamic time-history analysis

A simple finite element (FE) model was used to study the general dynamic behaviour of the box-frame. The FE model (Fig. 12) consists of four times 40 horizontal construction elements of 2.73 m length and 0.15 × 0.33 m cross-section. Vertically the beams are connected by 0.33 m long beams of 0.15 × 0.15 m to account for the reduced cross-section size at the junctions (Fig. 7). Total height of the structure is about 13 m. Compressive and tensile strengths of 23 and 18 GPa parallel to the fibres are assumed, respectively, following DIN EN 338 (1996) taking class C30 for oak wood. The strain at peak stress is 0.04 m/m. Two mass elements of 35 kg each are assigned to each horizontal element giving a total mass of 11.2 t.

In a first step, an eigenvalue analysis of the box-frame was made, assuming it is free-standing without any restraints on the sides. Three major eigenmodes with frequencies below 30 Hz exist in the direction parallel to the horizontal ground motion (Fig. 12). The major eigenmode in the vertical direction was found at 29.5 Hz with 80% of the total mass.

In a second step, the seismograms form the composite source model for the two earthquake scenarios were used as dynamic loading of the FE model of the box-frame. In order to account for

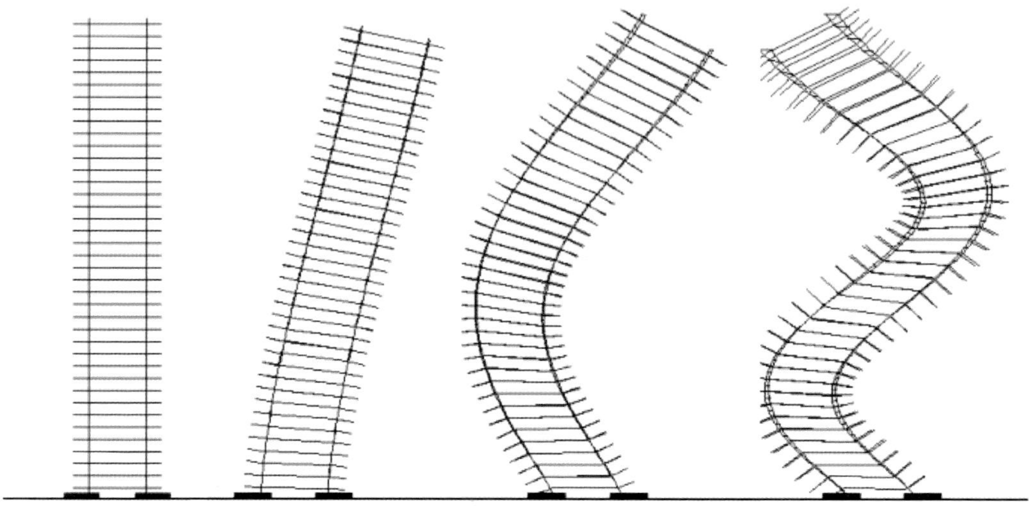

Fig. 12. Shape of the structural model of Kückhoven Neolithic well casing 1 and the first three basic eigenmodes in the horizontal x-direction. From left to right, the frequencies of the eigenmodes are 3.73 (66.3%), 14.2 (19.7%) and 28.9 Hz (4.6%): numbers in parentheses give the percentage of total mass of that eigenmode. Exaggeration of the deformation with respect to the size of the well is 5000.

the interaction with the surrounding subsurface material, dashpot-dampers were added to the corner nodes of the model. Two calculations were made for each scenario: (A) common damping of 0.5 at all nodes, and (B) smaller damping of 0.25 at the nodes above 4 m measured from the bottom. This is a rather arbitrary assumption and an oversimplification to account for the soil structure interaction; however, it provides a rough simulation of the dynamic effect of the wider section with backfill material in the upper part of the pit, as indicated in Figure 3, to test for the proposed damage hypothesis.

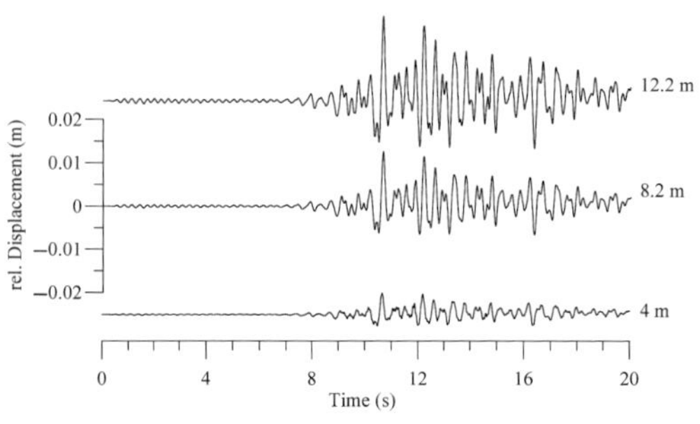

Fig. 13. Relative horizontal displacement from the FE model at three different levels from the bottom of the model. The dynamic load is the synthetic seismogram from Figure 9 for earthquake scenario 2. Maximum displacement is less than 2 cm.

The small relative displacements of less than 2 cm (Fig. 13) from the FE model calculation and the deformation of the sediments in the range of the well of less than 2 mm on a length corresponding to the width of the wooden box-frame of well I indicate that the earthquake scenarios cannot explain the failure of the wooden elements at a depth of *c.* 7.5 m if it is assumed that the elements were of first quality oak wood, free of any major flaws. In this range of deformation the wooden structure would have been in the elastic range.

Discussion and conclusion

The uncertainty as to the cause of structural damage of the Neolithic wooden well casing at Erkelenz Kückhoven and the close vicinity to one of the active faults in the Lower Rhine Embayment led to the question of whether these damages could have been seismogenic in origin. The assumption of a 'seismological worst case scenario' for the ground movements at the well site during an earthquake and the application of an engineering seismic model show, however, that the relative displacements which can be expected within the wooden structure of the well are too small to explain the damage. This result holds true if all beams can be assumed to be without any pre-damage. There is no reason to assume that the LBK well builders would have used either wood of poor quality or incorrectly worked construction elements. They did have more than 400 years of well-building experience, proven by several older well-known wooden LBK wells, and the precision of the beam dimensions (Fig. 7) is another argument against poor construction.

This result should not be regarded as a negative outcome of the archaeoseismic study; far from it, the exclusion of a seismogenic cause for archaeologically described structural damage of any construction has equal importance to the opposite outcome. Generally, we should be careful that archaeoseismology is utilized in a way commensurate with the scientific method, i.e. the testing of a hypothesis, and acceptance of results that either prove or disprove damage of a seismogenic nature. In accordance with this principle, the close vicinity of an active fault in this case study certainly justified the use of quantitative models to test the earthquake hypothesis.

Concerning the seismic hazard of the region, it was important to analyse the Neolithic well. The historic and instrumental earthquake record of NW Europe lists 14 events since 1350 with MS 5 or larger (Camelbeeck *et al.* 2007). Currently, the youngest palaeoseismic earthquake described in the area was found at the Bree Fault (Camelbeeck *et al.* 2007). This normal faulting earthquake with an average surface displacement of 0.55 m is dated 2970–8000 years BP. The hypothesized but very well datable Kückhoven event would have helped to close the gap between the historic and palaeoseismic record.

Considering the more than 400 km of active faults in the Lower Rhine Embayment and the very few (ten) palaeoseismic trenches which presently serve to document the Holocene displacement history at five locations, any corroborating evidence for a possible damaging earthquake, historical or archaeological, should be thoroughly studied in the future. Three other objects in the Lower Rhine Embayment have thus far been analysed with archaeoseismological methods: the Roman fortification of Tolbiacum (Hinzen 2005*a*), a Roman villa in Kerkrade (Hinzen 2005*b*) and the Praetorium of Cologne (Hinzen & Schütte 2003). Damage in the Kerkrade villa is possibly of seismogenic origin; however, clear evidence could not be established. The Tolbiacum city wall was probably damaged by an earthquake; however, the archaeological work to evaluate a date is still in progress. The proposed secondary earthquake effects on the Cologne Praetorium, which might have happened around AD 800, are the target of an ongoing study in connection with further excavation of the building and its surroundings in the Cologne Archaeological Zone.

The model used in this study distinguishes the buried wooden well casing as not susceptible enough to 'record' ground movements from a nearby damaging earthquake. This might have implications for similar situations found elsewhere. Nevertheless, the Lövenicher and Kaster faults, striking *c.* 100° and connecting the almost parallel, 20 km separated, Rurand and Peel Fault with the Erft Fault system (Fig. 1) should be a target for future palaeoseismological studies.

The results of this study confirm the generally accepted experience that underground structures are less vulnerable to earthquake-induced ground movements than above-ground structures (e.g. Chen & Scawthorn 2003). However, without quantitative analysis, the seismogenic damage hypothesis for the Neolithic well could not have been excluded. The outcome is important for further interpretation of the archaeological findings at the LBK site in Kückhoven. Further studies should reveal whether an oscillating water table around *c.* 4 m above the well bottom caused rotting of the wood to a stage at which the sediment pressure finally broke the necks of several oak beams, or whether the possible shrinkage and swelling of scenario 5 was the true cause.

We are grateful to R. Zähringer, Niederzier, for providing oak wood, which was used in physical testing. The testing of material strength was done at the Institut für Werkstoffkunde, RWTH Aachen, and we thank P. Beiss and E. El-Magd for their efforts. S. Schubert of ETH Zürich and D. Liberatore of University Basilicata, Potenza, gave valuable hints on mechanical and dynamic wood properties. Finite element calculations were made with the SeismoStruct code, which was provided by SeismoSoft (www.seismosoft.com). S. Reamer carefully read the manuscript and gave valuable hints to improve the text. The very helpful comments and suggestions by M. Sintubin and K. Reicherter, who reviewed the original manuscript, are gratefully acknowledged.

References

AHORNER, L. 1962. Untersuchungen zur quartären Bruchtektonik in der Niederrheinischen Bucht. *Eiszeitalter und Gegenwart*, **13**, 24–105.

AHORNER, L. 2001. Abschätzung der statistischen Wiederkehrperiode von starken Erdbeben im Gebiet von Köln auf Grund von geologisch-tektonischen Beobachtungen an aktiven Störungen. *DGG Mitteilungen*, **2**, 2–9.

CAMELBEECK, T. & MEGHRAOUI, M. 1998. Geological and geophysical evidence for large paleoearthquakes with surface faulting in the Roer Graben (northwest Europe). *Geophysical Journal International*, **132**, 347–362.

CAMELBEECK, T., VANNESTE, K. *ET AL.* 2007. Relevance of active faulting and seismicity studies to assessments of long-term earthquake activity and maximum magnitude in intraplate northwest Europe, between the Lower Rhine Embayment and the North Sea. *In*: STEIN, S. & MAZZOTTI, S. (eds) *Continental*

Intraplate Earthquakes; Secience, Hazard, and Political Issues. Geological Society of America, Special Papers, **425**, 193–224.

CAMPEN, I. 2000. Zwei weitere bandkeramische Brunnen aus dem Tagebau Zwenkau. *Archäologie aktuell im Freistaat Sachsen*, **6**(1998/99), 42–47.

CHEN, W.-F. & SCAWTHORN, C. 2003. *Earthquake Engineering Handbook (New Directions in Civil Engineering)*. CRC Press, Boca Raton, Florida.

CODUTO, D. P. 1998. *Geotechnical Engineering: Principles and Practices*. Prentice-Hall, New York.

COULOMB, C. A. 1776. *Sur une application des régles, de maximis et minimis à quelques problèmes de statique, relatifs à l'architecture*. Paris, L'Imprimerie Royale, Mémoires de Mathématique et de Physique, 343–384.

DECKER, K., GANGL, G. & KANDLER, M. 2006. The earthquake of Carnuntum in the fourth century A.D. – archaeological results, seismologic scenario and seismotectonic implications for the Vienna Basin fault, Austria. *Journal of Seismology*, **10**, 479–495.

DIN 52184. 1979. *Prüfung von Holz – Bestimmung der Quellung und Schwindung*. Deutsches Institut für Normung, Berlin.

DIN EN 338, 1996. *Bauholz für tragende Zwecke, Festigkeitsklassen*. Beuth Verlag, Berlin.

FÄH, D., STEIMEN, S. *ET AL*. 2006. The earthquake of 250 A.D. in Augusta Raurica, a real event with a 3D site-effect? *Journal of Seismology*, **10**, 459–477.

GALADINI, F., HINZEN, K.-G. & STIROS, S. 2006. Archaeoseismology: methodological issues and procedure. *Journal of Seismology*, **10**, 395–414.

GUIDOBONI, E., MUGGIA, A. & VALENSISE, G. 2000. Aims and methods in territorial archaeology: possible clues to a strong fourth-century A.D. earthquake in the strait of Messina (southern Italy). *In*: MCGUIRE, W. J., GRIFFITH, G. D., HANCOCK, P. L. & STEWART, I. S. (eds) *The Archaeology of Geological Catastrophe*. Geological Society, London, Special Publications, **171**, 45–70.

HINZEN, K.-G. 2003. Stress field in the northern Rhine area, Central Europe, from earthquake fault plane solutions. *Tectonophysics*, **377**, 325–356.

HINZEN, K.-G. 2005*a*. The use of engineering seismological models to interpret archaeoseismological findings in Tolbiacum, Germany, a case study. *Bulletin Seismological Society America*, **95**, 521–539.

HINZEN, K.-G. 2005*b*. Archaeoseismological case study of the main building of Kerkrade-Holzkuil. *In*: TICHELMAN, G. (ed.) Het villacomplex Kerkrade Holzkil, ADC ArchaeoProjekten, rapport **155**, 407–420.

HINZEN, K.-G. & REAMER, S. K. 2007. Seismicity, seismotectonics, and seismic hazard in the Northern Rhine Area. *In*: STEIN, S. & MAZZOTTI, S. (eds) *Continental Intraplate Earthquakes; Science, Hazard, and Political Issues*. Geological Society of America, Special Papers, **425**, 225–242.

HINZEN, K.-G. & SCHÜTTE, S. 2003. Evidence for earthquake damage on Roman buildings in Cologne, Germany. *Seismological Research Letters*, **74**, 124–140.

KEATON, J. R. 1999. *Synthetic seismograms for normal-faulting earthquakes using the composite source model*. EERI-FEMA Professional Fellowship Report, **PF99-8**.

KLOSTERMANN, J. 1988. Quartär. *In*: *Geologie am Niederrhein*. Geologisches Landesamt Nordrhein-Westfalen, Krefeld, 40–63.

MATWEB. 2005. Online material data sheet, European white oak wood. World Wide Web address: http://www.matweb.com (accessed 5.12.07).

NIKONOV, A. A. 1988. On the methodology of archaeoseismic research into historical monuments. *In*: MARINOS, P. G. & KOUKIS, G. C. (eds) *The Engineering Geology of Ancient Works, Monuments and Historical Sites, Preservation and Protection*, Proceedings of an International Symposium organized by the Greek National Group of IAEG, Athens, 19–23 September 1988, Rotterdam (The Netherlands) 1315–1320.

PELZING, R. 2002. *Seismizität und Erdbebengefährdung in der Niederrheinischen Bucht*. Jahrestagung Deutsche Geophysikalische Gesellschaft, Berlin.

RADBRUCH, D. H. 1969. *Areal engineering geology of the Oakland East quadrangle*. US Geological Survey Map **GQ-769**.

REAMER, S. K. & HINZEN, K.-G. 2004. An earthquake catalog for the Northern Rhine Area, Central Europe (1975–2002). *Seismological Research Letters*, **74**, 575–582.

SCHMIDT, B. 1992. Holzfunde des Jahres 1991. Ein dendrochronologischer Bericht. *In*: *Archäologie im Rheinland 1991*. Landschaftsverband Rheinland, Köln, 166–168.

SCHMIDT, B. & GRUHLE, W. 2003. Wuchshomogenität als ein neues Analyseverfahren zur Verbesserung der dendrochronologischen Datierungsmethode. Die Hölzer der neolithischen Brunnen von Erkelenz-Kückhoven, Zwenkau und Mohelnice sowie vom Fundplatz Kaster. *In*: *Archäologische Perspektiven. Analysen und Interpretationen im Wandel*. Festschrift für Jens Lüning zum 65, Geburtstag, Internationale Archäologie, Studia Honoraria, Rahden/Westf, **20**, 49–60.

SCHMIDT, B., HÖFS, E., KHALESSI, M. & SCHEMAINDA, P. 1998. Dendrochronologische Befunde zur Datierung des Brunnens von Erkelenz-Kückhoven in das Jahr 5090 vor Christus. *In*: *Brunnen der Jungsteinzeit*, Internationales Symposium Erkelenz, 27. bis 29. Oktober 1997, Landschaftsverband Rheinland, Rheinisches Amt für Bodendenkmalpflege, Materialien zu Bodendenkmalpflege im Rheinland, Bonn, **11**, 279–289.

SPELTER, M. 1998. Hydrologische Aspekte zum Befund eines bandkeramischen Holzbrunnens von Erkelenz-Kückhoven. *In*: *Brunnen der Jungsteinzeit*, Internationales Symposium Erkelenz, 27. bis 29. Oktober 1997, Landschaftsverband Rheinland, Rheinisches Amt für Bodendenkmalpflege, Materialien zu Bodendenkmalpflege im Rheinland, Bonn, **11**, 223–228.

STÄUBLE, H. & CAMPEN, I. 1998. 7000 Jahre Brunnenbau im Südraum von Leipzig. *In*: *Brunnen der Jungsteinzeit*, Internationales Symposium Erkelenz, 27. bis 29.

Oktober 1997, Landschaftsverband Rheinland, Rheinisches Amt für Bodendenkmalpflege, Materialien zu Bodendenkmalpflege im Rheinland, Bonn, **11**, 51–71.

STEWART, I., SINTUBIN, M. & SIMILOX-TOHON, D. 2007. Archaeoseismology: a new standardised methodology using logic trees. *Geophysical Research Abstracts*, **9**, 01886.

STIROS, S. & JONES, R. E. (eds) 1996. *Archaeoseismology*. British School at Athens, Fitch Laboratory Occasional Paper **7**.

VAN DEN BERG, M. W. 1994. Neotectonics of the Roer Valley rift system. Style and rate of crustal deformation inferred from syn-tectonic sedimentation. *Geologie en Mijnbouw*, **73**, 143–156.

VANNESTE, K., VERBEECK, K. *ET AL.* 2001. Surface-rupturing history of the Bree fault scarp, Roer Valley graben: evidence for six events since late Pleistocene. *Journal of Seismology*, **5**, 329–359.

WEBER, B. & HINZEN, K.-G. 2006. Bodenverstärkungen in der südlichen Niederrheinischen Bucht. *Bauingenieur*, **81**, S9–S15.

WEINER, J. 1991. Nur in der Tiefe gab es Wasser. Die Entdeckung und Interpretation eines außergewöhnlichen bandkeramischen Befundes. *In*: *Archäologie im Rheinland 1990*. Landschaftsverband Rheinland, Köln, 21–22.

WEINER, J. 1992. The Bandkeramik Wooden Well of Erkelenz-Kückhoven. *NewsWARP*, **12**, 3–11.

WEINER, J. 1994. Well on my back – an update on the Bandkeramik Wooden Well of Erkelenz-Kückhoven. *NewsWARP*, **16**, 5–17.

WEINER, J. 1997. A Bandkeramic Settlement with Wooden Well from Erkelenz-Kückhoven, Northrhine Westphalia (FRG). *In*: *Actes du 22e Colloque Interrégional sur le Néolithique*, Strasbourg, 27–29 octobre 1995. Supplément aux Cahiers de l'Association pour la Promotion de la Recherche Archéologique en Alsace, 401–405.

WEINER, J. 1998. Drei Brunnenkästen, aber nur zwei Brunnen: Eine neue Hypothese zur Baugeschichte des Brunnens von Erkelenz-Kückhoven. *In*: *Brunnen der Jungsteinzeit*, Internationales Symposium Erkelenz 27. bis 29. Oktober 1997. Landschaftsverband Rheinland, Rheinisches Amt für Bodendenkmalpflege, Materialien zur Bodendenkmalpflege im Rheinland, Köln, **11**, 95–112.

WEINER, J. & LEHMANN, J. 1998. Remarks concerning Early Neolithic Woodworking: The Example of the Bandkeramic Well of Erkelenz-Kückhoven, Northrhine-Westfalia, Germany. *In*: Introduzione all'Archeologia degli Spazi Domestici. Atti del seminario – Como, 4–5 novembre 1995. *Archeologia dell'Italia Settentrionale Como*, **7**, 35–55.

WELLS, D. L. & COPPERSMITH, K. J. 1994. New empirical relationships among magnitude, rupture length, rupture width, rupture area, and surface displacement. *Bulletin of the Seismological Society of America*, **84**, 974–1002.

ZENG, Y., ANDERSON, J. G. & YU, G. 1994. A composite source model for computing realistic synthetic strong ground motions. *Geophysical Research Letters*, **21**, 725–728.

ZIEGLER, P. A. 1992. European Cenozoic rift system. *Tectonophysics*, **208**, 91–111.

ZIEGLER, P. A. 1994. Cenozoic rift system of western and central Europe: an overview. *Geologie en Mijnbouw*, **73**, 99–127.

ZIEGLER, P. A. & CLOETINGH, S. 2004. Dynamic processes controlling evolution of rifted basins. *Earth-Science Reviews*, **64**. 1–50.

Speleoseismology and palaeoseismicity of Benis Cave (Murcia, SE Spain): coseismic effects of the 1999 Mula earthquake (m_b 4.8)

R. PÉREZ-LÓPEZ[1]*, M. A. RODRÍGUEZ-PASCUA[1], J. L. GINER-ROBLES[2], J. J. MARTÍNEZ-DÍAZ[3], A. MARCOS-NUEZ[4], P. G. SILVA[5], M. BEJAR[2] & J. P. CALVO[1]

[1]*Área de Riesgos Geológicos, Instituto Geológico y Minero de España, C/Ríos Rosas 23, Madrid 28003, Spain*

[2]*Dpto. de CCAA y RRNN, Universidad CEU San Pablo, Madrid, Spain*

[3]*Dpto. de Geodinámica, Facultad de Ciencias Geológicas, Universidad Complutense de Madrid, Spain*

[4]*Grupo Especial de Rescate en Altura, GERA, Cuerpo de Bomberos de la Comunidad de Madrid, Spain*

[5]*Dpto. de Geología, Universidad de Salamanca, Escuela Politécnica Superior de Ávila, Spain*

**Corresponding author (e-mail: r.perez@igme.es)*

Abstract: This work describes the coseismic ceiling block collapse within Benis Cave (−213 m; Murcia, SE Spain), associated with the 1999 Mula earthquake ($m_b = 4.8$, MSK VII). The collapse occurred at −156 m into the Earthquake Hall, and as a consequence one small gallery became blind. We studied the geology, topography and active tectonic structures relevant to the cave. In addition, we carried out a seismotectonic analysis of the focal mechanism solutions, and also a fault population analysis on slickensides measured in fault planes in the cave. The stress and strain regime is interpreted as being congruent with the palaeoseismic evidence, and agrees with the fault kinematics established for cave galleries developed within fault planes and growth anomalies of coral flowstone. Our analysis suggests that one active segment (NNE–SSW) determined the morphology and topography of the Benis Cave, where strong to moderate palaeoearthquakes ($6 \leq M \leq 7$) took place. As a consequence of this intense seismic activity a small gallery collapsed. A new palaeoseismic structure, or seismothem, has been recognized, namely the effect of palaeoearthquakes affecting the pattern of development of the spatial coral flowstone distribution located at the bottom of the cave.

Endokarstic terrains represent favourable environments for the preservation of palaeoseismic evidence associated with shaking and faulting produced by earthquakes (Postpischl *et al.* 1991; Gilli *et al.* 1999; Cadorin *et al.* 2001; Kagan *et al.* 2005). This terrain results from a delicate equilibrium between dissolution–precipitation processes affecting soluble rocks (carbonates, evaporites, lime sandstones, etc.). The equilibrium reflects the climatic conditions of the area (atmospheric content of CO_2, temperature, humidity, etc.), the tectonic setting of the rock mass (fault and joint geometry), the hydrogeologic constraint (phreatic level) and the chemical composition of the source area (water and rock).

A rhythmic and constant velocity of the dissolution–precipitation rate controls the genesis of the endokarst (Bauer *et al.* 2003). This fact implicates quiet periods, slow variations of the phreatic level and the presence of permanent water-sheets and pools. In this context, the occurrence of earthquakes perturbs the short-term dynamic of the karst, for example causing broken and tilted speleothems, growth anomalies in carbonate layers and block collapses (Gilli 2005).

Speleoseismology represents a new branch of the palaeoseismology that analyses the earthquake record in caves (Kagan *et al.* 2005; Becker *et al.* 2006). Earthquake shaking and faulting in caves can be deduced from two different features.

(A) *Destructive features:* broken speleothems, fallen stalactites, severed stalagmites (Kagan *et al.* 2005), block collapse (incasion), blind galleries, deformed cave sediment structures and coseismic fault displacement affecting cave deposits. These types of features are classified as *seismothems* (Delaby 2001).

From: REICHERTER, K., MICHETTI, A. M. & SILVA, P. G. (eds) *Palaeoseismology: Historical and Prehistorical Records of Earthquake Ground Effects for Seismic Hazard Assessment*. The Geological Society, London, Special Publications, **316**, 207–216. DOI: 10.1144/SP316.13 0305-8719/09/$15.00

(B) *Constructive features:* growth anomalies in carbonate layers and sudden changes in the pattern of precipitation layers produced by tilting or sharp migration of the phreatic surface.

However, processes other than tectonic shaking and faulting may produce similar structures, either destructive and/or constructive, e.g. creep movement of saturated silt sediments, ice filling in caves (glacitectonics), gravity failure and thermo-hydromecanics (Gilli 2005; Becker *et al.* 2006).

Previous work has identified and characterized active faults from the analysis of coseismic effects in caves, assuming a relationship between earthquakes and cave dynamics (e.g. Gilli *et al.* 1999). These authors concluded that one earthquake of 5.2 magnitude was responsible for breaking several soda-straw structures located at the Saint-Paul-de-Fenouillet caves (eastern Pyrenees). The main orientation of these broken soda-straws was east–west, agreeing with the proposed seismogenic fault, the North Pyrenean Fault. Gilli *et al.* (1999) also observed previous palaeoseismic damage. Gilli (2005) and Becker *et al.* (2006) pointed out the scarcity of direct observation of coseismic earthquake damage in caves.

The main question regarding speleoseismology is the speculative relationship between the endokarst dynamic and the earthquake effect. Accordingly, we propose a protocol for the first approach to the relationship between active faults and the long-term evolution of speleothems:

(1) *detailed description of the karst dynamic:* the cave topography, tube sections, geology, phreatic level evolution and principal speleothems;
(2) *systematic description of seismothems:* destructive and constructive features;
(3) *a complete seismotectonic analysis:* mapping of active faults, Quaternary sediments, study of focal mechanism solutions of instrumental earthquakes (located at the proximity of ca. 50 km.), historical seismic databases, palaeoseismic evidence and tectonic geomorphology studies near to the cave;
(4) *numerical dating:* (U–Th series) age of the structures to correlate with the time of occurrence of seismic events in the area (taken from seismic catalogues or palaeoseismic dating in active faults close to the cave) (Kagan *et al.* 2005; Gilli 2005).

In this work we present the first conclusion obtained from the morphotectonic and seismotectonic analysis of Benis Cave, in order to provide new descriptions of structures formed during an instrumental seismic event. In addition, a new type of palaeoseismic feature or seismothem – a migrating coral flowstone – may be used to identify and quantify at least two palaeoearthquakes.

Benis Cave, located in a Palaeocene to early Eocene carbonate massif in the Prebetic Zone of the Betic Cordillera, suffered a massive collapse of blocks at a depth of −156 m, coeval with the 1999 Mula earthquake (m_b 4.8). From this earthquake several blocks larger than 1 m^3 collapsed in a small gallery. These observations encouraged us to study the morphology and structure of this cave in detail. This study provided several structures that may be interpreted as palaeoseismic evidence.

The geological mapping of the study area (Jerez-Mir *et al.* 1972) shows the existence of a NNE–SSW strike-slip fault with normal component, which determines the morphostructure and topography of the cave. This fault appears within the cave, and in the collapsed gallery with impressive slickensides on the fault plane. We performed a kinematic analysis of these slickensides, and have compared the results with the present stress field responsible for the recent seismicity in the area. Hence, focal mechanism solutions were also analysed.

Antonio Salmerón, President of the Murcian Federation of Speleology (Spain) and topographer of the cave, climbed down to the abyss a few days after the Mula earthquake. From his observations, and the original topography, he concluded that one gallery and a small cave room (−156 m) had collapsed during the main shaking event of the earthquake. At this location in the cave we measured striations on fault planes to obtain the kinematic model of the fault movement, in order to make comparisons with focal mechanism solutions of the Mula earthquake.

Geographic location and geology of the cave

Benis Cave is located within the province of Murcia (SE Spain) (Fig. 1), close to the border with the province of Albacete, 10 km SE of Cieza Town. The UTM (30N) coordinates are X 645,220, Y 4,243,248 and Z 405 m high.

Benis Cave is developed within a Mesozoic–Palaeocene–Early Eocene carbonate massif of the Internal Prebetic autochthonous unit of the Betic Cordillera (Fig. 1) (Jerez-Mir *et al.* 1972). This massif consists of an anticline affecting deposits from Upper Cretaceous limestone (Caenomanian) to Eocene massive limestone. This anticline is flanked by synclines filled by Tortonian and Quaternary deposits (Fig. 2a, c).

The stratigraphic sequence of Benis Cave is composed of two principal units (Jerez-Mir *et al.* 1972): (1) concordant Cretaceous limestone and (2) Tertiary massive carbonates (Fig. 2b), also concordant with the Upper Cretaceous. The Cretaceous

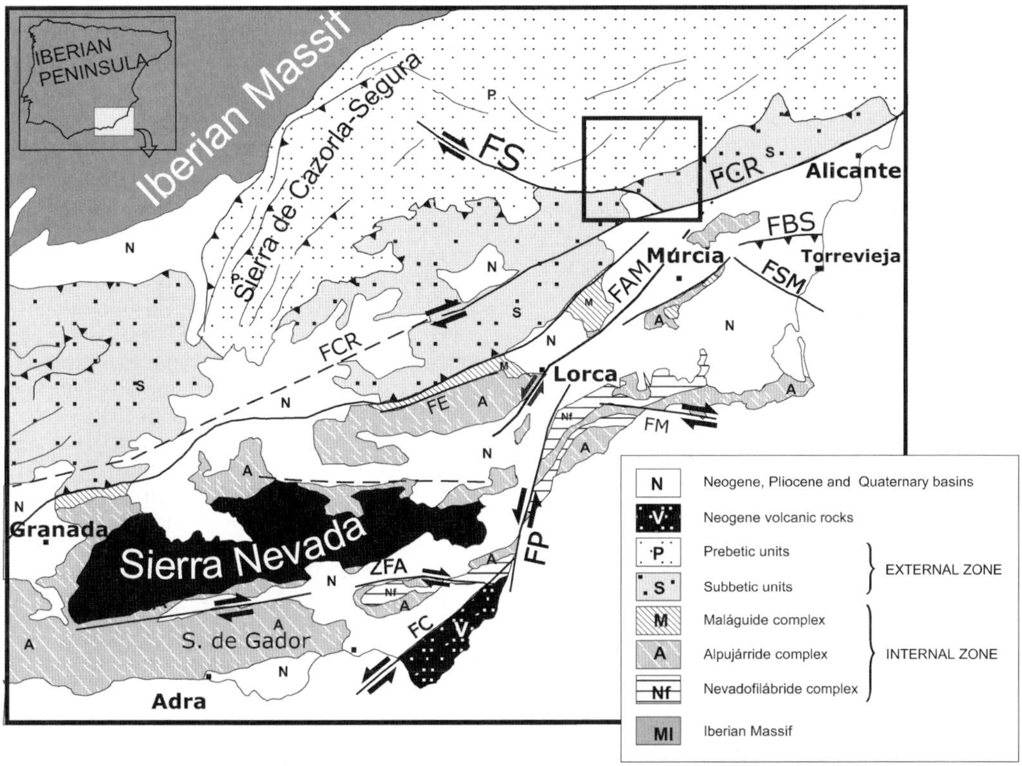

Fig. 1. Geographical location and major geological units of Benis Cave (black rectangular area). Major active faults of the area are indicated: FAM, Alhama–Murcia Fault; FBS, Bajo Segura Fault; FC, Carboneras Fault; FCR, Crevillente Fault; FE, Las Estancias Fault; FM, Las Moreras Fault; FP, Palomares Fault; FS, Socovos-Calasparra Fault; FSM, San Miguel Fault; ZFA, Alpujarras Fault.

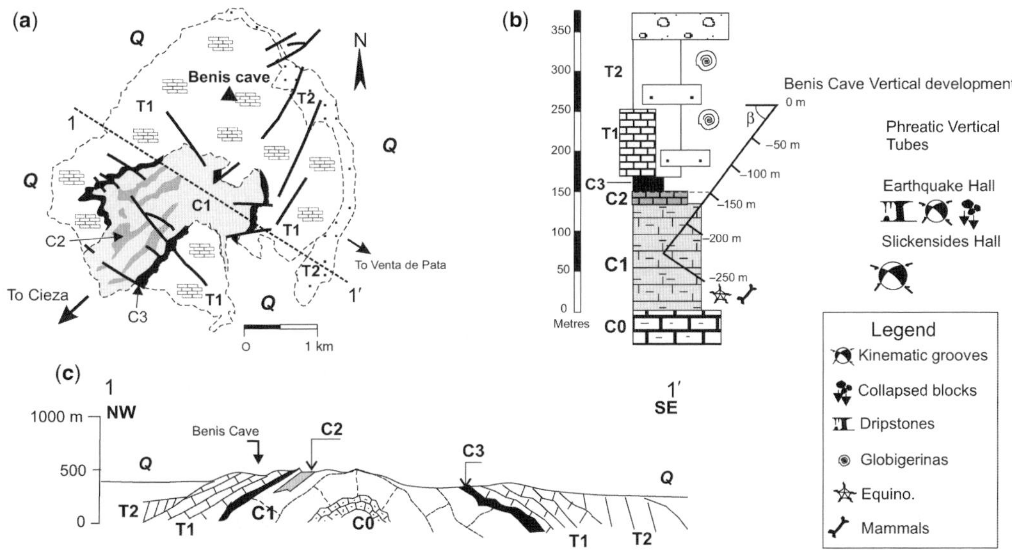

Fig. 2. (**a**) Geological sketch of Benis Cave. Key: C0, Cretaceous carbonate; C1, massive Upper Cretaceous Dolomite; C2, concordant massive white limestone; C3, marls and marly limestone; T1, Tertiary massive carbonate; T2, Tertiary. (**b**) Detailed stratigraphic log of the geology of Benis Cave. Kinematic data and earthquake evidence appear below *c*. 156 m. (**c**) Cross section 1–1′ of the geological map (a). Benis Cave is located in a dome-shaped anticline (after Jerez-Mir *et al.* 1972).

sequence is formed by 40 m of massive dolomite (Fig. 2b, C1), 20 m of massive white limestone with scarce fauna (e.g. *Equinocorix vulgaria*) (Fig. 2b, C2) and 25 m of white marls and marly limestone with *globigerinas* (Fig. 2b, C3). Tertiary carbonates are almost 150 m in width (Fig. 2b, T1).

The vertical topographic section of Benis Cave is developed to a depth of *c.* 213 m, conforming to a quasi-vertical cave (Fig. 3). The surface of the cave begins with a small and narrow window section to the Railing Handy (after Ferrer 2004). At this location there appears a sequence of

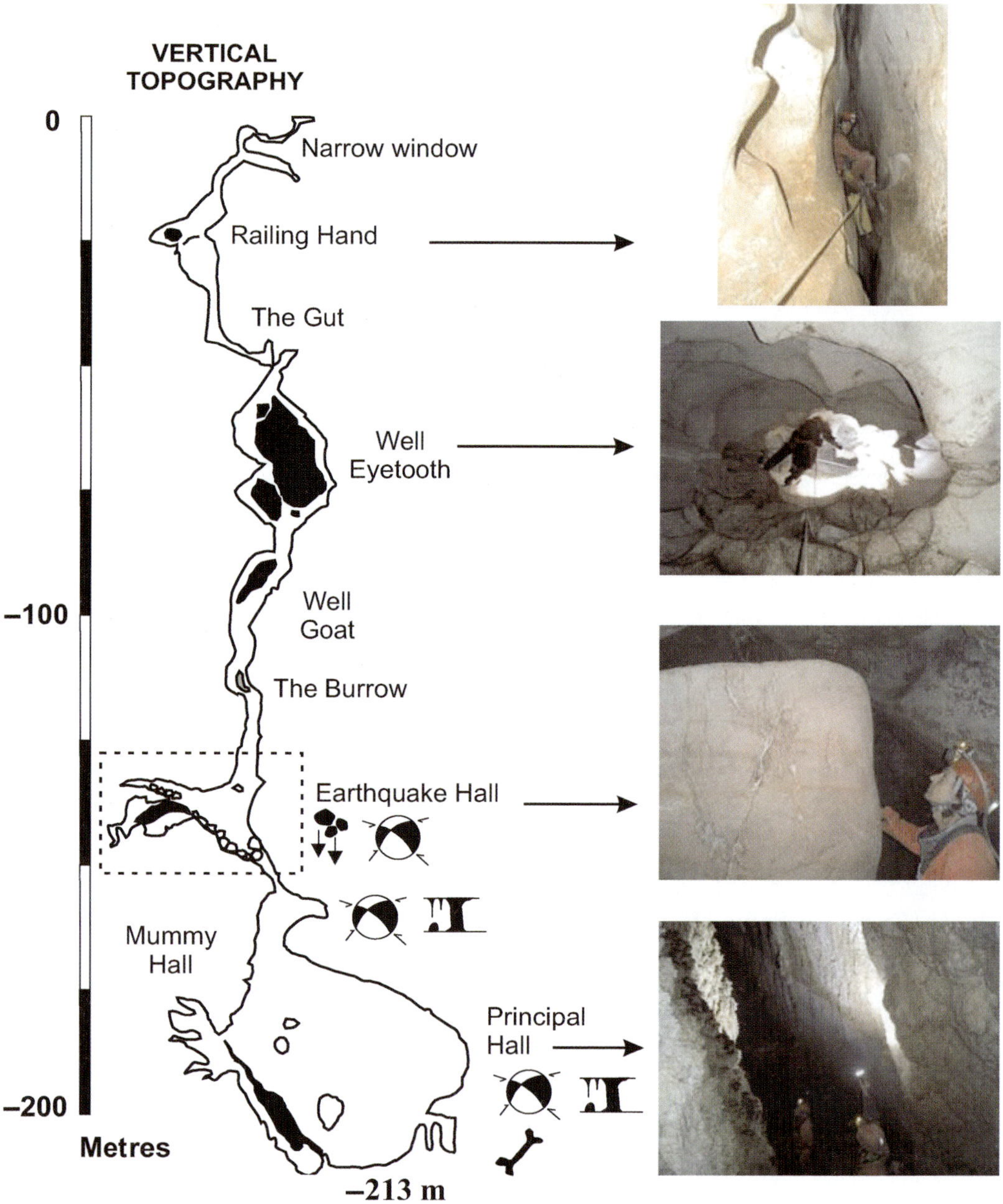

Fig. 3. Topographic sketch of Benis Cave (after Ferrer 2004). Symbols are described in the legend of Figure 2. See text for further explanation.

various deep wells, affecting the Tertiary limestone (Gut Gallery, Eye-tooth Well, Goat Well and Burrow Gallery) (Figs 2 and 3). These galleries are phreatic tubes, eroded and showing great scallops (metric size). The last well ends in the so-called Chaos Hall (*c.* 156 m) and a Deep Well head. It is in this part of the cave that the ceiling collapsed as a result of the Mula earthquake (1999). For this reason the collapsed gallery was renamed the Earthquake Hall (A. Salmerón, pers. comm.) (Fig. 3). The bottom of Benis Cave is the Principal Hall (Fig. 3), developed within a fault plane, and at its bottom small relict pools appear. This cave is preserved as an outcrop of Tertiary macro-mammals, containing bones of *Felix* (*Lynx*) *spelaea* and *Ursus spelaea*, covered by a thin carbonate sheet.

Palaeoseismic study

Earthquake Hall (−156 m)

Earthquake Hall (formerly Chaos Hall) exhibits several blocks from different collapses of the ceiling (Fig. 3) at this section of the cave. After the Mula earthquake, one small gallery collapsed, and new blocks appeared in the bottom of this hall. Based on a speleological survey, we reported great marl blocks with extensional calcite veins (Fig. 3) and fault breccia at this level (Fig. 4). These blocks range between 1 m^3 to centimetric size, although a systematic analysis of the size distribution is needed to quantify a relationship between the collapse and the earthquake event. Both halls developed coincidently with the Upper Cretaceous contact between limestone and marls (Figs 3 and 4).

Several slickensides were measured on fault planes within the Earthquake Hall (Fig. 5a), and we used these data to obtain the strain field associated with these faults.

The Earthquake and Chaos Halls show poor decoration development and small calcite concretions without great speleothems. From the Chaos Hall to the bottom of the cave the Deep Well is developed along a fault plane (orientated NNE–SSW), with a vertical topographic section of a phreatic tube. This tube is tilted according to the dip of the fault (75°NW). The decoration on both sides of the fault plane is different. The footwall exhibits flowstone and carbonate corals (pop-corn), whereas small stalactites (centimetric size) decorate the hanging wall. We have interpreted this fact to mean that the water-sheet flux is dominant, with the precipitation of thin laminated carbonates on the surface. In contrast, the hanging wall is decorated with dripstone structures (stalactites). The length of the fault plane reaches more than 80 m (Fig. 5b).

Principal Hall (−213 m)

The morphology of this hall is controlled by the NNE–SSW fault plane dipping 75° to the east (Fig. 5b), with dip-slip striations (>70 cm long.). These slickensides are analysed in the next section.

Coral flowstone (pop-corns) of centimetric size decorates the dip-slip plane of the hall. We have interpreted the spatial distribution of three coral structures in relation to potential palaeoearthquakes (Fig. 6), and suggest a sudden migration of the coral flowstone towards the SW (Fig. 6, p1, p2 and p3). This pattern was deduced from the apical point migration of the coral flowstone towards the SW. We speculate that episodic movements of the fault, due to at least two potential palaeoearthquakes, displaced the source of water flow responsible of the coral precipitation (Fig. 7).

Bearing in mind that the fault plane is striated, we can reconstruct the kinematics of this fault.

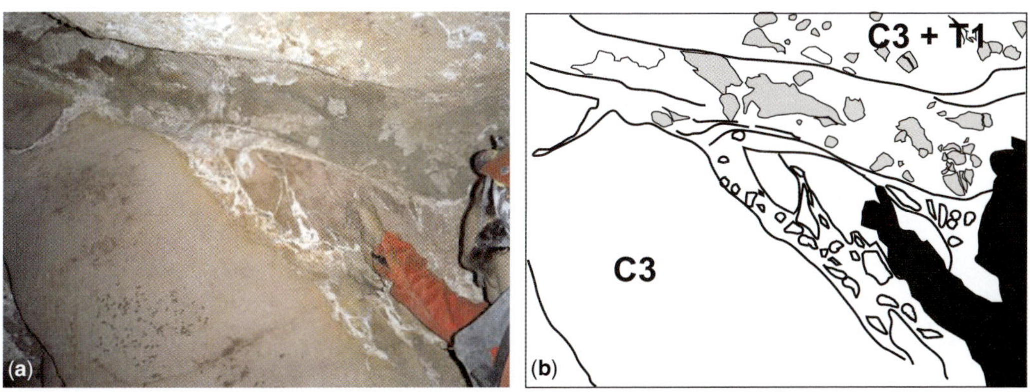

Fig. 4. (**a**) Detailed photograph of the contact between the Tertiary limestone (T1) and Upper Cretaceous marls (C3), within the Earthquake Hall of Benis Cave (−156 m). The contact exhibits a fault breccia. See Figure 2 for the stratigraphic sequence of the cave. (**b**) Interpretation of the contact.

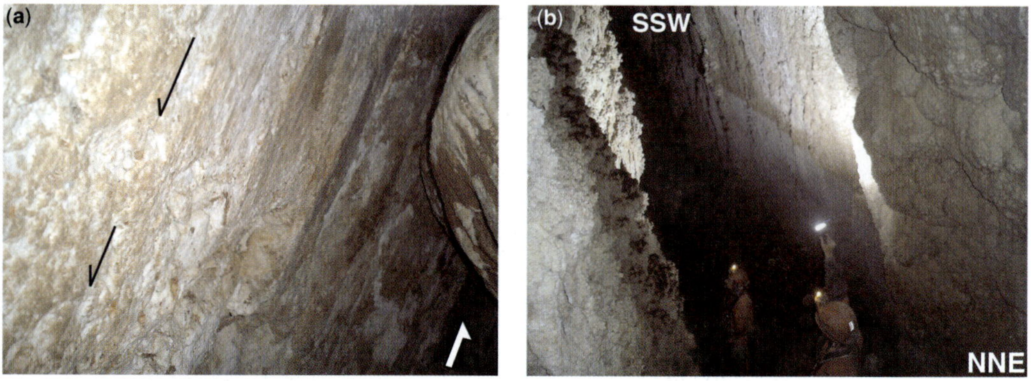

Fig. 5. Photographs of the fault plane showing slickensides (from −156 m to the bottom of the cave). (**a**) White arrow indicates the movement direction of the uplifted block. (**b**) Fault plane of the bottom of the cave with the hanging wall covered by dripstones, developed across large slickensides.

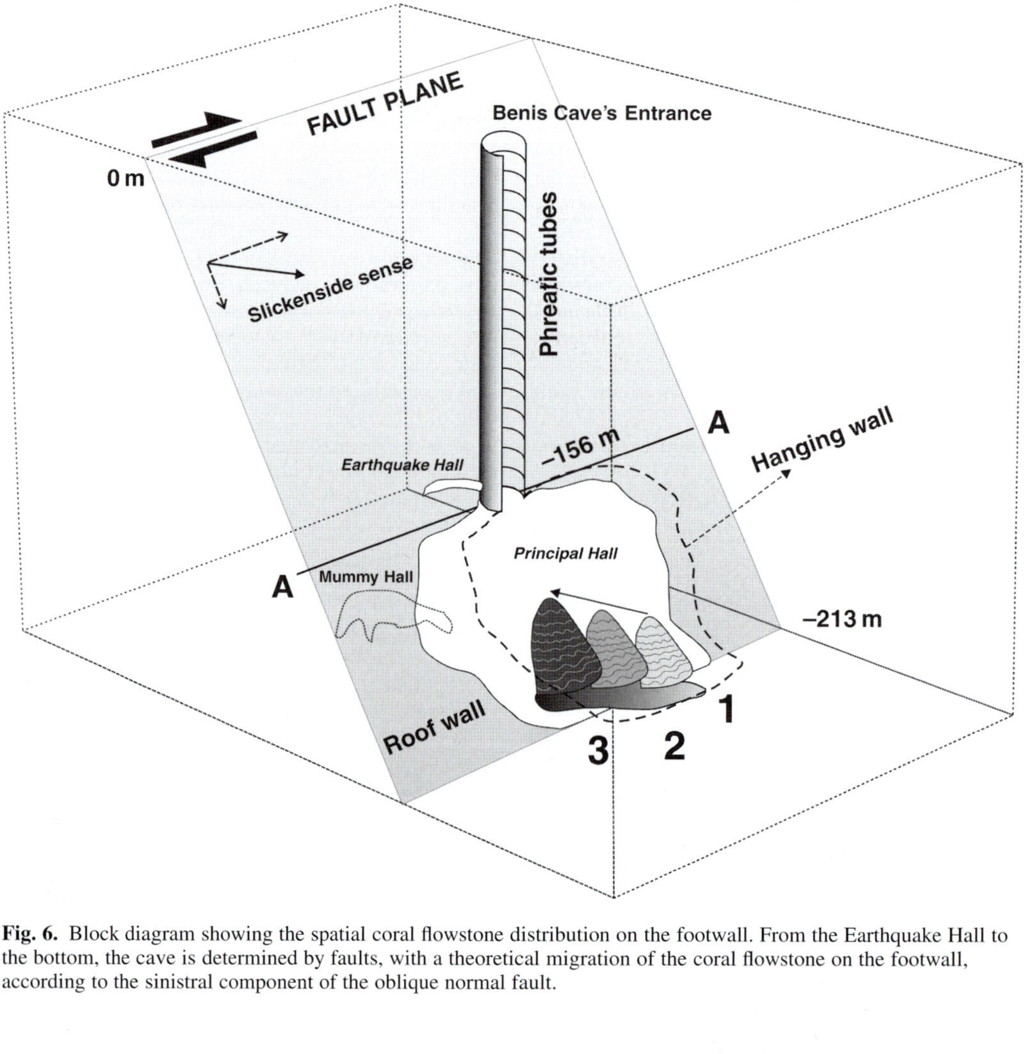

Fig. 6. Block diagram showing the spatial coral flowstone distribution on the footwall. From the Earthquake Hall to the bottom, the cave is determined by faults, with a theoretical migration of the coral flowstone on the footwall, according to the sinistral component of the oblique normal fault.

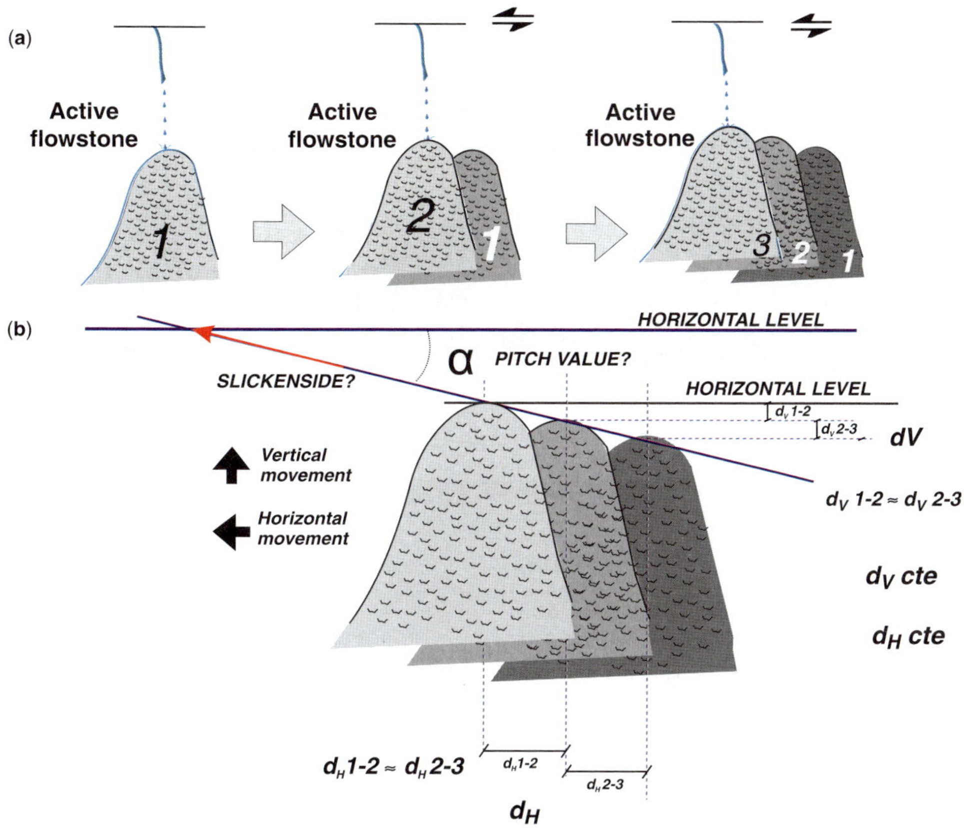

Fig. 7. (**a**) Front view of the coral pattern and fault plane through time. Vertical and horizontal movements correspond with the relative movements with the roof of the fault plane as the fixed tip. (**b**) Geometric reconstruction of the theoretical pitch value associated with the migration of the flowstone tip point. See text for further explanation.

The direction of the coral flowstone migration agrees with the rake of the striations on the fault plane estimated above. We measured the averaged vertical and horizontal fault throw, estimated from the apical point migration, at 1.2 m and 0.8 m respectively. This means a net fault displacement of 1.44 m, and normal movement with a sinistral strike-slip component (rake = 58°).

Fault population analysis in the cave

Using the structural data obtained from slickensides we have performed a fault population analysis, in order to determine the strain regime. These slickensides were principally measured on two fault planes, at the Earthquake Hall (−156 m) and the hanging wall of the Principal Hall (−213 m).

The strain analyses applied here were the slip method (Reches 1983), and the right dihedral method (Angelier & Mechler 1977). The slip method (SM) allows us to establish the fault plane from both nodal planes (Capote *et al.* 1991), thus indicating the orientation of the maximum horizontal shortening (*ey*) of each analysed fault.

Right dihedral (RD) is a qualitative method based on the stereographic projection of areas with similar behaviour, dilatation or compression (Reches 1983). RD analysis indicates the approximate orientation of the main axes of the strain ellipsoid: maximum shortening (*ey*) and minimum shortening axes (*ex*).

Figure 8 shows the results obtained for both strain analyses, with *ey* trending NNW–SSE (Fig. 8a). The right dihedral diagram indicates a strike-slip strain regime with a normal component, compatible with the main orientation of the fault plane of the Principal Hall (NNE–SSW) (Fig. 8b). Accordingly, we have four sinistral strike-slip faults (Fig. 8c), in agreement with the sinistral strike-slip movement interpreted above from the coral growth (Fig. 7), and with similar pitch values.

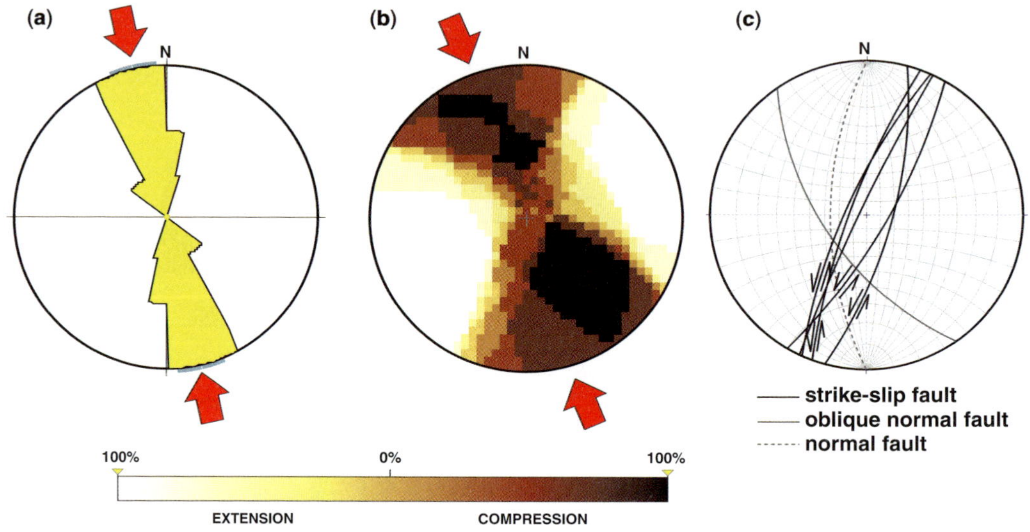

Fig. 8. Right dihedral diagram obtained from the slickensides, measured on fault planes from the Earthquake Hall to the bottom of the cave. This diagram defines a strike-slip strain field with the maximum horizontal shortening (*ey*) trending NNW–SSE.

Seismotectonics of the Mula earthquake

The Mula earthquake (2 February 1999) occurred close to the city of Mula and 28 km SW of Benis Cave (Fig. 9). The estimated magnitude was $m_b = 4.8$ with MSK intensity of VI–VII (IGN 1999). In the area of the cave the estimated intensity was less than V (Martínez-Díaz *et al.* 2002). In spite of the low population and infrastructures around the epicentral area, the cost of the damage reached almost 40 million Euros.

The epicentral area is located to the SE of the Betic Cordillera (Fig. 9), an area of moderate to low seismic activity with earthquakes of less than $m_b = 5$. Nevertheless, the historic seismicity shows high intensity values such as the Torrevieja earthquake (MSK X) and several earthquakes with MSK close to VIII within the Segura Basin (Mézcua & Martín-Solares 1983).

Two focal mechanism solutions have been obtained for the Mula seismic crisis, with different interpretations. Firstly, Buforn & Sanz de Galdeano (2001) obtained a reverse focal mechanism solution for the mainshock and also for the aftershocks (by using seismic wave polarity analysis) (Fig. 9). In contrast, Mancilla *et al.* (2002) proposed a strike-slip focal mechanism at a depth of 12.5 km, by using the inversion technique of the seismic moment (Fig. 9). The hipocentral depth calculated by the National Geographic Institute of Spain ranged between 4 and 5 km. According to the fault population analyses applied above, our result agrees with the focal mechanism solution of a strike-slip obtained by Mancilla *et al.* (2002) (Fig. 9).

However, the seismogenic source of the Mula earthquake can only be estimated from the focal mechanisms described above due to its low magnitude. Geological data suggest a possible NE–SW seismogenic source with high dip ($\beta > 70°$) (Martínez-Díaz *et al.* 2002). Within the epicentral area several faults with similar strike also show neotectonics evidence (Fig. 9). These faults present geometric features compatible with active segments, able to trigger earthquakes of such a magnitude ($m_b = 5$) (Martínez-Díaz *et al.* 2002). The epicentral position of the Mula earthquake suggests that the Crevillente Fault could be the seismic source, although the aftershock swarm appears displaced to the SE zone of this fault.

The map of isosist (lines of equal seismic intensity) obtained by Martínez-Díaz *et al.* (2002) also exhibits a NE–SW elongation. Bejar *et al.* (2006) carried out a relevant interferometric analysis related to the Mula earthquake, and suggested that it produced a maximum surface deformation of 12 mm. In this sense, these authors proposed a focal mechanism of strike-slip with foci greater than 8 km in depth, although such strike-slip deformation is difficult to detect using this kind of analysis.

The result obtained from the fault population analysis indicates a sinistral strike-slip sense for NNE–SSW faults. One focal mechanism agrees with this fault character, as described by Mancilla *et al.* (2002) and other neotectonic evidence for

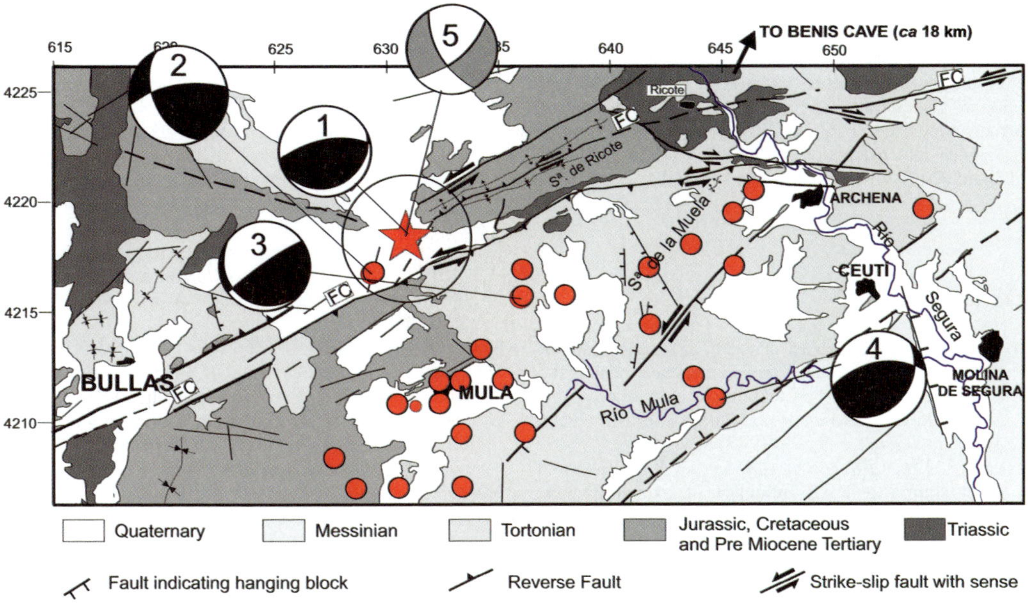

Fig. 9. Seismotectonic sketch of Benis Cave (FC, Crevillente Fault). Focal mechanisms obtained by Buforn & Sanz de Galdeano (1999): 1, mainshock; 2, 3, 4, aftershocks. The focal mechanism number 5 corresponds to Mancilla *et al.* (2002) (after Martínez-Díaz *et al.* 2001).

the area (Silva *et al.* 1993, 1996; Martínez-Díaz *et al.* 2002). The palaeoseismic evidence described from the coral flowstone spatial distribution at the bottom of the cave agrees with a dextral strike-slip movement across a fault plane orientated in a NNE–SSW direction (Fig. 9).

Conclusions

Caves represent a powerful tool as potential palaeoseismic indicators of recent tectonic activity. The present work reports a coseismic rock fall as a consequence of the Mula earthquake ($m_b = 4.8$, MSK VII, 1999). As a result, several blocks of marls collapsed in one small gallery (metre size) at a depth of −156 m. This represents one of the more severe damaging effects for a deep cave, in relation to an instrumental earthquake, described by a field survey.

We have therefore performed a morphotectonic, structural and palaeoseismic study of Benis Cave. The morphotectonic analysis indicates that the galleries and vertical phreatic tubes developed along an active fault with a NNE–SSW trend. The main galleries, below −150 m depth, were determined by the geometry of this fault and the contact between Tertiary limestones and Upper Cretaceous marls.

In addition, a new seismothem is described as a palaeoseismic indicator. The spatial distribution of the coral flowstone (pop-corn) at the bottom of the cave (−213 m) could be controlled by palaeoearthquake activity. From this pattern growth, two potential palaeoearthquakes showing normal fault movement with dextral strike-slip component were established. By using the Wells & Coppersmith (1994) empirical relationship for normal seismogenic faults, the palaeoearthquake size corresponds to approximately $6.5 < M < 7$ in both cases.

Several fault planes in the cave exhibit slickensides and normal-directional movement grooves. The strain analysis reveals that both the right dihedral diagram and the slip model method are in agreement with the focal mechanism obtained by Mancilla *et al.* (2002) for the Mula earthquake. These authors obtained a shallow sinistral strike-slip seismic source (15–8 km depth) that is well orientated with respect to the regional stress field defined by Stich *et al.* (2006).

Nonetheless a more exhaustive work, and numerical dating, is also required in order to establish the complete temporal dynamic of the Quaternary seismicity in the area, although the coseismic block collapse is valuable geological evidence for palaeoseismicity in caves.

We would like to record our sincere thanks to Elisa Kagan and Ives Quinif for their constructive and helpful reviews. Our great thanks also go to Antonio Salmerón, President of the Murcian Speleological Federation of Spain, for communicating to us the coseismic effect on Benis Cave a few days after the Mula earthquake (1999). We also thank Eva González, head speleologist, who recently has suffered an awful accident into the Tibia Fresca cave (−500 m, Cantabria), and Emilio Usaola. This work was partially supported by the Spanish Projects of the Ministry of Science and Education ACTISIS, CGL2006-05001/BTE and TECTO2, CGL2006-28134-E/CLI.

References

ANGELIER, J. & MECHLER, P. 1977. Sur une méthode graphique de recherche des contraintes principales également utilisables en tectonique et en sismologie: la méthode des dièdres droits. *Bulletin Société Géologique de la France*, **9**, 1309–1318.

BAUER, S., LIEDL, R. & SAUTER, M. 2003. Modelling of karst aquifer genesis: Influence of exchange flow. *Water Resources Research*, **39**(10), 1285–1309.

BECKER, A., DAVENPORT, C. A., EICHENBERGER, U., GILLI, E., JEANNIN, P. Y. & LACAVE, C. 2006. Speleoseismology: A critical perspective. *Journal of Seismology*, **10**(3), 371–388.

BEJAR, M., HERRÁIZ, E., MARTÍNEZ-DÍAZ, J. J., LÓPEZ, C., CAPOTE, R. & TSIGE, M. 2006. Seismotectonic interpretation of the 2002 Mw 4.8 Gérgal seismic sequence using seismological data and RADAR interferometry (InSAR). *Geogaceta*, **39**, 67–70 [in Spanish].

BUFORN, E. & SANZ DE GALDEANO, C. 2001. Focal Mechanism of Mula (Murcia, Spain) earthquake of February 2, (1999). *Journal of Seismology*, **5**, 277–280.

CADORIN, J. F., JONGMANS, D., PLUMIER, A., CAMELBEECK, T., DELABY, S. & QUINIF, Y. 2001. Modelling of speleothems failure in the Hotton Cave (Belgium). Is the failure earthquake induced? *Netherlands Journal of Geoscience* **80**(3–4), 315–321.

CAPOTE, R., DE VICENTE, G. & GONZÁLEZ-CASADO, J. M. 1991. An application of the slip model of brittle deformations to focal mechanism analysis in three different plate tectonic situations. *Tectonophysics*, **191**, 339–409.

DELABY, S. 2001. Paleoseismic investigations in Belgium Caves. *Cahier du Centre Européen de Geodynamique et de Sismologie*, **18**, 45–48.

FERRER, V. 2004. *Grandes Cuevas y Simas del Mediterráneo*. Editorial Ferrer Rico, Barcelona.

GILLI, E. 2005. Review on the use of natural cave speleothems as palaeoseismic or neotectonics indicators. *Geosciencie*, **337**, 1208–1215.

GILLI, E., LEVRET, A., SOLLOGOUB, P. & DELANGE, P. 1999. Research on the February 18, 1996 earthquake in the caves of Saint-Paul-de-Fenouillet area (eastern Pyrenees, France). *Geodinamica Acta*, **12**(3–4), 143–158.

IGN. 1999. *Serie sísmica de Mula (Murcia)*. Segundo Informe General. Edita Subdirección General de Geodesia y Geofísica. Madrid.

JEREZ MIR, L., JEREZ MIR, J. & GARCÍA-MONZÓN, G. 1972. *Mapa geológico de España E. 1:50.000*. Serie MAGNA (IGME). Hoja de Mula, **912**.

KAGAN, E. J., AGNON, A., BAR-MATTHEWS, M. & AYALON, A. 2005. Dating large, infrequent earthquakes by damaged cave deposits. *Geology*, **33**(4), 261–264.

MANCILLA, F. L., AMMON, C. J., HERRMANN, R. B. & MORALES, J. 2002. Faulting parameters of the 1999 Mula Earthquake, Southeastern Spain. *Tectonophysics*, **354**, 139–155.

MARTÍNEZ-DÍAZ, J. J., MASANA, E., HERNÁNDEZ-ENRILE, J. L. & SANTANACH, P. 2001. Evidence for co-seismic events of recurrent prehistoric deformation along the Alhama de Murcia fault, southeastern Spain. Monografía: Paleosismicidad en España. *Acta Geológica Hispánica*, **36**(3–4), 315–327.

MARTÍNEZ-DÍAZ, J. J., RIGO, A., LOUIS, L., CAPOTE, R., HERNÁNDEZ-ENRILE, J. L., CARREÑO, E. & TSIGE, M. 2002. Caracterización geológica y sismotectónica del terremoto de Mula (febrero de 1999, Mb:4,8) mediante la utilización de datos geológicos, sismológicos y de interferometría de RADAR (INSAR). *Boletín Geológico y Minero* **113**(1), 23–33.

MÉZCUA, J. & MARTÍNEZ-SOLARES, J. M. 1983. *Sismicidad del área Ibero-Magrebí*. Instituto Geográfico Nacional. Madrid.

POSTPISCHL, D., AGOSTINI, S., FORTI, P. & QUINIF, Y. 1991. Palaeoseismicity from karst sediment: the "Grotta del Cervo" cave case study (Central Italy). *Tectonophysics*, **193**, 33–44.

RECHES, Z. 1983. Faulting of rocks in three-dimensional strain fields, II. Theoretical analysis. *Tectonophysics*, **95**, 133–156.

SILVA, P. G., GOY, J. L., SOMOZA, L., ZAZO, C. & BARDAJÍ, T. 1993. Landscape response to strike–slip faulting linked to collisional settings: Quaternary tectonics and basin formation in the Eastern Betics, Southeast Spain. *Tectonophysics*, **224**, 289–303.

SILVA, J. P., MATHER, J. L., GOY, J. L., ZAZO, C. & HARVEY, A. M. 1996. Controles en el desarrollo y evolución del drenaje en zonas tectónicamente activas: el caso del río Mula (Región de Murcia, SE España). *Revista de la Sociedad Geológica de España*, **9**(3–4), 269–285.

STICH, D., SERPELLONI, E., MANCILLA, F. L. & MORALES, J. 2006. Kinematics of the Iberia–Maghreb plate contact from seismic moment tensors and GPS observations. *Tectonophysics*, **426**(3–4), 295–317.

WELLS, D. L. & COPPERSMITH, K. J. 1994. New empirical relationships among magnitude, rupture length, rupture width, rupture area, and surface displacement. *Bulletin of the Seismological Society of America*, **84**(4), 974–1002.

Tsunami deposits in the western Mediterranean: remains of the 1522 Almería earthquake?

KLAUS REICHERTER[1]* & PETER BECKER-HEIDMANN[2]

[1]*Institute of Neotectonics and Natural Hazards, RWTH Aachen University, Lochnerstr. 4-20, 52056 Aachen, Germany*

[2]*Institut für Bodenkunde, Universität Hamburg, Allende-Platz 2, 20146 Hamburg, Germany*

**Corresponding author (e-mail: k.reicherter@nug.rwth-aachen.de)*

Abstract: Shallow drilling in the lagoon of the Cabo de Gata area proved sedimentary evidence for a palaeo-tsunami along that part of the Spanish Mediterranean coast. Several coarse-grained intervals form fining-up and thinning-up sequences that are interpreted as tsunamites. Inland-extending sand sheets are used to identify tsunamigenic inundations. Other indicative features found are erosive bases, rip-up clasts, broken shells of bivalves and benthic/planktic foraminifera. The coarse-grained intervals consist of up to three sequences separated from each other by a silty mud drape. These intervals are interpreted as deposits of a tsunami train and correspond to three individual waves. Radiocarbon dating reveals evidence that these layers can be ascribed to deposition during the 1522 Almería earthquake.

The 1522 Almería earthquake (M > 6.5) affected large areas in the western Mediterranean and caused more than 1000 casualties. The epicentral area was offshore in the Gulf of Almería (southern Spain) along the Carboneras Fault Zone and seismic shaking triggered submarine slides in the Gulf of Almería, which may have caused tsunami waves.

We have also found another intercalation of tsunamites downhole, which are interpreted as either an expression of repeated earthquake activity or tsunami-like waves induced by submarine slides triggered by seismic shaking in the Gulf of Almería. Our evidence suggests a definite tsunami potential and hazard for offshore active and seismogenic faults in the western Mediterranean region.

During the last 20 years, several tsunami and palaeotsunami have been described, mainly in the circum-Pacific region, e.g. Washington and Oregon (Atwater 1992; Atwater & Yamaguchi 1991; Atwater & Moore 1992; Kelsey *et al.* 2005; Atwater *et al.* 2005), Chile (Cisternas *et al.* 2005), Kamchatka (Pinegina *et al.* 2003), Japan (Minoura & Nakaya 1991; Nanayama *et al.* 2003). Generally, coarse-grained or blocky deposits known from the Caribbean, Alaska or Australia are distinguished from fine-grained sandy layers, which are distributed spatially (e.g. Minoura & Nakaya 1991; Dabrio *et al.* 1998; Gianfreda *et al.* 2001; Luque *et al.* 2002; Pinegina *et al.* 2003; Tuttle *et al.* 2004). Along the coasts of the Mediterranean Sea, the remains of tsunami waves have also been described, e.g. Italy (Mastronuzzi & Sanso 2000; Pantosti *et al.* 2008), Greece (Dominey-Howes 1996; Dominey-Howes *et al.* 1999, 2000, 2006; Vött *et al.* 2008), Cyprus (Kelletat & Schellmann 2002), Spain (Bartel & Kelletat 2003; Whelan & Kelletat 2005). The western Mediterranean region lacks these studies, with the exception of the preliminary report of Becker-Heidmann *et al.* (2007).

The great tsunami of Banda Aceh on 26 December 2004 demonstrated that catastrophic tsunami events are rare and that sometimes the historical reports do not necessarily help to document this natural hazard for coastal zones. However, geological archives such as lagoons or estuaries serve as valuable documents of palaeotsunami activity (i.e. tsunamites) and help to close the gap between historic descriptions and prehistoric events. To study palaeotsunami, an interdisciplinary team of geologists, seismologists, modellers, stratigraphers, geochemists and geomorphologists is necessary, in order to establish a coastal hazard assessment. The research of palaeotsunamis has also to be supplemented by coastal and social engineers for emergency plans, in terms of civil protection. Sedimentological studies of tsunamites allow the calculation of run-up heights and distances, estimation of wave velocities and the localization of the tsunamigenic source and process. If dating of the sediments is possible, the time of occurrence and interval of recurrence can be delineated. Studies of coastal morphology, including ecological impacts, complement palaeotsunami research. The coast of southern Spain is a touristic hot-spot in western Mediterranean Europe and is densely populated, but it lacks these investigations and plans. Hence, the impact on coastal vulnerability from destructive

From: REICHERTER, K., MICHETTI, A. M. & SILVA, P. G. (eds) *Palaeoseismology: Historical and Prehistorical Records of Earthquake Ground Effects for Seismic Hazard Assessment*. The Geological Society, London, Special Publications, **316**, 217–235. DOI: 10.1144/SP316.14 0305-8719/09/$15.00

earthquakes and their secondary effects, such as tsunami, is of great concern for society and economy, especially in holiday and recreation areas in the western Mediterranean region.

In this paper, we present geological and stratigraphical evidence by percussion drilling for two tsunamite layers in the saline area of Cabo de Gata in the Gulf of Almería, southern Spain, pointing to repeated tsunami wave action in the western Mediterranean during the last 1000 years.

Geological setting

Generally, the eastern Betic Cordilleras are dominated by intense active strike-slip faulting along the sinistral Carboneras (CFZ) and Palomares Fault Zones (Fig. 1; Montentat & Ott d'Estevou 1995; Stapel *et al.* 1996; Jonk & Biermann 2002). This part of the Betics is underlain by an attenuated continental crust of 15–22 km thickness, in contrast to the western parts, which are characterized by crustal thicknesses of around 35 km separated by the Trans-Alborán Shear Zone (Montenat & Ott d'Estevou 1995). The CFZ represents a major sinistral strike-slip fault in the Betic Cordilleras of south-eastern Spain accompanied by pressure ridges and pull-apart basins, and several fault strands entering the Gulf of Almería (Figs 1 and 2). The onshore segment of the CFZ is striking approximately NE–SW and is anastomosing in east–west striking

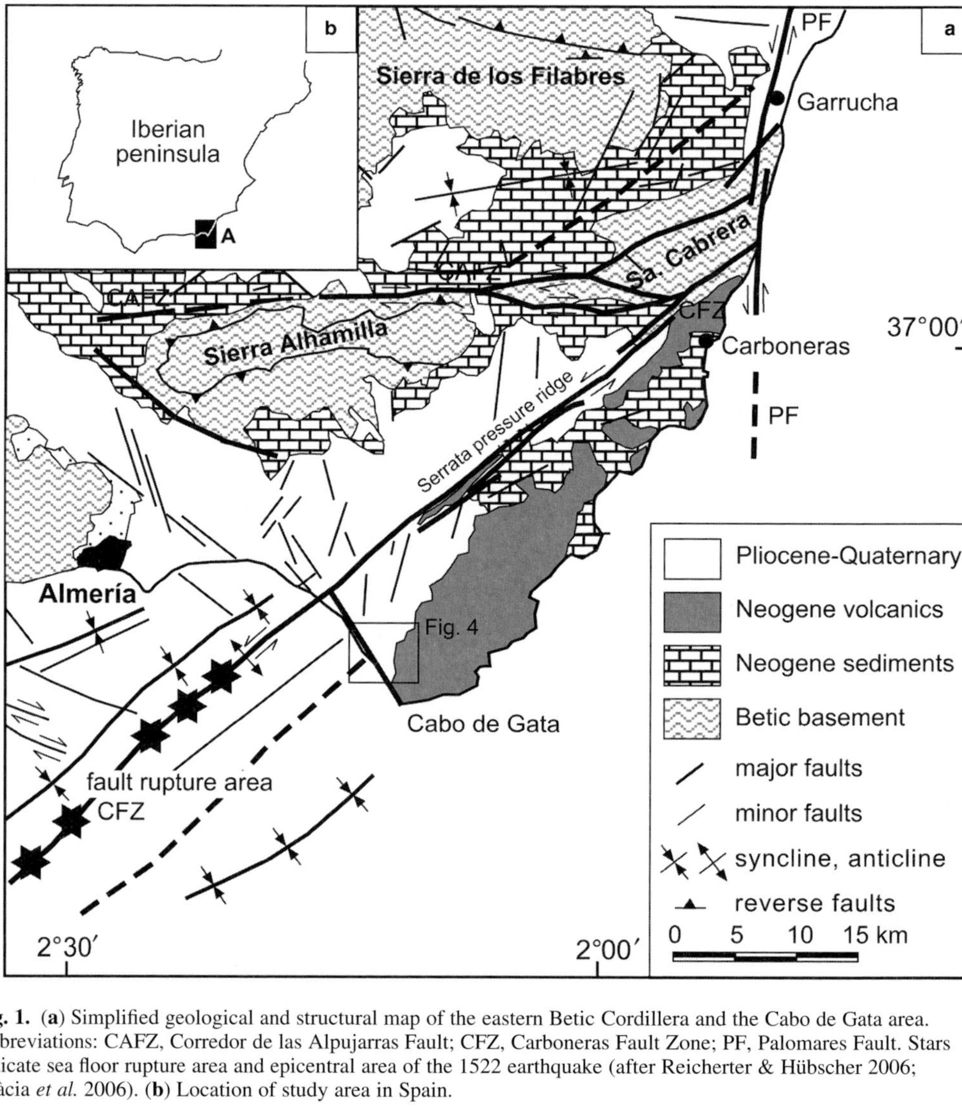

Fig. 1. (**a**) Simplified geological and structural map of the eastern Betic Cordillera and the Cabo de Gata area. Abbreviations: CAFZ, Corredor de las Alpujarras Fault; CFZ, Carboneras Fault Zone; PF, Palomares Fault. Stars indicate sea floor rupture area and epicentral area of the 1522 earthquake (after Reicherter & Hübscher 2006; Gràcia *et al.* 2006). (**b**) Location of study area in Spain.

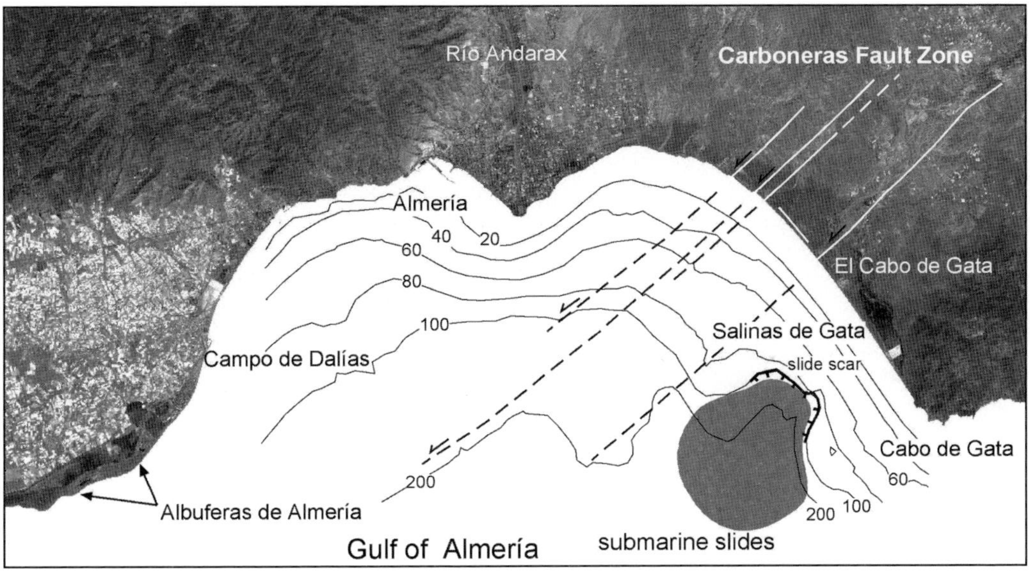

Fig. 2. Satellite image of the Gulf of Almería with localities and fault strands of the Carbon eras Fault Zone. Bathymetry (in metres) from Sanz *et al.* (2003) and unpublished data from Hoernle *et al.* (2003). Position of the different strands of the Carboneras Fault Zone; note a submarine silde scar in front of the saline area of the Cabo de Gata.

minor faults of about 50 km length. The northeastern termination of the CFZ (Fig. 1) connects with the sinistral NNE–SSW trending Palomares Fault. The southwestern continuation of the CFZ extends offshore into the Gulf of Almería for at least 50 to 100 km (Gràcia *et al.* 2006; Reicherter & Hübscher 2006).

The CFZ is seismically active. The Almería earthquake of 22 September 1522 affected large areas in the western Mediterranean. The earthquake, of magnitude $M > 6.5$ (IAG 2005), was followed by several aftershocks: the province capital of Almería was almost completely destroyed. A contemporaneous wood cut of 1523, by an anonymous German artist mentioned in Varela Hervías & von Waldheim (1948), displays drowning people, ships in distress, and inundations along the coastline (in Reicherter & Hübscher 2006). The text in German gives details about the earthquake accompanied by flooding in the western Mediterranean. Reicherter & Hübscher (2006) interpreted the picture as follows: ground shaking destroyed the city of Almería and its harbour and tsunami waves occurred. Arguments for an offshore epicentre for the 1522 Almería earthquake relatively close to the coast along the CFZ are reasonable (Fig. 3). Based on seismic data, they proposed the epicentre precisely at the observed sea floor rupture area at 36°42 N, 2°23 W in the Gulf of Almería.

Onshore investigations were concerned with Quaternary tectonics and earthquake deformation: Bell *et al.* (1997), and Reicherter & Reiss (2001) found evidence for Quaternary deformation, but were not able to relate those to the 1522 Almería earthquake. Martín *et al.* (2003) studied the long-term uplift of the Cabo de Gata area from the Neogene to the Recent and found that most of the uplift took place well before the Pliocene.

The remarkably straight coastline in the southeastern part of the Gulf of Almería (Figs 2 and 4) is controlled by several parallel and NW–SE trending normal faults. These faults played an important role in the formation of the lagoon and later the saline areas. Also, footwall uplift caused the formation of a 2 to 4 m high beach ridge (Fig. 4). The location of the individual strands of the CFZ has been confirmed by ground-penetrating radar studies (Reicherter & Reiss 2001); also the straight coastline is displaced in two places sinistrally by the CFZ (Fig. 3). Horizontal and vertical slip rates of the CFZ have been estimated on the order of 1 mm/a, and 0.1 mm/a, respectively (Reicherter & Reiss 2001; Silva *et al.* 2004).

Cabo de Gata lagoon

The Gulf of Almería is framed by the Cabo de Gata consisting of Neogene volcanics and carbonates in the east and by the Plio-Quaternary sediment complex of the Campo de Dalías in the west (Fig. 2). Both sides developed coastal lowlands,

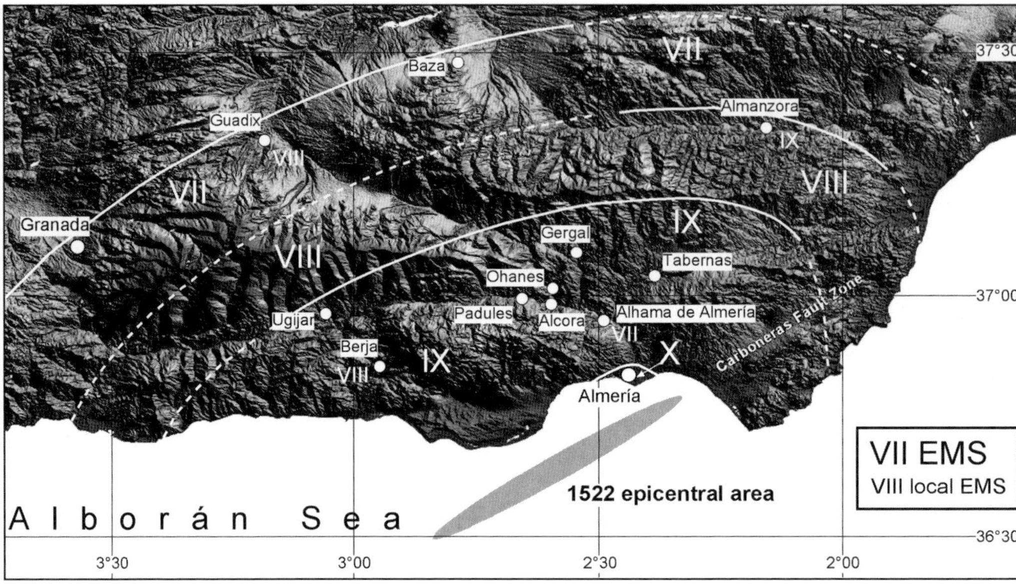

Fig. 3. Digital elevation model of the eastern Betic Cordilleras and the Cabo de Gata block based on SRTM90 data. Isoseismals constructed after damage reports (López Marinas 1985; Martínez Solares 1995). European Macroseismic Scale (EMS), intensity varies locally as in Almanzora, where EMS IX has been delineated from building damage. Offshore epicentral area of the 1522 Almería earthquake (after Reicherter & Hübscher 2006; Gracia *et al.* 2006).

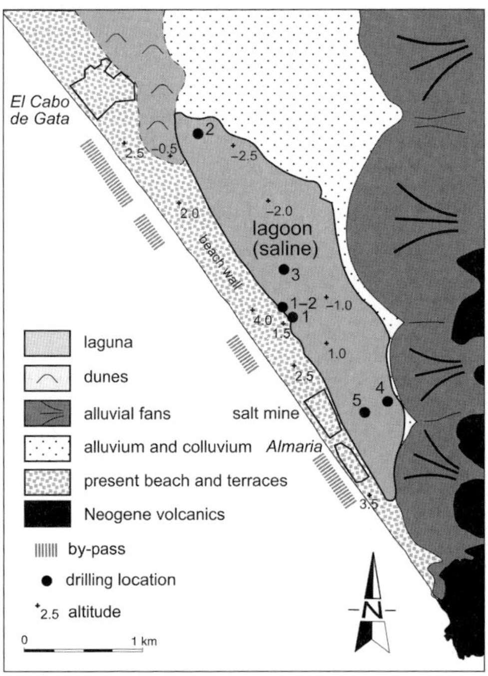

Fig. 4. Geological sketch map of the saline area of the Cabo de Gata, including the drilling positions (large numbers, which refer to drill cores), height of the beach wall (small numbers in metres), and position of possible by-passes in the beach wall.

the lagoon of the Cabo de Gata (2°13′875 W, 36°45′346 N), which has been used as a saline source since Phoenician times (Ruiz-Gálvez Friego 1993), and the Albuferas de Almería, a swampy area in the Campo de Dalías. The central part is characterized by the delta of the Río Andarax.

We focused our investigations on the saline area of the Cabo de Gata (Fig. 4). The lagoon is situated at the foot of the San Miguel volcanic hills. Salt production facilities occupy 500 ha, 300 of which are flooded, favouring the entrance of the sea by gravity. The beach and lagoon have a depositional history of about 6000 years (Goy *et al.* 1996; Jalut *et al.* 2000). The sediment cores allowed us to distinguish three periods in the evolution of the lagoon. The initial stage is a predominantly alluvial fan phase, which commenced during Pleistocene times (Harvey *et al.* 1999), followed by an intermediate beach phase from approximately 6000 to 3000 years BP, and from then on a marsh lagoon developed (Fig. 5). The early period is characterized by alluvial fan deposits (Harvey *et al.* 1999), which consist of reddish, poorly sorted, coarse-grained gravels, sands and intercalated palaeosols. The sedimentary source is the volcanic Cabo de Gata range; hence clastic material is exclusively made up of volcanic rocks and some Neogene carbonates. The alluvial fan sequences are cyclic and show fining-up cycles, mostly terminated by soil development. Up-section, sands with intercalated clays are

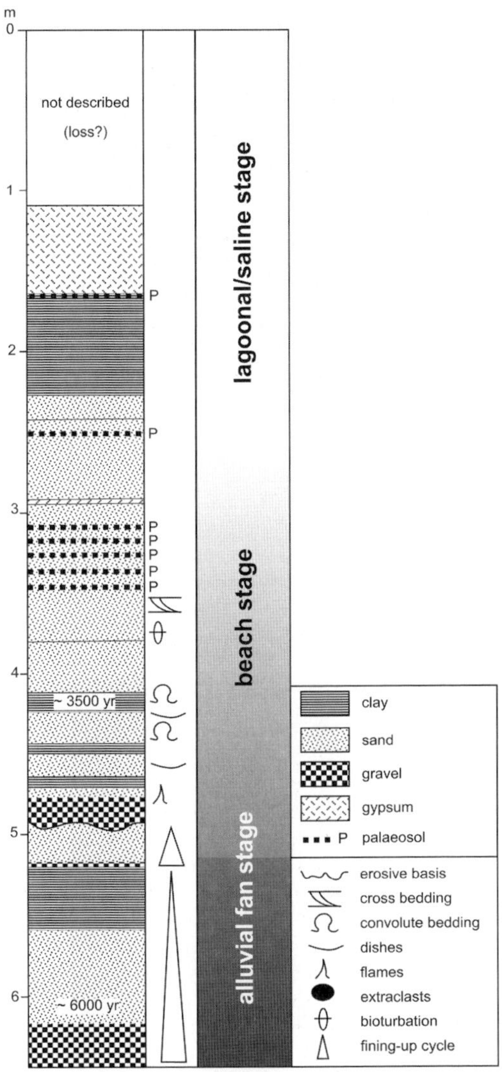

Fig. 5. Lithostratigraphic log of the saline area of the Cabo de Gata, showing the development of the lagoon and the Holocene stratigraphy of the sediments (after Jalut *et al.* 2000; and unpublished data). Note that the upper part is missing.

developed, which point to open marine, and hence beach-like conditions between 6000 and 3000 years BP. The sands are partly cross-bedded, but they also have convolute bedding, flames and dishes, interpreted as liquefaction (Fig. 5). The beach sands and, partly, aeolian sands are composed of volcanic lithic fragments, carbonates and metamorphic rocks (mica schists and amphibolites) of the Internal Betic Zone. Typical Mediterranean marine faunal such as mollusc shells are also found. The metamorphic rocks crop out in the Sierras de Alhamilla and Gador nearby (Fig. 1), and are transported by several ephemeral ramblas (Almoladeras, Morales and Retamar) and permanent rivers (Rio Andarax; Fig. 2) into the Mediterranean.

A series of palaeosols and dune sands are preserved in the upper part of the beach period. After approximately 3000 years BP a hypersaline environment, typical for a lagoon, developed with organic-rich clayey and evaporitic layers (i.e. gypsum). In these well-stratified cyclic deposits of the lagoon stage several sandy and coarse-grained layers are intercalated; partly well-sorted sand layers are interpreted to be aeolian dunes. Several other coarse-grained intervals with fining-up and thinning-up sequences contain rip-up clasts, shells debris and foraminifera. The uppermost 50 cm of the section yields organic-rich dark clays intercalated with light grey strata; some evaporitic layers also occur. These sediments are attributed to normal lagoonal sedimentary conditions, deposited during the last 500 years.

Methods

During the fieldwork in September 2004, we drilled in the lagoon and saline area of the Cabo de Gata (Fig. 4). A total of seven sites were selected for drilling to get sediments along a north–south transect. Sites 4 and 5 were drilled deeper inland, close to the distal alluvial Cabo de Gata fan in the salt-mining area. Sites 6 and 7 will not be described here; the drill samples contained variegated gypsum to >3 m depth, and no clastic sediments. Applying an open window sampler we reached maximum depths of 7.42 m (CDG 4; Fig. 6a). After core description, the same site was drilled with a sampler with PVC liners for laboratory research and sampling. A total of 45 m of cores in liners was obtained. Sampling was carried out in the field and laboratory. In the laboratory, cores were split lengthwise, logged and subsampled for grain size, organic content, micro- and macrofossils, geophysical and radiocarbon analyses (AMS dating results).

Core description

At site 1 (02°13.148 W, 36°45.365 N) two cores were obtained relatively close to the beach wall, which is here only about 1.5 to 2 m high and forms a possible by-pass area for larger tsunami waves (Fig. 4). Core CDG 1 was taken close to the shallowest part of the beach wall. Core CDG 1-2 (2°13.183 W, 36°45.415 N; Fig. 7) was drilled about 100 m north of CDG 1, in order to obtain information on the extent of the fan-like sand sheet. CDG 1-2 was taken on the northern and

Fig. 6. (**a**) Drilling in the saline area of the Cabo de Gata. (**b**) Detail of core CDG-1-2 (71 to 81 cm depth); a sequence with erosive base, shell debris, fining-up and thinning-up is interpreted as a tsunami deposit (three cycles – tsunami train); age of the layer below the deposits is 680 ± 30 years BP. (**c**) Detail of core CDG-2 (50 to 81 cm depth), erosive base. (**d**) Detail of core CDG-2 (92 to 121 cm depth); erosive base, shell debris and rip-up clasts are interpreted as a tsunami deposit; age of the layer below the deposits is 850 ± 35 years BP.

Fig. 7. Core CDG-1-2, 0–100 cm section, and core CDG-2, 0–300 cm section. Boxed parts B, C and D, are shown in Figure 6 in detail. c = caliche horizon.

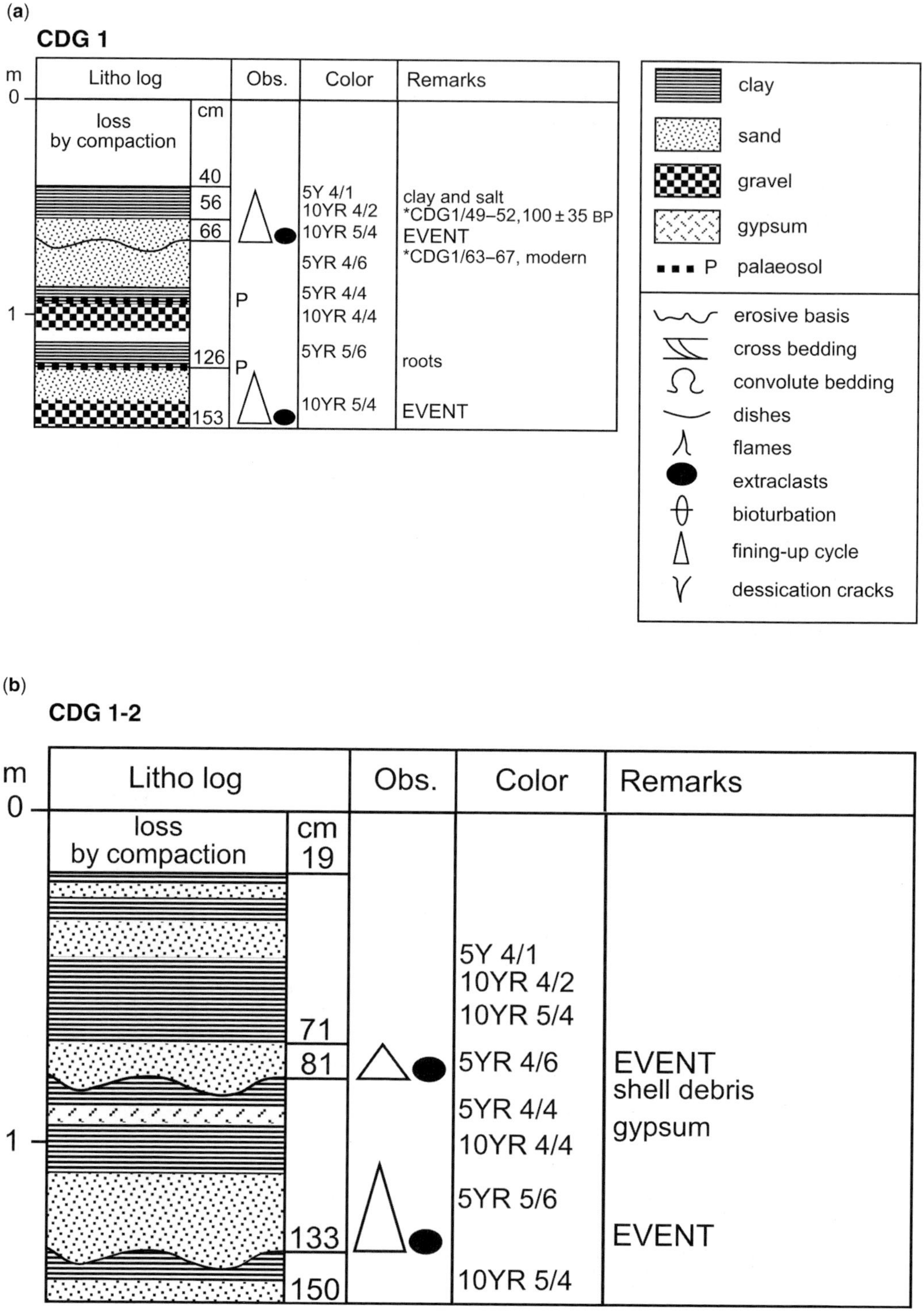

Fig. 8. (**a–d**) Lithostratigraphic logs of the drill cores (see Fig. 4 for locality). Colours refer to the Munsell colour code.

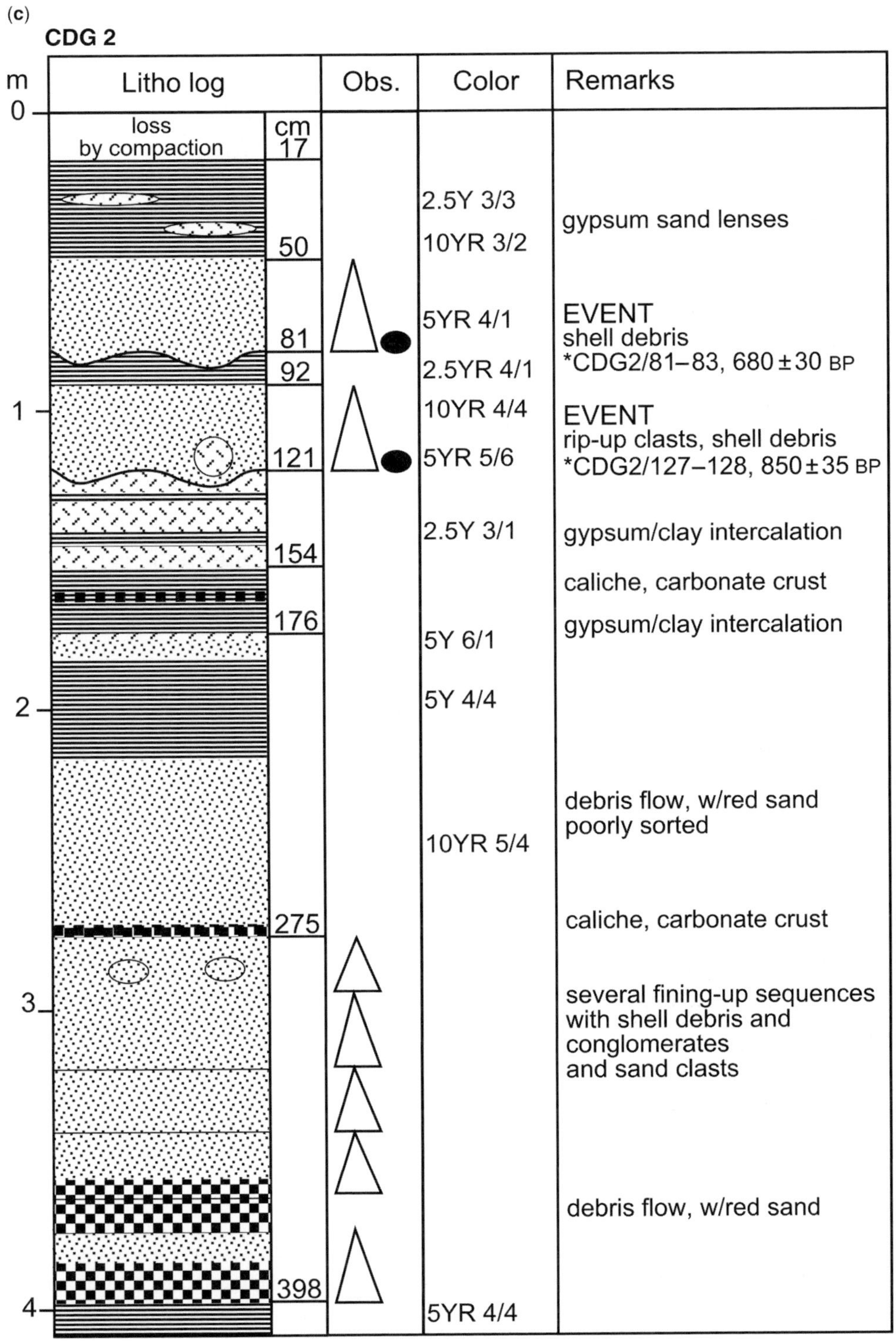

Fig. 8. (*Continued*).

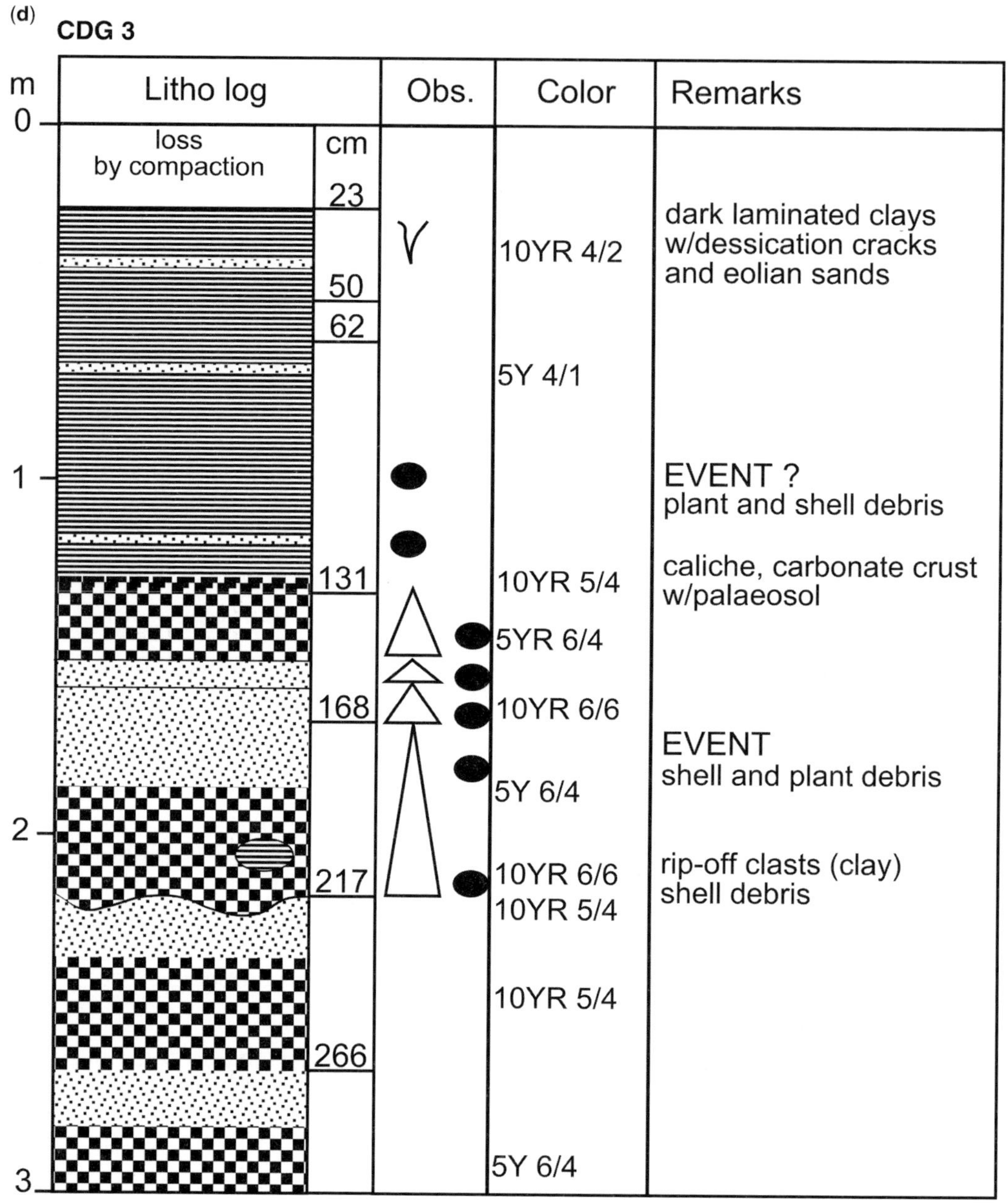

Fig. 8. (*Continued*).

distal part of the wash-over fan. At site CDG 1 (Fig. 8a) a 1.53 m long core yielded two coarse-grained (conglomeratic and sandy) layers, which form a fining-up sequence at depth of 66 cm below surface (b.s.) and 153 cm b.s. These layers are intercalated in the lagoonal clayey and evaporitic unit. The topmost lagoonal layer has been dated as 100 ± 35 years BP (Table 1). There is also a significant amount of compaction by percussion drilling of these unconsolidated sediments. Dating of the uppermost clay/sand layer gave an unexpectedly high ^{14}C activity, respectively young age, probably due to ground-water modifications. The sand layer was heavily oxidized and modified. Two palaeosols with root rests have been found in the core, one on top of the second fining-up

Table 1. *Results of radiocarbon dating*

Lab. no.	Core (interval)	Depth and correction	Age $^{14}C \pm$ err.	pmC $\pm$ err.	eq. cal. age/2σ
HAM3857/Poz-12977	CDG1 whole rock	49–52 cm	100 ± 35	98.75 ± 0.41	1681–1764 cal AD 1801–1939 cal AD
KIA 32694	CDG1 bivalve shell	49–52 cm	2330 ± 25	74.83 ± 0.24	–
HAM3858/Poz-13095	CDG1 whole rock	63–67 cm	modern	162.27 ± 0.64	1963.30 (Apr)–1963.34 (May) 1966.77 (Oct)–1967.97 (Dec) 1968.59 (Aug)–1968.65 (Aug)
HAM3859/Poz-13037	CDG2 whole rock	81–83 cm	680 ± 30	91.88 ± 0.36	1272–1314 cal AD 1357–1388 cal AD
HAM3860/Poz-12978	CDG2-2 whole rock	127–128 cm	850 ± 35	89.94 ± 0.37	1050–1083 cal AD 1125–1137 cal AD 1152–1262 cal AD
HAM3861/Poz-13097	CDG2-5 palaeosol	498–500 cm	modern	106.18 ± 0.42	1955.50 (Jul)–1957.32 (Apr)
HAM3862/Poz-13098	CDG4-5 palaeosol	424–429 cm (283–287)	4585 ± 40	56.52 ± 0.27	3499–3429 cal BC 3379–3312 cal BC 3237–3106 cal BC
HAM3863/Poz-13099	CDG4-6 palaeosol	525–531 cm (383–389)	4755 ± 40	55.31 ± 0.28	3638–3499 cal BC 3428–3379 cal BC
HAM3864/Poz-13100	CDG4-7 palaeosol	638–641 cm (496–499)	2590 ± 70	72.46 ± 0.67	902–510 cal BC
HAM3865/Poz-13101	CDG4-8 palaeosol	729–733 cm (587–591)	modern	158.27 ± 0.45	1963.20 (Mar)–1963.32 (Apr) 1967.00 (Jan)–1967.07 (Jan) 1967.35 (May)–1969.43 (Jun) 1969.77 (Oct)–1969.82 (Oct) 1970.59 (Aug)–1970.64 (Aug)
HAM3866/Poz-12980	CDG4-8 palaeosol	768–771 cm (626–629)	modern	108.91 ± 0.44	1956.96 (Dec)–1957.84 (Nov) 1995.23 (Mar)–1995.31 (Apr) 1995.91 (Nov)–1997.41 (May)

Performed at laboratories of University of Hamburg (HAM), Poznan (Poz), Poland and University of Kiel (KIA), Germany. The depth was corrected to account for an artificial dyke and compaction of the core. pmC, percent modern carbon; eq. cal. age/2σ, equivalent calibrated age given in decimal format (see Stuiver & Polach 1997 for details).

sequence at 126 cm. Both fining-up sequences have an erosive base, where considerable lagoonal sediments have been eroded. Also, both layers yield extra-clasts of the lagoonal clays (rip-up clasts) and shell debris. Core CDG 1-2 yielded comparable sequences (Figs 6b and 8b), of finer-grained 'event' layers. The event layers show up to three sequences separated from each other by a small clayey layer. These intervals are interpreted as tsunamites. These event layers express extraordinary sedimentary conditions, e.g. tsunamites. No conglomerates have been found.

Core CDG 2 (02°13.951 W, 36°46.726 N) was drilled in the northernmost part of the actual lagoon. Here, a core longer than 4 m was obtained (Fig. 8c). Again, up to 121 cm depth two sandy layers with fining-up sequences, erosive bases and rip-up clasts (Figs 6d and 7) have been found. ^{14}C dating of the lagoonal sediments directly below the upper sandy intercalation yielded an age of 680 ± 30 years BP (Fig. 6c). The sediments below the second sand layer downhole turned out to have an age of 850 ± 35 years BP (Fig. 6d). Down-section more lagoonal sediments, claystones and evaporitic layers have been drilled, partly with intercalated palaeosols and caliche crusts. The basal core section is made up of sand layers with reddish colours; they contain some shell debris and form fining-up cycles with conglomeratic bases. This base is not erosional, pointing to a distinct depositional process. The sediments of the pre-lagoonal stage are interpreted as poorly sorted debris flows mixed with marine beach layers with shell debris.

Drilling in the centre of the lagoon between the bird-watchers hide and close to the central pumping station (02°13.145 W, 36°45.621 N) at the CDG 3 site ended with a 3 m long core (Fig. 8d). Generally, this core is much more coarse-grained than those previously described. However, at 217 cm b.s. shell debris, rip-up clasts and fining-up sequences have been encountered. This 86 cm thick sequence also has an erosive base and is topped by a caliche. However, the thickness of the layer is exceptional for one single tsunami layer, which is usually on the order of 30–50 cm (Morton *et al.* 2007). Up-section clays with desiccation cracks and intercalations of aeolian sands follow. Around 100 cm b.s. plant and shell debris, as well as extra-clasts, have been found. This interval is not as clearly developed as in the other cores, but may correspond to the upper event layer. We tried to correlate the so-called 'event layers' in Figure 9: the younger event occurred later than 680 ± 30 years BP and the older event should be younger than 850 ± 35 years BP. With the exception of core CDG 3 the thickness of the layers is relatively constant; CDG 3 yields possibly a thicker deposit of the older event, but the younger one is relatively difficult to identify.

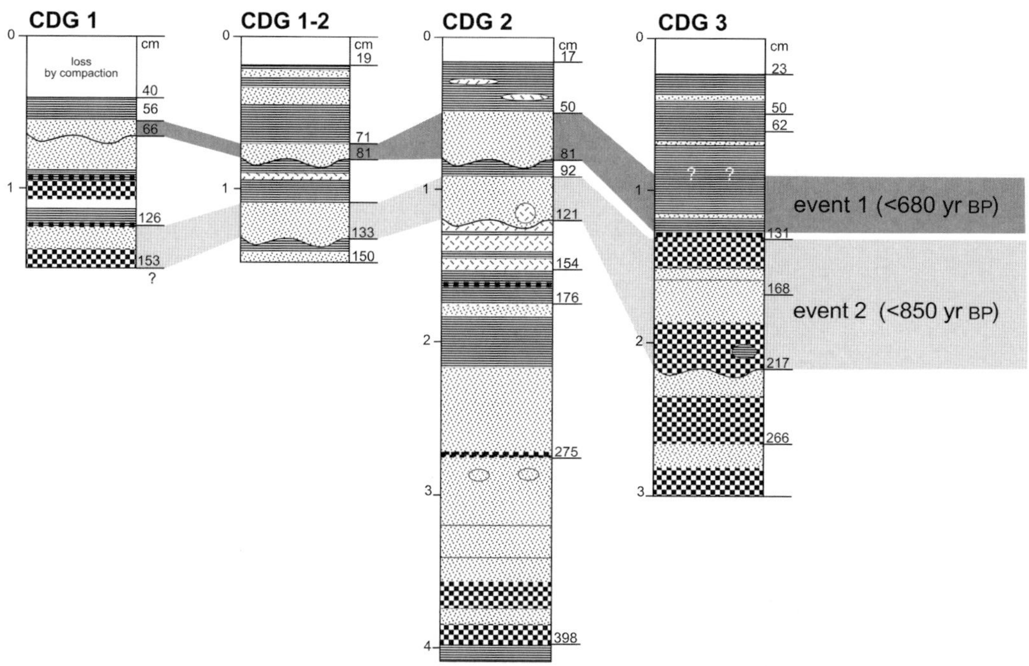

Fig. 9. Core logs of holes 1–3 with parallelization of the event layers. For legend see Figure 8.

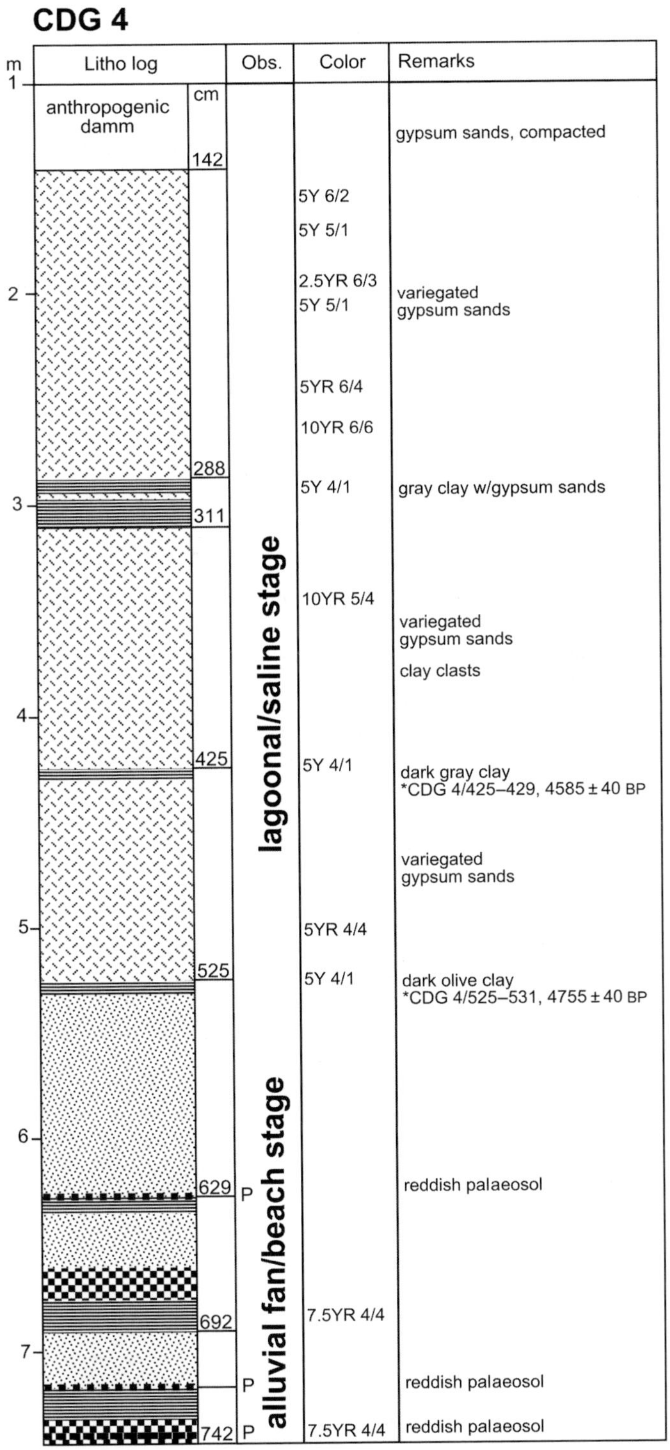

Fig. 10. Core log of drill hole 4 (see Fig. 4 for locality). Colours refer to the Munsell colour code. Compared to Figure 5, the most distal drill hole does not contain evidence for tsunamites, but delineates the development of the lagoon. Absolute ages were used to reconstruct the sedimentation rates.

As alternative interpretation is that the younger layer may have eroded the lower tsunamite, and is now not clearly distinguishable from it.

To get an idea of mean sedimentation rates and the depositional history of the sediments in the lagoon, we drilled another site CDG 4 (02°12.159 W, 36°45.026 N), where we obtained a core of 7.70 m length close to the gypsum basin of the saline area (Fig. 4). The drill core lithology (Fig. 10) is dominated by variegated gypsum deposits up to 525 cm depth b.s., intercalated with some grey clays. We dated two of these clay intercalations and obtained ages of 4585 ± 40 years BP at 425 cm b.s., and 4755 ± 40 year BP at 525 cm b.s. (Table 1). Below a sand layer and reddish palaeosols, coarse-grained conglomerates and sand/clay of the alluvial fan stage and the beach stage have been encountered. Probably due to contamination and fresh-water influences, again the results of ^{14}C dating were unexpectedly low in the deeper parts of the core. Core CDG 5 (2°12.634 W, 36°44.90 N; Fig. 4) was drilled on an artificial dam (+1.5 m a.s.l.) in the saline area, and we obtained 1.8 m anthropogenic filling and 2.2 m of variegated gypsum, but no clastic sediments, which could be used for the development of the lagoon.

Sedimentation rates in the lagoon of the Cabo de Gata

According to Zazo *et al.* (2008), the lower part of the core sections falls into a period of increased aridity. First reddish conglomerates from the Cabo de Gata fan system were deposited. The radiocarbon dates of all core segment samples we analyzed lie between the establishment of the lagoonal conditions at about 4755 years BP, which is in quite good agreement with the data of Goy *et al.* (1996), and very recent years. The $\delta^{13}C$ values scatter in the normal range for organic samples and marine carbonate and validate the ^{14}C dates (Table 1). A fragment of an articulated bivalve from core CDG 1 at 49–52 cm depth yields significantly older ages (2,330 ± 25 years BP) than expected for the sediments; this suggests reworking and redeposition of older shell fragment by tsunami or storm activity.

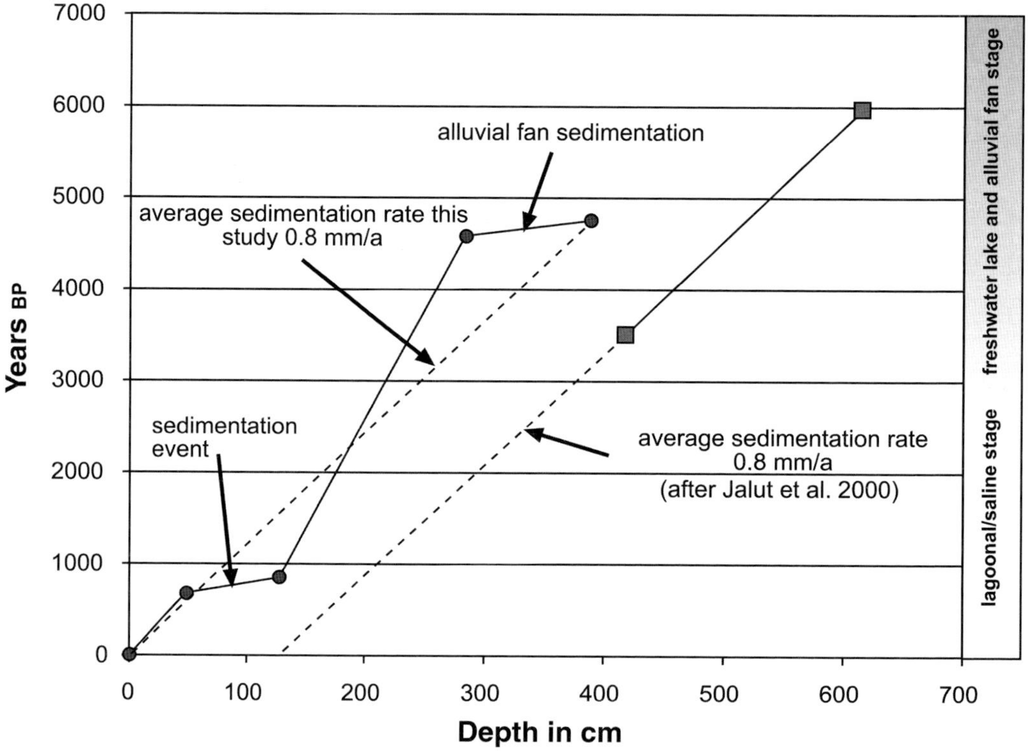

Fig. 11. Average sedimentation rates calculated from ^{14}C dates (see Table 1); note comparable sedimentation rates to Jalut *et al.* (2000). The alluvial fan stage (up to 4000 years BP) and the last 1000 years are characterized by sedimentation events.

The 'modern' dates of samples CDG 1 63–67 cm, CDG 2 498–500 cm, CDG 4 729–733 cm and CDG 4 768–771 cm can be interpreted as post-deposition modification, e.g. by fresh water. The occurrence of layers containing modern carbon in greater depth of sediments and soils is not unusual and has been explained by rapid percolation and accumulation at abrupt texture steps (Becker-Heidmann & Scharpenseel 1986). Sivan *et al.* (2002) discuss in detail the downward flux of dissolved modern carbon in saline environments. We consider the younger age as unreliable on the basis of the sedimentary history of the lagoon (Jalut *et al.* 2000), which should be older than 6000 years. For the calculation of the sedimentation rate, we took the age data of Table 1 and compared them to the data of Jalut *et al.* (2000). Also, we have to take into account compaction while drilling, which may modify the calculation of sedimentation rates. Based on that, the mean sedimentation rate is about 80 cm to 1 m per 1000 years (0.8–1 mm a^{-1}), and is relatively stable (Fig. 11); such sedimentation rates are quite common along the western Mediterranean shore (Hoffmann 1988). On the other hand, sedimentation events are rare and form outliers in the calculated regression, such as during the alluvial fan stage or the observed sandy intervals in the last 1000 years. The shift of the data of Jalut *et al.* (2000) in the plot may be explained by the incompleteness of the core description (see Fig. 5).

Sedimentological evidence for tsunamites

In contrast to the European Atlantic coast, the western Mediterranean lacks description of tsunamigenic sediments. Seismic events, volcanic explosions, impacts of extraterrestrial bolides and submarine slides may generate tsunami in the Mediterranean, for example the 1693 earthquake in eastern Sicily or 1908 in the Straits of Messina. In particular coastal situations, meteo-tsunami occur (e.g. Rissaga, Marrubio, Stigazzi), triggered by a sudden drop of pressure. Submarine slides caused the 1979 Nizza tsunami, accompanied by a 3 m high wave (comparable to an earthquake of M = 6; Tinti 1996). The last tsunami in the western Mediterranean region was observed during the earthquake of 21 May 2003 along the coast of northern Algeria. Significantly damaging waves in the ports along the southern coasts of the Balearic Islands north of the epicentral region were reported. Waves of 1 to 3 m in amplitude damaged many vessels in the harbours (e.g. modelling of Borrero 2005).

Major difficulties arise in distinguishing between storm and tsunami deposits (Minoura & Nakaya 1991; Tuttle *et al.* 2004; Morton *et al.* 2007; Tappin 2007), because of sedimentological similarities of inundation and tsunamigenic layers. Usually, sea sands from the beach or the littoral and neritic part of the shelf are transported landward by the wave (seaward derived tsunami sand sheet; see Nelson *et al.* 1996), which are landward-fining. Also, wash-over sand sheets have been described in lagoons or run-up sediments in estuaries (e.g. Kortekaas & Dawson 2007); beaches and barriers are washed over by the wave, bringing open-marine fauna into marshy and salty or terrigenous environments. However, the back-wash or back-flow brings material from the hinterland to the beach and shelf. Molluscs, foraminifera, diatoms and ostracodes provide useful tools for recognizing tsunamites (Dominey-Howes 1996; Dominey-Howes *et al.* 1999; Dawson 2007).

Generally, these lagoons form a potential reservoir for marine ingressions, like strong winter storms or tsunami. However, the Cabo de Gata saline area was never entered by marine water during a storm in the last 40 years (saline workers pers. comm.). The beach wall is up to 4 m high and very straight, and presently no active wash-over fans or by-passes exist (Fig. 5). We regard present-day storm surges and storm waves as incapable of overtopping the beach barrier and dunes. Hence, we interpret our coarse-grained event intervals as tsunamites.

The coarse-grained intervals of the Cabo de Gata lagoon have erosive bases and show up to three sequences separated from each other by a small clayey layer. These intervals are interpreted as tsunamites, and the sequence as 'tsunami train' deposits (Fig. 6b, c). Other indicative sedimentary features are rip-up clasts (gypsum, sand and clay extra-clasts; Fig. 6d) and fossil content. Bivalve and gastropod shells, mainly fragmented, and benthic and planktic foraminifera from shallow to deeper seawater habitats (>20 m water depth), such as *Ammonia beccarii*, *Elphidium crispum* and *Bolivina* sp., with some porcellanoid foraminifera such as *Quinqueloculina* sp. and *Triloculina* genera, have also been found in cores CDG 1, 1-2 and 2. These are typical faunal assemblages of eelgrass (*Zostera marina*) meadows, which are frequently found in front of the beaches in the Gulf of Almería.

Discussion

We suspect the layers found in the saline area of the Cabo de Gata are likely to be tsunamigenic because: (1) the drilling site is located at the back of the tectonically risen beach wall and back dune area exceeding altitudes of 3 m a.s.l.; (2) the two event

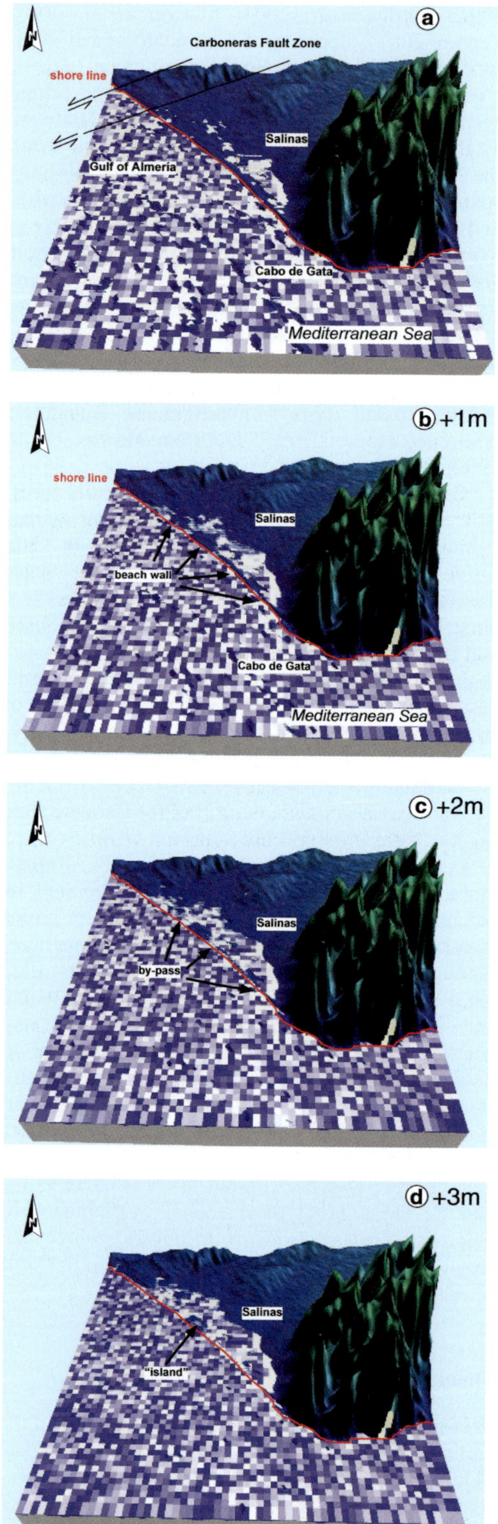

Fig. 12. (**a**) DEM of the Cabo de Gata area, with present-day sea level; red line represents the present-day coastline. Note that an *c.* 10-fold vertical exaggeration has been used to visualize topographic features along the beach and beach barrier. (**b**–**d**) Inundation models for the Cabo de Gata area. Modelling of the wave height: (b) 1 m; (c) 2 m; (d) 3 m. All based on SRTM data.

layers match exactly the sedimentology, lithology and stratigraphy expected for a tsunami, and these layers are unique in the drill cores (CDG 1 to 3, see Fig. 9); (3) the layers have sharp and erosive contact with the layers below; we found three individual layers separated by a clayey-silty drape from each other, representing a 'tsunami train' deposit (in cores CDG 1 and 1-2); (4) the layers can be parallelized in several cores along a north–south transect; (5) we found open-marine planktic and benthic foraminifers as well as mollusc fragments (gastropods and bivalves); (6) large storms and waves are regarded as incapable of overtopping the beach barrier and dunes, and storm layers (tempestites) should be more frequent.

The organic matter of the stratified clayey sediments directly below the tsunamites in CDG 2 was dated with ^{14}C-AMS as 680 ± 30 years BP for the upper layer and 850 ± 35 years BP for the lower layer. Taking into account the findings of rip-up clasts and an erosive base, the upper tsunamite may well correspond to the 1522 Almería earthquake. Inundations in coastal zones are supported by historical reports and drawings. The lower intercalation of those coarse-grained layers is quite difficult to interpret by means of historical events, and is explained as either an expression of repeated earthquake activity or tsunamigenic waves induced by submarine slides triggered by seismic shaking in the Gulf of Almería. The earthquake catalogue provides evidence for a regional strong earthquake, which occurred in AD 1013–1014 along the east coast of Andalusia (Reicherter 2001; IAG 2005). However, more detailed information is not available, neither precise locality nor intensity. But if the seismic source is located along the CFZ, the major active tectonic fault in the area, we assume recurrence periods on the order of 500–1000 years for major and tsunamigenic earthquakes in the Gulf of Almería region (because of the recurrence of tsunamigenic events). Recurrence periods for major earthquakes ($M_S > 6.5$) were estimated to be on the order of 10 ka for the Almería corridor and the CFZ (Muñoz & Udías 1985). Our new findings suggest significantly shorter return periods of destructive seismic events.

We used digital elevation models to test the extent of inundations in the lagoon of the Cabo de Gata (Fig. 12a). As a topographic base, the SRTM data with a horizontal resolution of 3 arcsec (90 m × 90 m), but a significantly higher vertical resolution (<1 m in that area) have been used. We increased the sea level step-wise in order to model and identify inundation areas as well as topographic barriers, like the beach wall. Results show that by raising the sea level 1 m above the present-day level (Fig. 12b), the beach barrier is still a straight ridge and is not overtopped by seawater. The situation also reflects winter storm wave height of *c.* 1–1.5 m (saline workers pers. comm.), but only when westerly winds hit the coast (Poniente situation). Raising the sea level to +2 m causes bypasses to develop (Fig. 12c), and this may help explain the wash-over fan formation in the area of site 1. Here, drill cores CDG 1 and 1-2 showed sedimentological differences, which are interpreted as proximal (or central) for CDG 1 and lateral fan for CDG 1-2, respectively. Other parts of the lagoon are flooded. Raising the sea level to +3 m, the entire lagoon and the beach wall are flooded (Fig. 12d), and isolated patches of topographic heights develop. The run-up height and run-up distances were not modelled; however, the static model provides good evidence that wave heights between 2 and 3 m are sufficient and necessary to bring beach sediments into the lagoon. Then, accounting for the absence of storm sediments in the cores, we have to attribute the 'event layers' to a different high-energy wave action – a tsunami.

The Carboneras Fault Zone is the most prominent and most active tectonic structure in the Gulf of Almería, and is characterized by its historic and prehistoric earthquakes (e.g. the 1522 earthquake). Generally, strike-slip faults are not considered to produce large tsunamis. Therefore, we have to take into account earthquake-triggered submarine mass movements as a possible source for tsunami generation in the Gulf of Almería. Evidence for submarine slides has already been described by Reicherter & Hübscher (2006) on the basis of high-resolution parametric echo sounding north of the Cabo de Gata spur, between the CFZ and the cape. Directly in front of the lagoon and the saline area, the bathymetric map of the Spanish coastal platforms (Sanz *et al.* 2003) shows a significant 3 km long slide scar (Fig. 3), which may be the possible source zone for the tsunami found in the lagoonal sediments. The age dating of the sediments directly below the tsunamite is slightly older than the historic 1522 Almería event due to erosion. Submarine topography revealed a slide scar directly in front of the lagoon; this suggests most probably earthquake-triggering of a submarine slide and provides a plausible scenario for the generation of tsunamigenic waves in the Gulf of Almería and sediments in the lagoon of the Cabo de Gata.

Conclusions

Drilling in the lagoon and saline area of the Cabo de Gata (Province of Almería, Spain) provided sedimentary evidence for a palaeotsunami along the Spanish Mediterranean coast. Several coarse-grained intervals with fining-up and thinning-up sequences, rip-up clasts, shells of lamellibranchs and foraminifera show erosive bases. The coarse-grained intervals show up to three sequences, each separated from the next one by a small clayey layer. These intervals are interpreted as tsunamites, a tsunami train deposit. The next coarse-grained intercalation downhole is found at 1.2 m depth, and is interpreted as either an expression of repeated earthquake activity or tsunami-like waves induced by submarine slides triggering seismic shaking in the Gulf of Almería. The age of the uppermost tsunamites is almost coeval to the 1522 Almería earthquake, hence we suggest tsunami action during that earthquake. However, taking into account that the Carboneras Fault is a strike-slip fault, mass wasting due to earthquake shaking may provide a plausible scenario for the tsunami. A nearby slide scar in the marine platform supports this.

Sedimentary evidence of a Holocene tsunami has been found at a very suitable location in a lagoon on the coast of southern Spain; the layers suggest a non-negligible tsunami and hazard potential for offshore active and seismogenic faults in the western Mediterranean region. As the Costa de Sol is one of the touristic hot-spots in Mediterranean Europe and is very densely populated, the impact on the vulnerability is of great concern for society and economy. Therefore, further and more detailed investigations along the Mediterranean coast are necessary, e.g. in shallow marshy coastal areas like the Albuferas de Almería (Campo de Dalías), the Albuferas de Valencia and south of Alicante (Santa Pola and Torrevieja), and the saline areas of Majorca Island.

This study was financially supported by the German Research Foundation (DFG-project Re 1361/3). Bathymetry is from the Meteor M51-1 cruise. AMS dating of the Cabo de Gata drilling core samples was conducted by T. Goslar (Poznan Radiocarbon Lab). A. Kaiser, C. Grützner, A. Schmidt and C. Scur helped during the field campaign. Daniel Stich is thanked for excavating the data mine of the IAG for historical earthquake reports. We thank Javier Lario and Alessandra Smedile for thoughtful and helpful reviews.

References

ATWATER, B. F. 1992. Geologic evidence for earthquakes during the past 2000 years along the Copalis River, Southern Coastal Washington. *Journal of Geophysical Research*, **97**(B2), 1901–1919.

ATWATER, B. F. & MOORE, A. L. 1992. A tsunami about 1000 years ago in Opuget Sound, Washington. *Science*, **258**, 1614–1623.

ATWATER, B. F. & YAMAGUCHI, D. K. 1991. Sudden, probably coseismic submergence of Holocene trees and grass in coastal Washington state. *Geology*, **19**, 706–709.

ATWATER, B. F., MUSUMI-ROKKAKU, S., SATAKE, K., TSUJI, Y., UEDA, K. & YAMAGUCHI, D. K. 2005. *The orphan tsunami of 1700: Japanese clues to a parent earthquake in North America*. US Geological Survey, Professional Paper **1707**.

BARTEL, P. & KELLETAT, D. 2003. Erster Nachweis holozäner Tsunamis im westlichen Mittelmeergebiet (Mallorca, Spanien) mit einem Vergleich von Tsunami- und Sturmwellenwirkung auf Festgesteinsküsten. *Berichte Forschungs- und Technologiezentrum Westküste der Universität Kiel, Büsum*, **28**, 93–107.

BECKER-HEIDMANN, P. & SCHARPENSEEL, H. W. 1986. Thin layer ^{13}C and $D^{14}C$ monitoring of "Lessivé" soil profiles. *Radiocarbon*, **28**, 383–390.

BECKER-HEIDMANN, P., REICHERTER, K. & SILVA, P. G. 2007. ^{14}C dated charcoal and sediment drilling cores as first evidence of Holocene tsunami at the Southern Spanish coast. *Radiocarbon*, **49**(2), 827–835.

BELL, J. W., AMELUNG, F. & KING, G. C. P. 1997. Preliminary Late Quaternary slip history of the Carboneras Fault, southeastern Spain. *Journal of Geodynamics*, **24**, 51–66.

BORRERO, J. C. 2005. Preliminary Simulations of the Algerian Tsunami of 21 May, 2003 in the Balearic Islands by Jose C. Borrero, University of Southern California; Tsunami Research group. World wide web address: http://www.usc.edu/dept/tsunamis/ALGERIA/

CISTERNAS, M., ATWATER, B. F. *ET AL*. 2005. Predecessors to the giant 1960 Chile earthquake. *Nature*, **437**, 404–407.

DABRIO, C. J., GOY, J. L. & ZAZO, C. 1998. The record of the tsunami produced by the 1755 Lisbon earthquake in Valdelagrana spit (Gulf of Cádiz, southern Spain). *Geogaceta*, **23**, 31–34.

DAWSON, S. 2007. Diatom biostratigraphy of tsunami deposits: Examples from the 1998 Papua New Guinea tsunami. *Sedimentology*, **200**(3–4), 328–335.

DOMINEY-HOWES, D. 1996. *The Geomorphology and Sedimentation of Five Tsunamis in the Aegean Sea Region, Greece*. PhD thesis, Coventry University.

DOMINEY-HOWES, D., DAWSON, A. G. & SMITH, D. E. 1999. Late Holocene coastal tectonics at Falasarna, western Crete (Greece): a sedimentary contribution. STEWART, I. S. & VITA-FINZI, C. (eds) *Coastal Tectonics*. Geological Society, London, Special Publications, **146**, 343–352.

DOMINEY-HOWES, D., CUNDRY, A. & CRUDACE, I. 2000. High energy flood deposits on Astypalaea Island, Greece: possible evidence for the AD 1956 southern Aegean tsunami. *Marine Geology*, **163**, 303–315.

DOMINEY-HOWES, D., HUMPHREYS, G. S. & HESSE, P. P. 2006. Tsunami and palaeotsunami depositional signatures and their potential value in understanding the late-Holocene tsunami record. *The Holocene*, **16**, 1095–1107.

GIANFREDA, F., MASTRONUZZI, G. & SANSÓ, P. 2001. Impact of historical tsunamis on a sandy coastal barrier: an example from the northern Gargano coast, southern Italy. *Natural Hazards and Earth System Sciences*, **1**, 213–219.

GOY, J. L., ZAZO, C., DABRIO, C. J., LARIO, J., BORJA, F., SIERRO, F. J. & FLORES, J. A. 1996. Global and regional factors controlling changes of coastlines in southern Iberia (Spain) during the Holocene. *Quaternary Science Reviews*, **15**, 773–780.

GRÀCIA, E., PALLÀS, R. *ET AL*. 2006. Active faulting offshore SE Spain (Alboran Sea). Implications for earthquake hazard assessment in the Southern Iberian Margin. *Earth and Planetary Science Letters*, **241**, 734–749.

HARVEY, A. M., SILVA, P. G., MATHER, A. E., GOY, J. L., STOKES, M. & ZAZO, C. 1999. The impact of Quaternary sea-level and climatic change on coastal alluvial fans in the Cabo de Gata ranges, southeast Spain. *Geomorphology*, **28**, 1–22.

HOERNLE, K. A., DUGGEN, S., GELDMACHER, J., KLÜGEL, A. & Scientific Party 2003. METEOR Cruise No. 51, Leg 1, Vulkosa: Vulkanismus Ostatlantik-Alboran. *In*: *Meteor Berichte 03-1 Ostatlantik-Mittelmeer-Schwarzes Meer, Cruise No. 51, 12 September to 28 December 2001*. GEOMAR Research Centre, Kiel, 1–38.

HOFFMANN, G. 1988. *Holozänstratigraphie und Küstenlinienverlagerung an der andalusischen Mittelmeerküste*. Berichte, FB Geowissenschaften, Universität Bremen, **2**.

IAG 2005. Instituto Andaluz de Geofísica. World wide web address: http://www.ugr.es/~iag/

JALUT, G., AMAT, A. E., BONNET, L., GAUQUELIN, T. & FONTUGNE, M. 2000. Holocene climatic changes in the Western Mediterranean, from south-east France to south-east Spain. *Palaeogeography, Palaeoclimatology, Palaeoecology*, **160**, 255–290.

JONK, R. & BIERMANN, C. 2002. Deformation in Neogene sediments of the Sorbas and Vera Basins (SE Spain): constraints on simple-shear deformation and rigid body rotation along major strike-slip faults. *Journal of Structural Geology*, **24**, 963–977.

KELLETAT, D. & SCHELLMANN, G. 2002. Tsunamis on Cyprus: field evidences and ^{14}C dating results. *Zeitschrift für Geomorphologie, NF*, **46**(1), 19–34.

KELSEY, H. M., NELSON, A. R., HEMPHILL-HALEY, E. & WITTER, R. C. 2005. Tsunami history of an Oregon coastal lake reveals a 4600 yr record of great earthquakes on the Cascadia subduction zone. *Geological Society of America Bulletin*, **177**, 1009–1032.

KORTEKAAS, S. & DAWSON, A. G. 2007. Distinguishing tsunami and storm deposits: An example from Martinhal, SW Portugal. *Sedimentology*, **200**(3–4), 208–221.

LÓPEZ MARINAS, J. L. 1985. El terremoto catastrofico de 22 de Septiembre de 1522, en Almería. *Seminario sobre Sismicidad y Riesgo Sísmico Comunicaciones y ponencias*. Instituto Geografico Nacional, **1**, 51–60.

LUQUE, L., LARIO, J., CIVIS, J., SILVA, P. G., ZAZO, C., GOY, J. L. & DABRIO, C. J. 2002. Sedimentary record of a tsunami during Roman times, Bay of

Cádiz, Spain. *Journal of Quaternary Science*, **17**(5–6), 623–631.

MARTIN, J. M., BRAGA, J. C. & BETZLER, C. 2003. Late Neogene – Recent uplift of the Cabo de Gata volcanic province, Almería, SE Spain. *Geomorphology*, **50**, 27–42.

MARTÍNEZ SOLARES, J. M. 1995. Catalogo sismico. *In*: *Sismicidad historica del Reino de Granada (1487–1531)*. Instituto Geografico Nacional, Monografia, **12**, 9–36.

MASTRONUZZI, G. & SANSÓ, P. 2000. Boulders transport by catastrophic waves along the Ionian coast of Apulia (southern Italy). *Marine Geology*, **170**, 93–103.

MINOURA, K. & NAKAYA, S. 1991. Traces of tsunami preserved in inter-tidal lacustrine and marsh-deposits– Some examples from northeast Japan. *Journal of Geology*, **99**, 265–287.

MONTENAT, C. & OTT D'ESTEVOU, P. 1995. Late Neogene basins evolving in the Eastern Betic transcurrent fault zone: an illustrated review. *In*: FRIEND, P. F. & DABRIO, C. J. (eds) *Tertiary Basins of Spain*. Cambridge University Press, Cambridge, 372–386.

MORTON, R. A., GELFENBAUM, G. & JAFFE, B. E. 2007. Physical criteria for distinguishing sandy tsunami and storm deposits using modern examples. *Sedimentary Geology*, **200**, 184–207.

MUÑOZ, D. & UDÍAS, A. 1985. Estudios recientes del riesgo sismico en España. *Seminario sobre Sismicidad y Riesgo Sísmico Comunicaciones y ponencias*. Instituto Geografico Nacional, **1**, 105–114.

NANAYAMA, F. K., SATAKE, K., FURUKAWA, R., SHIMOKAWA, K., ATWATER, B., SHIGENO, K. & YAMAKI, S. 2003. Unusually large earthquakes inferred from tsunami deposits along the Kuril Trench. *Nature*, **424**, 660–663.

NELSON, A., SHENNAN, I. & LONG, A. J. 1996. Identifying coseismic subsidence in tidal-wetland stratigraphic sequences at the Cascadia subduction zone of western North America. *Journal of Geophysical Research*, **101**(B3), 6115–6135.

PANTOSTI, D., BARBANO, M. S., SMEDILE, A., DE MARTINI, P. M. & TIGANO, G. 2008. Geological evidence of palaeotsunamis at Torre degli Inglesi (northeast Sicily). *Geophysical Research Letters*, **35**, L05311. DOI: 10.1029/2007/GL032935.

PINEGINA, T., BOURGEOIS, J., BAZANOVA, L., MELEKESETEV, I. & BRAITSEVA, O. 2003. A millennial-scale record of Holocene tsunamis on the Kronotskiy Bay coast, Kamchatka, Russia. *Quaternary Research*, **59**, 36–47.

REICHERTER, K. 2001. Paleoseismological advances in the Granada Basin (Betic Cordilleras, southern Spain). *Acta Geologica Hispanica*, **36**(3–4), 267–281.

REICHERTER, K. & HÜBSCHER, C. 2006. Off-shore evidence for the 1522 Almería earthquake (M $>$ 6.5) in the Alborán Sea (southern Spain). *Journal of Seismology*, **11**, 15–26.

REICHERTER, K. & REISS, S. 2001. The Carboneras Fault Zone (southeastern Spain) revisited with Ground Penetrating Radar – Quaternary structural styles from high-resolution images. *Geologie en Mijnbow*, **80**(3–4), 11–20.

RUIZ-GÁLVEZ FRIEGO, M. 1993. El Occidente de la Península la Iberica, Punto de encuentro entre el Mediterraneo y el Atlantico a fines de la edad del Bronce. *Complutum*, **4**, 41–68.

SANZ, J. L., TELLO, O., *ET AL*. 2003. *Estudio de La Plataforma Continental Española*. Hoja MC 050 – Cabo de Gata, Serie C, Modelos digitales y geomorphologia, Promontorio de Cabo de Gata (Almeria), Instituto Español de Oceanografia.

SILVA, P., MASANA, E. *ET AL*. 2004. Active tectonics in the Mediterranean Sector of the Iberian Peninsula (east Spain). *32nd International Geological Congress, Florence 2004, Field Trip Guide*, **B10**.

SIVAN, O., LAZAR, B., YECHIELI, Y., BOARETTO, E., HEINEMEIER, J. & HERUT, B. 2002. ^{14}C excess in deep-sea sediments porewater driven by diffusion — Southeast Mediterranean. *Limnology and Oceanography*, **47**, 565–570.

STAPEL, G., MOEYS, R. & BIERMANN, C. 1996. Neogene evolution of the Sorbas basin (SE Spain) determined by paleostress analysis. *Tectonophysics*, **255**, 291–305.

STUIVER, M. & POLACH, H. A. 1977. Discussion; reporting of C-14 data. *Radiocarbon*, **19**, 355–363.

TAPPIN, D. R. 2007. Sedimentary features of tsunami deposits — Their origin, recognition and discrimination: An introduction. *Sedimentary Geology*, **200**, 151–154.

TINTI, S. 1996. Genesis and impact of tsunamis on the European coasts – GITEC – A European effort to foster tsunami research. *In*: GHAZI, A. & YEROYANNI, M. (eds) *Seismic Risk in the European Union* (Vol. 1). European Commission on Environment and Climate Programme and natural hazards, Brussels, 135–151.

TUTTLE, M. P., RUFFMAN, A., ANDERSON, T. & JETER, H. 2004. Distinguishing tsunami deposits from storm deposits along the coast of northeastern North America: Lessons learned from the 1929 grand Banks tsunami and the 1991 Halloween storm. *Seismology Research Letters*, **75**, 117–131.

VARELA HERVÍAS, E. & VON WALDHEIM, G. 1948. *Una relacíon alemana sobre el terremoto de Andalucía, Marruecos y Azores del año 1522*. Artes Gráficas Municipales, Madrid.

VÖTT, A., BRÜCKNER, H., MAY, M., LANG, F., HERD, R. & BROCKMÜLLER, S. 2008. Strong tsunami impact on the Bay of Aghios Nikolaos and it environs (NW Greece) during Classical-Hellenistic times. *Quaternary International*, **181**, 105–122.

WHELAN, F. & KELLETAT, D. 2005. Boulder deposits on the southern Spanish Atlantic coast: possible evidence for the 1755 AD Lisbon tsunami? *Science of Tsunami Hazards*, **23**(3), 25–38.

ZAZO, C., DABRIO, C. J. *ET AL*. 2008. The coastal archives of the last 15 ka in the Atlantic- Mediterranean Spanish linkage area: Sea level and climate changes. *Quaternary International*, **181**, 72–87.

Palaeoseismology of the Vilariça Segment of the Manteigas-Bragança Fault in northeastern Portugal

THOMAS ROCKWELL[1], JOÃO FONSECA[2], CHRIS MADDEN[1], TIM DAWSON[1,3], LEWIS A. OWEN[4], SUSANA VILANOVA[5] & PAULA FIGUEIREDO[5,6]

[1]*Earth Consultants International, 1642 East 4th Street, Santa Ana, CA 92701, USA*

[2]*Physics Department, IST, Av Rovisco Pais 1, 1049-001 Lisbon, Portugal*

[3]*California Geological Survey, 345 Middlefield Road, MS520, Menlo Park, CA 94025, USA*

[4]*Department of Geology, University of Cincinnati, PO Box 210013, Cincinnati, OH 45221, USA*

[5]*ICIST, IST, Av Rovisco Pais 1, 1049-001 Lisbon, Portugal*

[6]*LATTEX, Geology Department, University of Lisbon, Campo Grande, Lisbon, Portugal*

**Corresponding author (e-mail: trockwell@geology.sdsu.edu)*

Abstract: The Manteigas-Bragança fault is a major, 250-km-long, NNE-striking, sinistral strike-slip structure in northern Portugal. This fault has no historical seismicity for large earthquakes, although it may have generated moderate (M5+) earthquakes in 1751 and 1858. Evidence of continued left horizontal displacement is shown by the presence of Cenozoic pull-apart basins as well as late Quaternary stream deflections. To investigate its recent slip history, a number of trenches were excavated at three sites along the Vilariça segment, north and south of the Douro River. At one site at Vale Meão winery, the occurrence of at least two and probably three events in the past 14.5 ka was determined, suggesting an average return period of about 5–7 ka. All three events appear to have occurred as a cluster in the interval between 14.5 and 11 ka, or shortly thereafter, suggesting a return period of less than 2 ka between events within the cluster. In the same area, a small offset rill suggests 2–2.5 m of slip in the most recent event and about 6.1 m after incision below a *c*. 16 ka alluvial fill event along the Douro River. At another site along the Vilariça River alluvial plain, NE of the Vale Meão site, several trenches were excavated in late Pleistocene and Holocene alluvium, and exposed the fault displacing channel deposits dated to between 18 and 23 ka. In a succession of closely spaced parallel cuts and trenches, the channel riser was traced into and across the fault to resolve *c*. 6.5 m of displacement after 18 ka and *c*. 9 m of slip after *c*. 23 ka. These observations yield a slip rate of 0.3–0.5 mm/a, which is consistent with earlier estimates. Combining the information on timing at Vale Meão winery and displacement at Vilariça argues for earthquakes in the M7+ range, with coseismic displacements of 2–3 m. This demonstrates that there are potential seismic sources in Portugal that are not associated with the 1755 Lisbon earthquake or the Tagus Valley, and, although rare, large events on the Vilariça fault could be quite destructive for the region. This work provides an analogue for the study of active faulting in intracontinental settings and supports the view that earthquakes within intracontinental settings tend to cluster in time. In addition, this study highlights the usefulness and application of multiple field, remote sensing and geochronological techniques for seismic hazard mitigation.

We investigated the late Quaternary (*c*. last 30 000 years) rupture history of the Vilariça segment of the Manteigas-Bragança fault (called here Vilariça fault for simplicity) in the vicinity of the Douro River in northern Portugal (Fig. 1) as part of a seismic hazard investigation for a proposed dam across the Sabor River. This provided an opportunity to study the history and nature of intracontinental earthquakes in an area that is generally considered tectonically stable, and with low seismic hazard (see Vilanova & Fonseca (2007) for an assessment of seismic hazard in Portugal).

For this study, we analysed stereo-paired aerial photography of about 25 km of the Vilariça fault to assess the overall expression of the fault, and to map landforms that are typically associated with active faults. We also used photography to select potential trench sites to date the timing of past surface ruptures. We then evaluated potential sites in the field for both access and to assess their likelihood for the presence of suitable stratigraphy and datability. Finally, we excavated six trenches at three sites to expose the fault and to study its late Quaternary history, and conducted additional

From: REICHERTER, K., MICHETTI, A. M. & SILVA, P. G. (eds) *Palaeoseismology: Historical and Prehistorical Records of Earthquake Ground Effects for Seismic Hazard Assessment*. The Geological Society, London, Special Publications, **316**, 237–258. DOI: 10.1144/SP316.15 0305-8719/09/$15.00

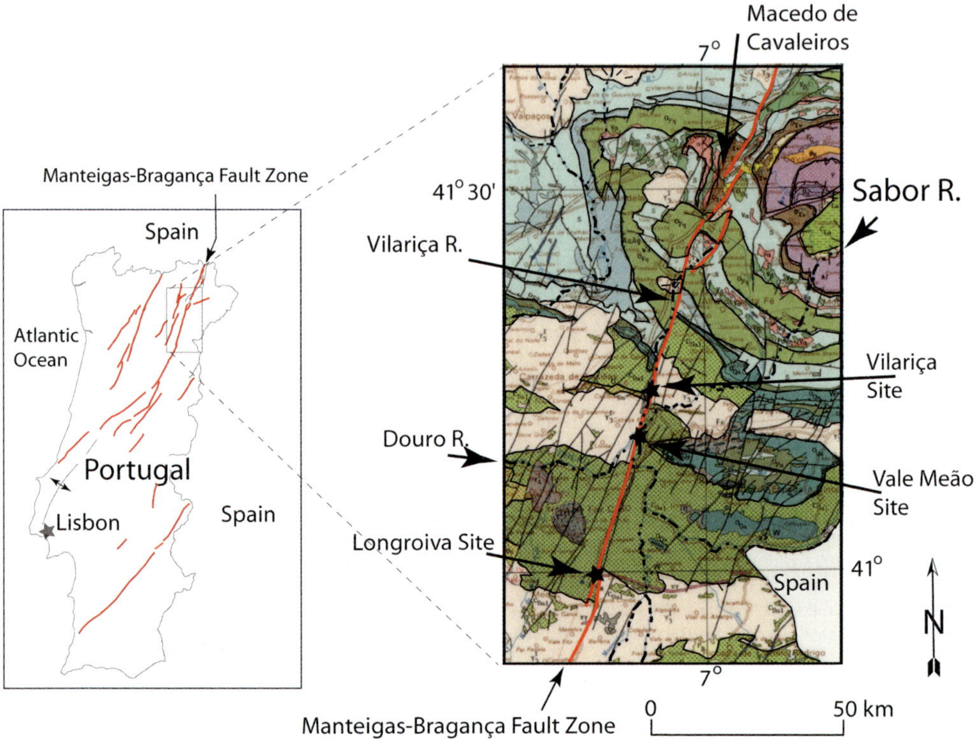

Fig. 1. Geological map of the study area in northeastern Portugal along the Manteigas-Bragança fault zone. The background map is scanned and modified from the 1992 version of the Carta Geológica de Portugal, produced by the Servicos Geológicos de Portugal. Note the left-lateral offset of the folded Palaeozoic strata along the Manteigas-Bragança and other parallel faults. Please refer to original map for description of units.

three-dimensional trenching at one of the sites to resolve displacement of a buried channel margin. This paper summarizes the results of these field studies, the first of this type along the Vilariça fault, and clearly demonstrates that there are large earthquake sources in northern Portugal that have been recurrently active in the late Quaternary. We begin with a brief background into the tectonic setting and current kinematic framework.

Regional tectonic overview and current kinematic framework

The tectonic setting of Portugal reflects a long history of deformations that extend back to the early Palaeozoic. Variscan basement rocks underlie most of western Iberia, with the Hesperian massif consisting of the most continuous fragment of Variscan basement in Europe (Ribeiro 1974, 1981; Ribeiro *et al.* 1979). The basement in the study area comprises primarily folded Precambrian to Devonian greywackes, schists and quartzites, intruded by granites (Cabral 1989).

The fold and thrust belt, within which this study is sited, formed in the Variscan Orogeny during the Carboniferous and into the Permian, with the early development of strike-slip faulting across the fold belt in late Variscan time. The Late Variscan NNE–SSW trending faults were considered as primary left-lateral faults by many workers (Ribeiro 1974, 1981; Arthaud & Matte 1975; Ribeiro *et al.* 1979), although Marques *et al.* (2002) suggest that Variscan faulting may have been dextral, and later became sinistral during reactivation in the Alpine Orogeny. In Late Cretaceous to late Eocene, Iberia moved *c.* 120 km northward, producing the Pyrenean orogenic belt (Grimaud *et al.* 1982). The region experienced continued shortening in the Miocene (Ribeiro *et al.* 1990), with some activity persisting to the present on structures such as the Vilariça fault in northern Portugal (Cabral 1989) and other structures in Portugal (Cabral & Ribeiro 1988).

Based on analysis of current seismicity, and structural indicators onshore and offshore of Portugal, Fonseca & Long (1991) suggest that western Iberia is being extruded westward towards the Atlantic. This extrusion may have begun as early as the late Eocene but apparently was a major factor in the Alpine collision during the Miocene. Fonseca & Long (1991) interpret the current activity on many faults as 'a subdued continuation of the upper Miocene evolution'. The extrusion model is significant because it: (1) allows for the reactivation of many basement strike-slip faults in Portugal, including the Vilariça fault; (2) explains the apparent activity of folds and thrusts offshore to the west; and (3) explains the current presence of seismic activity in and around Portugal.

The amount of Quaternary slip associated with these faults is yet to be defined, although at least the Vilariça fault has experienced moderate earthquakes in 1751 and 1858 (Vilanova & Fonseca 2007). Furthermore, it is likely that fault reactivation has a long history. There is a total of *c.* 8 km of left slip on the Vilariça fault, as estimated from offset Palaeozoic folds measured from the geological map of Portugal (Fig. 1) (for the estimation of total lateral offset, we used sheets 11C, Torre de Moncorvo, and 15A, Vila Nova de Foz Côa, published by the Serviços Geológicos de Portugal, both at a scale of 1:50 000). However, this deformation has been accruing since the Palaeozoic, and if Marques *et al.* (2002) are correct, the sinistral displacement may have been superposed on some component of dextral slip. In any case, the fault has sufficient displacement to have developed a relatively straight, localized trace over a length scale of more than 100 km.

Cabral (1985, 1989) was the first to delineate direct evidence of neotectonic activity along the Manteigas-Bragança fault, and he attributes the formation of the Vilariça basin as a pull-apart structure resulting from reactivation of the fault. Based on an inferred deflection of the Douro River, Cabral (1985) also attributes a slip rate of 0.2–0.5 mm/a to this structure, which he states is 'compatible with its regional geomorphic expression and historical seismicity'. Using this rate, only 0.4–1 km of slip would have accrued in the Quaternary. Thus, an assessment of the late Quaternary movement history of the Vilariça fault is important in quantifying the likely hazard posed to planned and existing dams in the Douro River region, and to the population in general. Towards that end, the focus of much of our work has been the establishment of the timing and size of past surface ruptures along the Vilariça segment of the fault, resolving its late Quaternary slip rate, and estimating the size and likelihood of future large earthquakes.

Geomorphic analysis

Black and white stereo aerial photographs (1:18 000 scale) purchased from the Instituto Geográfico de Portugues in Lisbon were examined to identify active structures and map tectonic landforms, following the work of Cabral (1985, 1989) who mapped this same area at a scale of 1:25 000. The photographs were scanned at 800 dpi (*c.* half-metre pixels), and mosaics were constructed to compile the observations, as presented in Figures 2–4. We divided the study area into three subsections for ease of presentation. The northernmost section

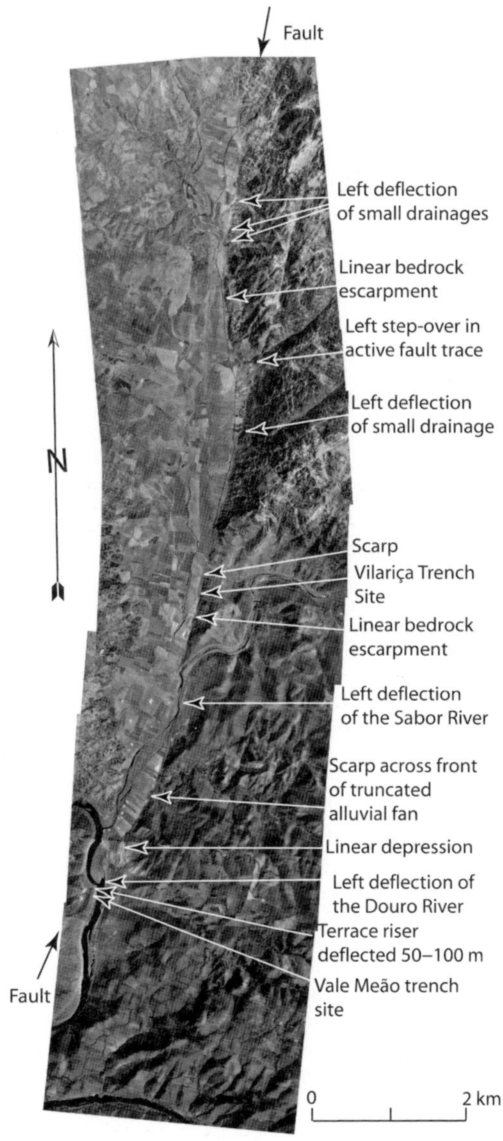

Fig. 2. Photomosaic of the fault zone in the northern area of detailed study along the Vilariça River southward to the Douro River, detailing the tectonic geomorphology in Vilariça basin.

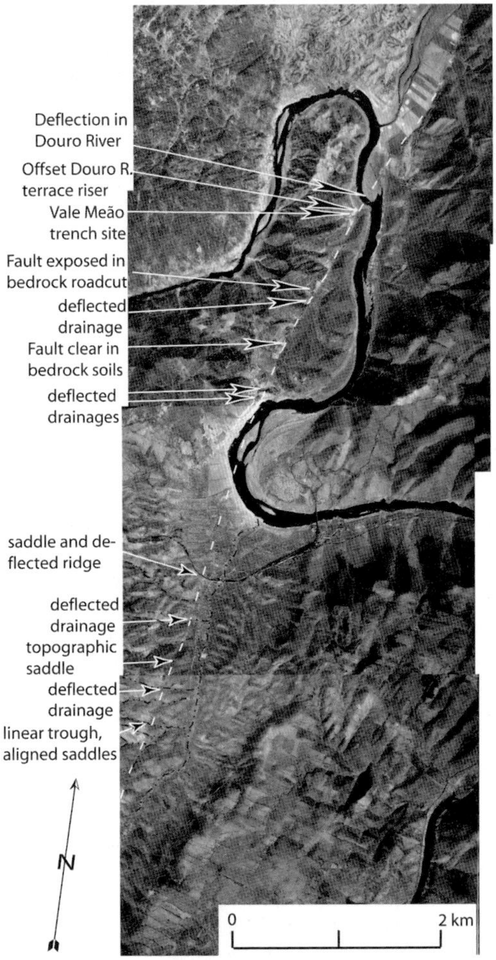

Fig. 3. Photomosaic of the fault zone in the central area, showing the Vilariça fault segment south of the Douro River.

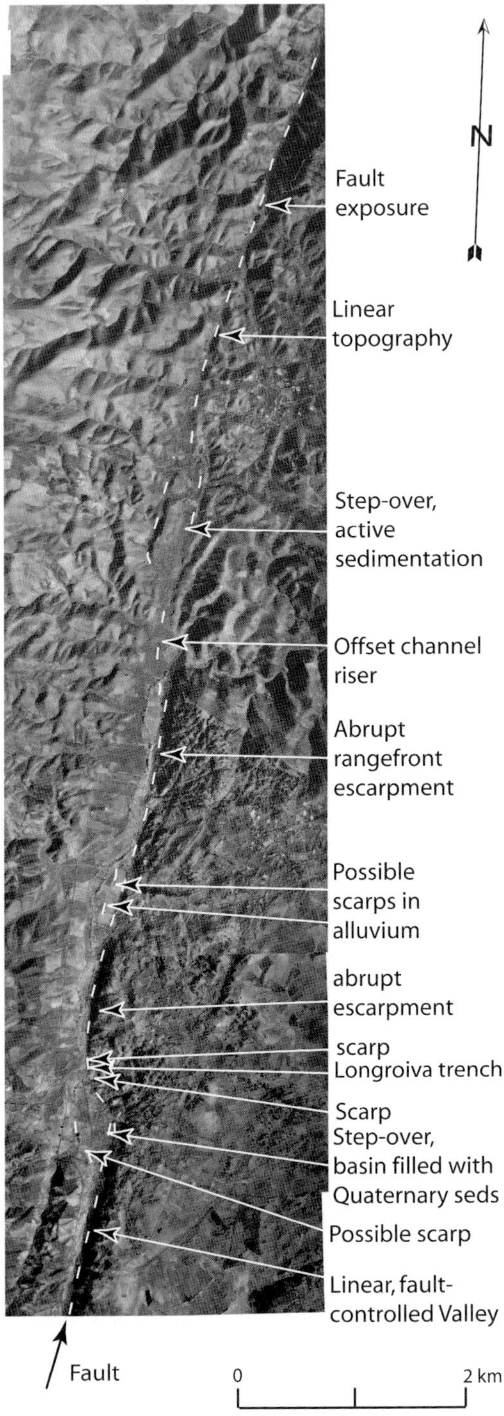

Fig. 4. Photomosaic of the fault zone in the southern area along the Vilariça fault segment, showing the tectonic geomorphology southward to the Longrovia area.

includes the Vilariça basin southward to the northernmost bend of the Douro River (Fig. 2). The central section includes the northward bend of the Douro River and the Vale Meão winery (Quinta do Vale Meão) southward into the Ribeira do Vale da Vinha area (Fig. 3), whereas the south section extends down to Longroiva (Fig. 4). A short section in the upper Ribeira do Vale da Vinha/Vale do Escudeiro is not covered by any of the mosaics presented in Figures 2 to 4, but most of this area is undergoing erosion and features associated with recent activity are sparse.

Northern section: Vilariça Basin

The Vilariça fault is well expressed north of the Douro River in the area that Cabral (1985, 1989) called the Vilariça basin (Fig. 2). Much of the area along the fault has undergone Holocene deposition, with accumulations of young sediments against the bedrock escarpment that was produced by the fault. This has resulted in a very pronounced and linear escarpment that is clearly seen in Figure 2. The surface trace of the fault is mostly buried by late Holocene alluvium along much of the length of the Vilariça basin, both north and south of the basin's confluence with the Sabor River, as exposed in the Vilariça trenches discussed later in this paper. Nevertheless, the escarpment generally marks the location of the fault, although it is locally trimmed by the Vilariça and Sabor River margins.

Landforms specifically attributed to fault movement include offset or deflected drainages. These are well seen where they cross the fault in the northern Vilariça basin area (upper part of Fig. 2). Most of these channels are incised suggesting that erosion has dominated during the late Quaternary. There is a scarp that truncates an alluvial fan south of where the Sabor River enters the Vilariça basin. This scarp is related to either fault movement or erosion by the Sabor River. The Sabor River is also deflected to the left, but this is down the fluvial gradient towards the Douro River so the magnitude of the deflection is likely enhanced by this effect. Farther south, the course of the Douro River is strongly left-deflected at the fault. This deflection is pronounced and not likely the result of simple differential erosion. The rocks east of the fault are predominantly schist and should erode more rapidly than the granitoid rocks west of the fault. Nevertheless, the schist is shunted northward across the Douro River. We consider it likely that at least part of this deflection is related to slip on the Vilariça fault, especially considering that this is the largest (highest order) stream in this area and it would have been hard to overcome its erosive force if the fault had not been tectonically active. The deflection observed in Figure 2 is enhanced by ponding of the Douro River behind a hydroelectric dam across the river downstream from the fault. Hence, the deflection represents offset of the channel walls rather than the Holocene channel itself and probably represents slip that accrued after a major phase of incision during the Pleistocene. On the north margin of the Douro River, the deflection is also likely enhanced by deposition from the Sabor River, the confluence of which is very close to where the fault crosses the Douro.

Central area south of the Douro River

The area SW of the major northward bend in the Douro River, which includes the Vale Meão winery that hosts one of our trench sites, has several features that are likely the direct result of slip on the fault. The fault is well-expressed in the Vale Meão area, as shown in Figure 3. The southern terrace riser of the Douro River is left-deflected by *c.* 50–100 m, which we interpret as largely the result of fault offset.

Farther south, the location of the fault is very clear in the aerial photography because the surface topography is undergoing erosion and the soils have developed very differently across either side of the fault. On the northwestern side, the granitic bedrock soil is much lighter in colour than the soil developed in the schist SE of the fault. The fault surface is well-exposed in at least one road-cut exposure, and striae on gouge of the active fault surface are nearly horizontal, indicating nearly pure horizontal motion, which is consistent with the surface morphology in this area.

There are several deflected or offset stream channels that incise the bedrock in the Vale Meão area. We interpret these as the result of long-term motion on the fault. The smallest of these deflections, located adjacent to the Douro River where the fault intersects the river on the SW edge of the Vale Meão winery, shows *c.* 2.2 m of near-field left-lateral deflection and *c.* 6 m of far-field deflection on a small rill, as discussed later when we estimate slip per event. The next stream to the NE is deflected *c.* 50 m and the fault coincides precisely with the location of these deflections. There are also drainage basins within the Vale Meão winery that do not currently align with their principal channels and are likely offset.

To the SW of the Douro River, the location of the fault is inferred for several hundred metres, where it is buried beneath late Quaternary alluvium of Ribeira do Vale da Vinha. Farther south, the fault traverses an incised area with minor deflections of the principal channels (Fig. 3). The drainage courses are delineated on Figure 3 to illustrate that although deflections are present, they are relatively subtle. We interpret this to reflect the relatively low slip rate associated with the fault, because the overall geomorphology represents long-term incision.

Southern section to Longroiva

The fault makes several small, left (releasing) stepovers with intervening areas of late Quaternary sedimentation south of the drainage divide that separates Ribeira do Vale da Vinha from the Longroiva Valley. The Longroiva trench site is at the northern edge of the largest of these basins.

The fault is principally expressed as aligned linear escarpments and small scarps in bedrock and possibly alluvium (Fig. 4). An offset channel riser with several metres of deflection was mapped in the field, but this feature is not very apparent in the aerial photography (Fig. 4). In general, the surface expression of the fault along this southern section is subdued with less evidence for direct fault motion (channel offsets). This suggests that channel aggradation is younger in this area as compared to the other areas to the north.

Summary

The overall expression of the Vilariça fault in the area covered by Figures 2–4 indicates that the principal sense of late Quaternary motion is left-lateral strike-slip, consistent with the sinistral displacement observed for the basement rocks. In areas undergoing erosion, channel and drainage basins are consistently deflected to the left. In general, the deeper incisions express greater amounts of deflection, and the smallest deflection of *c*. 2.2 m is associated with a small rill. In areas where the fault steps to the left, young deposits bury the active trace of the fault, consistent with basin formation due to a releasing step-over. The general lack of scarps in Holocene alluvium suggests that either the sedimentation is generally younger than the most recent fault activity in most areas, that the evidence for young activity is generally not recognizable at the scale of the photos, or that recent agricultural activity has obliterated scarps and other youthful landforms associated with active faulting. In any case, the general expression of the geomorphology of the fault in the study area argues for an overall low late Quaternary slip rate, but one that has remained left-lateral.

Trenching investigations

Two of the principal tasks that we undertook in this study were to establish the timing of the most recent surface ruptures, and to assess the likely magnitude of these prehistorical earthquakes. Towards this end, we explored three sites in the subsurface to establish the recent slip history of the fault. At one of the sites, we conducted three-dimensional trenching to resolve displacement on a buried channel.

The success of a palaeoseismic site hinges on exposing good stratigraphy in an area of active sedimentation embedded with datable material, preferably detrital charcoal or peat. The good stratigraphy is required so as to recognize evidence for individual events. If the strata are too coarse, it is likely that some events will be unrecognized. Furthermore, coarse strata, such as channel gravel, are commonly associated with periods of erosion and may remove evidence for surface rupturing events. In an ideal site, there is sufficient sedimentation to separate evidence for discrete surface ruptures, but not so much to deeply bury the fault and make it difficult to study. Also, coarse strata tend to have much sparser concentrations of detrital charcoal and are harder to directly date with other methods, such as thermoluminescence or optically stimulated luminescence (OSL) dating. For all of these reasons, the best sites to date past earthquakes tend to be in areas of structural releasing step-overs, or in areas where sediment is ponded behind a scarp, as both environments tend towards moderate rates of sediment accumulation which are typically better at preserving palaeoseismic records.

For this study, we found no young sag ponds or other small-scale depressions that our experience has shown to be the best for palaeoseismic studies. Nevertheless, there did appear to be active sedimentation at all three of the sites we chose, although the stratigraphy at each site did pose its own unique challenges. The southernmost site at Longrovia, however, yielded low-resolution information as the main fault ruptures bedrock to the surface and is not overlain by an alluvial sequence. Hence, this site was abandoned to focus on the more promising Vale Meão winery and Vilariça sites. We discuss each of these two northern sites beginning with a brief description of the site and the rationale for choosing each trenching location. The work at Vale Meão is discussed first, as that site yielded the best data on timing of past events. The Vilariça site, in contrast, provided the best information on fault displacement.

Vale Meão winery site

The Vale Meão trench site was originally chosen to explore the possibility of determining a long-term slip rate. Figure 5 shows the detailed geomorphology in the Vale Meão area. In particular, it appears in the aerial photography that the Douro River channel wall, the riser to a major terrace, and potentially the terrace itself, are likely laterally offset by several tens of metres. The Douro River itself is likely deflected by up to a hundred metres. To explore the possibility of resolving a slip rate, we excavated a long trench (VM T-1) from the east side of the fault near the edge of the terrace westward across a drainage and up onto the terrace surface to the west. The intent was to cross the fault, resolve the locations of the fault and terrace edges on each side of the fault, and set up for trenches parallel to the fault to resolve slip on the terrace. Based on the location of the fault exposed in trench VM T-2, it is likely that we terminated the trench just east of the fault. Nevertheless, the trench exposed sediments that required rethinking of using this site to resolve a long-term slip.

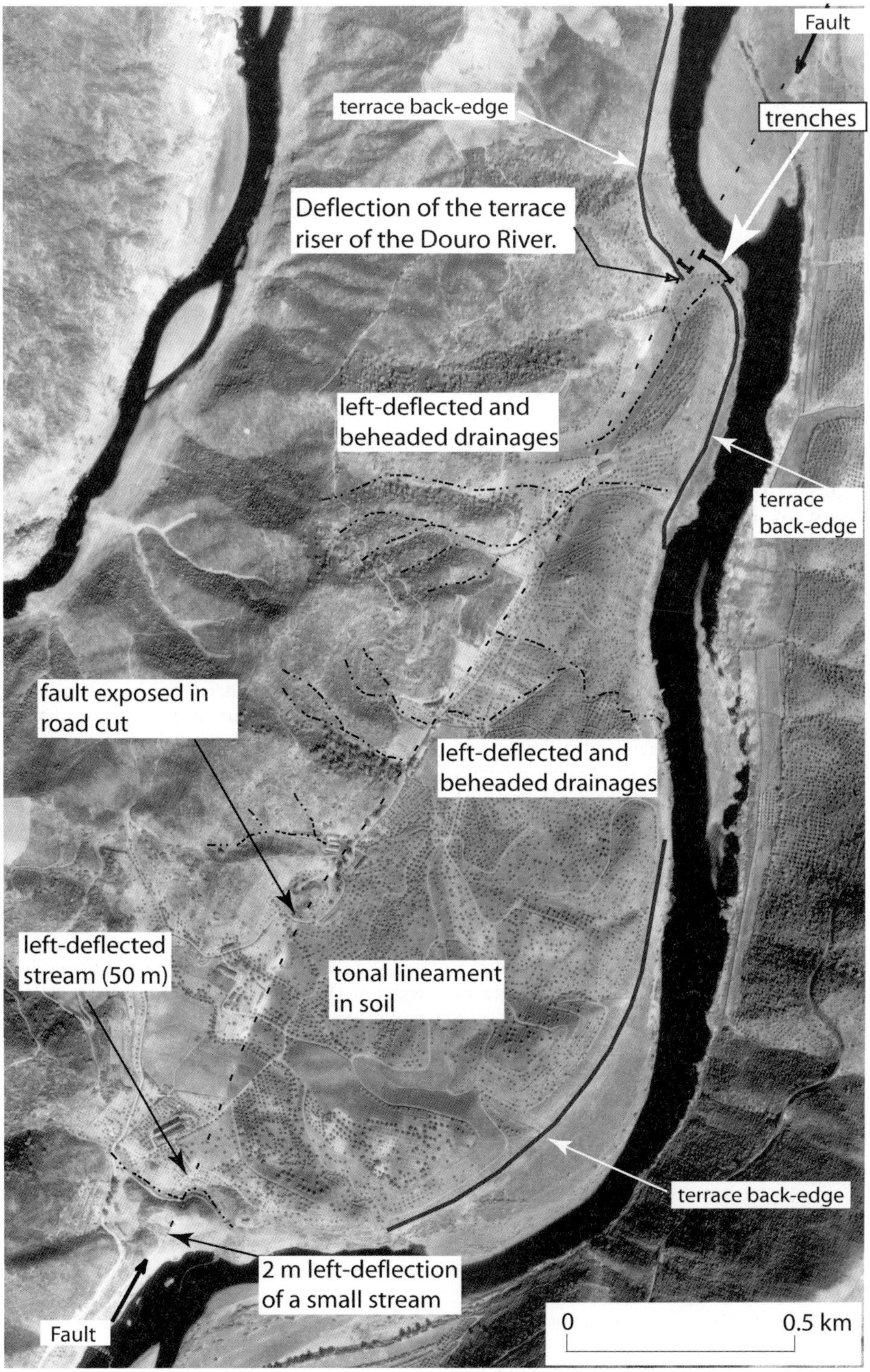

Fig. 5. Detail of the Vale Meão winery area showing the deflection of the Douro River and the trench site. The fault is clear in the geology and geomorphology. The interpreted location of the back-edge of the fill terrace exposed in trench T-1 at Vale Meão winery, dated at about 16 ka, is shown.

In trench T-1, the terrace comprised a massive fill event, rather than a strath (cut) surface, and was composed of very poorly bedded silt and sandy silt. On the stable terrace surface, the soil has a 1–2 m thick strong brown (7.5YR 4/5d) argillic horizon with common to many thin to moderately thick clay films. In southern California, which has a similar xeric climate and fluvial deposits with comparable parent material, this soil would be interpreted as having started to form in the late Pleistocene (Rockwell *et al.* 1985). In contrast, across the river to the NW in a road-cut, there is a strath terrace exposed with a much stronger soil (2.5–5YR 5/8 colour, plugged with clay) that is at a slightly higher elevation (by *c.* 10 m) than the Vale Meão terrace. It is clear from trench T-1 and comparison of the soils that these are different terraces of very different ages, with the older and redder soil likely at least 100 ka in age.

We collected two samples for OSL dating from the silty alluvial fill of this fill terrace, with resulting ages of 25.0 $\pm$ 2.3 ka and 16.0 $\pm$ 1.1 ka. These dates do not overlap at 2σ and suggest that at least the older date has a significant component of inheritance, especially as it is stratigraphically above the younger date (see discussion of OSL dating below). Nevertheless, the dates do support the soil observations that the alluvial fill event that comprises this terrace formed in the late Pleistocene. Considering that OSL dates are maximum ages because of the inheritance issue (Fuchs & Lang 2008), the likely age of the fill event is *c.* 16 ka or younger.

We interpret the geomorphic expression of the apparently offset terrace as simply reflecting deposition across the left-deflected riser to the Douro River. Thus, the fill does not date the deflection, but rather the fill drapes the deflection. In any case, sediments in trench VM T-1 show that the terrace was not the right type of terrace deposit to resolve a good, long-term slip rate.

Trench VM T-2, however, exposed the fault cutting late Quaternary deposits and yielded valuable information on the timing of the past two (and possibly three) surface ruptures. In the trench, Zedes Granite on the west was juxtaposed against phyllite of the Desejosa Formation on the east. A sequence of alluvial and colluvial deposits overlies the bedrock fault, and two strands of the fault extended upward to two different stratigraphic levels, indicating multiple events, as discussed in more detail below (Fig. 6).

Colluvial sediment dominated the stratigraphy in the trench. At least the upper part of the stratigraphically highest and youngest unit (Col_1) was found to bury a stone wall that was likely built in *c.* 1890 during the early development of the winery. The wall was exposed in the eastward continuation of the trench, but this portion was back-filled for safety reasons. The modern A horizon soil caps this colluvial deposit.

Units Col_{2b} and Col_{3a} are likely middle Holocene colluvial deposits, based on the radiocarbon ages of two detrital samples (VM-08 and VM-02, *c.* 4.8 and 4.65 ka, respectively; Table 1) collected from these units (Fig. 6). Unit Col_{3a} grades upward into the base of Col_2, but they were differentiated in the field because Col_2 apparently fills a fairly well-defined palaeogulley, whereas the contact between Col_{3a} and Col_{3b} was very poorly defined and difficult to resolve. Col_2 is capped by a buried A horizon that has a slightly darker appearance, indicative of slightly greater organic matter content, which in turn suggests some surface stability for a significant period of time.

Col_3 represents a colluvial fill event that backfills a channel eroded into units Col_4 to Qa_6. The two radiocarbon age (*c.* 4.80 ka and 4.65 ka; Table 1) indicate that Col_3 and all overlying units are Holocene, and that Col_2 and Col_{3a} were probably deposited in the middle Holocene and represent a period of colluvial aggradation.

Below the relatively massive middle to late Holocene colluvial deposits which tend to be organic-rich, units Col_4 to Qa_6 represent at least two phases of aggradation and incision, with some component of fluvial deposition, probably from the small drainage that runs through the site. Notably, these older units are essentially devoid of organic material, either as detrital charcoal or humus, and suggest accumulation under dryer conditions than are present today.

Unit Qa_6 comprises stratified silty sand to gravelly sand and is likely a mix of colluvium and fluvial sediment. The sloping character of the overall unit, however, points more towards colluvial processes. Qa_6 is cut out by an incision event, the channel of which is then backfilled by Col_{5b}, stratified alluvium of Qa_5 and more colluvium of Col_{5a}. Col_{4b} is interpreted as a wedge or sheet of colluvium derived from the fault, as discussed below, and Col_{4a} is more colluvium with minor stratified sand and gravelly sand deposits embedded within it. The presence of water-laid sediments within the colluvium suggests that the system aggraded at least up to the level of the alluvial components.

We initially collected six OSL samples from units Col_4 to Qa_6, but the highest (youngest) and two lowest (oldest) samples disintegrated in transport to the laboratory in the USA. These three samples were re-collected along with an additional four samples, so ten OSL samples were altogether dated from this trench (Table 2).

OSL dating is used to determine the time elapsed since a sediment sample was exposed to daylight. This technique has been successfully applied to

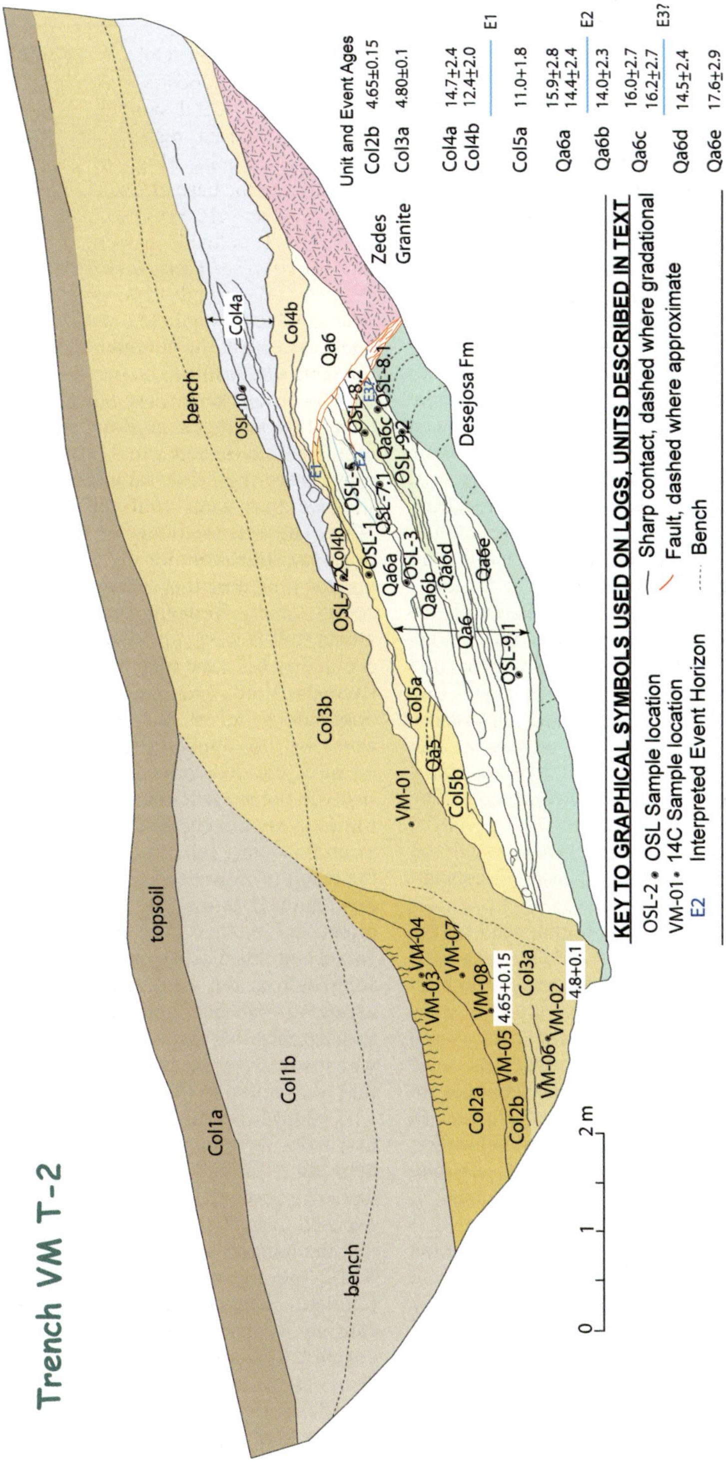

Fig. 6. Log of the VM T-2 trench at Vale Meão winery. Units and event interpretations are generally described in the text.

Table 1. *Radiocarbon ages for detrital charcoal samples recovered from sediments in the Vilariça (VR) and Vale Meão (VM) trenches*

CAMS #	Sample name	$\delta^{13}C$	Fraction modern	Error (±)	$D^{14}C$	Error (±)	^{14}C age	Cal. age (ka BP)
115427	VR3-02	−25	0.9705	0.0045	−29.5	4.5	240 ± 40	0.05–0.5
115428	VR3-04	−25	0.9431	0.0040	−56.9	4.0	470 ± 35	0.55 ± 0.05
115429	VM-02	−25	0.5933	0.0028	−406.7	2.8	4195 ± 40	4.8 ± 0.1
115430	VM-08	−25	0.5988	0.0024	−401.2	2.4	4120 ± 35	4.65 ± 0.15

Sample names correlate to those on the respective trench logs.

dating deformed sediments for palaeoseismic studies in the western USA (e.g. Machette *et al.* 1992; Crone *et al.* 1997; Rockwell *et al.* 2000; Lee *et al.* 2001; Kent *et al.* 2005; Wesnousky *et al.* 2005) and elsewhere in the world (e.g. Owen *et al.* 1999; Washburn *et al.* 2001). The technique relies on the interaction of ionizing radiation with electrons within semi-conducting minerals resulting in the accumulation of charge in metastable location within minerals. Illuminating the minerals and detrapping the charge that combines at luminescence centres can determine the population of this charge. This results in the emission of photons (luminescence). Artificially dosing subsamples and comparing the luminescence emitted with the natural luminescence can determine the relationship between radiation flux and luminescence. The equivalent dose (D_E) experienced by the grains during burial therefore can be determined. The other quantity needed to calculate the age is the ionizing radiation dose rate, which can be derived from direct measurements or measured concentrations of radioisotopes.

The uncertainty in the age is influenced by the systematic and random errors in the D_E values and the possible temporal changes in the radiation flux. The quoted error is the deviation of the D_E values on multiple subsamples and the error in measured ionizing radiation dose rate or the concentration of radioisotopes. Determining temporal changes in the dose rate that is a consequence of changes in water content and the growth and/or translocation of minerals within the sediment is not possible. The dose rate is therefore generally assumed to have remained constant over time.

For most deposits derived from a common source, the dose rate determined directly from those sediments commonly falls within a fairly narrow range as long as there are not secondary factors such as the accumulation of pedogenic carbonate or other authigenic deposits that have affected the post-deposition radiogenic content of the sample. The Vale Meão trench samples, however, produced dose rates that varied from *c.* 4 to 7 mGy/a and yielded apparent ages that are stratigraphically reversed, with one of the youngest apparent ages at the bottom of the section (Table 2). Furthermore, there are no secondary minerals apparent in the sediments for which to attribute post-deposition changes in dose. We reanalysed the samples for dose and found surprisingly large variability from the same sample, as well as between samples that were collected very close to each other. These results indicated to us that there are surprisingly large, very local variations in dose rate. In that the final dose that a sample has been exposed to comes largely from the surrounding mass of sediments (out to *c.* 30 cm or so from the sample), we decided to take the mean of all of the dose results from the Vale Meão OSL samples and apply that dose rate to all of the samples to calculate their ages. We also applied the standard deviation in the mean of the dose rate samples as an added uncertainty in the age calculations. In so doing, the resulting dates are mostly in stratigraphic order, with only a couple of very minor age inversions that are likely the result of inheritance. We present all of the dose data, as well as the calculated ages using both the direct and average dose rates, in Table 2. For the balance of this discussion, we use only the ages calculated from the average dose rate, as those ages appear to provide good stratigraphic coherence. For samples that are age-reversed, we assume that part of this reversal is due to the inheritance effect that is common in OSL dates.

The OSL results indicate that all of the Col_4 to Qa_6 units were deposited in the latest Pleistocene between *c.* 11 and 18 ka, with the lower part of Qa_6 dating to about the same age as the fine-grained fill exposed in trench VM-1. This observation suggests that the alluvial and colluvial fill exposed in trench VM-2 may have been forced by the same factors (climate change, tectonics, allocyclic processes) causing aggradation of the main Douro River channel. The absence of organic material (humus, charcoal) in the latest Pleistocene section, and its abundance in the late Holocene section, is consistent with development under dryer climate conditions, as have been suggested for this region during the latter part of the last glacial by Sobrino *et al.* (2007).

Table 2. *Summary of OSL dating results from quartz extracted from sediment matrices*

Sample number	Unit	Location (°N/°W)	Altitude (m a.s.l)	Depth (cm)	U[a] (ppm)	Th[a] (ppm)	K[a] (%)	Rb[a] (ppm)	Cosmic dose rate[b,c] (mGy/a)	Total dose rate[b,d] (mGy/a)	N[e]	Mean D_E[f] (Gy)	Age[g] (ka)
VM-10	Col4a	41.168/7.113	115	180	5.2	11.8	2.2	214	0.17	3.99 ± 0.23	24(30)	72.0 ± 8.5	18.5 ± 0.9
					4.5	*10.4*	*2.3*	*215*		*3.81 ± 0.23*			*14.7 ± 2.4*
										3.90 ± 0.16			
VM-7.2	Col4b	41.168/7.113	115	248	5.7	9.3	3.4	420	0.16	5.01 ± 0.32	17(30)	60.7 ± 8.0	14.2 ± 0.8
					4.6	*4.6*	*2.6*	*348*		*3.89 ± 0.24*			*12.4 ± 2.0*
										4.29 ± 0.19			
VMT1-OSL1[h]	Col5a	41.168/7.113	115	290	7.1	10.9	1.6	123	0.15	4.04 ± 0.24	20(28)	53.9 ± 10.1	13.8 ± 1.2
													11.0 ± 1.8
VMT1-OSL5[h]	Qa6a	41.168/7.113	115	300	6.3	7.5	3.0	328	0.15	4.89 ± 0.31	11(28)	77.9 ± 32.2	15.9 ± 2.4
													15.9 ± 3.2
VMT1-OSL3[h]	Qa6a	41.168/7.113	115	320	6.0	8.9	2.5	235	0.14	4.40 ± 0.27	20(28)	70.6 ± 13.1	16.1 ± 1.4
													14.4 ± 2.4
VM-7.1	Qa6b	41.168/7.113	115	305	11.4	8.8	2.7	272	0.15	5.55 ± 0.33	16(24)	68.8 ± 8.3	14.0 ± 0.7
					8.0	*6.9*	*2.5*	*243*		*4.50 ± 0.27*			*14.0 ± 2.3*
										4.92 ± 0.21			
VM-8.2	Qa6c	41.168/7.113	115	300	9.5	11.4	3.4	363	0.15	5.90 ± 0.36	21(24)	78.8 ± 14.0	14.6 ± 0.8
					9.0	*10.4*	*2.6*	*312*		*5.07 ± 0.30*			*16.0 ± 2.7*
										5.41 ± 0.23			
VM-8.1	Qa6c	41.168/7.113	115	317	7.4	13.0	3.9	420	0.15	6.01 ± 0.37	28(30)	79.6 ± 15.3	14.3 ± 0.8
					7.0	*7.0*	*3.2*	*369*		*5.23 ± 0.32*			*16.2 ± 2.7*
										5.56 ± 0.24			
VM-9.1	Qa6e	41.168/7.113	115	395	8.2	13.5	3.2	268	0.13	5.56 ± 0.33	27(30)	86.4 ± 16.5	16.6 ± 0.9
					9.4	*11.3*	*2.3*	*210*		*4.93 ± 0.29*			*17.6 ± 2.9*
										5.20 ± 0.22			
VM-9.2	Qa6e	41.168/7.113	115	340	14.3	14.8	3.1	277	0.14	6.95 ± 0.41	24(30)	71.2 ± 9.7	11.0 ± 0.5
					12.2	*13.1*	*2.8*	*269*		*6.13 ± 0.36*			*14.5 ± 2.4*
										6.49 ± 0.27			
VR3-2[h]	Qa2	41.215/7.095	120	275	6.4	17.2	2.8	194	0.15	5.38 ± 0.32	19(28)	102.2 ± 28.3	18.2 ± 1.6
VR 3-1	Qa2	41.212/7.094	120	315	5.2	15.5	2.6	195	0.15	4.54 ± 0.27	24	>108[j]	>25
					4.9	*14.8*	*2.2*	*191*		*4.05 ± 0.23*			
										4.26 ± 0.18			
VM PLEISTOCENE		41.168/7.113	112	250	5.2	14.6	2.1	127	0.16	4.05 ± 0.23	18(24)	64.9 ± 10.3	16.3 ± 0.9
					5.3	*14.8*	*1.9*	*124*		*3.90 ± 0.22*			
										3.97 ± 0.16			

(*Continued*)

Table 2. *Continued*

Sample number	Unit	Location (°N/°W)	Altitude (m a.s.l)	Depth (cm)	U[a] (ppm)	Th[a] (ppm)	K[a] (%)	Rb[a] (ppm)	Cosmic dose rate[b,c] (mGy/a)	Total dose rate[b,d] (mGy/a)	N[e]	Mean D_E^f (Gy)	Age[g] (ka)
VM-UPPERP[h]		41.168/7.113	115	150	5.6	18.0	2.1	123	0.18	4.57 ± 0.26	20(28)	134.8 ± 28.2	25.0 ± 2.3
VROSL3/VR3V		41.212/7.094	120	290	3.4	6.5	1.7	247	0.15	2.83 ± 0.17	12(24)	91.3 ± 18.7	30.4 ± 2.4
					4.0	*8.1*	*1.9*	*198*		*3.21 ± 0.19*			
										3.00 ± 0.13			
VR2 OSL4/VR 2-4	Qt	41.212/7.094	120	290	3.9	9.9	2.6	178	0.15	3.86 ± 0.24	8(12)	132.4 ± 8.4	38.6 ± 2.7
					3.2	*7.0*	*2.4*	*156*		*3.32 ± 0.12*			
										3.43 ± 0.11			

[a]Elemental concentrations from NAA of whole sediment measured at Becquerel Laboratories, Lucas Heights, NSW, Australia, for samples shown in normal text, and at USGS Nuclear Facility in Denver for samples shown in italics. Uncertainty taken as ±10%. Elemental concentrations shown in italics are duplicate samples.
[b]Estimated fractional water content from whole sediment (Aitken 1998). Uncertainty taken as 10 ± 5%.
[c]Estimated contribution to dose rate from cosmic rays calculated according to Prescott & Hutton (1994). Uncertainty taken as ±10%.
[d]Total dose rate from beta, gamma and cosmic components. Beta attenuation factors for U, Th and K compositions incorporating grain size factors from Mejdahl (1979). Beta attenuation factor for Rb arbitrarily taken as 0.75 (cf. Adamiec & Aitken 1998). Factors utilized to convert elemental concentrations to beta and gamma dose rates from Adamiec & Aitken (1998) and beta and gamma components attenuated for moisture content. Dose rates shown in italics are based on the elemental concentrations for the duplicate samples, while those in bold are the weighted mean doses based on original and duplication dose rates.
[e]Number of replicated D_E estimates used to calculate mean D_E. The number in parentheses is the total number of aliquots measured. These are based on recuperation error of <10%.
[f]Mean equivalent dose (D_E) determined from replicated single-aliquot regenerative-dose (SAR; Murray & Wintle 2000) runs. Errors are 1-sigma incorporating error from beta source estimated at about ±5%.
[g]Errors incorporate dose rates errors and 1-sigma standard errors (i.e. $\sigma_{n-1}/n^{\frac{1}{2}}$) for D_E using the weighted mean of dose rates determined at Becquerel Laboratories and the USGS Nulcear facility in Denver. Values shown in italics are calculated using a dose rate of 4.91 ± 0.79 Gy/ka determined from the mean of the fault trench samples with the error being the standard deviation of the dose rates.
[h]Samples measured on a Daybreak OSL reader. All other samples were measured on a Riso OSL reader.

Interpretation and timing of faulting events

We see clear evidence of two surface ruptures, and probably a third surface rupture, exposed in trench VM T-2. The youngest rupture breaks up through unit Col_{5a} and is overlain by unit Col_{4b}. Col_{4b} is interpreted as a colluvial deposit that was shed directly off of the fault and resulted from this event, designated as E1 in Figure 6. Units Col_{4a} and Col_3 are unbroken and must post-date the most recent surface rupture. A sample for OSL dating from the highest displaced unit (Col_{5a}) yielded an age of 11.0 ± 1.8 ka. The OSL age from Col_{4b} that caps the event had a similar age of 12.4 ± 2.0 ka, which overlaps within uncertainty with the sample from Col_{5a}. The sample from the stratigraphically higher Col_{4a} had a slightly discordant age of 14.7 ± 2.4 ka and apparently has some inherited signal, although even this date barely overlaps with the youngest of the three OSL ages. At face value, these data suggest that the most recent event (E1) occurred in the latest Pleistocene soon after 11 ka.

The most conservative assessment of the timing of E1 is provided by assuming that the OSL dates are maximum ages and by using the two radiocarbon samples that we dated from unfaulted units Col_{3a} and Col_{2b}, which yielded calibrated ages of *c.* 4.65–4.80 ka. Using these data, the timing of the most recent event is constrained to have ruptured the Vilarica fault segment sometime between 4.8 and 11 ka. However, considering that OSL dates are quite consistently in the late Pleistocene, along with the lack of organic matter in this part of the section which is consistent with a dryer latest Pleistocene climate, we argue that the best estimated age for event E1 is more likely close to 11 ka.

The penultimate event, designated as E2 in Figure 6, ruptured the eastern (lower) fault strand up through the stratified sediment of Qa_{6b}, and is overlain by more sediment of Qa_{6a}. The fault rolls over to the palaeosurface, much like the upper, western strand, and is clearly truncated by unfaulted alluvium of Qa_{6a}. We dated two samples from unit Qa_{6a} and one sample from unit Qa_{6b}, which place constraints on the timing of this event. All three dates are slightly discordant in that the oldest is apparently above the youngest, but considering the uncertainties, they all overlap and are about the same age (Table 2). As they are maximum ages, event E2 is younger than *c.* 14 ka but older than the capping unit Qa_{5a} at *c.* 11 ka. We therefore infer the age of E2 as being 11–14 ka.

A probable third event, designated as E3 on Figure 6, is indicated by the presence of a wedge of colluvial (unit Qa_{6c}) that thickens towards and is terminated by the lower fault. The observation that this deposit thickens into the fault is diagnostic of fault-generated colluvial wedges. Unfortunately, this inference is based on only a stratigraphic observation and there are no separate faults that ruptured in this event, although subsequent activity (E2 and E1) may have obscured the structural evidence for this event. Nevertheless, with a lesser degree of certainty, we infer that a surface-rupturing event may have produced this colluvial wedge.

We dated two OSL samples from the wedge sediments, both of which yielded dates around 16 ka (Table 2). However, the sediments from unit Qa_{6d} directly below the wedge yielded an age of 14.5 ± 2.4 ka, which appears slightly younger but overlaps with the dates from Qa_{6d}. Taken together, and assuming that the wedge represents the occurrence of a real event, E3 occurred at *c.* 14.0–14.5 ka, assuming no significant inheritance in the OSL signal from the deepest sample and minor inheritance from the two samples from the wedge.

In summary, we observed evidence for two and probably three surface ruptures on the Vilariça fault segment at Vale Meão between *c.* 14.5 and 4.8 ka, with a high likelihood that all three occurred between 14.5 and 11 ka, or soon thereafter. Although this yields an average late Quaternary return period of *c.* 5 ka if all three events are real, and closer to 6–7 ka if only E1 and E2 are real, the events appear clustered with interevent times during the cluster of less than 2 ka. It is clear that there is an open interval at 4.8 to 11 ka, but intervals between clusters in intracontinental regions are poorly known and could range to many tens of thousand of years or longer.

Vilariça site

The Vilariça site lies on the floodplain of the Vilariça River *c.* 2 km north of its confluence with the Sabor River (Fig. 1). The site was chosen because there is an apparent scarp, with bedrock cropping out to the east. The escarpment approximately follows the trace of the fault, although the active trace of the fault is buried by very young alluvium and the base of the scarp represents the area where the young sediments are ponded against rock.

Four trenches were initially opened at the Vilariça site to explore the location of the fault and the timing of past slip events (Fig. 7). Trench VR-1, which is the longest of all of the exposures and was excavated as the locator trench, exposed the main fault east of the sequence of Holocene alluvial deposits (Qal) in which we had hoped to capture the faulting history, so we do not include the log. The main fault juxtaposed Cambrian chloritic phyllite of the Pinhao Formation against very old Quaternary alluvial deposits (Qoa) at the base of the trench. We also observed several minor

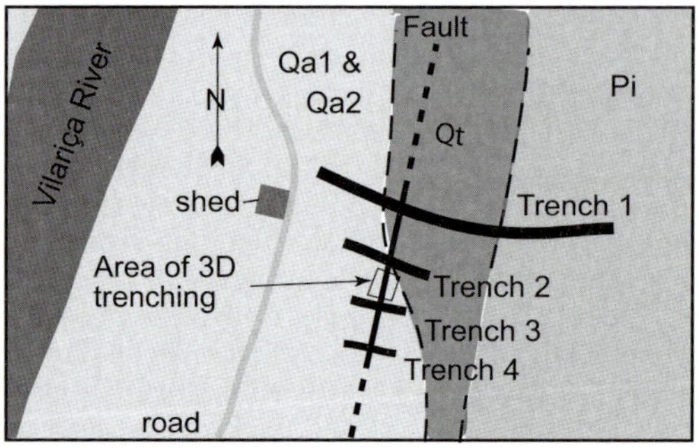

Fig. 7. Schematic map of the Vilariça trench site showing the spatial relations between trenches. The 3D trenching was conducted between trenches 2 and 3.

secondary shears in the phyllite itself, just east of the main fault. The age of the Qoa is inferred to be middle Quaternary or older based on its very red (10R 7/8d) colour and degree of cementation, but could well pre-date the Quaternary altogether. The Qal deposits in the western part of the trench were reasonably well-stratified and contained abundant detrital charcoal, two pieces of which were dated from trench VR-3 that yielded late Holocene ages (after *c.* AD 1400; Table 1).

Stratigraphy was lacking within the fault zone, which made the assessment of the timing of past events impossible. We therefore decided to excavate additional trenches to the south where we expected the younger alluvium to cross the fault. Towards that end, trench VR-2 was excavated parallel to VR-1, but *c.* 10 m south followed by VR-3 and VR-4 (Fig. 7). Each successive trench exposed the fault in deeper and thicker fluvial deposits such that the fault was at or below the level of groundwater in VR-4. The margin of a channel deposit (Qa2) in trench VR-2 (Fig. 8, south face), however, was *c.* 2 m west of the fault, whereas the correlative deposit was in fault contact on the east side in both the northern and southern trench face exposures. In trench VR-3, the same channel deposit was in fault contact on both sides of the fault (Fig. 9). This implied that the channel's western margin intersected the fault in the volume of earth between the two trenches. Similarly, the equivalent eastern margin must intersect the fault between trenches VR-1 and VR-2. These set up the 3D exercise to resolve displacement on the channel margin, as none of the critical piercing points were removed in the initial trenching.

Age of deposits

We collected and dated four samples for OSL dating from the Vilariça trenches with the intent to test the ages of deposits that could potentially yield information on timing and displacement. We also used radiocarbon dating of detrital charcoal to date the young unfaulted channel alluvium (Qa1) that buries the older, faulted deposits.

The radiocarbon dating showed that the Qa1 alluvium is very young, on the order of a few hundred years (Table 1), and was not useful in constraining the timing of the most recent event. The Qa2 channel was sampled in two exposures for OSL dating, and both the channel alluvium and the overlying overbank (fine sand) deposits yielded results. We analysed up to 28 aliquots per sample and found that there was significant spread among the individual aliquot ages, which suggests a strong inherited signal. We also re-ran sample splits for dose rate and got generally similar

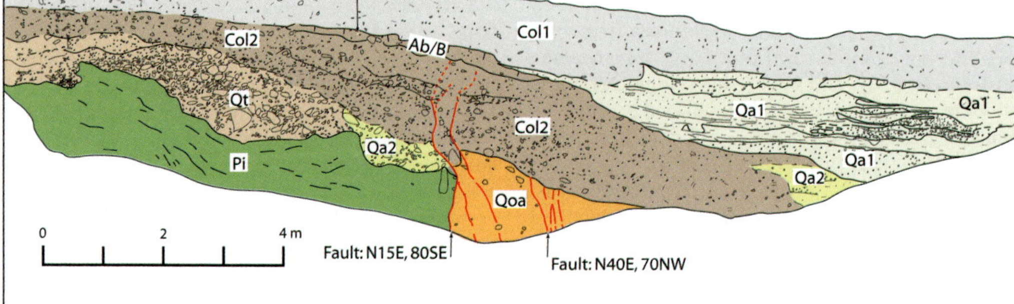

Fig. 8. Log of the Vilariça VR-2 trench, south wall. The Qa2 alluvium, which was used to resolve displacement, interfingers with the colluvium of Col2, part of which is derived from the gravelly deposits of an older terrace deposit (Qt). The primary discriminator was whether the gravel is matrix-supported (colluvial) or grain-supported (fluvial).

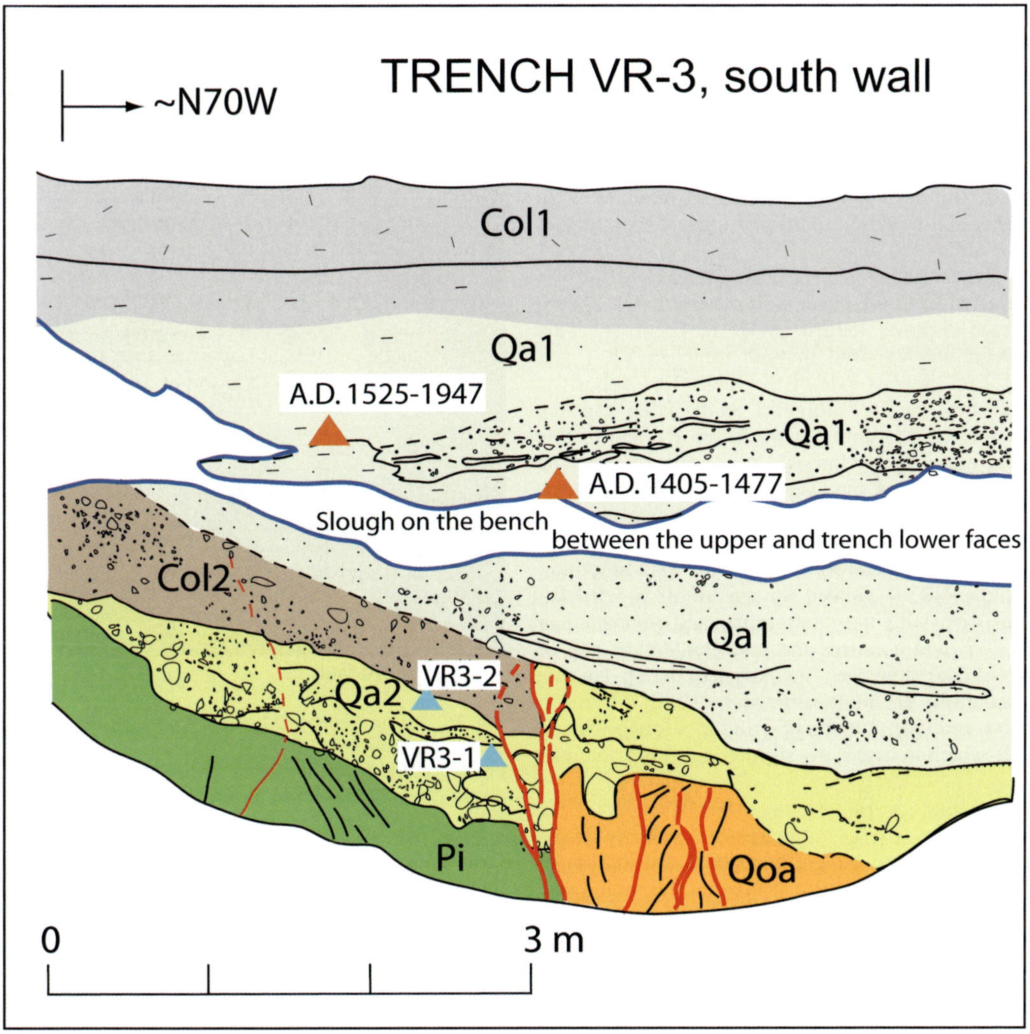

Fig. 9. Log of Vilariça trench VR-3, south wall. Note that the Qa2 alluvium is in complete fault contact, indicating that the channel margin lies between trenches VR-2 and VR-3. The eastern margin was exposed both in VR-2 and VR-3 and is the edge of the Qa2 deposits on the left side of the log. Also note the position of two of the OSL samples used to constrain the age of the Qa2 alluvium.

results (within *c.* 0.5 mGy). For the age calculations, we present results from both dose rates, as well as their mean (Table 2).

For the channel deposits, sample VR3/3 from slice 4 of the 3D trenches yielded an average age of 30.4 ± 2.4 ka. However, the lowest 20% of aliquots averaged only 23.1 ± 1.3 ka, again indicating a strong inheritance component. Sample VR3-1 yielded an apparent minimum age of *c.* 24 ka, as the grains are OSL-saturated. However, considering the likely effect of partial bleach (inheritance) on this sample too, this may well be a maximum age. In any case, although it generally corroborates the results from VR3/3, the fact that the OSL signal is saturated makes this not very useful. Based on these results, and heavily weighting the results from VR3/3, we infer the maximum age of the channel to be *c.* 23 ka.

Sample VR3-2 was collected from a fine-grained sand unit that directly overlies the gravel component of the channel (Fig. 9). There was no evidence of weathering or surface exposure between these units, and they fill the same channel form, so we expected the results to be similar. The OSL analyses yielded an age of 18.2 ± 1.6 ka, several thousand years younger than the underlying coarse channel alluvium. Taken together, along with the lack of weathering or erosion between units, we suspect that the *c.* 18 ka age is closer to the true age of the channel complex. If correct, then the channel is similar in age to the base of the alluvium at the Vale Meão palaeoseismic site. In any case, it appears that the channel should record slip from all of the two or three events exposed at Vale Meão, and possibly additional ones if the channel is as old as *c.* 23 ka.

The fourth sample for OSL dating (VR2-4) was collected from the north wall of trench VR-2 from terrace deposits logged as Qt in Figure 8. This deposit was exposed in both trenches VR-1 and VR-2 on the east side of the fault. This deposit is older than the Qa2 alluvium, as the latter deposits are incised into the Qt deposits, so the age of this unit provides additional age control on the maximum age of the Qa2 alluvium that we used to resolve slip. Furthermore, it may provide a source of inheritance if a significant amount of the Qt deposits were incorporated into the Qa2 alluvium. Sample VR2-4 yielded an age of 38.6 ± 2.7 ka, consistent with the stratigraphy and younger age of Qa2, and confirms the late Quaternary age of the alluvial sequence. The margin of the Qt deposits must cross the fault somewhere north of trench VR-1 and will potentially provide a longer period over which to assess the slip rate in future studies.

Fig. 10. Set-up for the 3D trenching, looking from trench VR-3 towards VR-2. Note that the fault zone is highly localized, which is ideal for resolution of displacement. The units are described in the text.

Resolution of slip

We explored the three-dimensional distribution of the Qa2 alluvium where it crosses the fault zone, with special attention to using the channel margin as a piercing point. The channel margin, as seen in the south wall of trench VR-2, has two elements that looked useful: the tread/riser intersection; and the top edge of the channel gravel itself. We mapped both of these features in 3D across the fault. (Note that only the top edge of the channel is evident on Fig. 8, as we deepened and lengthened trench VR-2 prior to starting the 3D exercise.)

The Qa2 alluvium was generally *c.* 2.5–3.0 m below the modern ground surface, and 3D trenching at these depths can be hazardous without adequate trench bracing (shoring). Instead, we removed the upper 1.5–2 m of sediment from the site over the entire area between trenches VR-2 and VR-3, making a flat surface from which to work, and keeping the depths of the trenches to *c.* 1 m (Fig. 10). We then opened two trenches parallel to the fault, one on each side and located 1–2 m from the fault. This left an intact block of soil that contains the fault, the Qa2 alluvium near the fault, and the expected western piercing point (Figs 11 and 12a). We set a grid system on the surface of the block by establishing an orthogonal system of large nails connected by string: this new grid was tied to the original grid reference frame in trenches VR-2 and VR-3. We also continued the grid into the fault-parallel trenches so that all significant features could be located to within a few centimetres. Although surveying would have been preferable for accuracy of locations, this method allowed us to plot everything in the field as each cut was made, and a few centimetres of imprecision is negligible when considering offsets at the metre scale. This also allowed for rapid decisions on the next cuts.

We then began to trace the western channel margin into the fault by cutting successive slices across the fault in 'the block' (Figs 11 and 12). Each exposure was photographed and logged in the field, and the location of each cut determined with the grid reference system (Fig. 12).

After resolving the location of the piercing point west of the fault, we turned to the north face of trench VR-2, where both the riser (R) and channel margin (C) piercing points were preserved (Fig. 13a). We cut the trench face back *c.* 0.5 m, and in so doing removed the piercing point, so we know its location to ± 25 cm (half the cut distance) (Fig. 11).

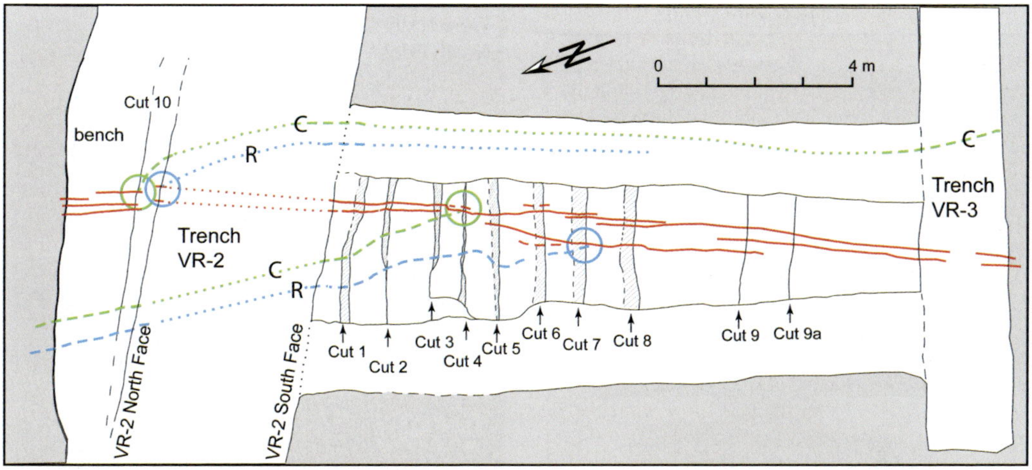

Fig. 11. Detailed map of the area between trenches VR-2 and VR-3 showing the locations of each trench and cut, the fault strands, the location of the channel edge and tread/riser contact, and the piercing points at the fault.

Figure 11 shows the location of both piercing lines as they were traced from exposure to exposure. Figure 14 is a simplification of the offset piercing lines and shows that the riser is laterally offset *c.* 8.9 m whereas the channel margin is displaced *c.* 6.5 m. As the exposures were closely spaced, we know the locations of the piercing point/fault intersection to within *c.* 25 cm for each side of the fault, so we assign an uncertainty of *c.* 0.5 m for these displacements. Thus, we assign displacement values of 6.5 ± 0.5 m for the channel edge and 8.9 ± 0.5 m for the channel bed and riser.

It is interesting that the channel tread/riser and gravel lag are apparently displaced more than the top channel edge. Our preferred interpretation is that the channel was active during a large displacement event, so the difference represents slip in that event. If true, this suggests a displacement of *c.* 2.4 m for an event that is constrained to between *c.* 18 and 23 ka if the OSL ages read true, or younger than 18 ka if the OSL ages are maximum ages and the channel alluvium is all *c.* 18 ka or less.

Other constraints on slip per event

Although we have resolved slip for the past *c.* 18–23 ka at the Vilariça site, we also make inferences from the deflection of a small channel at Vale Meão winery that may represent slip after incision of the Douro River below the level of the *c.* 16 ka fill event.

The fault is exposed in the river riser on the southwestern margin of Vale Meão winery where the fault cuts into the Douro River gorge. In the vicinity of the small channel at the base of the slope, colluvium covers the fault but bedrock outcrops upslope allow for placement of the fault to within ± 3 m close to the channel. At this location, the small rill is deflected exactly where the fault projects through the rill, suggesting that the deflection is the result of fault slip (Fig. 15). The small rill was surveyed with a total station, and the map generated from that survey is presented as Figure 15. In the field, we estimated *c.* 2 m of deflection in the near-field, and that estimate is refined to *c.* 2.0–2.5 m, as shown in Figure 15. However, the far-field deflection of the rill is *c.* 6.1 m, as measured along the strike of the fault. The smaller deflection reflects the fairly abrupt jog in the channel at the fault, whereas the larger deflection appears to reflect the total offset of the drainage. Taken together, we interpret the 6.1 m to represent slip in the past two or three events, with the most recent event 'freshening' the deflection in the vicinity of the fault by *c.* 2–2.5 m. The other possibility is that the 6.1 m deflection is the result of only one event and that this reflects slip in the most recent event with substantial off-fault bending, as has been documented in recent earthquakes (Rockwell *et al.* 2002; Treiman *et al.* 2002). However, because the near-field deflection appears to be superposed on the farther-field deflection, and because there is no thick cover of alluvium as is typical in areas of off-fault bending, we prefer an interpretation where the overall 6.1 m deflection has resulted from at least two and possibly three events. Furthermore, because of the uncertainty in interpretation of the deflected channel, we use the range of possible values to infer slip per event (2–3 m), assuming that the far-field deflection has resulted from at least two but not more than three events.

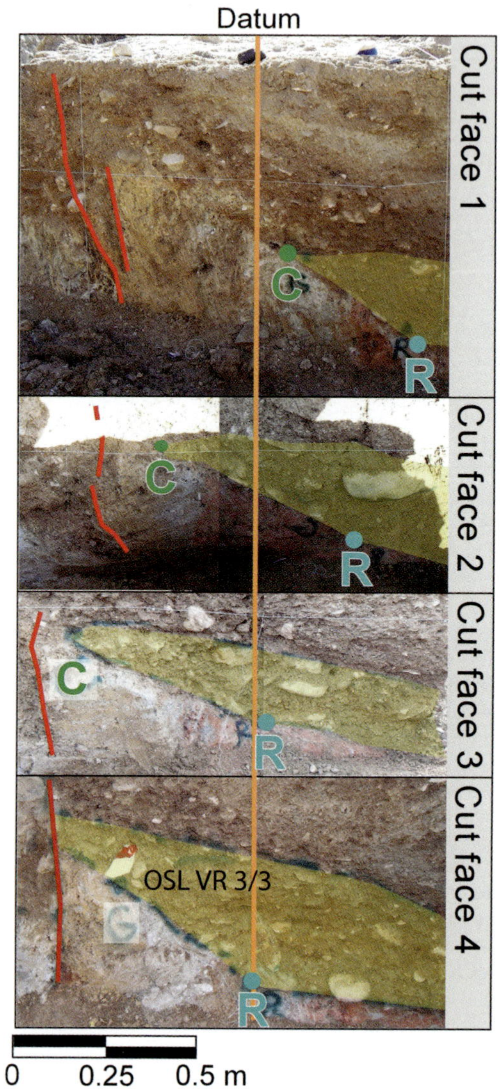

Fig. 12. First four cuts or slices into the block used to trace the location of the channel edge (C) and the tread/riser (R) into the fault. Cuts 5 to 7 determined the location of R at the fault (not shown). Note the location (flagging) of OSL sample VR 3/3.

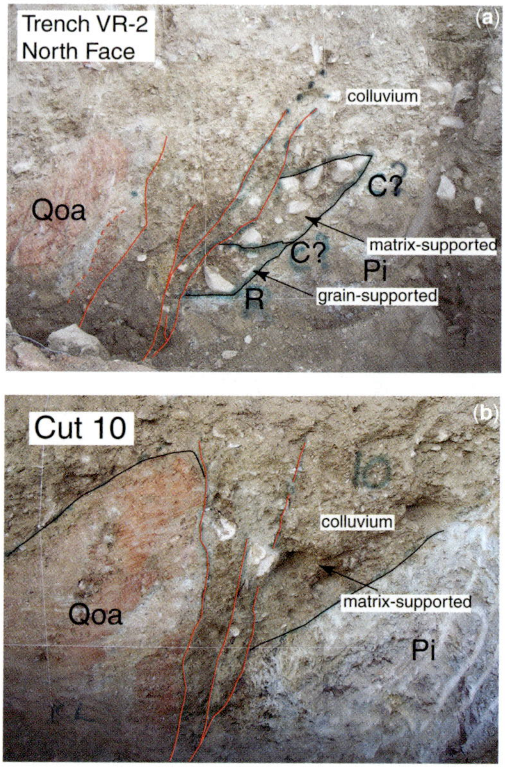

Fig. 13. Remnant of both piercing points exposed in the north face of trench VR-2. Note that we are uncertain whether the actual channel margin is represented by the upper or lower edge of the coarse, bouldery deposits, as the upper portion is partially matrix-supported and could be interpreted as colluvium. In either case, both contacts intercepted the fault between the VR-2 face and cut 10, as shown in (b), so their location at the fault is known to within ±25 cm.

The age of the rill must be younger than the *c.* 16 ka fill event represented by the prominent terrace along the Douro River, as the rill is topographically below this level. Consequently, it can only represent slip that has occurred after 16 ka, or only the past two to three events as determined at the Vale Meão trench site. The 6.1 m of deflection is very similar to the *c.* 6.5 m of offset of the channel margin at the Vilarica site which has occurred in a similar timeframe of <18 ka. Taken together, these observations make a strong argument for relatively large displacements on the Vilariça fault in the late Quaternary, with evidence preserved both at the surface and in the subsurface.

Estimation of slip rate

We have determined that two and probably three surface ruptures have occurred on the Vilariça fault in the past 14.5 ka at the Vale Meão site. We have also inferred *c.* 6 m of slip in the same time-frame from deflection of a small rill at Vale Meão. Taken together, these observations argue for 2–3 m of displacement every 5–7 ka, suggesting a rate of *c.* 0.3–0.6 mm/a (2–3 m events with a recurrence of 7 ka is *c.* 0.29–0.43 mm/a; 2–3 m events every 4.5 ka is *c.* 0.44–0.66 mm/a; 6.1 m in <16 ka is a minimum rate of 0.38 mm/a).

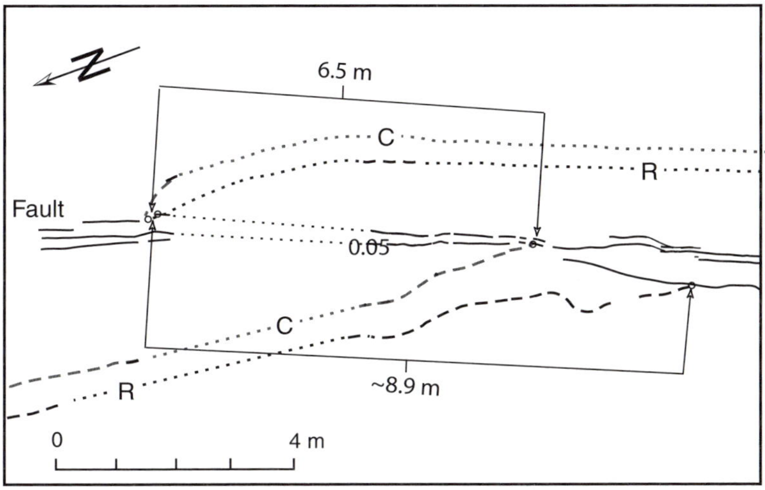

Fig. 14. Simplification of Figure 11 showing the estimates of displacement for the channel margin (C) and riser (R).

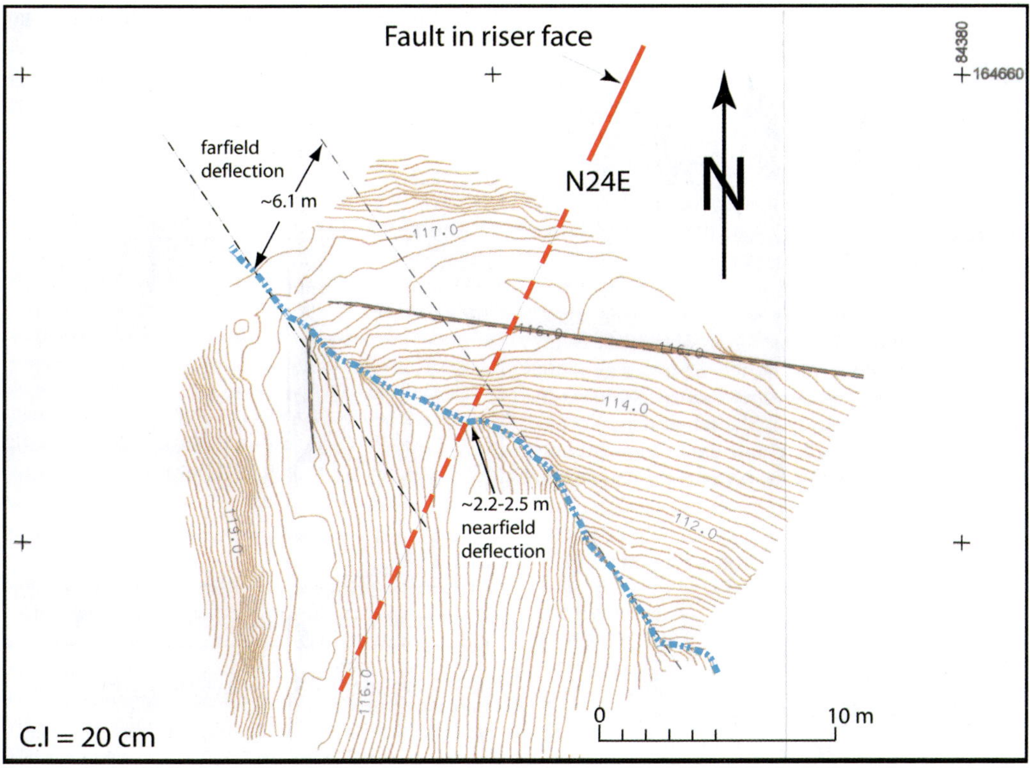

Fig. 15. Surveyed map of a small deflected channel on the SW edge of the Vale Meão winery. Contour interval is 20 cm. The field estimate of deflection was about 2 m, similar to the surveyed estimate of near-field deflection. The far-field deflection probably represents slip from multiple events.

Similarly at the Vilariça site, we resolved *c.* 6.5 m of displacement in less than the past 18 ka (>0.35 mm/a) and *c.* 8.9 m in less than 23 ka (>0.38 mm/a) with the likelihood that both offsets are <18 ka (0.5 mm/a). The absolute maximum age of the 8.9 m offset is the age of the Qt deposits (38.6 ± 2.7 ka) into which the Qa2 alluvium is incised, yielding an absolute minimum longer-term rate of 0.23 mm/a.

All of these observations at both Vale Meão and Vilariça are consistent with each other and make a robust set of data that argues for a preferred slip rate in the range of 0.3 to 0.5 mm/a for the late Quaternary. The uncertainty is larger than this range suggests because we do not know the variability in long-term seismic production rate. Our data suggest that the events were clustered, and that by sampling the cluster in our estimation of rate, we may have overestimated the rate. Nevertheless, this rate agrees favourably with the long-term Quaternary rate suggested by Cabral (1989) based on his interpretation of deflection of the Douro River.

Estimate of earthquake magnitude

The size of an earthquake is directly proportional to the average amount of slip and the surface area of rupture. Wells & Coppersmith (1994) compiled historical surface ruptures and regressed a number of properties (average and maximum slip, rupture length) against magnitude. Using the equations for strike-slip faults ($M = 7.04 + 0.89 \log(AD)$ and $M = 6.81 + 0.78 \log(MD)$, where AD = average displacement and MD = maximum displacement), and using the displacement values of 2.0–2.5 m for average displacement in individual earthquakes, we estimate the magnitude of these events is on the order of Mw 7.3. If these are maximum slip values, then the magnitude range is closer to Mw 7.1. In either case, it is likely that the palaeoevents that we identified on the Vilariça fault represent slip in relatively large earthquakes in the M7+ range.

As an independent check on our magnitude estimates, we assessed the Vilariça fault segment in terms of its segmentation or continuity. The fault is virtually straight for a distance of about 75 km between Longrovia northward to the vicinity of Macedo de Cavaleiros (Fig. 1), where the fault makes a complex restraining bend, several kilometres wide. At Longrovia, the fault makes a relatively minor 2-km-wide releasing step. A 75 km length is consistent with earthquakes at the upper end of our magnitude range and supports the idea that this fault is capable of producing relatively large events.

Conclusions

We have demonstrated that the Vilariça fault in northeastern Portugal has sustained multiple surface ruptures in the late Quaternary (over last *c.* 30 000 years), each likely producing multiple metres of slip (Fig. 16). The most recent surface slip occurred between *c.* 4.8 and 11 ka, but is most likely close to the older end of this range, indicating that there is an open interval of *c.* 5–11 ka. At least two and probably three large slip events occurred on the fault in the past 14.5 ka, suggesting an average return period in the range of 5 to 7 ka but with the high likelihood that the events were clustered in time. Clustering is common in intracontinental settings (Crone & Wheeler 2000) and suggests that the Vilariça segment of the Manteigas-Bragança fault is behaving as a typical intracontinental fault.

In the same timeframe as the occurrence of the two or three events, we resolved that at least 6.5 m, and possibly as much as 9 m, of displacement accompanied these events (Fig. 16), arguing that these earthquakes were in the magnitude range of M7+. Combining information on displacement and the age of the alluvial units, we suggest that the slip rate is on the order of *c.* 0.4 mm/a and is not likely more than 0.5 mm/a. Provided that the open interval is at least 5 ka, there has likely been at least 2 m of accumulated elastic strain on the Vilariça fault, similar to our minimum estimate of slip in the past events. If the Vilariça fault behaves in a clustered mode, however, it could still be several thousand years (or longer) before the next cluster begins. We could also have overestimated the slip rate by sampling a cluster of events if the intercluster period ranges into tens of thousands of years. It will be important to try to establish both a

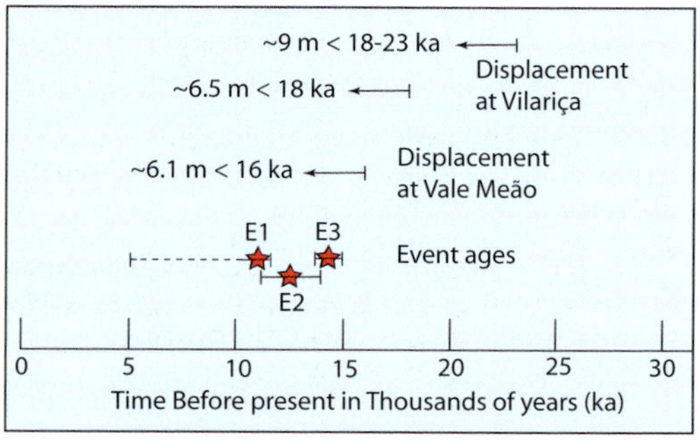

Fig. 16. Summary of palaeoseismic data from the Vilariça segment of the Manteigas-Bragança fault zone. The two or three documented surface ruptures apparently cluster in time between *c.* 14.5 and 11 ka, and produced 6.1–6.5 m of left-lateral slip. The slightly older Qa2 terrace tread at the Vilariça site is offset *c.* 9 m and may represent the occurrence of an additional event, or it may be a better measure of displacement in all three events.

better date for the most recent Vilariça event, as well as to extend the record back to include events associated with the previous cluster. Only then can we increase the reliability of any forecast that uses time-based conditional probabilities.

Finally, our studies demonstrate that there are potential seismic sources in western Iberia that are not associated with the 1755 Lisbon earthquake or the Tagus Valley and, although rare, large events on the Vilariça fault could be quite destructive for the region. As there are other major faults with evidence of Quaternary activity in Portugal (Cabral & Ribeiro 1988), these observations collectively argue for continued Alpine shortening in western Iberia, albeit at low rates.

We thank Electricidade de Portugal (EDP) for the opportunity to work on this project and permission to publish the results. We also thank the many landowners who granted access and allowed the subsurface investigations. We especially appreciate the cooperation of Mr Francisco Olazabal of Quinta de Vale Meão, who not only allowed access to his land but graciously introduced us to his secrets of making and tasting among the best wine in Portugal. We thank Tim Debey at the USGS Nuclear Reactor Facility in Denver for making the NAA measurements for the OSL dating. Finally, we thank João Cabral and Matthieu Ferry for their excellent reviews and meticulous attention to detail.

References

ADAMIEC, G. & AITKEN, M. 1998. Dose-rate conversion factors: update. *Ancient TL*, **16**, 37–50.

AITKEN, M. J. 1998. *An Introduction to Optical Dating*. Oxford University Press, Oxford.

ARTHAUD, F. & MATTE, P. 1975. Les decrochements tardi-Hercyniens du sud-ouest de L'Europe. Geometrie et essai de reconstitution des conditions de la deformation. *Tectonophysics*, **25**, 139–171.

CABRAL, J. 1989. An example of intraplate neotectonic activity, Vilariça basin, Northeast Portugal, *Tectonics*, **8**, 285–303.

CABRAL, J. & RIBEIRO, A. 1988. *Carta Neotectónica de Portugal Continental, escala 1/1 000 000*. Serviços Geológicos de Portugal, Lisboa.

CRONE, A. J. & WHEELER, R. L. 2000. *Data for Quaternary faults, liquefaction features, and possible tectonic features in the central and eastern United States, east of the Rocky Mountain front*. US Geological Survey Open-File Report, **00–0260**.

CRONE, A. J., MACHETTE, M. N., BRADLEY, L.-A. & MAHAN, S. A. 1997. *Late Quaternary surface faulting on the Cheraw Fault, Southeastern Colorado*. Geologic Investigations Map, USGS Map **I-2591**.

FONSECA, J. F. B. D. & LONG, R. E. 1991. Seismotectonics of SW Iberia: A distributed plate margin? *In*: MOSTENA, T. & UDIAS, A. (eds) Instituto Geografico National, Madrid, Memoria **8**.

FUCHS, M. & LANG, A. 2008. Luminescence dating of hillslope deposits – a review. *Geomorphology* (in press).

GRIMAUD, S., BOILLOT, G., COLLETTE, B. J., MAUFFRET, A., MILES, P. R. & ROBERTS, D. B. 1982. Western extension of the Iberian–European Plate boundary during the early Cenozoic (Pyrenean) convergence: A new model. *Marine Geology*, **45**, 63–77.

KENT, G. M., BABCOCK, J. M. *ET AL*. 2005. 60 k.y. record of extension across the western boundary of the Basin and Range province: Estimate of slip rates from offset shoreline terraces and a catastrophic slide beneath Lake Tahoe. *Geology*, **33**, 365–368.

LEE, J., SPENCER, J. Q. & OWEN, L. A. 2001. Holocene slip rates along the Owens Valley fault, California: Implications for the recent evolution of the Eastern California Shear Zone. *Geology*, **29**, 819–822.

MACHETTE, M. N., PERSONIUS, S. F. & NELSON, A. R. 1992. The Wasatch Fault Zone, USA. *Annales Tectonicae*, **6**, 5–39.

MARQUES, F. O., MATEUS, A. & TASSINARI, C. 2002. The Late-Variscan fault network in central-northern Portugal (NW Iberia): A re-evaluation. *Tectonophysics*, **359**, 255–270.

MEJDAHL, V. 1979. Thermoluminescence dating: beta attenuation in quartz grains. *Archaeometry*, **21**, 61–73.

MURRAY, A. S. & WINTLE, A. G. 2000. Luminescence dating of quartz using an improved single-aliquot regenerative-dose protocol. *Radiation Measurements*, **32**, 57–73.

OWEN, L. A., CUNNINGHAM, W. D., RICHARDS, B., RHODES, E. J., WINDLEY, B. F., DORNJAMJAA, D. & BADAMGARAV, J. 1999. Timing of formation of forebergs in the northeastern Gobi Altai, Mongolia: Implications for estimating mountain uplift rates and earthquake recurrence intervals. *Journal of the Geological Society*, **156**(3), 457–464 (Erratum: *Journal of the Geological Society*, **156**, Contents xii).

PRESCOTT, J. R. & HUTTON, J. T. 1994. Cosmic ray contributions to dose rates for luminescence and ESR dating. Large depths and long-term time variations. *Radiation Measurements*, **23**, 497–500.

RIBEIRO, A. 1974. *Contribution a l'etude de Trás-os-Montes Oriental*. Memorias Servicos Geologicas de Portugal, **24**.

RIBEIRO, A. 1981. A geotraverse through the Variscan fold belt in Portugal. *In*: ZWART, H. J. & DORNSIEPEN, U. F. (eds) *The Variscan Orogen in Europe. Geologie en Mijnbouw*, **60**, 41–44.

RIBEIRO, A. *ET AL*. 1979. *Introduction a la geologie générale du Portugal*. Serviços Geológicos de Portugal, Lisbon.

RIBEIRO, A., KULLBERG, M. C., KULLBERG, J. C., MANUPPELLA, G. & PHIPPS, S. 1990. A review of Alpine tectonics in Portugal: Foreland detachment in basement and cover rocks. *Tectonophysics*, **184**, 357–366.

ROCKWELL, T. K., JOHNSON, D. L., KELLER, E. A. & DEMBROFF, G. R. 1985. A late Pleistocene-Holocene soil chronosequence in the central Ventura Basin, Southern California, U.S.A. *In*: RICHARDS, K., ARNETT, R. & ELLIS, S. (eds) *Geomorphology and Soils*. George Allen and Unwin, London, 309–327.

ROCKWELL, T. K., LINDVALL, S., HERZBERG, M., MURBACH, D., DAWSON, T. & BERGER, G. 2000.

Paleoseismology of the Johnson Valley, Kickapoo and Homestead Valley faults of the Eastern California Shear Zone. *Bulletin of the Seismological Society of America*, **90**, 1200–1236.

ROCKWELL, T. K., LINDVALL, S., DAWSON, T., LANGRIDGE, R. & LETTIS, W. 2002. Lateral offsets on surveyed cultural features resulting from the 1999 Izmit and Duzce earthquakes, Turkey. *Bulletin of the Seismological Society of America*, **92**, 79–94.

SOBRINO, C. M., REGO, P. R. & GOMEZ-ORELLANA, L. 2007. Late Wurm and early Holocene in the mountains of northwest Iberia: Biostratigraphy, chronology and tree colonization. *Vegetation History and Archaeobotany*, **16**, 223–240.

TREIMAN, J. A., KENDRICK, K. A., BRYANT, W. A., ROCKWELL, T. K. & MCGILL, S. F. 2002. Primary surface rupture associated with the Mw7.1 16 October, 1999 Hector Mine earthquake, San Bernardino County, California. *Bulletin of the Seismological Society of America*, **92**, 1171–1191.

VILANOVA, S. & FONSECA, J. F. B. D. 2007. Probabilistic seismic-hazard assessment for Portugal. *Bulletin of the Seismological Society of America*, **97**, 1702–1717.

WASHBURN, Z., ARROWSMITH, J. R., FORMAN, S. L., COWGILL, E., WANG, X., ZHANG, Y. & CHEN, Z. 2001. Late Holocene earthquake history of the central Altyn Tagh fault, China. *Geology*, **29**(11), 1051–1054.

WELLS, D. L. & COPPERSMITH, K. J. 1994. New emperical relationships among magnitude, rupture length, rupture width, rupture area, and surface displacement. *Bulletin of the Seismological Society of America*, **84**, 974–1002.

WESNOUSKY, S. G., BARRON, A. D., BRIGGS, R. W., CASKEY, S. J., KUMAR, S. & OWEN, L. 2005. Paleoseismic transect across the northern Great Basin. *Journal of Geophysical Research*, **110**, B05408. DOI: 10.1029/2004JB003283.

ZIEGLER, P. 1982. *Geological Atlas of Western and Central Europe*. Shell Internationale Petroleum Maatschappij, B.V.

Recent seismic activity in the NW Himalayan Fold and Thrust Belt, Pakistan: focal mechanism solution and its tectonic implications

MONALISA

Department of Earth Sciences, Quaid-i-Azam University, Islamabad, Pakistan (e-mail: lisa_qau@yahoo.com)

Abstract: The Himalayas in northern Pakistan have been the site of several disastrous earthquakes of moderate to high intensity. The 8 October 2005 Muzaffarabad earthquake, with magnitude M_w 7.6, occurred in the NW Himalayan Fold and Thrust Belt at 08:50:38 local time. The epicentre of the main shock was located 19 km NE of Muzaffarabad. This earthquake took a death toll of more than 80 000 human lives and caused widespread destruction in Kashmir and north Pakistan, particularly in the towns of Muzaffarabad, Bagh, Rawalakot, Mansehra, Balakot, Abbottabad and Batgram. Based on the information obtained from print and electronic media (and for some areas from field studies), an intensity of X (MMI scale) has been assigned at the epicentral location including the localities of Muzaffarabad and Balakot. Epicentral distribution of 300 aftershocks indicates that more than one tectonic subdivision of the fold belt have experienced instability. Focal depths indicate that most activity is confined to a narrow depth range (5–20 km). Further extension of the Indus Kohistan Seismic Zone in the Hazara–Kashmir syntaxial area and activation of more than one fault seem to be the cause of this seismic activity, as suggested by the focal mechanism of the main event and depth distribution of the aftershocks. About 100 large landslides caused by active faulting have been observed in the rupture zones near Balakot, Muzaffarabad, Kardalla, Hattian Bala, Sarain, Sunddangali and Bagh, through field studies and satellite images.

On the morning of 8 October 2005 (08:50 local time), a large portion of northern Pakistan (about 11 000 km^2) was violently shaken by an earthquake of magnitude M_w 7.6 near the city of Muzaffarabad (Fig. 1). The towns of Muzaffarabad, Balakot, Bagh, Alai, Rawalakot, Mansehra and Abbottabad were severely damaged. The collapse of some buildings and minor damage were reported from Islamabad (one 12-storey tower collapsed), Lahore, Sialkot and Gujranwala located 100 to 200 km from the epicentre. The earthquake was also felt in central Afghanistan and in most parts of northern India. More than 80 000 people in Pakistan were killed and millions affected by this earthquake. Due to the widespread destruction caused, it is considered to be the worst of all the earthquakes that have occurred in the region. The seismicity of the area is related to the collision between the Indian and Eurasian plates. Active tectonic features (Figs 1 and 2) within the area, e.g. Main Boundary Thrust, Jhelum Fault, Panjal Thrust (MonaLisa and Jan, 2007*a*), Indus Kohistan Seismic Zone (IKSZ; Armbruster *et al.* 1978) and surface expression of the IKSZ, i.e. Muzaffarabad or Balakot-Bagh Fault (BBF), provide evidence for the ongoing collision. In the present work, an attempt has been made to understand the seismicity and seismotectonics of the area in the light of the 8 October 2005, Muzaffarabad earthquake. Fault plane solution of the main event and its aftershock distribution suggest that the NW–SE trending IKSZ is currently active. However, more fault plane solutions are required to support this contention.

Seismotectonics

The active tectonic features in northern Pakistan are the result of northward movement of the Indian Plate (about 4 cm/a) with respect to the Eurasian Plate. Collision between the two plates has resulted in intense folding and faulting. The Main Karakorum Thrust, the Main Mantle Thrust (MMT) and the Main Boundary Thrust (MBT) are well-known faults of regional extent. Along with these compressional features, transpression is also prevalent along the Thakot, Puran and Jhelum faults (Fig. 1).

Kazmi & Jan (1997) introduced the term NW Himalayan Fold and Thrust Belt for the western portion of the Himalayas. This irregularly shaped mountainous belt forms part of an active zone of convergence with the Kohistan Island Arc in the north. The MMT and Salt Range Thrust are its northern and southern extremities, whereas the Panjal-Khairabad Fault further divides it into northern Hinterland Zone and southern Foreland Zone (Fig. 1). The area is characterized by a large number of mostly east–west trending thrust faults orientated perpendicular to the principal stress directions (NW and NE). At the same time, the

From: REICHERTER, K., MICHETTI, A. M. & SILVA, P. G. (eds) *Palaeoseismology: Historical and Prehistorical Records of Earthquake Ground Effects for Seismic Hazard Assessment*. The Geological Society, London, Special Publications, **316**, 259–267. DOI: 10.1144/SP316.16 0305-8719/09/$15.00

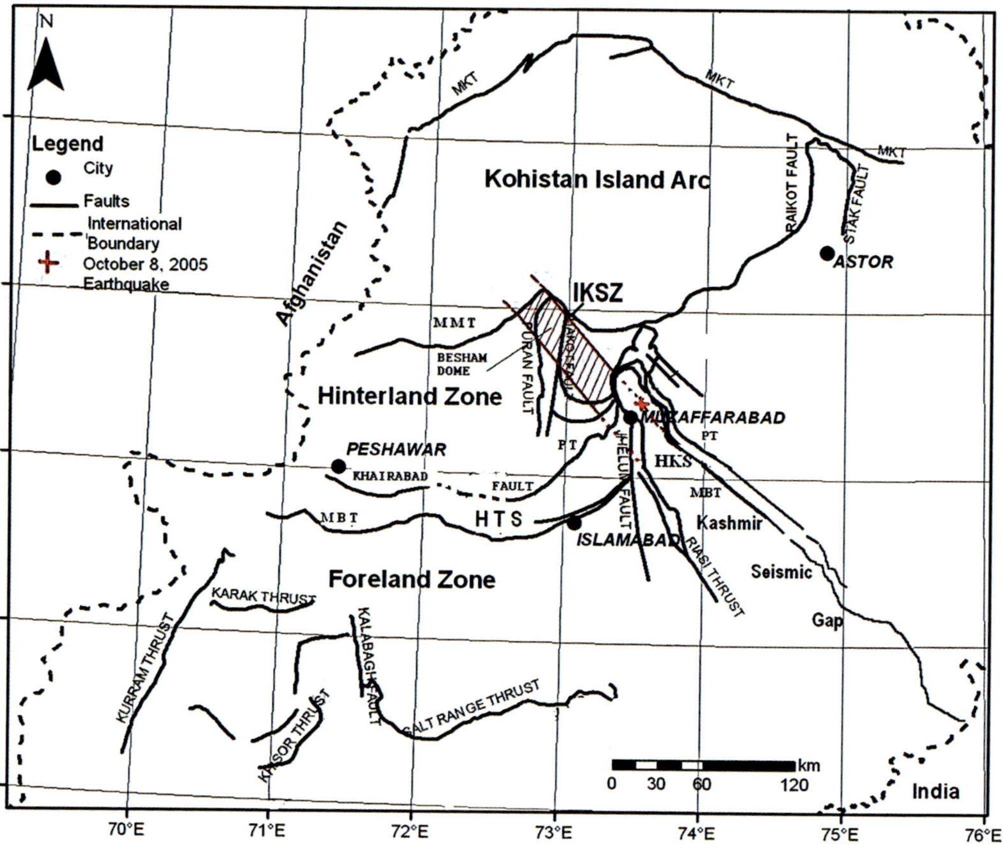

Fig. 1. Tectonic setting of NW Himalayan Fold and Thrust Belt. Abbreviations: HKS, Hazara-Kashmir Syntaxis; HTS, Hazara Thrust System; IKSZ, Indus Kohistan Seismic Zone; MBT, Main Boundary Thrust; MKT, Main Karakoram Thrust; MMT, Main Mantle Thrust; PT, Panjal Thrust (adapted from MonaLisa *et al.* 2007*b*).

western boundary of the fold belt that is also the Indian plate boundary is marked by the 860- km-long left-lateral strike-slip Chaman Fault. Its influence has resulted in the development of a zone of transpression. Further complexity is due to differential rates of uplift within the fold belt thereby resulting in strike-slip faulting. The complex deformation affecting the region resulted recently in the devastating Muzaffarabad earthquake ($M_w = 7.6$) of 8 October 2005 (Fig. 1) that claimed more than 80 000 lives. An area of about 30 000 km^2 within the fold belt experienced severe damage.

The historical and instrumental documentation of seismicity data (AD 25 to present) show that the area is seismically active (Figs 2 and 3). More than 100 earthquakes of magnitude greater than 5.5 and depth ranges of less than 75 km in most cases and 80–100 km in some have been documented. More recently two significant earthquakes with magnitudes m_b 6.0 and 5.6 occurred in Pattan (1974) and Kaghan (2004). The city of Muzaffarabad, at a distance of 19 km from the 8 October 2005 earthquake (Figs 2 and 3), itself had no history of large earthquakes before this event even though it is situated at the apex of a hairpin structure called the Hazara-Kashmir Syntaxis (HKS) (Wadia 1931) (Figs 1 and 3). The HKS is a complex tectonic feature encompassing active faults, such as MBT, Panjal Thrust, Jhelum Fault, BBF (Figs 2 and 3) and IKSZ (Fig. 1) etc. The HKS is a tectonic subdivision of the NW Himalayan Fold and Thrust Belt (Kazmi & Jan 1997), which although a compressional belt has some transpressional features as well. Among the above-mentioned tectonic features, the IKSZ is considered to be seismically the most active structure (Gahalaut 2006). The IKSZ is a 100-km-long, 50-km-wide, NW-trending wedge-shaped thrust blind fault zone, which lies between the MMT and HKS (Fig. 1); it has a NE-dipping lower surface and is parallel to the general trend of the MBT (Fig. 1). The 28 December 1974, Pattan earthquake (m_b 6), another destructive

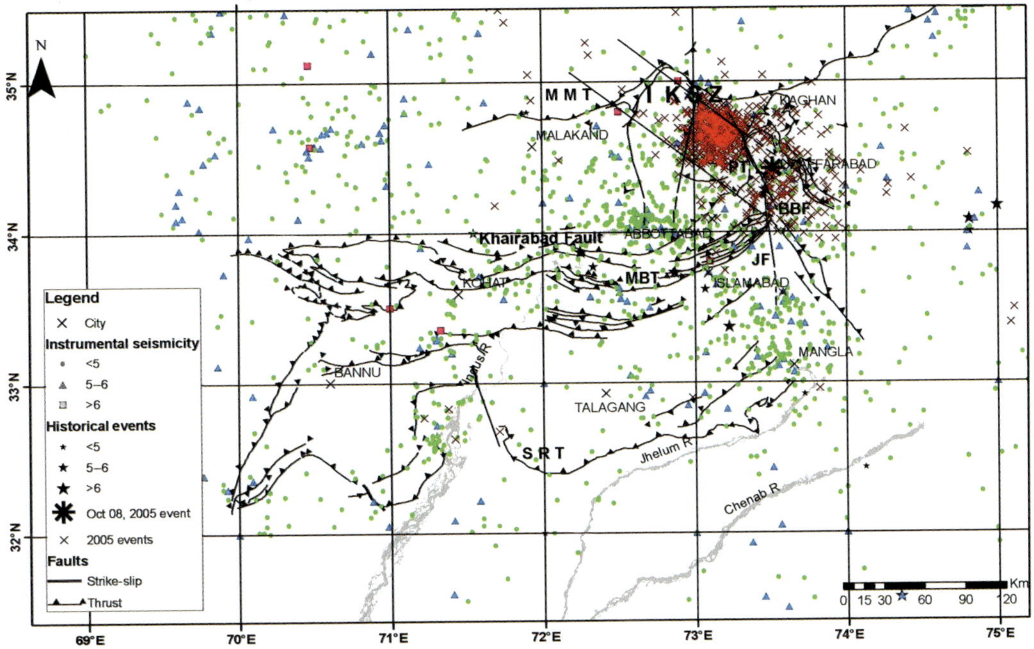

Fig. 2. Seismotectonic map of the area showing seismicity, structure and aftershock distribution of 8 October 2005 Muzaffarabad earthquake. Abbreviations: BBF, Balakot-Bagh Fault; IKSZ, Indus Kohistan Seismic Zone; JF, Jhelum Fault; MBT, Main Boundary Thrust; MMT, Main Mantle Thrust; PT, Panjal Thrust; SRT, Salt Range Thrust.

earthquake that occurred before the 2005 Muzaffarabad earthquake was also due to the IKSZ. This zone was first identified by Armbruster *et al.* (1978) using micro-earthquake survey data (1973–1974) and is further confirmed by Seeber & Armbruster (1979) and Ni *et al.* (1991). The IKSZ is believed to be divided into two zones, i.e. the upper shallow zone (8–10 km depth) and a mid-crustal zone (12–25 km depth), by a decollement surface. Based upon epicentral distribution of the aftershocks and the focal mechanism solution of the 8 October 2005, Muzaffarabad earthquake, we propose that the IKSZ is the source of this earthquake and it further reactivated the BBF (surface trace of IKSZ, having right-lateral strike-slip component) and several other faults in the area (Mahajan *et al.* 2006; Singh *et al.* 2006).

Seismic observations for the 8 October 2005 earthquake

Source parameters and aftershock distribution

The local seismic observatory provides the following source parameters of the main shock: date, 8 October 2005; time, 08:52:20 PST; magnitude, 7.6 Mw; epicentral location, 73.52 longitude, 34.42 latitude; focal depth, 13 km.

According to the local network, the main event occurred inside the HKS, near its eastern limb close to the MBT (Figs 2 and 3). As already described, the HKS is a complex tectonic feature dominated by several active thrust faults. Baig & Lawrence (1987) reported the presence of a NW–SE trending thrust fault at the site of the mainshock and named it the Muzaffarabad Fault (currently known as the Balakot-Bagh Fault, BBF) as shown in Figure 3. The earthquake was followed by thousands of aftershocks. The distribution of 300 selected aftershocks (magnitude $\geq$4.0 obtained from local observatory) for the period of October 2005 to October 2006 shows some concentration along the BBF (Fig. 3). However, most aftershocks were distributed between the MMT and HKS in a NW–SE direction (Fig. 3). This shows that more than one tectonic division has been activated. A lot of the aftershocks were also concentrated in the Batagram and Alai area, the major damaged sites. The majority of the aftershocks were in the depth range of 5–20 km (Fig. 4). Thus both the aftershock pattern and depth distribution clearly indicate activation of the IKSZ.

Focal mechanism solution

The focal mechanism solution (FMS) of the 13-km-deep mainshock was determined using the

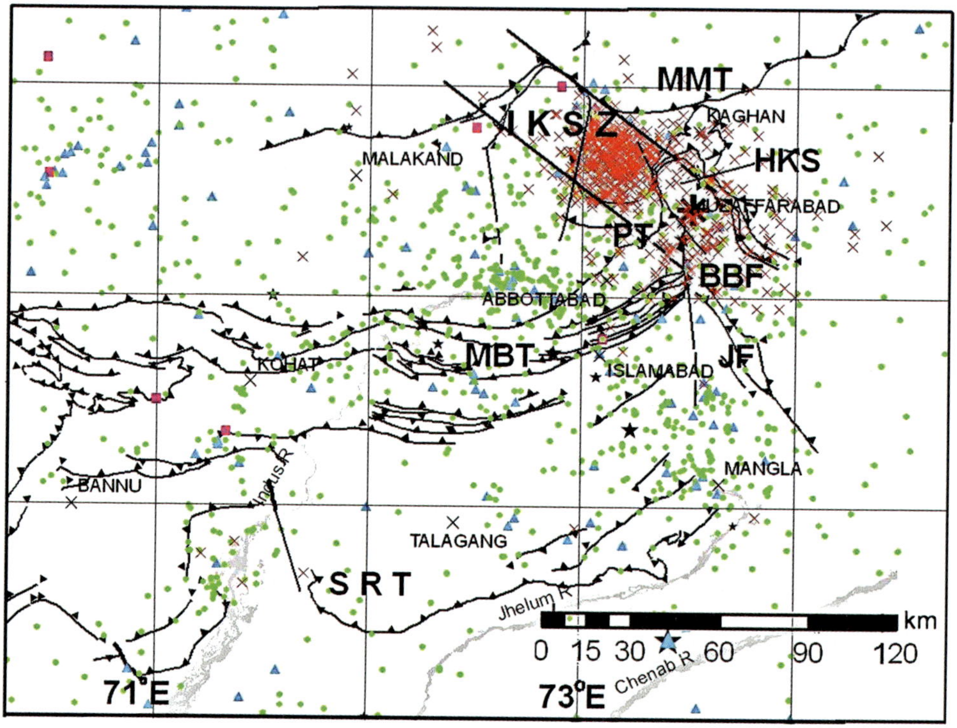

Fig. 3. Closer view of the active seismotectonic features, seismicity (AD 25 to present) and aftershock distribution of the 2005 Muzaffarabad earthquake. For legend and abbreviations see Figure 2.

first motion data of the local observatory as well as data of the International Seismological Centre (ISC) and United States Geological Survey (USGS). The software AZMTAK and PMAN (Suetsugu 1997) were used in order to obtain the 'beach-ball' diagram (Fig. 5). The FMS shows thrust with minor strike-slip component (Fig. 5). The USGS and Harvard network also obtained the thrust FMS with minor strike-slip solution as shown in Figure 6. Structurally this earthquake is surrounded by a number of active faults such as the MBT, Panjal Thrust, Jhelum fault and BBF (Fig. 3). A blind thrust seismic zone, i.e. the IKSZ, with the same trend (NW–SE) and depth (its upper layer above decollement lies at a depth of 8–10 km) also exists. Thus, based upon the trend and depth of the IKSZ, the nodal plane striking NW–SE and dipping NE is considered to be the rupture plane.

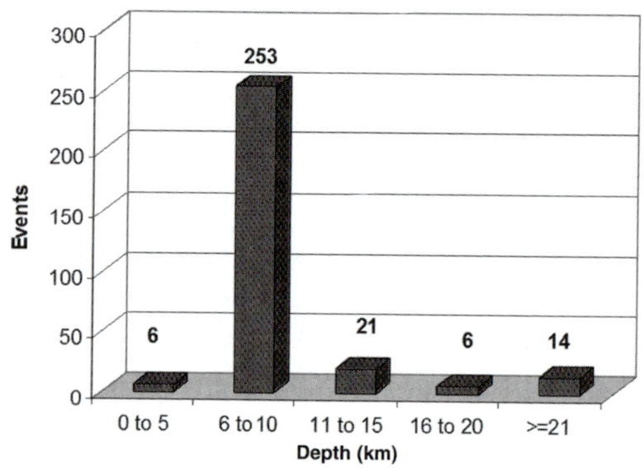

Fig. 4. Depth distributions of 300 aftershocks of the Muzaffarabad earthquake.

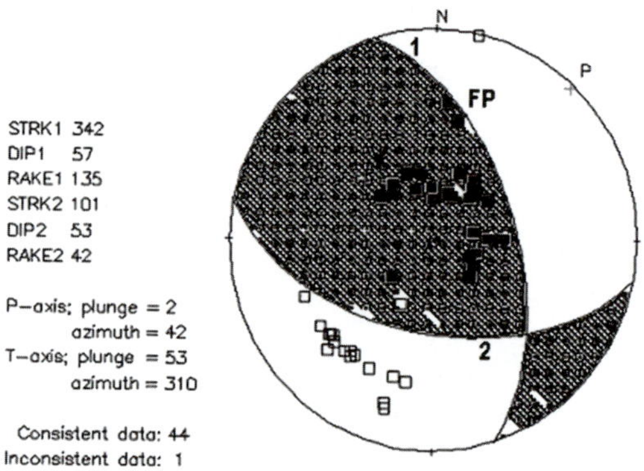

Fig. 5. Focal mechanism solution of the Muzaffarabad earthquake.

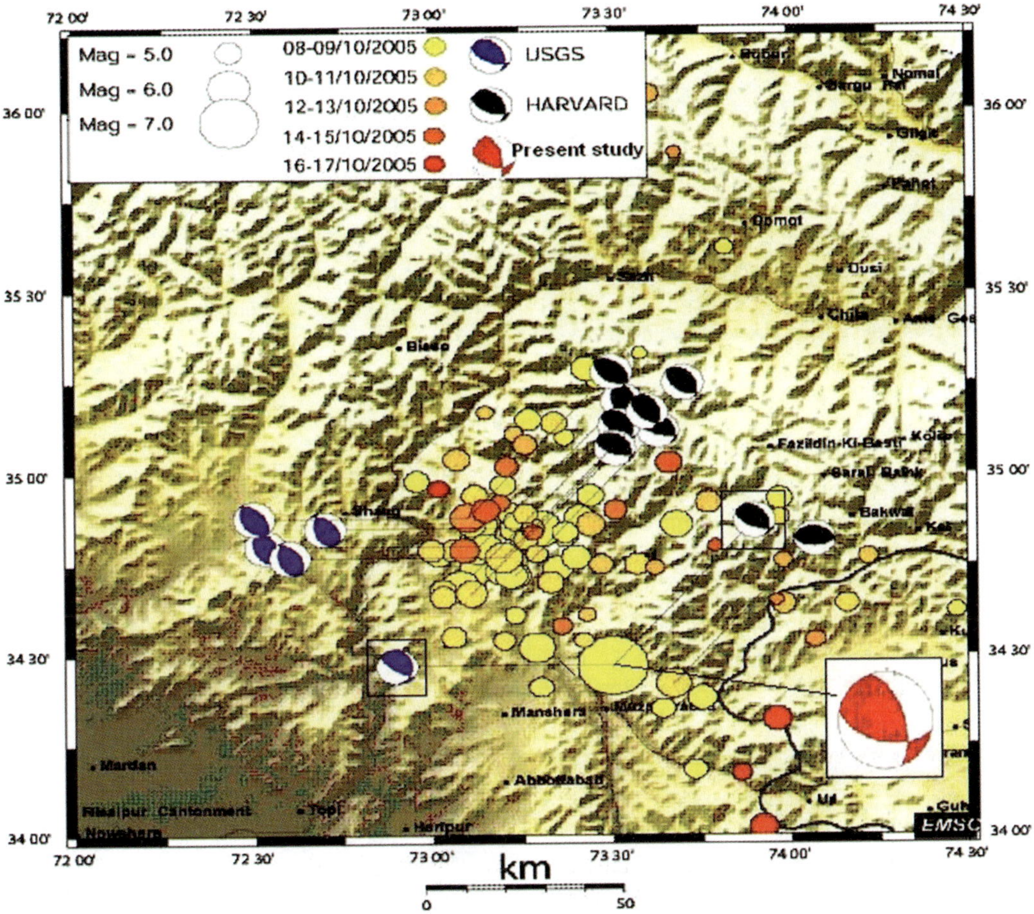

Fig. 6. Focal mechanism solutions (FMS) of the Muzaffarabad earthquake and its aftershocks by USGS (blue), Harvard (black; courtesy of USGS) and the present study (red; main event only).

Fig. 7. Total collapse of Balakot city.

Fig. 8. Collapse of Sangam Hotel, a five-storey reinforced concrete frame building in Domel, Muzaffarabad.

Fig. 9. Total collapse of Balakot city.

Intensity and damage distribution

The 8 October 2005, Muzaffarabad earthquake is considered to be the worst of the Himalayan earthquakes due to the widespread destruction it caused. The damage extended to a radius of 150 km. Muzaffarabad, Bagh and Rawalakot in Kashmir, and Mansehra, Balakot, Abbottabad and Batgram in NW Frontier Province (NWFP) were the most affected towns. Damage to buildings and other structures in general corresponded with the intensity of ground shaking observed at that location. Within a radius of 30 km from the main epicentral location, nearly a quarter of the buildings collapsed and the rest were badly damaged (MonaLisa *et al.* 2009). Based on the information obtained from the print media, an intensity of X on the MMI scale has been assigned near the rupture zone. The shallow focal depth (13 km), high magnitude (7.6 M_w) and poorly constructed buildings are also indicative of this intensity of X (MMI scale). Some of the damage can be seen in photographs displayed in Figures 7–9.

Fault rupture

The Muzaffarabad earthquake originated due to the Muzaffarabad or BBF (a surface expression of the IKSZ) that runs parallel to Main Boundary Thrust Fault Zone in the HKS between Bagh and Balakot (Fig. 3). This information is based upon several surface ruptures reported between Bagh and Balakot (Kausar *et al.* 2006). Between Balakot and Muzaffrabad this fault reactivates the MBT, and southward it follows the northern bank of the

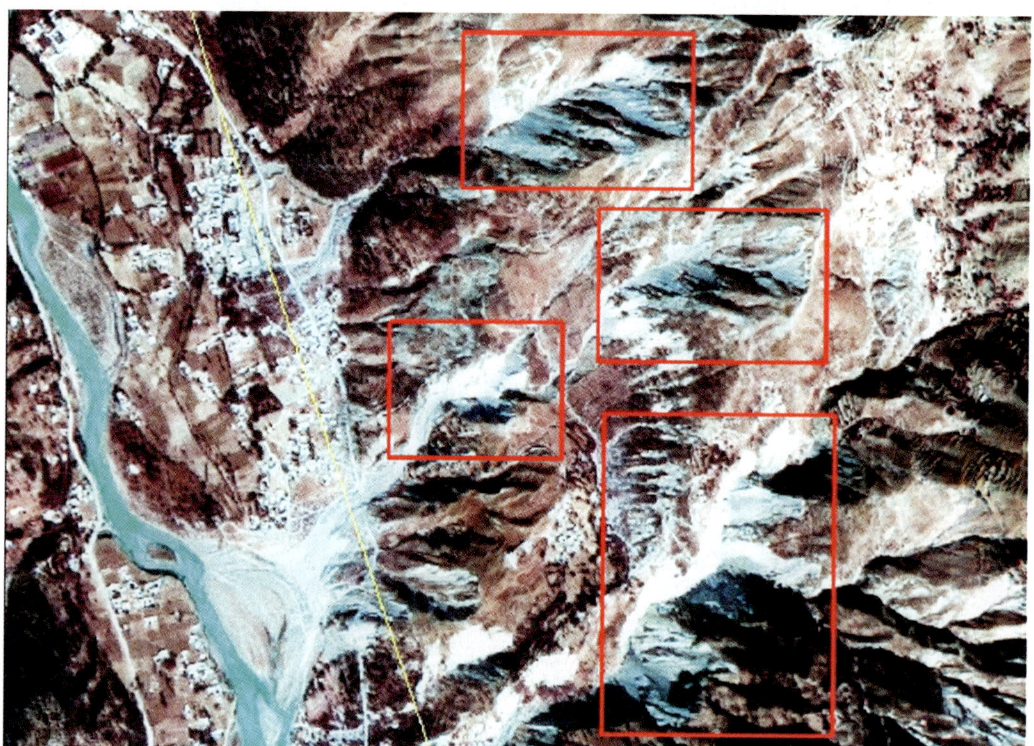

Fig. 10. Large-scale landslides occurred in the north of Muzaffarabad.

Jhelum River passing near the villages of Garhi Dupatta, Chakar and Bagh, where its surface trace ends (Hussain & Yeats 2006). These villages were some of the most heavily damaged by the earthquake. Based upon the landslides (Figs 10–12), surface evidence of the coseismic ruptures, the emergence of new springs, and some field evidence, it is interpreted that this *c.* 110-km-long fault is a thrust with strike-slip component, striking in a NNW direction and dipping 30–70° NE. The damage distribution and depth of the aftershocks (5–20 km) suggest that the BBF may extend at depth and coincide with the IKSZ.

Landslides

Active faulting in any tectonically unstable area may cause huge landsliding. In the present case, landslides on a moderate to large scale occurred at numerous locations. The most prominent of them occurred in the north of Muzaffarabad city, due to which a large portion of the mountainous range was lost (Sato *et al.* 2007 and present study) as shown in Figures 10–12. About 100 of the landslides were observed in the active rupture zones near Balakot, Muzaffarabad, Kardalla, Hattian Bala, Sarain, Sunddangali and Bagh, through field studies and satellite images (Figs 10–12). NW-trending linear landslides are common in the hanging wall and footwall of the Muzaffarabad Fault. These landslides have been formed due to ground shaking, gravity collapse and NE–SW late extension in the area. The landsliding not only caused the rivers to flood by damming with excess sedimentary load, but also blocked the roads which badly affected communications and rescue work. Using the 2.5 m resolution images (courtesy of National Engineering Services of Pakistan, NESPAK), the landslides are identified as bright white slopes on the eastern and northwestern portion of the Muzaffarabad Fault (between Bagh and Balakot) as shown in Figures 10–12.

Fig. 11. Closer view of one of the largest landslides that occurred in Hattian Bala.

Fig. 12. Massive landslide occurrence in the north of Muzaffarabad.

Conclusions

Based on this preliminary seismotectonic study of the 8 October 2005, Muzaffarabad earthquake, it is hypothesized that the IKSZ is seismically very active and the real cause of this event. The epicentre of the 1974 Pattan earthquake was also located in this zone. This earthquake was of magnitude 6.0, with focal depth of 10–20 km. It is concluded that the IKSZ trends NNW–SSE and extends for about 100 km from the centre of the HKS to the MMT in the vicinity of Pattan. This is evident from the aftershock pattern, their depth distribution and focal mechanism. The FMS of the mainshock (8 October 2005 earthquake), aftershock pattern and their depth distribution are clearly indicative of a NW–SE striking, NE-dipping thrust fault, following the IKSZ (BBF, surface expression of IKSZ) from the nose of the HKS to the MMT. The IKSZ has also been considered the source of the second major earthquake of the area, i.e. the 1974 Pattan earthquake (15 km depth). The FMS of this earthquake was also a NW–SE trending thrust with minor strike-slip component (Pennington 1979). The FMS of the Muzaffarabad earthquake (USGS, Harvard and present study) is in agreement with those of the previous earthquakes of the IKSZ.

On the one hand, this event has released high seismic energy in a NW–SE direction along the IKSZ, and on the other it has increased the tectonic stresses on the northern and SW directions, which may cause high seismicity in the near future. It is also worth noting that such a big event in the area has triggered many offshoots and splays of the larger faults (data to be presented separately). Thus the possibility of large earthquakes in future, causing serious damage in the populated and poorly constructed areas cannot be excluded (Parsons *et al.* 2006, and present study). It is therefore suggested that an effective National Seismic Hazard Study programme should be initiated with emphasis on precise delineation of the high risk zones.

The author is indebted to Dr M. Qasim Jan and Dr Azam A. Khwaja for their critical comments. The provision of data from the local observatory and landslide images from Yawer S. Ansari of National Engineering Services Pakistan (NESPAK) is also gratefully acknowledged.

This work is partially supported by Higher Education Commission (HEC) Projects Nos 20-749/R & D/07/336 and 20-600/R & D/06/1761.

References

ARMBRUSTER, J. G., SEEBER, L. & JACOB, K. K. 1978. The northwest termination of the Himalayan mountain front: active tectonics from micro earthquakes. *Journal of Geophysical Research*, **83**, 269–282.

GAHALAUT, V. K. 2006. 2005. Kashmir earthquake: not a Kashmir Himalaya seismic gap event. *Current Science*, **90**(4), 507–508.

HUSSAIN, A. & YEATS, R. S. 2006. The Balakot-Bagh fault that triggered the October 8 earthquake and other active faults in the Himalayan foreland region, Pakistan. *In*: KAUSAR, A. B., KARIM, T. & KHAN, T. (eds) *Extended Abstracts, International Conference on 8 October 2005 Earthquake in Pakistan: Its Implications & Hazard Mitigation*. Geological Survey of Pakistan, 115–116.

KAUSAR, A. B., HUSSAIN, S. H., KANEDA, H., KONDO, H., AWATA, Y. & JOUANNE, F. 2006. Hazara Kashmir Syntaxis & Seismic Hazard of 8 October 2005 earthquake. *In: Abstracts, International Workshop on Seismology, Seismotectonics and Seismic Hazard in the Himalayan Region*. Geological Survey of Pakistan, Islamabad, 21.

KAZMI, A. H. & JAN, M. Q. 1997. *Geology and Tectonics of Pakistan*. Graphic Publishers, Karachi.

MAHAJAN, A. K, KUMAR, N. & ARORA, B. R. 2006. Quick look isoseismal map of 8 October 2005 Kashmir earthquake. *Current Sciences*, **91**(3), 356–361.

MONALISA, KHWAJA, A. A. & JAN, M. Q. 2007*a*. Seismic hazard assessment of the NW Himalayan Fold-and-Thrust Belt, Pakistan using probabilistic approach. *Journal of Earthquake Engineering*, **11**, 257–301.

MONALISA, KHWAJA, A. A., JAN, M. Q., JAN, M. Q., YEATS, R. S. & HUSSAIN, A. 2007*b*. 8 October 2005 Muzaffarabad Earthquake: New data on the Indus Kohistan Seismic Zone and its extension into the Hazara Kashmir Syntaxis, NW Himalayas of Pakistan. *Abstracts of American Geophysical Union (AGU) Fall meeting*, 10–14 December 2007. San Francisco, USA.

NI, J., IBENBRAHIM, A. & ROCKER, S. W. 1991. 3 dimensional velocity structure and hypocenters of earthquakes beneath the Hazara arc, Pakistan: Geometry of the under thrusting Indian plate. *Journal of Geophysical Research*, **96**(19), 865–877.

PARSONS, T., YEATS, R. S., YAGI, Y. & HUSSAIN, A. 2006. Static stress change from the 8 October 2005 $M = 7.6$ Kashmir earthquake. *Geophysical Research Letters*, **33**. DOI: 10.1029/2005GL025429.

PENNINGTON, W. D. 1979. A summary of field and seismic observations of the Pattan earthquake—28 December 1974. *In*: FARAH, A. & DEJONG, K. A. (eds) *Geodynamics of Pakistan*. Geological Survey of Pakistan, Quetta, 143–147.

SATO, H. P., HASEGAWA, H., FUJIWARA, S., TOBITA, M., KOARAI, M., UNE, H. & IWAHASHI, J. 2007. Interpretation of landslide distribution triggered by the 2005 Northern Pakistan earthquake using SPOT 5 imagery. *Landslides*, **4**(2), 113–122.

SEEBER, L. & ARMBRUSTER, J. 1979. Seismicity of Hazara Arc in northern Pakistan: decollement vs. basement faulting. *In*: FARAH, A. & DEJONG, K. A. (eds) *Geodynamics of Pakistan*. Geological Survey of Pakistan, Quetta, 131–142.

SINGH, S. K., IGLISIAS, A. *ET AL*. 2006. Muzaffarabad earthquake of 8 October 2006 (M_w 7.6): A preliminary report on source characteristics and recorded ground motion. *Current Science*, **91**(5), 689–695.

SUETSUGU, D. 1997. Source Mechanism Practice. *In*: *Training Course in Seismology and Earthquake Engineering II*. Japan International Cooperation Agency (JICA), Ibaraki-ken, Japan, 13–48.

WADIA, D. N. 1931. The syntaxis of the northwest Himalaya: Its rocks, tectonics and orogeny. *Geological Survey of India Records*, **63**(1), 129–138.

Palaeoearthquake surface rupture in a transition zone from strike-slip to oblique-normal slip and its implications to seismic hazard, North Island Fault System, New Zealand

VASILIKI MOUSLOPOULOU[1,2]*, ANDREW NICOL[3], TIMOTHY A. LITTLE[1] & JOHN G. BEGG[3]

[1]*School of Earth Sciences, PO Box 600, Victoria University of Wellington, Wellington, New Zealand*

[2]*Present address: Department of Mineral Resources Engineering, Technical University of Crete, Chania 73100, Greece*

[3]*GNS Science, PO Box 30368, Lower Hutt, New Zealand*

**Corresponding author (e-mail: vasiliki@mred.tuc.gr)*

Abstract: The North Island Fault System (NIFS) is the longest and highest slip-rate active strike-slip fault system within the Hikurangi subduction margin in New Zealand, accommodating up to 10 mm/a of the margin-parallel plate motion. Displacement of landforms over the last *c.* 30 ka indicates a gradual northward change from right-lateral strike-slip to oblique-normal slip along the northern NIFS and within 60 km of its intersection with the active Taupo Rift. This change is expressed by a *c.* 60° increase in the pitch of the slip vectors. We use fault data from 20 trenches and displacements along active traces to explore whether changes in late Quaternary fault kinematics principally arise due to earthquake rupture arrest and/or variations in slip vector pitch during individual earthquakes that span the kinematic transition zone. Results show that earthquake rupture arrest occurs along the strike of the NIFS, with at least 60–80% of all events during the last 10–13 ka terminating across the zone of late Quaternary (*c.* 30 ka) transition from strike-slip to oblique-normal slip. The strike of the faults across the kinematic transition is unchanged, and we suggest that rupture was arrested there due to a 20–30° northward shallowing of fault-dip across this zone. Rupture arrest limits earthquake lengths and magnitudes which, when combined with recurrence intervals from trenching, locally decreases the seismic hazard in the region of the faults. Simple kinematic earthquake slip models, which simulate the addition of slip vectors during individual earthquakes, suggest that rupture arrest was accompanied by a northward steepening of slip vectors during individual earthquakes. Changes in coseismic slip vectors may arise due to the northward decrease in fault dip and associated steepening of the principal compressive stress axis (σ_1) which, in turn, is due to fault interactions between the NIFS and the adjacent active Taupo Rift.

Some strike-slip earthquake ruptures are characterized by slip vectors that change their pitch significantly along-strike to include oblique motion (e.g. Yoshida *et al.* 1996; Eberhart-Phillips *et al.* 2003; Muller & Aydin 2004). In many cases these changes appear to correlate with corresponding changes in the strike of the fault plane as occurs, for example, across fault bends or step-overs (i.e. Freund 1974; Barka & Kandinsky-Cade 1988; Baljinnyan *et al.* 1993; Wald & Heaton 1994; Aydin & Du 1995; Muller & Aydin 2004) (Fig. 1a), or where large strike-slip faults terminate against reverse or normal faults (Bayasgalan *et al.* 1999; Hreinsdottir *et al.* 2003) (Fig. 1b). Changes in the slip vector pitch along a fault or fault system may also accompany along-strike partitioning of the strike-slip and dip-slip components across a fault system (Begg & Johnston 2000) (Fig. 1c) or up-dip partitioning of slip along a single fault (Ide *et al.* 1996; Nicol & Van Dissen 2002). In this paper we examine how cumulative slip in individual earthquake ruptures may accrue to produce late Quaternary spatial changes in the pitch of slip vectors, along fault systems that neither change in strike nor partition their coseismic slip into strike-slip and dip-slip.

Globally some faults change their kinematics along-strike yet do not vary significantly in strike (e.g. Beroza 1991; Johnson & Segall 2004; Oglesby 2005). These changes presumably reflect local variations in the orientations and/or magnitudes of the principal stress axes. Along-strike

From: REICHERTER, K., MICHETTI, A. M. & SILVA, P. G. (eds) *Palaeoseismology: Historical and Prehistorical Records of Earthquake Ground Effects for Seismic Hazard Assessment*. The Geological Society, London, Special Publications, **316**, 269–292. DOI: 10.1144/SP316.17 0305-8719/09/$15.00

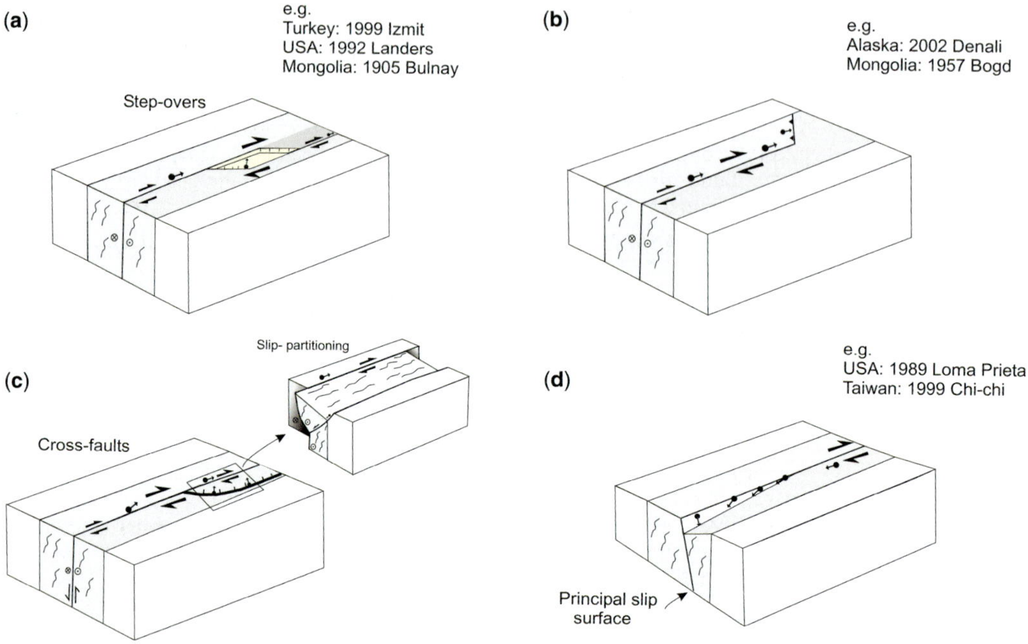

Fig. 1. Schematic block diagrams illustrating cases in which strike-slip and dip-slip faulting is produced coseismically. (**a**) The strike of a strike-slip fault swings and coseismic slip propagates through fault bends and/or step-overs (i.e. 1905 Bulnay; 1992 Landers; 1999 Izmit earthquakes). (**b**) Coseismic slip on strike-slip faults terminates at high-angle reverse or normal faults (1957 Bogd; 2002 Denali earthquakes). (**c**) Slip is partitioned along the strike of the fault into strike-slip and dip-slip components (1855 Wairarapa earthquake). (**d**) The coseismic slip vector pitch changes spatially along-strike (1989 Loma Prieta; 1999 Chi-chi earthquakes).

changes in the stress regime require that: (1) earthquakes rupture through the kinematic transition zone and undergo a corresponding change in their slip-vector pitch (Figs 1d and 2a); (2) isolated earthquake ruptures have different slip vectors on either side of the transition zone into which they terminate (Knuepfer 1989) (Fig. 2c); or (3) a combination of the two slip rupture models, with a number (but not all) of earthquakes being arrested within the kinematic transition zone (Fig. 2b).

Support for the first model (Fig. 2a) is provided by the 1989 Loma Prieta and 1999 Chi-chi

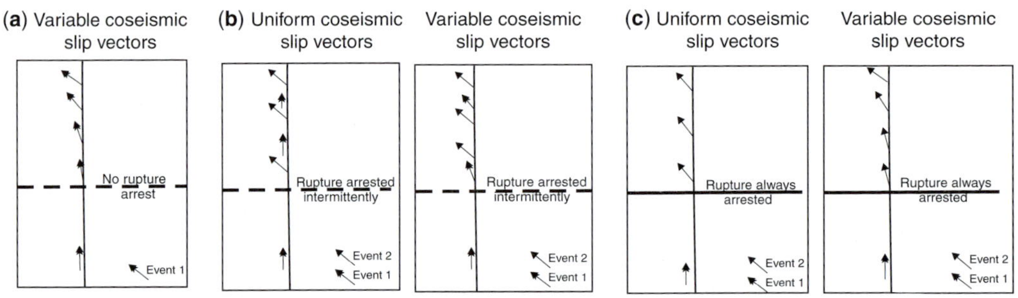

Fig. 2. End-member models of earthquake surface rupture propagation for the northern NIFS with respect to the kinematic transition from strike-slip to oblique-normal faulting (thick dashed and solid lines). (**a**) There is no arrest of surface-rupturing earthquake across the transition zone. (**b**) The kinematic transition zone arrests all oblique-slip but not all strike-slip earthquakes, which intermittently rupture through the transition zone, resulting in a mixture of strike-slip and oblique-slip earthquake ruptures north of the transition zone. (**c**) Strike-slip and oblique-slip earthquake ruptures are always arrested within the kinematic transition zone. For all scenarios we consider the possibility that earthquakes rupture with either spatially variable or uniform slip-vector pitch. To simplify the model, the segment boundary is drawn as a line although it represents a zone.

earthquakes where the rake of coseismic slip changed by up to 70° along sections of the faults with relatively uniform strike (Beroza 1991; Oglesby & Day 2001). These two examples of extreme changes in slip vector orientations are thought to have resulted from a combination of geometric (i.e. change in the fault-dip) and static (spatial variations in the pre-earthquake magnitudes and orientations of the principal stress axes) effects (Árnadóttir & Segall 1994; Guatteri & Cocco 1996; Oglesby & Day 2001).

Displacements of ≤30 ka old landforms by the North Island Fault System (NIFS), New Zealand, suggest that the kinematics of the 500-km-long strike-slip fault system change from dominantly strike-slip (horizontal to vertical ratio *c.* 10) to dominantly oblique-normal slip (horizontal to vertical ratio *c.* 0.8) near its northern termination against the Taupo Rift. Fault strike across this kinematic transition zone remains relatively uniform. The question we address in this paper is whether, in the case of the NIFS, changes in fault kinematics along the system arise due to variable slip-vector pitch during individual earthquakes (model 1) or to fault-rupture arrest within the kinematic transition zone (model 2). Which of these two models best accounts for the kinematics of the northern NIFS, has implications for the seismic hazard of the Bay of Plenty region. To test these models we compile fault-trench data from a total of 20 excavations to characterize the timing and coseismic slip vectors of prehistoric earthquakes, during the last 10–13 ka, on faults either side of the kinematic transition zone. We investigate the way individual earthquakes, with either a uniform or variably orientated coseismic slip vector, aggregate to account for the late Quaternary (*c.* 30 ka) finite slip vector orientations in northern NIFS (Figs 2 and 3).

The NIFS in New Zealand provides an excellent opportunity to conduct this study as the change in the kinematics of these faults is well documented (Beanland 1995; Wallace *et al.* 2004; Mouslopoulou *et al.* 2007), there are numerous trenches along the fault system and across the kinematic transition zone and also because tephra layers in the trenches provide, together with radiocarbon dating, excellent age control on the timing of past earthquakes (Manning 1995; Mouslopoulou 2006). Collectively these data, combined with simple kinematic earthquake-slip modelling, suggest that the late Quaternary change in the kinematics along-strike the NIFS arises due to a combination of earthquake rupture arrest and variations in the orientation of the coseismic slip vectors. Rupture arrest may arise due to a 20–30° shallowing of fault-dips across the kinematic transition zone, while spatial variations in the pitch of the earthquake vectors may reflect a regional northward steepening of the principal stress axis (σ_1) with increasing proximity to the Taupo Rift. Rupture arrest in the kinematic transition zone of the northern NIFS results in a decrease of the earthquake magnitude in the northern NIFS by *c.* 0.5 compared with data included in the New Zealand national seismic hazard model (Stirling *et al.* 2002) and, therefore, to a reduction of the seismic hazard in the Bay of Plenty region.

Definition of kinematic zones in the NIFS

Along the Hikurangi margin in the North Island of New Zealand, the Pacific Plate is being subducted beneath the Australian Plate (Ballance 1975; Walcott 1978, 1984; Rait *et al.* 1991; Beavan & Haines 2001). The relative motion between the two plates is oblique with >80% of the margin-normal motion accommodated on the plate interface itself and >50% of the margin-parallel motion (i.e. strike-slip and vertical axis fault-block rotations) accommodated in the upper plate (Nicol & Beavan 2003; Wallace *et al.* 2004; Nicol & Wallace 2007).

The NIFS is the principal strike-slip fault system in the upper plate of the Hikurangi margin and comprises two main strands, which are typically referred to as the western and the eastern strands (inset in Fig. 3). This study focuses on the western strand of this fault system which hereafter is referred to as the NIFS.

The NIFS is *c.* 500 km long and traverses the entire North Island, from Wellington to Bay of Plenty, with an average strike of NE–SW (i.e. 30°) and mainly dips steeply (i.e. >80°) either to the east or west (Fig. 3, inset). The fault system is well located by its active traces. The rate of strike-slip in the NIFS increases in a southward direction, from *c.* 4 to 10 mm/a, and accommodates up to about 30% of the margin-parallel relative plate motion near Wellington (Beanland 1995; Mouslopoulou *et al.* 2007; Nicol & Wallace 2007). The NIFS comprises two principal splays, the Wellington–Mohaka and Ruahine faults. The Wellington–Mohaka Fault bifurcates northwards into the Patoka–Rangiora, Whakatane, Waimana, Waiotahi and Waioeka faults, whilst the Ruahine Fault becomes the Waiohau Fault which has several secondary fault splays (e.g. Wheo Fault) (Fig. 3). Near the Bay of Plenty coast, where the faults in the NIFS are oblique-normal, the NIFS is intersected by the NE-striking active Taupo Rift (Fig. 3).

The geometry of the faults in the NIFS varies along-strike. The NIFS has a constant NE–SW strike along its southern 400 km and swings abruptly 25° anticlockwise to a north–south strike about 100 km south of the Bay of Plenty coast (Fig. 3). This change in fault strike is accompanied

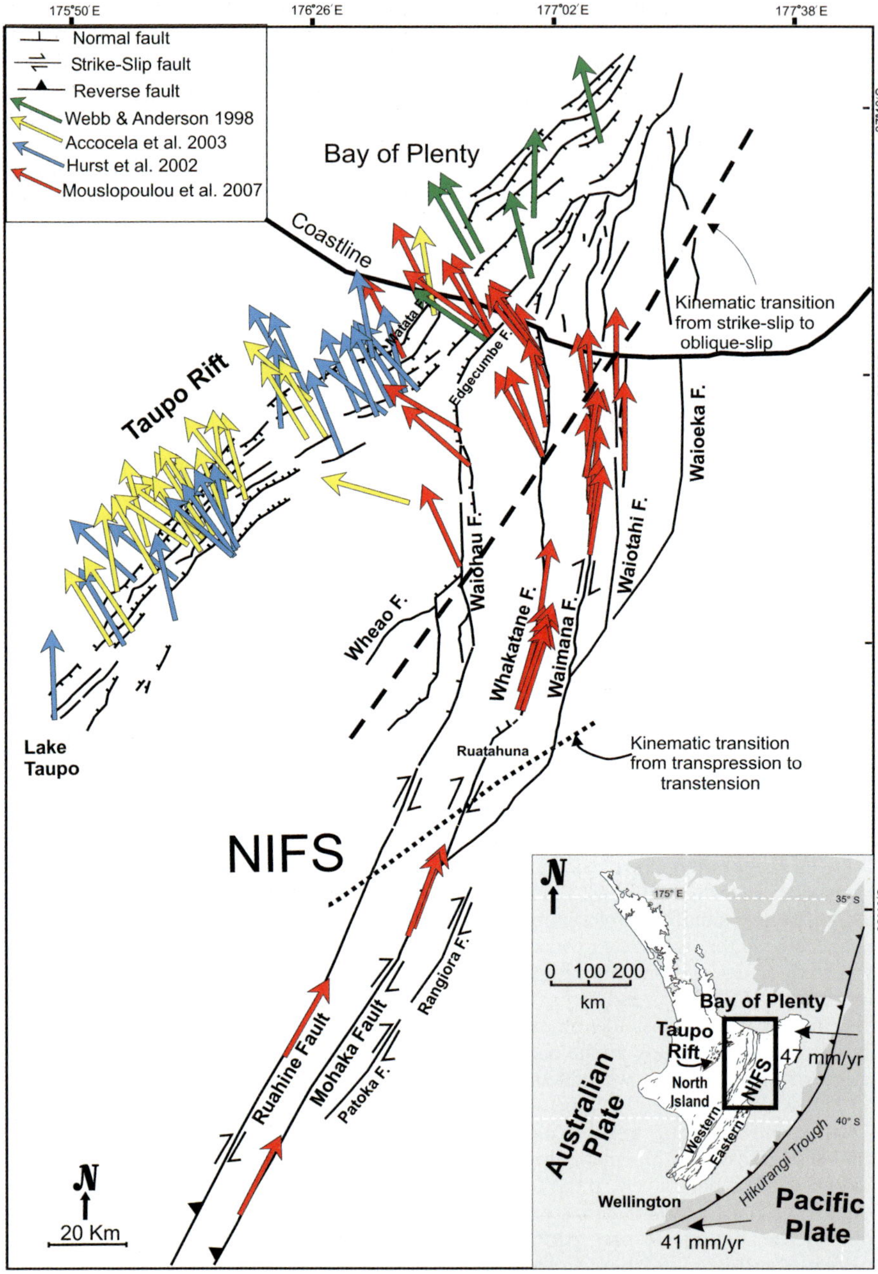

Fig. 3. Map of the NIFS–Taupo Rift fault intersection showing the location of the main active faults. The azimuths of the net slip vectors on the faults, which derive from a combination of outcrop geology (Acocella *et al.* 2003; Mouslopoulou *et al.* 2007) and focal mechanism solutions (Webb & Anderson 1998; Hurst *et al.* 2002), and the two kinematic transition zones across the northern NIFS (dashed lines) are indicated. These transition zones are drawn as lines although they represent zones (of uncertain width). The overall kinematic pattern mirrors the way displacement vectors on the faults in the NIFS deflect gradually anticlockwise to accommodate some of the NW–SE orientated extension which is distributed outside the main Taupo Rift. Displacement vectors in the NIFS represent a number of cumulative earthquakes, ranging from one to 17 events, during the last *c.* 30 ka. Inset: The NIFS, on the upper plate of the Hikurangi Margin, accommodates the margin-parallel component of the westward oblique subduction of the Pacific Plate beneath the Australian Plate. Rates of relative plate motion from De Mets *et al.* (1994).

by bifurcation of the fault system into five splays. In this north–south striking section of the fault, dips decrease northwards from almost vertical to *c*. 60°W (Mouslopoulou *et al.* 2007). Within 10–15 km of their intersection with the Taupo Rift, the north–south striking faults of the NIFS swing 20–25° clockwise in strike (Fig. 3).

In conjunction with the observed strike and dip changes, the faults in the NIFS experience significant along-strike changes in their kinematics, as measured from displacements of ≤30 ka old landforms. In the south, where the faults dip steeply (>80°), they function chiefly as strike-slip faults with typical slip vector pitch angles of 0–10° (Fig. 4a). Along this steeply dipping section of the NIFS, the slip azimuth is approximately constant and parallel to fault strike, although the principal downthrown side of the faults changes northwards, from east to west, to produce a gradual transition from minor transpression to minor transtension

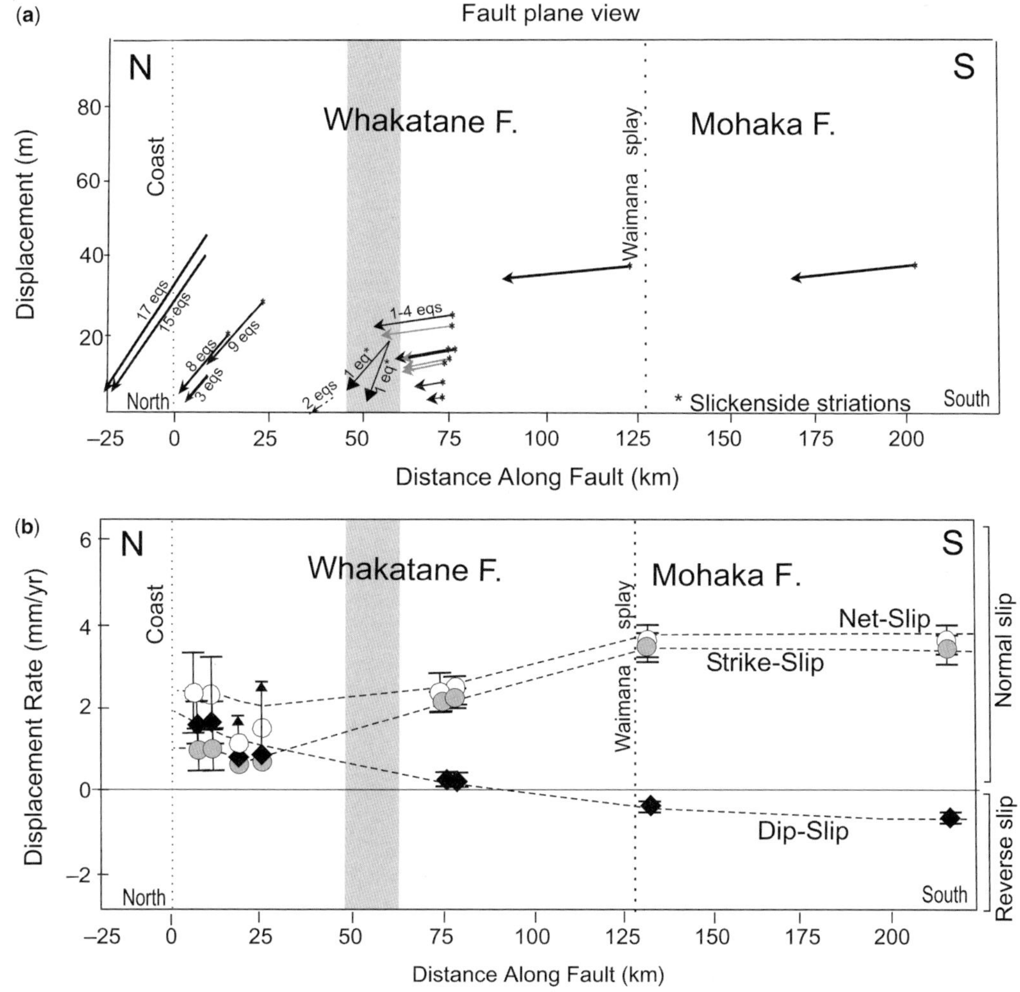

Fig. 4. (**a**) Fault plane view illustrating the pitch of the slip vectors along the length of the Whakatane–Mohaka Fault. Length of arrows is proportional (with the exception of two slickenside striations) to the magnitude and time period over which the slip was measured. Black arrows indicate slip down to the NW while grey arrows indicate slip down to the NE. The numbers of earthquakes adjacent to the arrows derive after combining the total displacement recorded at different locations along the strike of the faults with the average single event displacement measured at, or near to, these locations. (**b**) The three components of slip (i.e. dip-slip, strike-slip and net-slip) of the Whakatane–Mohaka Fault are plotted against distance along the strike of the fault. Data are consistent with a kinematic transition from strike-slip to oblique-slip northwards as the Whakatane Fault approaches the Taupo Rift. In both diagrams, this kinematic transition zone is shaded grey. Data are from Mouslopoulou *et al.* (2007).

(Figs 3 and 4b). Approximately 50–60 km south of the Bay of Plenty coast, the late Quaternary rate of normal-slip in the NIFS begins to increase significantly, while the corresponding rate of strike-slip on the same faults decreases (Fig. 4a and b). These changes in the rates of strike- and dip-slip are accompanied by a gradual northward steepening (up to 60°) of the slip vector (Fig. 4a) and thus an anticlockwise rotation in the trend of the slip azimuth (Fig. 3) (Mouslopoulou *et al.* 2007). The ratio of horizontal to vertical slip decreases northwards from *c.* 10 to <1. At the northern end of the NIFS, slip vector orientations on faults in the NIFS are comparable to those on faults in the Taupo Rift and subparallel to the lines of intersection between the two fault systems (Fig. 3) (Mouslopoulou *et al.* 2007).

Temporal and spatial distribution of surface rupturing earthquakes

To test whether the observed changes in late Quaternary fault kinematics could be achieved by fault rupture segmentation and associated surface rupture arrest across, or within, the kinematic transition zone, we estimate the timing of palaeoearthquakes from 16 trenches excavated across faults in the NIFS (Fig. 5). The trenches are distributed along three of the main strands of the NIFS (Waiohau–Ruahine, Whakatane–Mohaka and Waimana faults) and record large-magnitude surface rupturing earthquakes during the past *c.* 10–13 ka (Fig. 5, Table 1). All trenches but two (i.e. sites 5 and 6 in Fig. 5) are excavated at localities where the fault comprises a single active trace and therefore the entire fault zone is trenched. To complement these data and to examine the possible interrelationship between palaeoearthquakes in the northern NIFS and the Taupo Rift, we draw on published palaeoearthquake information from trenching on the four principal faults of the northern Taupo Rift (Edgecumbe, Rotoitipakau, Braemar and Matata faults) (Fig. 5, Table 1). In this section, we examine the spatial relationships between palaeoearthquakes in the NIFS and the locations of kinematic transition zones on the faults (see previous section).

Individual strike-slip earthquakes are identified in trenches where stratigraphic horizons are truncated or deformed against the fault, where colluvial scarp-derived wedges abut the fault, or where slip surfaces within fault zones terminated upwards against younger less deformed units (Fig. 6a). In the case of oblique-slip palaeoearthquakes, offset stratigraphic units may be correlated across the fault and, in such cases, the timing of successive earthquakes can be constrained by downward increases in fault dip-separation (Fig. 6b). In most trenches, surface-rupturing earthquakes with displacements <0.1 m cannot be reliably resolved. The timing of the earthquakes was determined by a combination of tephrochronology (analysis of glass chemistry) and radiocarbon dating of selected organic-rich layers.

The temporal resolution provided by the available dating techniques does not allow discrimination of earthquakes that occurred closely in time (e.g. <500–1500 years). Thus, when the timing of fault rupture in two neighbouring trenches is approximately the same (e.g. 8–9 ka BP), we cannot differentiate whether the fault ruptured in each trench during one event, or in two events very closely spaced in time (e.g. <10 years) or in two events separated by 500–1000 years. Based on trench data, therefore, the occurrence of mechanical segment boundaries (i.e. locations where earthquake rupture is arrested) cannot be excluded even where surface rupturing events appear to occur at about the same time in both sides of that putative boundary. By contrast, when events of approximately the same age are not observed in contiguous trenches, we can infer that a rupture was arrested between those two sites.

The details of each trench, including source of data, grid reference and timing of the most recent event, are summarized in Table 1. The timing of the palaeoearthquakes revealed by the trenches of Table 1 are plotted in Figure 7, which summarizes the temporal and spatial distribution of surface-rupturing palaeoearthquakes for the northern *c.* 200 km of the NIFS over the last 10–13 ka. The black circles, located between trench sites, indicate the inferred locations of surface rupture terminations, signifying mismatches in the number of palaeoearthquakes recorded in neighbouring trenches for the same time span. The solid black line attached to each black circle shows the part of the fault that is inferred to have ruptured during each event, and points away from the location of inferred arrest (Fig. 7). To avoid biasing in favour of the notion of earthquake surface rupture segmentation at specific localities, we always assume that a nearby rupture propagated through trench sites for which there is no available temporal palaeoearthquake information of the same age (indicated by white boxes). Moreover, where the number of earthquake ruptures in a single site in uncertain (i.e. one or possibly two events), we adopt the maximum number of events, again to avoid biasing the data towards rupture arrest. The number of rupture tips indicated on Figure 7 is therefore the minimum required to account for the available data. For the data underpinning the timing of each earthquake event presented in Figure 7 see Mouslopoulou (2006, Chapter 4).

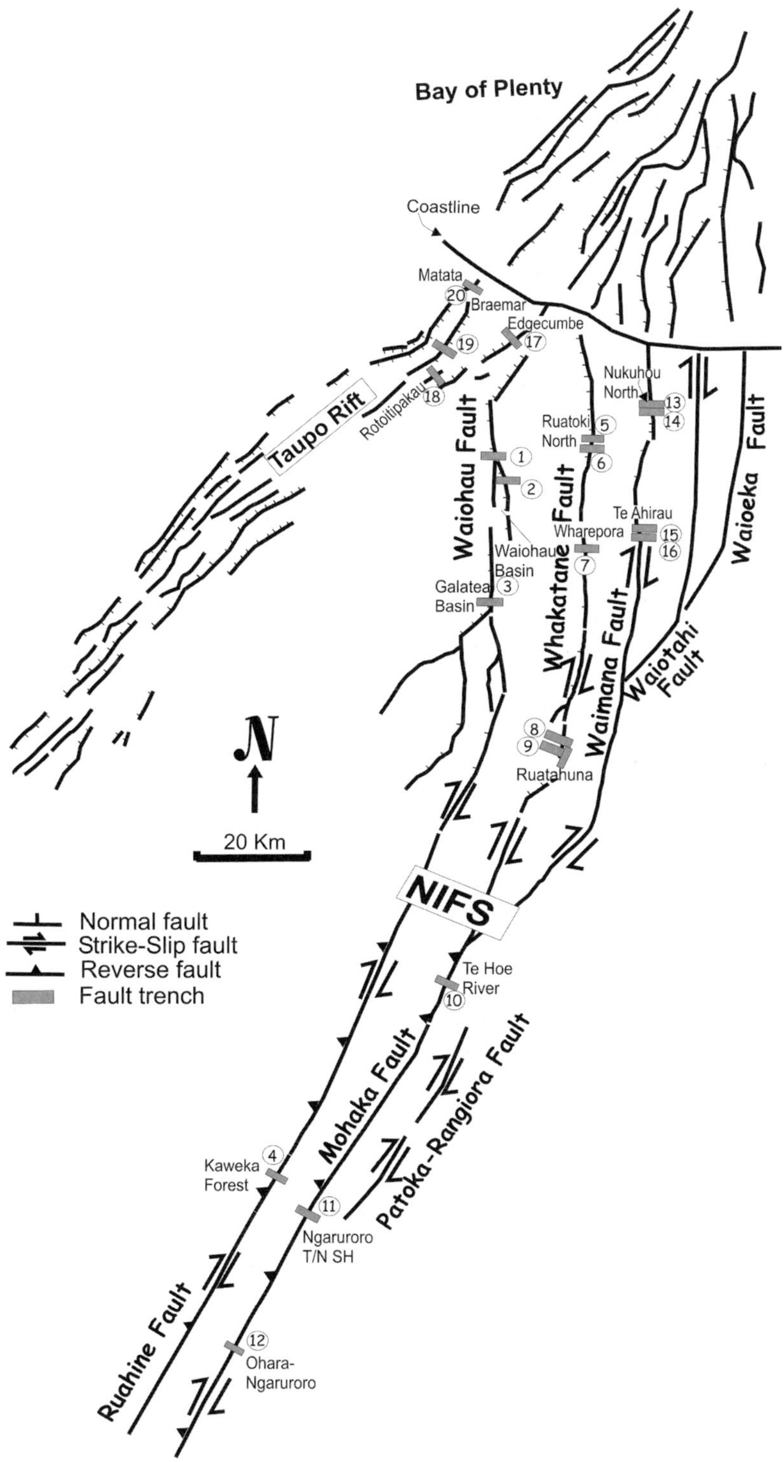

Fig. 5. Fault map showing the location of the 20 fault trenches across the northern NIFS (n = 16) and Taupo Rift (n = 4). The details of each trench are summarized in Table 1.

Table 1. *Summary of palaeoearthquake data from 20 trenches across the northern NIFS and Taupo Rift intersection*

Site no.	Trench	Grid reference	Fault	Location	Last earthquake recorded (ka BP)*	Oldest earthquake recorded (ka BP)*	Net Slip Rate (mm/yr)†	References
1	Tasman	V16/453298	Waiohau	Waiohau Basin	1.8 < eq. < 5.6	<17.6	0.7 ± 0.2	Woodward-Clyde 1998‡
2	Cornes	V16/472252	Waiohau	Waiohau Basin	>8	9.5 < eq.< 13.8	0.7 ± 0.2	Woodward-Clyde 1998
3	Troutebeck	V17/446057	Waiohau	Galatea Basin	1.8 < eq. < 9.5	13.8 < eq. < 17.6	0.7 ± 0.2	Beanland 1989
4	Davis	V20/092077	Ruahine	Kaweka Forest	<1.8	>13.8	–	Hanson 1998‡
5	Ruatoki/Sarah	W16/620330	Whakatane	Ruatoki North	0.62 < eq. < 0.627	–	–	Beanland 1989
6	Te Whetu	W16/619327	Whakatane	Ruatoki North	0.52 ± 0.2 < eq. < 4.7	<50	1.5 (min)	Mouslopoulou 2006
7	Te Marama	W16/603193	Whakatane	Wharepora	<0.794	–	–	Mouslopoulou 2006
8	Armyra	W17/556802	Whakatane	Ruatahuna	<1.8	<3.5	3 ± 1.1	Mouslopoulou 2006
9	Thalassa/Helios	W17/555802	Whakatane	Ruatahuna	0.93 < eq. < 1.8	8 < eq. < 9.5	3 ± 1	Mouslopoulou 2006
10	Te Hoe	V19/379395	Mohaka	Te Hoe River	<0.8	<9.5	3.5 ± 0.8	Hull 1983
11	Syme	V20/137986	Mohaka	Ngaruroro-T/N SH	<1.8	23 < eq. < 3.36	*c.* 6.5	Hanson 1998
12	McCool	U21/010746	Mohaka	Ohara-Ngaruroro	<1.8 or 0.6	>14	*c.* 4–5	Hanson 1998‡
13	Moana	W16/725379	Waimana	Nukuhou North	1.8 < eq. < 5.6	*c.* 13.8	1.2 ± 0.5 (max)	Mouslopoulou 2006
14	Moana-Iti	W16/725379	Waimana	Nukuhou North	1.8 < eq. < 3.36 ± 0.38	>10.1	1.2 ± 0.5 (max)	Mouslopoulou 2006
15	Ahirau1	W16/705161	Waimana	Te Ahirau	1.8 < eq. < 2.88	9.5 < eq. < 13.8	1.1 ± 0.6	Mouslopoulou 2006
16	Ahirau2	W16/705161	Waimana	Te Ahirau	>1.8	<13.8	1.1 ± 0.6	Mouslopoulou 2006
17	Edgecumbe	V15/477502	Edgecumbe	Edgecumbe	AD 1987	–	2.6 ± 1.4	Beanland *et al.* 1989
18	Rotoitipakau	V15/345436	Rotoitipakau	Rotoitipakau	AD 1987	<9.5	2.2 ± 0.25	Berryman *et al.* 1998
19	Braemar	V15/358480	Braemar	Braemar	<5.6	<9.5	0.7 ± 0.3	Beanland 1989
20	Matata	V15/414602	Matata	Matata	200 yr BP	<5.6	1.8 ± 0.2	Ota *et al.* 1988

*Calibrated age.
†Average slip rate of the section trenched.
‡Re-interpreted data.

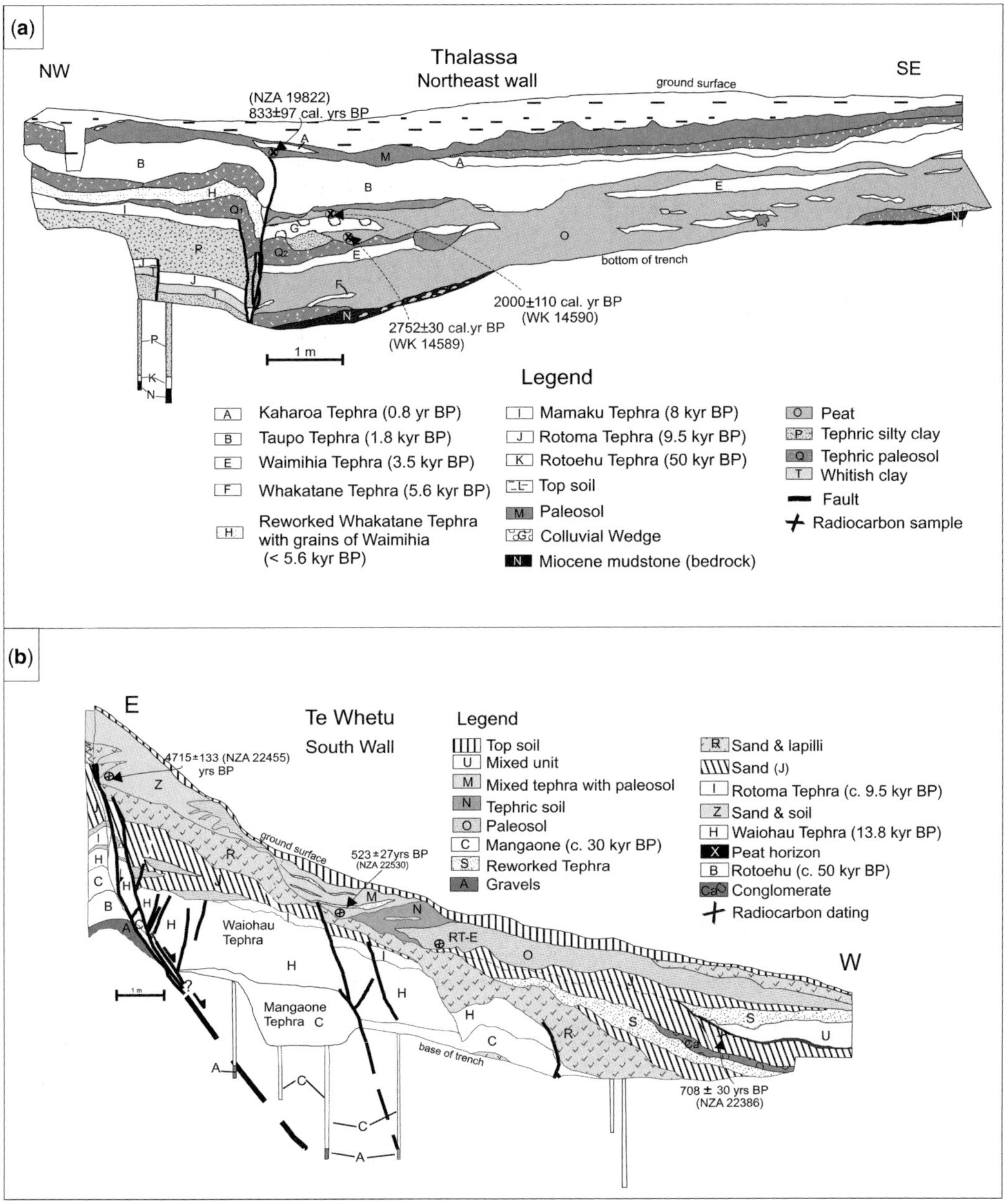

Fig. 6. Two simplified trench logs that derive from sections of the Whakatane Fault that most often ruptured during different earthquakes. The NE wall of Thalassa, at Ruatahuna, is characterized by confined faulting on a vertically dipping fault plane that juxtaposes dissimilar stratigraphic units (**a**). In contrast, displaced horizons can be correlated across the main, low angle (68°), distributed slip surfaces on the south wall of Te Whetu (*c*. 50 km north of Ruatahuna) (**b**). For detailed logs and description of the trench units see Mouslopoulou (2006).

Waiohau–Ruahine Fault

Figure 7 (left-hand panel) summarizes the temporal and spatial distribution of palaeoearthquakes recorded along the northernmost 150 km of the Waiohau–Ruahine Fault during the past *c*. 13 ka. These palaeoearthquakes were revealed by trenches at three locations along the Waiohau–Ruahine Fault (Fig. 5). During the last 13 ka the fault appears to have ruptured most recently and

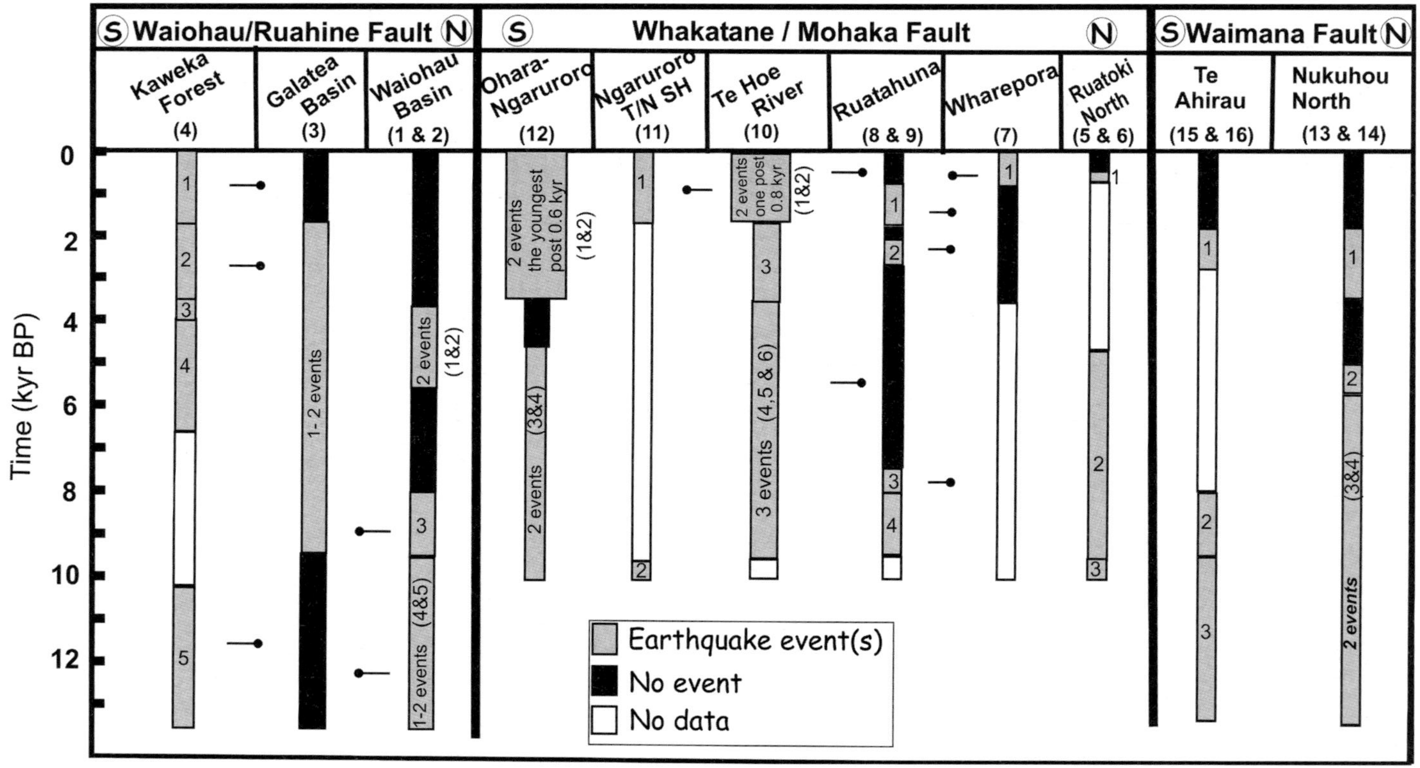

Fig. 7. Time–distance plot summarizing the timing of the palaeoearthquakes revealed by trenches along the northernmost 200 km of the NIFS during the last *c.* 10–13 ka. Faults are presented from west to east, and localities on each fault from south to north. The numbers in parentheses below each site correspond to trenches listed in Table 1 and Figure 5. Numbers within (or adjacent to) each grey box represent consecutive earthquakes. Lower and upper boundaries on each grey box represent maximum and minimum age for the earthquake, respectively. The timing of most earthquakes is constrained by tephrochronology. Black circles, located between trench sites, indicate the inferred locations of surface rupture terminations or indicate mismatches in the number of palaeoearthquakes recorded in neighbouring trenches for the same time span. Locations of repeated earthquake arrest are inferred on the Waiohau and Whakatane faults where neighbouring trenches appear to record different earthquake histories. Note that although the vertical position of the rupture tips (black solid lines) in some cases is non-unique, the number of rupture tips between sites at given time intervals is unique.

more frequently at the Kaweka Forest and Waiohau Basin sites than the Galatea Basin site (Fig. 7). The section of the fault at Kaweka Forest appears to have ruptured at least once within the last 1.8 ka whilst the section immediately to the north, in the Galatea Basin, did not accommodate any surface-rupturing earthquake during this period. The southern Kaweka Forest section has ruptured at least five times during the last *c*. 13 ka while the Galatea Basin section appears to have ruptured one or possibly two times during the same time interval (Fig. 7). A maximum of two events could have ruptured the entire fault between the Kaweka Forest and Waiohau Basin, while a minimum of three palaeoearthquakes ruptured the Kaweka Forest section of the fault but terminated before reaching the Galatea Basin site (trenches 3 and 4 in Figs 5 and 7 and Table 1).

The timing of palaeoearthquakes also differs on the two sections of the Ruahine–Waiohau Fault, further to the north of Kaweka Forest, at Galatea Basin and Waiohau Basin (sites 1–3 in Figs 5 and 7 and Table 1). During the last 13 ka, at least two earthquakes that ruptured the Waiohau Basin section of the fault did not extend as far south as the Galatea Basin (Fig. 7). The section of the fault in the Waiohau Basin appears to be overall more active seismically, with a minimum of four palaeoearthquakes during the last *c*. 13 ka, than the Galatea Basin section, which has accommodated a maximum of two palaeoearthquakes during the same time interval.

The non-uniform spatial and temporal distribution of palaeoearthquakes along the Waiohau–Ruahine Fault is consistent with the presence of two localities where palaeoearthquakes were repeatedly arrested. The southern arrest site occurs along the *c*. 100 km of the fault between the Kaweka Forest and Galatea Basin trench sites (Fig. 5 and Table 1) and is associated with a *c*. 20° change in the dip of the fault, a 25° change in the fault strike immediately south of the Galatea Basin and a northward anticlockwise rotation of the fault-slip azimuth on the faults (Figs 3 and 4). The northern arrest site occurs along the 25 km between the Galatea and Waiohau basins (Fig. 5, Table 1) where the fault trace is almost linear but the dip of the fault decreases northwards from 70° to 60°. The decrease in fault dip is accompanied by an increase in the dip-slip component on the fault (Mouslopoulou *et al*. 2007).

Additional support for earthquake rupture segmentation along the Waiohau–Ruahine Fault is provided by the recurrence intervals as estimated (independently of palaeoearthquake timing from the trenches) by the single event displacement and the slip rate (from offset landforms) at given points along the strike of the fault (Mouslopoulou 2006). The average recurrence intervals that typify the Kaweka Forest, Galatea Basin and Waiohau Basin sections of the fault, over the last 10–13 ka, are 2.3 ± 1.3 ka BP, 5 ± 1.3 ka BP and 3.6 ± 1.2 ka BP, respectively, and suggest different average return periods for each section of the fault (Mouslopoulou 2006).

Whakatane–Mohaka Fault

Six trenches summarize the earthquake events along the northernmost 200 km of the Whakatane–Mohaka Fault (Fig. 7, middle panel). Similar to the Waiohau–Ruahine Fault, the timing of palaeoearthquakes was not uniform along the Whakatane–Mohaka Fault (Fig. 7). In the southern (sites 10–12, Fig. 5) and northern (sites 5–7, Fig. 5) sections of the fault, for example, the most recent earthquake appears to have occurred within the last 800 years whereas at Ruatahuna (sites 8 and 9, Fig. 5), in between the southern and northern sections, the last earthquake is bracketed between *c*. 1.8–0.83 ka BP with no rupture during the last *c*. 770 years (Fig. 7).

In the southern section of the fault, between Ohara and Te Hoe River, the timing of palaeoearthquakes was comparable between sites (Fig. 7). Only one of the six Holocene palaeoearthquakes identified at Te Hoe River could not be correlated (within the temporal resolution of the data) with events at the Ohara-Ngaruroro and Ngaruroro-T/N SH sites further south (sites 11 and 12 in Figs 5 and 7). In addition, along the section of the fault between Te Hoe River and Ruatahuna (Fig. 5), where the fault kinematics change from minor transpression to minor transtension, four of the six palaeoearthquakes recorded in the Te Hoe Trench can be correlated with palaeoearthquakes identified in the trenches at Ruatahuna (Fig. 7). The termination of some Holocene palaeoearthquakes between Te Hoe River and Ruatahuna may have been related to the eastward bifurcation of the Whakatane–Mohaka Fault into the Waimana Fault, to the presence of a *c*. 1.5 km wide releasing bend at Ruatahuna and/or to the change from slight transpression to slight transtension (Figs 3, 4b and 5).

Within the northern section of the fault, between Ruatahuna and Ruatoki North, the timing of palaeoearthquakes is often different. The northern end of the fault, at Ruatoki North and at Wharepora sites (sites 5–7 in Fig. 5 and Table 1), appears to have ruptured at least once within the last 800 years in contrast to the southern end of this section, at Ruatahuna, which has not accommodated a surface-rupturing earthquake during that period (sites 8 and 9, Figs 5 and 7). Furthermore, the ultimate (1.8–0.83 ka BP) and penultimate (2.7–2 ka BP) earthquakes recorded at Ruatahuna do not appear to have ruptured the section of the fault immediately to the north, at Wharepora (site 7 in Figs 5 and 7). At least 60%, and possibly all, of the Holocene earthquakes recorded along the northernmost 70 km of the Whakatane Fault ruptured the Ruatoki North–Wharepora and Ruatahuna sections at different times. We infer, therefore, that most (if not all) fault rupture events terminated between the Ruatahuna and Wharepora sites (sites 7–9 in Fig. 5 and Table 1).

In summary, none of the palaeoearthquakes inferred from the trenches can be shown to have ruptured the entire 200 km sample length of the Mohaka–Whakatane Fault. Within the temporal resolution of the data, two palaeoearthquakes terminated somewhere within the kinematic transition zone from transtension (to the north) to transpression (to the south). By contrast, at least three (and perhaps all) palaeoearthquakes were arrested within the northern kinematic transition zone from strike-slip faulting to oblique-normal faulting. The presence of the latter transition zone is also indicated by the different fault geometries revealed in trenches located across the kinematic transition (Mouslopoulou 2006). The near-vertical strike-slip fault at Ruatahuna (Fig. 6a), for example, becomes a zone of distributed, predominantly normal faulting on a shallow (60–70°) dipping plane in Ruatoki North (Fig. 6b).

Waimana Fault

The Waimana Fault does not show significant variability in the timing of its Holocene palaeoearthquakes at two localities along its northernmost 35 km (sites 13–16 in Figs 5 and 7). The most recent earthquake, for example, is constrained at both localities to have ruptured between *c.* 1.8 ka and 3.3 ka BP. The penultimate earthquake event ruptured through the northern locality, at Nukuhou North, at *c.* 5.6 ka BP and no data can exclude its coeval rupturing through the southern locality, at Te Ahirau (Fig. 7, sites 13–16 in Fig. 5 and Table 1). A third event appears to have ruptured both trench locations at *c.* 7.5 ka BP (Fig. 7). The poorly dated oldest event at the northern locality (site 13 in Fig. 5 and Table 1) predated 7.5 ka BP, while a similarly poorly constrained event at the southern sites occurred prior to 9.5 ka BP.

Most (or all) of the palaeoearthquakes recorded at Nukuhou North therefore appear to have ruptured through the trenches located some 20 km to the south, at Te Ahirau (Figs 5 and 7, Table 1). The lack of evidence for rupture termination along the Waimana Fault is significant because neither the kinematics nor the geometries change considerably from Nukuhou North to Te Ahirau (Fig. 3) (Mouslopoulou *et al.* 2007). We anticipate, however, that the abrupt change in the fault's kinematics that occurs close to the coastline, some 10 km north of the Nukuhou North locality (Mouslopoulou *et al.* 2007), may be accompanied by rupture arrest, similar to that inferred for the Whakatane and Waiohau faults.

Earthquake rupture models

Fault rupture through the kinematic transition zone from strike-slip to oblique normal slip for all palaeoearthquakes during the last 10–13 ka is inconsistent with trench data for the Waiohau–Ruahine and Whakatane–Mohaka faults in northern NIFS. The observed cumulative late Quaternary displacements could, therefore, accrue in the northern NIFS during individual earthquakes (or earthquake cycles) in two ways: (i) mixed strike-slip and oblique-slip events occur, with the former intermittently rupturing, northwards, through the kinematic transition where they have a slip vector distribution that is either uniform or variable along-strike (Fig. 2b); or (ii) oblique-slip events are confined to fault segments north of the kinematic transition and strike-slip events confined to segments south of the zone (Fig. 2c). Given that faults north of the transition zone are associated with a gradual (10°/15 km) northward steepening in the pitch of the late Quaternary slip vectors, oblique-slip vectors in case ii (see Fig. 2c) must also involve a steepening in the pitch of the coseismic slip vectors in a northward direction along-strike.

Slickenside striations, which record the pitch of single-event slip vectors and would most likely permit differentiation between the two basic rupture scenarios (see i and ii above) are rare. Two sets of slickenside striations ($N = 5$) at a single site on the Whakatane Fault north of the transition zone (site 29 in Table 1 of Mouslopoulou *et al.* 2007), indicate oblique-slip motion (slip pitch 50° and 70°). These striations confirm oblique-normal slip on the faults north of the kinematic transition but do not exclude the possibility that large strike-slip events have also ruptured through the kinematic transition zone.

As we could not differentiate by first-order observations how slip during individual earthquakes accrued on the northern NIFS to produce its northward finite increase in late Quaternary dip-slip (Figs 3 and 4a), we use earthquake data to model earthquake cycles in order to explore which earthquake rupture scenario may account for this kinematic transition. In the next sections we compare earthquake rupture models for mixed strike-slip and oblique-slip ruptures (with uniform or variable coseismic slip vector pitches), with cumulative late Quaternary slip vector pitches (Fig. 8). Intraseismic variations in the pitch of the coseismic slip at given points on the faults (e.g. curved slickensides formed during single earthquakes), as discussed by Spudich *et al.* (1998) and Guaterri & Spudich (1998), are not considered in this paper.

Uniform earthquake-slip vector orientations

The uniform slip model requires that during individual earthquakes, slip vectors do not change orientation. For such uniform slip the observed spatial changes in fault kinematics arise due to the summation of differing slip vector orientations during non-coeval events. To test whether the observed

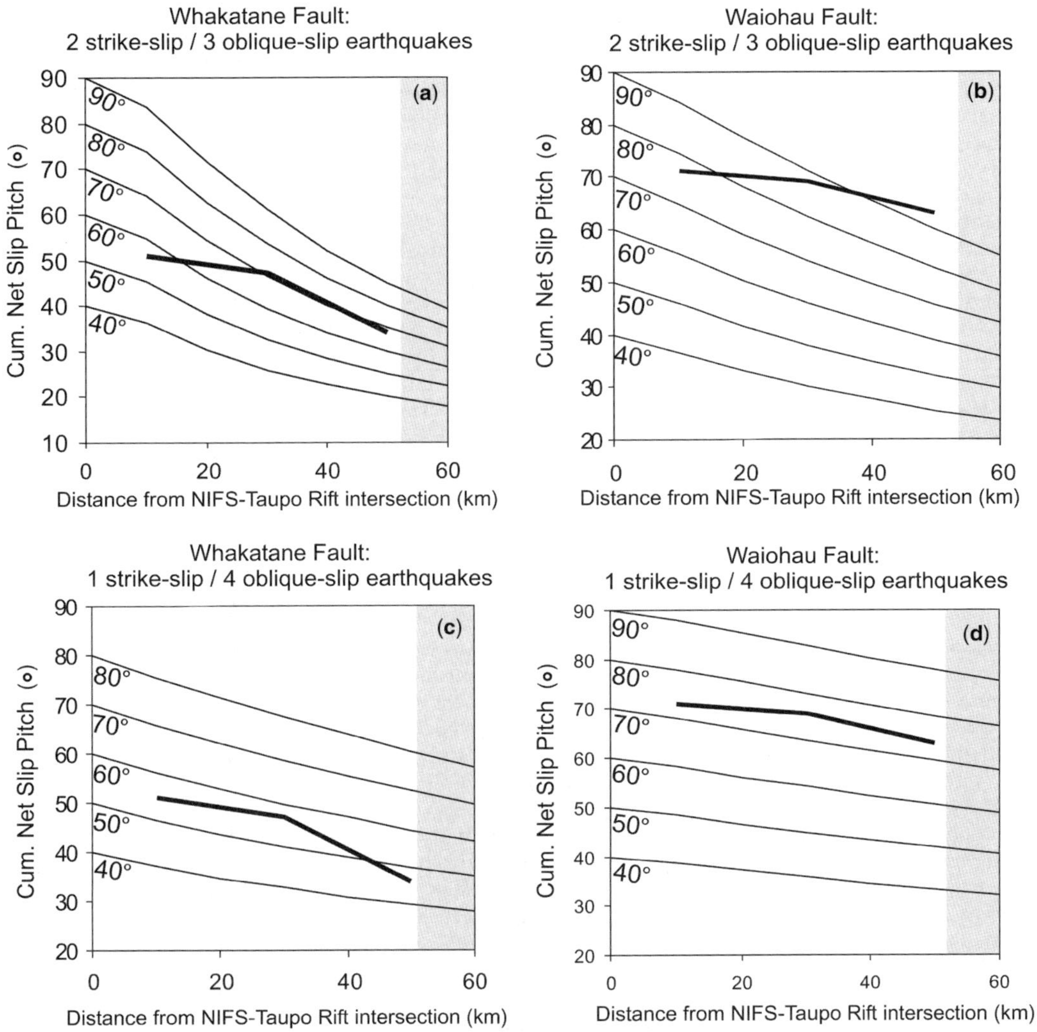

Fig. 8. Cumulative net-slip vector pitches reproduced by simulation of late Quaternary (0–13 ka) palaeoearthquakes that have ruptured the northern sections of the Whakatane and Waiohau faults plotted against distance from the NIFS–Taupo Rift fault intersection. Earthquakes for both the Whakatane and the Waiohau faults comprise five events, each of which comprises one (minimum) or two (maximum) strike-slip events aggregated with four and three oblique-slip earthquakes, respectively. The cumulative pitch for each earthquake cycle of five events ranges between 40 and 90° (thin black lines). The results are compared to observed cumulative net-slip vector pitches (thick black lines) from offset late Quaternary landforms. In all diagrams the kinematic transition zone is indicated by grey shading. See text for further discussion of model construction and input parameters.

changes in fault kinematics can be produced by aggregation of uniform slip vectors which vary between, but not during, individual earthquakes, a simple kinematic model has been constructed. In the absence of data on the distribution of slip vector orientations for individual earthquake ruptures along faults in the NIFS, we use site-specific data to constrain the input parameters (i.e. the incremental slip on earthquake ruptures, horizontal to vertical slip ratios and the number of earthquakes) for the models. The models simulate changes in slip vectors across, and north of, the kinematic transitions on the Waiohau and Whakatane faults during the last 10–13 ka.

In the slip models uniformly orientated strike-slip and oblique-slip events were aggregated to produce a cumulative net-slip pitch which has been compared to (cumulative) net-slip pitch measured from displaced late Quaternary landforms along the faults in the NIFS (Fig. 8 and Table 2) (Mouslopoulou *et al.* 2007). In all models, one or two strike-slip events rupture through the transition zone, while oblique-slip events are confined to the sections of the faults north of the transition. Occasional rupture of strike-slip events through the transition, and arrest of oblique-slip events at this boundary, are consistent with the inferred fault rupture lengths, which are generally larger (i.e. >50 km) south than north (i.e. ≤50 km) of the kinematic transition zone. Therefore, strike-slip events are likely to produce larger earthquake magnitudes which have greater potential to propagate through mechanical barriers (Aki 1979). Each oblique-slip and strike-slip event is taken to rupture the entire fault north of the transition zone, a distance of about 50 km (Fig. 8), and is arrested at the fault's intersection with the Taupo Rift. Rupture arrest at the fault intersection is supported by Figures 7 and 9 which indicate a lack of temporal correlation between the earthquakes in Taupo Rift and those in the northern NIFS. Holocene palaeoearthquakes on four faults (Edgecumbe, Rotoitipakau, Braemar and Matata faults; for references see Table 1) in the northern onshore Taupo Rift, suggest that these faults generally ruptured three or more times during the last 1.8 ka (Fig. 9, Table 1), while the faults in the northern NIFS have ruptured only once (Whakatane Fault) or not at all (Waiohau and Waimana faults) during the same period (Fig. 7). Thus, in most or all surface-rupturing earthquakes the NIFS–Taupo Rift fault junction appears to act as a barrier to rupture propagation.

Fault-trench data show that the Waiohau and Whakatane faults have ruptured five times during the last 13–10 ka with up to two strike-slip palaeoearthquakes that nucleated south of the kinematic transition to propagate through the zone (Fig. 7). Strike-slip events could be modelled using a range of possibilities between, and including, two end-member slip scenarios: strike-slip which decreases uniformly to zero at the NIFS–Taupo Rift fault intersection (Fig. 10a) or constant slip along the entire rupture length (Fig. 10b). Our models use only the former scenario, as this is most likely to produce the observed changes in pitch of the observed cumulative slip vectors. For each fault the single-event strike-slip at the transition zone is constrained by displaced landforms and is 5.5 and 3.5 m on the Whakatane and Waiohau faults, respectively. All strike-slip palaeoearthquakes have a uniform slip vector with a pitch of 0°.

Palaeoearthquakes restricted to the Whakatane and Waiohau faults north of the kinematic transition zone, carry oblique-slip in the models with no change in slip magnitude or orientation during individual events (Fig. 10, Table 2). The constant magnitudes are supported by the displacement (throw) accrued along the Whakatane Fault during the last event at two sites (Ruatoki North and Wharepora) which are located *c.* 20 km apart from one another, and which show a similar value of coseismic throw (1.7–1.9 m) (Mouslopoulou 2006). For each earthquake the pitch of the oblique-slip vector was constant along the length of the fault. The pitch also remains uniform for each earthquake cycle of three or four oblique-slip events.

For each fault, the pitch of the cumulative net slip was calculated for five events, with one or two strike-slip events, and the pitch of oblique-slip vectors set at 40°, 50°, 60°, 70°, 80° or 90° for each seismic cycle (the curves for each pitch are presented on each plot in Fig. 8). In the model, the strike-slip events are aggregated with oblique-slip events and the resulting model-derived cumulative net-slip pitch is directly compared to (cumulative) net-slip pitch measured from displaced late Quaternary landforms along the faults in the NIFS (Table 2) (Mouslopoulou *et al.* 2007). The input parameters in the models for hypothetical earthquake cycles are summarized in Table 2.

Figure 8 shows the simulation of the cumulative net-slip vector pitch produced by aggregating slip for each cycle of strike-slip and oblique-slip events (see Table 2). For example, when the cumulative strike-slip contribution for the Waiohau Fault is input at its maximum level (i.e. two strike-slip events ruptured through the kinematic transition zone during the last 10–13 ka), the pitch of each of the oblique-slip events ranges from *c.* 75–95° (Fig. 8b) in order to match the along-strike pattern of cumulative net-slip vector pitch (Fig. 3). Pure dip-slip events (i.e. ≥75°) are unlikely given the obliquity of the observed (albeit sparse) slickensides (Mouslopoulou *et al.* 2007) and the oblique-slip (pitches of 34–45°) measured from offset

Table 2. *Input data used to simulate earthquake cycles on the Waiohau and Whakatane faults*

Waiohau Fault (Earthquake cycle: 2 strike-slip and 3 oblique-slip events)						Whakatane Fault (Earthquake cycle: 2 strike-slip and 3 oblique-slip events)					
Input				Output	Real data	Input				Output	Real data
Distance (km)	Cumulative strike-slip (m)	Cumulative oblique-slip (m)	Oblique-slip pitch (°)	Cum. net-slip pitch (°)	Long-term net-slip pitch (°)	Distance (km)	Cumulative strike-slip (m)	Cumulative oblique-slip (m)	Oblique-slip pitch (°)	Cum. net-slip pitch (°)	Long-term net-slip pitch (°)
60	7	10	90	55		60	11	9	90	39	
50	5.8	10	90	60	63	50	9.2	9	90	44	34
40	4.6	10	90	65		40	7.4	9	90	51	
30	3.4	10	90	71	69	30	5.6	9	90	58	47
20	2.2	10	90	78		20	3.8	9	90	67	
10	1	10	90	84	71	10	1.6	9	90	80	51
0	0	10	90	90		0	0	9	90	90	
60	7	10	80	48		60	11	9	80	35	
50	5.8	10	80	53	63	50	9.2	9	80	39	34
40	4.6	10	80	57		40	7.4	9	80	45	
30	3.4	10	80	62	69	30	5.6	9	80	51	47
20	2.2	10	80	68		20	3.8	9	80	59	
10	1	10	80	74	71	10	1.6	9	80	70	51
0	0	10	80	80		0	0	9	80	80	
60	7	10	70	42		60	11	9	70	31	
50	5.8	10	70	46	63	50	9.2	9	70	35	34
40	4.6	10	70	50		40	7.4	9	70	39	
30	3.4	10	70	54	69	30	5.6	9	70	44	47
20	2.2	10	70	59		20	3.8	9	70	51	
10	1	10	70	65	71	10	1.6	9	70	61	51
0	0	10	70	70		0	0	9	70	70	
60	7	10	60	36		60	11	9	60	27	
50	5.8	10	60	39	63	50	9.2	9	60	30	34
40	4.6	10	60	42		40	7.4	9	60	33	
30	3.4	10	60	46	69	30	5.6	9	60	38	47
20	2.2	10	60	50		20	3.8	9	60	43	
10	1	10	60	55	71	10	1.6	9	60	52	51
0	0	10	60	60		0	0	9	60	60	
60	7	10	50	30		60	11	9	50	22	
50	5.8	10	50	32	63	50	9.2	9	50	25	34
40	4.6	10	50	35		40	7.4	9	50	28	
30	3.4	10	50	38	69	30	5.6	9	50	31	47
20	2.2	10	50	42		20	3.8	9	50	36	
10	1	10	50	46	71	10	1.6	9	50	43	51
0	0	10	50	50		0	0	9	50	50	
60	7	10	40	24		60	11	9	40	18	
50	5.8	10	40	26	63	50	9.2	9	40	20	34
40	4.6	10	40	28		40	7.4	9	40	22	
30	3.4	10	40	30	69	30	5.6	9	40	25	47
20	2.2	10	40	33		20	3.8	9	40	28	
10	1	10	40	37	71	10	1.6	9	40	34	51
0	0	10	40	40		0	0	9	40	40	

Waiohau Fault (Earthquake cycle: 1 strike-slip and 4 oblique-slip events)						Whakatane Fault (Earthquake cycle: 1 strike-slip and 4 oblique-slip events)					
Input				Output	Real data	Input				Output	Real data
Distance (km)	Cumulative strike-slip (m)	Cumulative oblique-slip (m)	Oblique-slip pitch (°)	Cum. net-slip pitch (°)	Long-term net-slip pitch (°)	Distance (km)	Cumulative strike-slip (m)	Cumulative oblique-slip (m)	Oblique-slip pitch (°)	Cum. net-slip pitch (°)	Long-term net-slip pitch (°)
60	3.5	13.5	90	75		60	5.5	12	80	57	
50	2.9	13.5	90	78	63	50	4.6	12	80	61	34
40	2.3	13.5	90	80		40	3.7	12	80	64	
30	1.7	13.5	90	83	69	30	2.8	12	80	68	47
20	1.1	13.5	90	85		20	1.9	12	80	71	
10	0.5	13.5	90	88	71	10	1	12	80	75	51
0	0	13.5	90	90		0	0	12	80	80	
60	3.5	13.5	80	66		60	5.5	12	70	50	
50	2.9	13.5	80	68	63	50	4.6	12	70	52	34
40	2.3	13.5	80	71		40	3.7	12	70	55	
30	1.7	13.5	80	73	69	30	2.8	12	70	59	47
20	1.1	13.5	80	75		20	1.9	12	70	62	
10	0.5	13.5	80	78	71	10	1	12	70	66	51
0	0	13.5	80	80		0	0	12	70	70	
60	3.5	13.5	70	57		60	5.5	12	60	42	
50	2.9	13.5	70	59	63	50	4.6	12	60	44	34
40	2.3	13.5	70	61		40	3.7	12	60	47	
30	1.7	13.5	70	64	69	30	2.8	12	60	50	47
20	1.1	13.5	70	66		20	1.9	12	60	53	
10	0.5	13.5	70	68	71	10	1	12	60	56	51
0	0	13.5	70	70		0	0	12	60	60	
60	3.5	13.5	60	49		60	5.5	12	50	35	
50	2.9	13.5	60	50	63	50	4.6	12	50	37	34
40	2.3	13.5	60	52		40	3.7	12	50	39	
30	1.7	13.5	60	54	69	30	2.8	12	50	41	47
20	1.1	13.5	60	56		20	1.9	12	50	44	
10	0.5	13.5	60	58	71	10	1	12	50	47	51
0	0	13.5	60	60		0	0	12	50	50	
60	3.5	13.5	50	40		60	5.5	12	40	28	
50	2.9	13.5	50	42	63	50	4.6	12	40	29	34
40	2.3	13.5	50	43		40	3.7	12	40	31	
30	1.7	13.5	50	45	69	30	2.8	12	40	33	47
20	1.1	13.5	50	47		20	1.9	12	40	35	
10	0.5	13.5	50	48	71	10	1	12	40	37	51
0	0	13.5	50	50		0	0	12	40	40	
60	3.5	13.5	40	32							
50	2.9	13.5	40	33	63						
40	2.3	13.5	40	34							
30	1.7	13.5	40	36	69						
20	1.1	13.5	40	37							
10	0.5	13.5	40	39	71						
0	0	13.5	40	40							

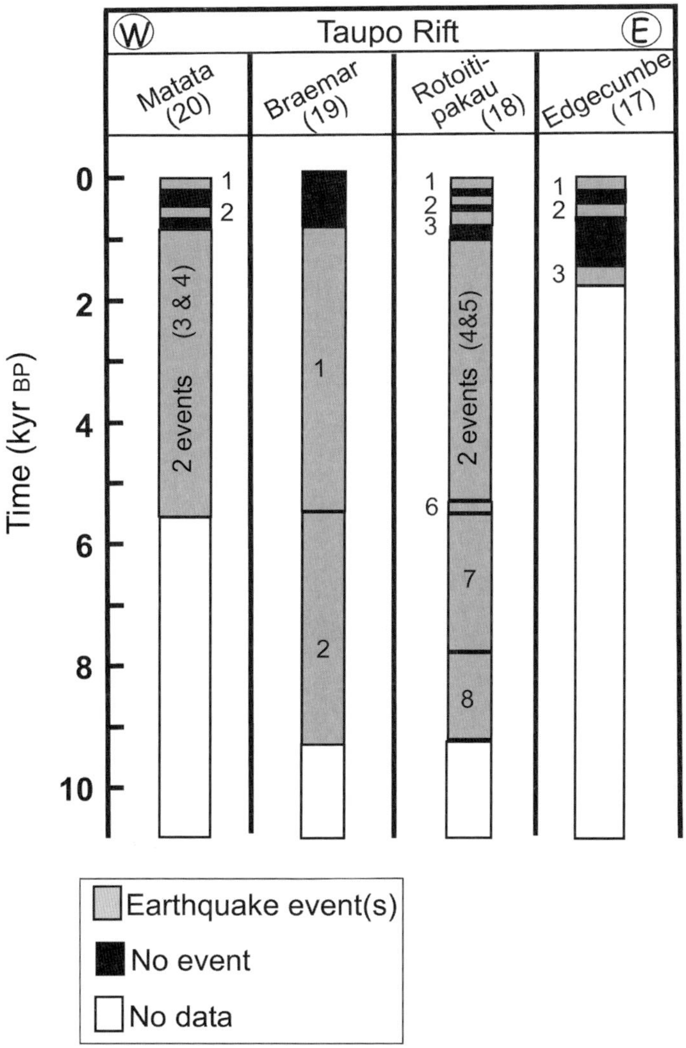

Fig. 9. Time–distance plot summarising the spatial and temporal distribution of palaeoearthquakes in the northern Taupo Rift. Note that most of the faults have ruptured several times within the last 2 ka. The numbers in parentheses below each site correspond to trenches listed in Table 1 and Figure 5. Numbers within (or adjacent to) each grey box represent consecutive earthquakes during given time period. Lower and upper boundaries on each grey box represent maximum and minimum age for the earthquake, respectively. The timing of most earthquakes is constrained by tephrochronology.

landforms (sites 19 and 28; Mouslopoulou *et al.* 2007, Table 1) which may have accrued in two or three events. For the remaining cases (Fig. 8a, c and d), the measured cumulative net-slip vector pitches can be reproduced for an oblique-slip pitch range of 45–75°. Although pitches of this range are consistent with slickenside data, the average late Quaternary northward gradient in net-slip vector pitch of *c.* 10°/15 km along the strike of the Whakatane and Waiohau faults (thick solid line on each plot of Fig. 8), cannot be accounted for with uniform slip orientations during individual earthquakes. The modelling suggests either a non-uniform oblique-slip magnitude (which is not supported by the measured data at Ruatoki North and Wharepora sites; Mouslopoulou 2006) or non-uniform slip vector orientations during each event (i.e. coseismic variation of the slip vector pitch).

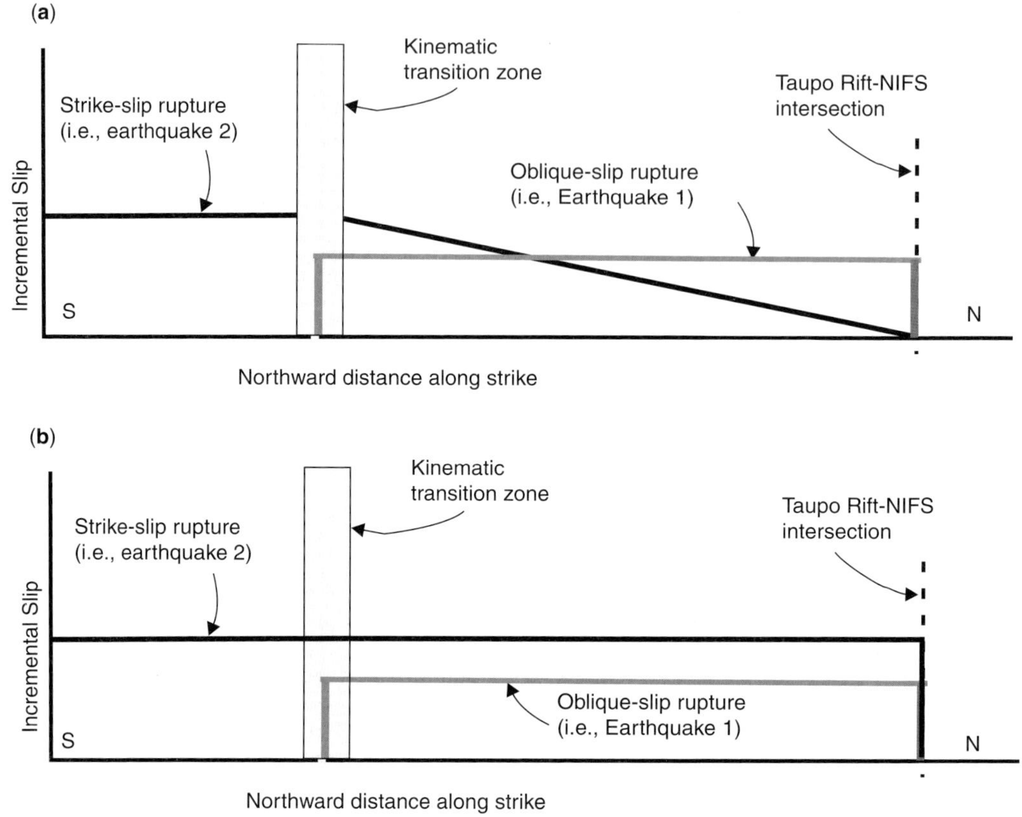

Fig. 10. Schematic diagram showing end-member coseismic slip models for the northern NIFS with respect to the inferred zone of earthquake rupture arrest for individual oblique-normal slip and strike-slip events. (**a**) Strike-slip is taken to decrease at a constant rate from a maximum at the transition to zero at the NIFS–Taupo Rift intersection. (**b**) Strike-slip remains constant along the entire fault rupture length. Oblique-slip events, which are restricted to the faults north of the kinematic transition zone, have uniform slip magnitudes in both cases (a and b).

Variable earthquake-slip vector orientations

An alternative means of achieving the observed late Quaternary along-strike gradient in finite slip vector orientations, is a case for which the pitch of the slip vector changes along-strike during individual earthquake ruptures. Large-magnitude historic earthquakes indicate that slip vector orientations can change significantly during individual events over a fault surface that does not change strike (Olson & Apsel 1982; Beroza 1991; Yoshida *et al.* 1996; Johnson & Segall 2004).

In Figure 2b (right diagram), where earthquake events are mostly, but not always, arrested in the kinematic transition zone, the section of the fault to the north of this zone experiences some strike-slip earthquakes with variable slip vector pitches in addition to the oblique-slip earthquake ruptures. In Figure 2c (right diagram) earthquake events are shown to be always arrested within the kinematic transition zone, with coseismic slip vectors changing in pitch during individual oblique-slip earthquakes to the north of the transition zone. Either scenario is consistent with the observed late Quaternary data if individual earthquakes carry a gradient in the pitch of the coseismic slip vectors of about $30^{\circ}/10$ km (Fig. 2b) and $10^{\circ}/15$ km (Fig. 2c). These values are less than the $70^{\circ}/13$ km observed during the 1989 Loma Prieta earthquake (Beroza 1991; Guatteri & Cocco 1996).

A property of the slip vector variations depicted in Fig. 2b and c (right diagrams) is that the cumulative and individual earthquake slip are parallel. The notion of parallel finite and incremental slip vectors is consistent with late Quaternary slip vectors, which formed in *c.* 2–17 palaeoearthquakes and yet are mainly parallel (Fig. 4a).

Spatial changes in coseismic slip vector rake angles of $>50^{\circ}$ are thought to result from a combination of geometric and static effects (Oglesby &

Day 2001). Geometric effects are usually associated with changes of the dip of the fault (Árnadóttir & Segall 1994), while the static effects are attributed to non-uniform pre-stress conditions proximal to the section of the fault that undergoes the spatial rake rotations (e.g. due to fault interactions) (Guatteri & Cocco 1996). We suggest that variable coseismic slip vector pitches could arise due to the 20–30° fault-dip changes observed across the kinematic transition. These variations in coseismic slip vectors are consistent with the along-strike gradient in the late Quaternary slip vectors in this part of the fault system (Fig. 3).

Earthquake magnitudes and hazard

Earthquake magnitudes for prehistoric events can be calculated from the seismic moment (Kanamori 1977; Hanks & Kanamori 1979) and estimated using empirical data sets from the dimensions of the rupture surface or the amount of slip per event (Wells & Coppersmith 1994; Stirling *et al.* 2002). The precise form of the relation between rupture length and magnitude varies between techniques and empirical data sets. Moment magnitude (M_w) of Hanks & Kanamori (1979) is $M_w = 2/3 \log M_o - 10.7$ (where M_o is the seismic moment, and $M_o = DA\mu$, D (cm) is the average displacement per event, A (cm^2) the rupture area, i.e. the product of the fault rupture length at the ground surface and the thickness of the seismogenic crust, and μ is the shear modulus of 3×10^{11} dyne/cm^2 (or 30 000 N/m^2), while the earthquake magnitude (M) of Wells & Coppersmith (1994) is $M = 3.98 + 1.02 \log A$ (A is again fault surface rupture area). Each technique produces comparable, although not identical, results (Table 3), with increases in fault rupture length raising the earthquake magnitude.

Estimating surface rupture length from the palaeoseismic record can be problematic in the absence of trench and displacement information at multiple sites along an active trace. Straight faults without steps or bends would, in the absence of data to the contrary, typically be assumed to rupture in their entirety. This approach produces maximum earthquake magnitudes and recurrence intervals. Trench data on faults in the NIFS suggest that, although the total lengths of the Waiohau–Ruahine and Whakatane–Mohaka faults exceed 150 and 200 km respectively, they do not rupture in their entirety. Rupture lengths for individual sections on the Waiohau–Ruahine and Whakatane–Mohaka faults are in the range 30–85 km and 30–70 km, respectively (Table 3, Fig. 11). These fault rupture lengths produce an average moment magnitude (M_w) and magnitude (M) of about 7 for events in the northern NIFS that do not rupture through the kinematic transition zone. These earthquake magnitudes are notably different to the M 7.4–7.6 of Stirling *et al.* (2002), whose data underpin the New Zealand National Seismic Hazard Model.

With the exception of the single event displacement on the Waiohau Fault in the Galatea Basin the single-event displacements in Table 3 are broadly consistent with the fault rupture lengths inferred

Table 3. *Earthquake magnitude estimates based on regression equations*

Fault	Section description	Rupture length (km)	Single event displacement (m)	Moment magnitude (Mw)*	Magnitude (M)	Average magnitude
Fault sections						
Waiohau	WR_N	*c.* 30	2.5 ± 0.5	7.0	6.6	6.8
Waiohau	WR_G	*c.* 30	3.4 ± 0.2	7.0	6.7	6.8
Ruahine	WR_S	85 (min)	3.5 ± 1.5	7.4	7.1	7.3
Whakatane	WM_N	*c.* 50	2.8 ± 0.2	7.2	6.9	7.1
Whakatane–Mohaka	WM_R	70	4.7 ± 0.7	7.4	7.1	7.2
Total fault length						
Waiohau–Ruahine	$WR_N + WR_G + WR_S$	*c.* 150	3.5 ± 1.5	7.5	7.4	7.4
Whakatane–Mohaka	$WM_N + WM_R + WM_T + WM_{Ng}$	200	5 ± 1	7.7	7.5	7.6

Magnitude estimates based on $M_w = 2/3 \log M_o - 10.7$ (Hanks & Kanamori 1979)* and $M = 3.98 + 1.02 \log A$ (Wells & Coppersmith 1994) regression equations for possible earthquake rupture lengths on the Waiohau–Ruahine, Whakatane–Mohaka and Waimana faults in northern NIFS. Single event displacements are from Mouslopoulou (2006). Rupture lengths represent a range of potential values and do not include all possible fault sections. For the full names of the fault sections refer to the caption of Figure 11.

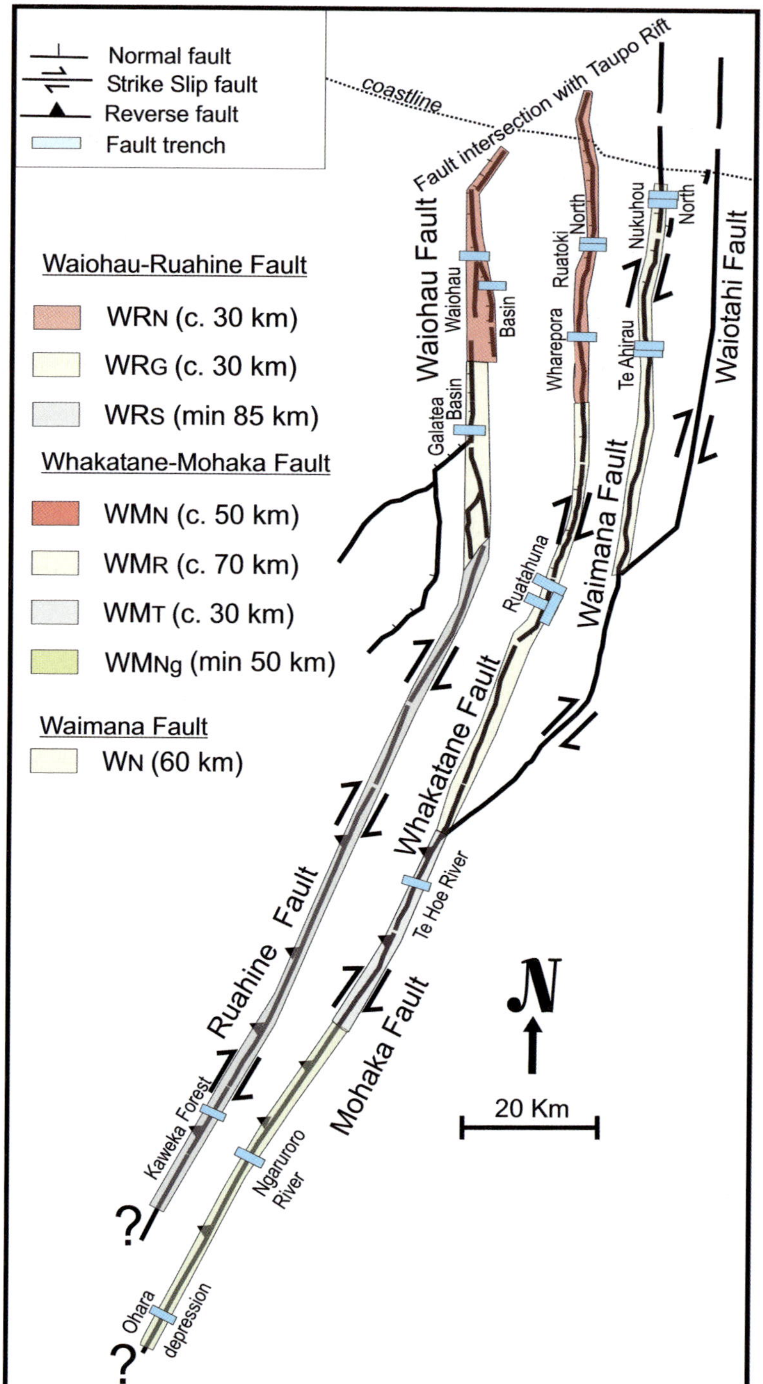

Fig. 11. Rupture lengths (km) assigned for the northern NIFS. Note that these rupture lengths represent a range of potential values and do not include all possible fault sections. Abbreviations: WR_N, northern section of the Waiohau–Ruahine Fault; WR_G, Galatea Basin section of the Waiohau–Ruahine Fault; WR_S, southern section of the Waiohau–Ruahine Fault; WM_N, northern section of the Whakatane–Mohaka Fault; WM_R, Ruatahuna section of the Whakatane–Mohaka Fault; WM_T, Te Hoe section of the Whakatane–Mohaka Fault; WM_{Ng}, Ngaruroro section of the Whakatane–Mohaka Fault; W_N, northern (onshore) section of the Waimana Fault.

from trenching for the displacement length relations of Wells & Coppersmith (1994). In addition, recurrence intervals calculated from the single event displacements and slip rates are, within the uncertainties, typically comparable to those determined from the trenches (Mouslopoulou 2006). If we adopt the timing of earthquakes from the trenches, then the average *c.* 0.5 decrease in M and M_w due to local rupture arrest on the Waiohau–Ruahine and Whakatane–Mohaka faults should locally decrease the seismic hazard in the Bay of Plenty region. The estimated decrease in magnitude is significant and reinforces the view that palaeoearthquake histories on long faults (e.g. >50 km) should be determined at multiple, widely separated (e.g. 10–20 km), locations.

Discussion and conclusions

The northern NIFS is characterized by northward changes in slip vector pitch and azimuth by up to 60° (Figs 3 and 4). A northward kinematic transition from strike-slip to oblique-normal faulting occurs along faults that maintain a nearly uniform strike. These variations in slip vector orientations arise due to regional changes in the kinematics of the North Island, from margin-parallel-dominated

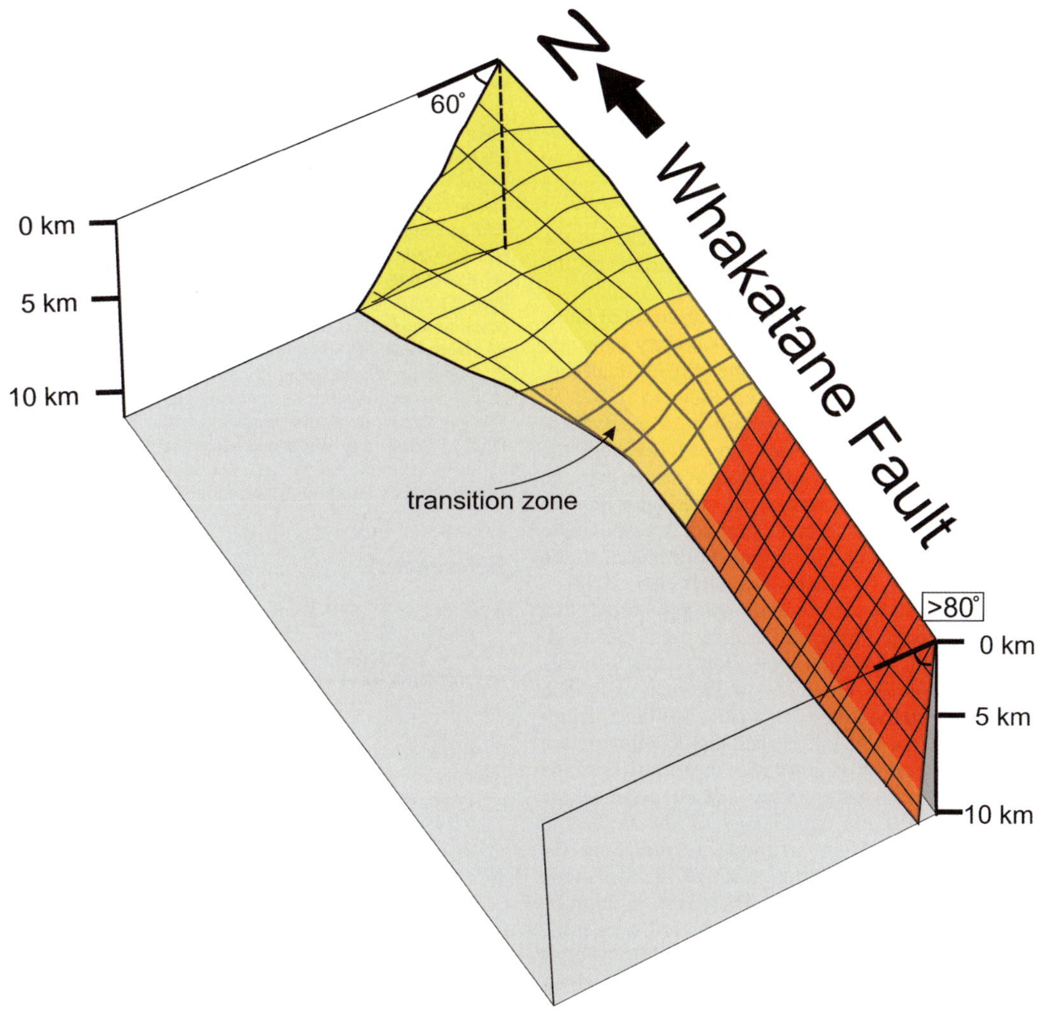

Fig. 12. Schematic diagram illustrating the gradual change in the fault dip on the Whakatane–Mohaka Fault, from near vertical in the south to *c.* 60° W in the north. The change in the fault dip may produce geometric changes during dynamic rupture that contribute to rupture arrest.

strike-slip in the south to oblique extension in the north within, and adjacent to, the Taupo Rift (Wallace *et al.* 2004). The observed late Quaternary changes in fault kinematics along-strike are accomplished by superimposition of earthquakes, involving a combination of rupture segmentation and rotation of the coseismic slip vector pitch.

Differences in the timing of palaeoearthquakes either side of the kinematic transition zone, from strike-slip to oblique-normal slip, indicate fault rupture arrest in at least 60–80% of events during the last 10–13 ka BP. Palaeoearthquake rupture arrest may contribute to the observed late Quaternary pattern of slip vectors by permitting faults north and south of the kinematic transition zone to rupture in separate events with different slip vector orientations (i.e. strike-slip in the south and oblique-slip in the north).

Earthquake rupture arrest in the northern NIFS results in rupture lengths of *c.* 30 to 85 km. For these rupture lengths average earthquake magnitudes are about 7. This estimate of earthquake magnitude is lower than the M 7.4–7.6 inferred for the same faults in the national seismic hazard model (Stirling *et al.* 2002). A decrease in the average magnitudes of surface-rupturing earthquakes in the NIFS may decrease the seismic hazard in the eastern Bay of Plenty.

There is no apparent change in the strike of faults in the NIFS across the segment boundary zone and we suggest that changes in the dip of the fault, from approximately 90° in the south to *c.* 60° W in the north (Fig. 12), contribute to a mechanical arrest of dynamic rupture propagation on the Waiohau and Whakatane faults. We interpret the gradual northward decrease in the fault-dips to be <1 Ma old and to have formed in response to the increase in NW–SE extension proximal to the Taupo Rift (Fig. 12) (Mouslopoulou *et al.* 2007).

Our data do not preclude the possibility that occasional large strike-slip events rupture through the kinematic transition zone from south to north. However, simple kinematic modelling, in which slip during individual earthquakes has been aggregated, indicates that if strike-slip events rupture through the transition zone with uniformly pitching coseismic slip vectors, they cannot produce the observed *c.* 10°/15 km change of the finite slip vector pitch (and slip azimuth) recorded by the 30 ka old offset landforms north of the kinematic transition zone (see above). Therefore, in addition to rupture arrest, the gradual northward steepening of the pitch on the late Quaternary slip vectors also requires co-seismic slip vector rotation across, and north of, the main transition zone.

Historical earthquakes provide support for variations in slip vector orientation during individual events. During the 1989 Loma Prieta earthquake, for example, rupture initiated as right-lateral strike-slip and the slip vector rake rotated up to 70° to become reverse dip-slip over a fault strike distance of 13 km (Beroza 1991; Guatteri & Cocco 1996). Similarly, during the 1999 Chi-chi earthquake, reverse dip-slip rotated more than 50° towards the horizontal to become oblique left-lateral at the rupture termination (Oglesby & Day 2001).

Non-uniform distribution of strength over a fault plane can cause a complex rupture process, including rotations of the coseismic slip vector (Mikumo & Miyatake 1978; Aki 1979). At the northern end of the NIFS the presence of the seismically active, rapidly extending (*c.* 10–15 mm/a) Taupo Rift may locally impact on static stresses. Changes in coseismic slip vectors in the NIFS may arise due to the northward decrease in fault dip and associated steepening of the principal compressive stress axis (σ_1) approaching the active rift.

This work is the result of a PhD study undertaken at Victoria University of Wellington, New Zealand, funded by the Earthquake Commission of New Zealand, the Foundation of Research, Science & Technology of New Zealand and GNS-Science. Shmulik Marco and Gerald Roberts are thanked for their constructive reviews. We are grateful to I. Nairn and J. Patterson for their invaluable help in tephra identification, to R. Robinson for helpful discussions and to D. Beetham, H. Seebeck, R. Langridge, K. Berryman, P. Villamor, D. Heron, B. Lukovic, P. Tamiana, G. Hughes, S. Toulmin, B. Savage and H. Brown for assisting in trenching and its interpretation. Special thanks go to the tangata whenua (people of the land) of Rahiri and Whakarae Marae in Waimana Valley, as well as Noti Teepa and the late Te Kiato Sonny Biddle for they taught us how poor people can be so 'rich'.

References

ACOCELLA, V., SPINKS, K., COLE, J. & NICOL, A. 2003. Oblique back arc rifting of Taupo Volcanic Zone, New Zealand. *Tectonics*, **22**(4), 1045. DOI: 10.1029/2002TC001447.

AKI, K. 1979. Characterization of barriers on an earthquake fault. *Journal of Geophysical Research*, **84**, 6140–6148.

ÁRNADÓTTIR, T. & SEGALL, P. 1994. The 1989 Loma Prieta earthquake imaged from inversion of geodetic data. *Journal of Geophysical Research*, **99**(B11), 21,835–21,855.

AYDIN, A. & DU, Y. 1995. Surface rupture at a fault bend: the 28 June 1992 Landers, California earthquake. *Bulletin of the Seismological Society of America*, **85**, 111–128.

BALJINNYAN, I. ET AL. 1993. Ruptures of major earthquakes and active deformation in Mongolia and its surroundings. *Memoirs of Geological Society of America*, **181**, 62.

BALLANCE, P. F. 1975. Evolution of the India-Pacific plate boundary in North Island, New Zealand. *Bulletin of the Australian Society of Exploration Geophysicists*, **6**(2/3), 58–59.

BARKA, A. A. & KADINSKY-CADE, K. 1988. Strike-slip fault geometry in Turkey and its influence on earthquake activity. *Tectonics*, **7**(3), 663–684.

BAYASGALAN, A., JACKSON, J., RITZ, J. F. & CARRETIER, S. 1999. Field examples of strike-slip fault terminations in Mongolia and their tectonic significance. *Tectonics*, **18**, 394–411.

BEANLAND, S. 1989. *Detailed mapping in the Matahina Dam region*. New Zealand Geological Survey Client Report **89/8**.

BEANLAND, S. 1995. *The North Island Dextral Fault Belt, Hikurangi Subduction Margin, New Zealand*. PhD Thesis, Victoria, University of Wellington.

BEANLAND, S., BERRYMAN, K. R. & BLICK, G. H. 1989. Geological investigations of the 1987 Edgecumbe earthquake. *New Zealand Journal of Geology and Geophysics*, **32**, 73–91.

BEAVAN, R. J. & HAINES, J. 2001. Contemporary horizontal velocity and strain rate fields of the Pacific-Australian plate boundary zone through New Zealand. *Journal of Geophysical Research, Solid Earth*, **106**(B1), 741–770.

BEGG, J. G. & JOHNSTON, M. R. 2000. *Geology of the Wellington area: scale 1:250,000*. Institute of Geological & Nuclear Sciences, Lower Hutt Map **10**.

BEROZA, G. C. 1991. Near-source modelling of the Loma Prieta earthquake: evidence for heterogeneous slip and implications for earthquake hazard. *Bulletin of the Seismological Society of America*, **81**(5), 1603–1621.

BERRYMAN, K. R., BEANLAND, S. & WESNOUSKY, S. 1998. Paleoseismicity of the Rotoitipakau Fault Zone, a complex normal fault in the Taupo Volcanic Zone, New Zealand. *New Zealand Journal of Geology and Geophysics*, **41**, 449–465.

DE METS, C. R., GORDON, R. G., ARGUS, D. & STEIN, S. 1994. Effect of recent revisions to the geomagnetic reversal time scale on estimates of current plate motions. *Geophysical Research Letters*, **21**, 2191–2194.

EBERHART-PHILLIPS, D., HAEUSSLER, P. J. ET AL. 2003. The 2002 Denali Fault earthquake, Alaska: A large magnitude, slip-partitioned event. *Science*, **300**, 1113–1118.

FREUND, R. 1974. Kinematics of transform and transcurrent faults. *Tectonophysics*, **21**, 93–134.

GUATTERI, M. & COCCO, M. 1996. On the variation of the slip direction during earthquake rupture: supporting and conflicting evidence from the 1989 Loma Prieta earthquake. *Bulletin of the Seismological Society of America*, **86**(6), 1935–1951.

GUATTERI, M. & SPUDICH, P. 1998. Coseismic temporal changes of slip direction: the effect of absolute stress on dynamic rupture. *Bulletin of the Seismological Society of America*, **88**, 777–789.

HANKS, T. C. & KANAMORI, H. 1979. A moment magnitude scale. *Journal of Geophysical Research*, **84**(B5), 2348–2350.

HANSON, J. A. 1998. *The neotectonics of the Wellington and Ruahine Faults between the Manawatu gorge and Puketitiri, North Island, New Zealand*. PhD Thesis, Massey University, New Zealand.

HREINSDOTTIR, S., FREYMUELLER, J. T., FLETCHER, H. J., LARSEN, C. F. & BURGMANN, R. 2003. Coseismic slip distribution of the 2002 M_W 7.9 Denali fault earthquake, Alaska, determined from GPS measurements. *Geophysical Research Letters*, **30**(13), 1670. DOI: 10.1029/2003GL017447.

HULL, A. G. 1983. *Trenching of the Mohaka Fault near Hautapu River, Hawkes Bay*. New Zealand Geological Survey Report, file **831/26**.

HURST, W. A., BIBBY, H. M. & ROBINSON, R. R. 2002. Earthquake focal mechanism in the Central Volcanic Zone and their relation to faulting and deformation. *New Zealand Journal of Geology and Geophysics*, **45**, 527–536.

IDE, S., TAKEO, M. & YOSHIDA, Y. 1996. Source process of the 1995 Kobe Earthquake: determination of spatio-temporal slip distribution by Bayesian Modeling. *Bulletin of the Seismological Society of America*, **86**(3), 547–566.

JOHNSON, K. M. & SEGALL, P. 2004. Imaging the ramp-decollement geometry of the Chelungpu fault using coseismic GPS displacements from the 1999 Chi-chi, Taiwan Earthquake. *Tectonophysics*, **378**, 123–139.

KANAMORI, H. 1977. Energy-release in great earthquakes. *Journal of Geophysical Research*, **82**(20), 2981–2987.

KNUEPFER, P. L. K. 1989. Implications of the characteristics of end-points of historical surface fault ruptures for the nature of fault segmentation. *In*: SCHWARTZ, D. P. & SIBSON, R. H. (eds) *Fault Segmentation and Controls of Rupture Initiation and Termination*. US Geological Survey, Open File Report, **89–315**, 193–228.

MANNING, D. A. 1995. *Late Pleistocene tephrostratigraphy of the eastern Bay of Plenty, North Island, New Zealand*. PhD Thesis, Victoria University of Wellington, New Zealand.

MIKUMO, T. & MIYATAKE, T. 1978. Dynamic rupture process on a three-dimensional fault with non-uniform frictions and near-field seismic waves. *Geophysical Journal of Royal Astronomical Society*, **54**, 417–438.

MOUSLOPOULOU, V. 2006. *Quaternary geometry, kinematics and palaeoearthquake history at the intersection of the strike-slip North Island Fault System and Taupo Rift, New Zealand*. PhD Thesis. Victoria University of Wellington, New Zealand (available at: http://researcharchive.vuw.ac.nz/handle/10063/502?show=full).

MOUSLOPOULOU, V., NICOL, A., LITTLE, T. A. & WALSH, J. J. 2007. Terminations of large strike-slip faults: an alternative model from New Zealand. *In*: CUNNINGHAM, W. D. & MANN, P. (eds) *Tectonics of Strike-Slip Restraining and Releasing Bends in Continental and Oceanic Settings*. Geological Society, London, Special Publications, **290**, 387–415.

MULLER, J. R. & AYDIN, A. 2004. Rupture progression along discontinuous oblique fault sets: implications for the Karadere rupture segment of the 1999 Izmit earthquake, and future rupture in the Sea of Marmara. *Tectonophysics*, **391**, 283–302.

NICOL, A. & BEAVAN, R. J. 2003. Shortening of an overriding plate and its implications for slip on a subduction thrust, central Hikurangi Margin, New Zealand. *Tectonics*, **22**(6), 1070. DOI: 10.1029/2003TC001521.

NICOL, A. & VAN DISSEN, R. 2002. Up dip partitioning of displacement components on the oblique slip

Clarence fault. *Journal of Structural Geology*, **24**, 1521–1535.

NICOL, A. & WALLACE, L. 2007. Temporal stability of deformation rates: comparison of geological and geodetic observations, Hikurangi subduction margin, New Zealand. *Earth and Planetary Science Letters*, **258**, 397–413.

OGLESBY, D. D. 2005. The dynamics of strike-slip stepovers with linking dip-slip faults. *Bulletin of the Seismological Society of America*, **95**(5), 1604–1622.

OGLESBY, D. D. & DAY, S. M. 2001. Fault geometry and the dynamics of the 1999 Chi-chi (Taiwan) earthquake. *Bulletin of the Seismological Society of America*, **91**(5), 1099–1111.

OLSON, A. H. & APSEL, R. J. 1982. Finite faults and inverse theory with applications to the 1979 Imperial Valley Earthquake. *Bulletin of the Seismological Society of America*, **72**(6), 1969–2001.

OTA, Y., BEANLAND, S., BERRYMAN, K. R. & NAIRN, I. A. 1988. The Matata Fault: active faulting at the north-western margin of the Whakatane graben, eastern Bay of Plenty. *New Zealand Geological Survey Record*, **35**, 6–13.

RAIT, G., CHANIER, F. & WATERS, D. W. 1991. Landward- and seaward-directed thrusting accompanying the onset of subduction beneath New Zealand. *Geology*, **19**, 230–233.

SPUDICH, P., GUATTERI, M., OTSUKI, K. & MINAGAWA, J. 1998. Use of fault striations and dislocation models to infer tectonic shear stress during the 1995 Hyogo-ken Nanbu (Kobe) earthquake. *Bulletin of the Seismological Society of America*, **88**(2), 413–427.

STIRLING, M. W., MCVERRY, G. H. & BERRYMAN, K. R. 2002. A new seismic hazard model for New Zealand. *Bulletin of the Seismological Society of America*, **92**, 1878–1903.

WALCOTT, R. I. 1978. Present tectonics and Late Cenozoic evolution of New Zealand. *Geophysical Journal of the Royal Astronomical Society*, **52**(1), 137–164.

WALCOTT, R. I. 1984. *An introduction to the recent crustal movements of New Zealand.* Royal Society of New Zealand, Wellington, Miscellaneous series, 7.

WALD, D. J. & HEATON, T. H. 1994. Spatial and temporal distribution of slip for the 1992 Landers, California, earthquake. *Bulletin of the Seismological Society of America*, **84**(3), 668–691.

WALLACE, L. M., BEAVAN, J., MCCAFFREY, R. & DARBY, D. 2004. Subduction zone coupling and tectonic block rotations in the North Island, New Zealand. *Journal of Geophysical Research*, **109**(B12), 2406. DOI:10.1029/2004JB003241.

WEBB, T. & ANDERSON, H. 1998. Focal mechanisms of large earthquakes in the North Island of New Zealand: slip partitioning at an oblique active margin. *Geophysical Journal International*, **134**, 40–86.

WELLS, D. L. & COPPERSMITH, K. J. 1994. New empirical relationships among magnitude, rupture length, rupture width, rupture area and surface displacement. *Bulletin of the Seismological Society of America*, **8**, 974–1002.

WOODWARD-CLYDE 1998. *Matahina Dam Strengthening Project.* Geological completion report to Electricity Corporation of New Zealand.

YOSHIDA, S., KOKETSU, K., SHIBAZAKI, B., SAGIYA, T., KATO, T. & YASHIDA, Y. 1996. Joint inversion of near- and far-field waveforms and geodetic data for the rupture process of the 1995 Kobe earthquake. *Journal of Physics of the Earth*, **44**, 437–454.

Pleistocene to Recent rejuvenation of the Hebron Fault, SW Namibia

STEPHEN WHITE[1], HARALD STOLLHOFEN[2]*, IAN G. STANISTREET[3] & VOLKER LORENZ[4]

[1]*Geological Survey of Western Australia, 100 Plain Street, East Perth WA 6004, Australia*

[2]*GeoZentrum Nordbayern, FG Krustendynamik, Universität Erlangen, Schlossgarten, 5, 91054 Erlangen, Germany*

[3]*Department of Earth and Ocean Sciences, University of Liverpool, Brownlow Street, PO Box 147, Liverpool, L69 3BX, UK*

[4]*Institut für Geologie, Universität Würzburg, Pleicherwall 1, D-97070 Würzburg, Germany*

**Corresponding author (e-mail: stollhofen@geol.uni-erlangen.de)*

Abstract: The Hebron Fault in SW Namibia is associated with a <1 m to 9.6 m high scarp displacing Proterozoic basement and Middle to Late Pliocene crystalline conglomerates. The young age of strata exposed in the fault scarp together with evidence for displacement of aeolian dunes, post-dating the Middle Stone Age, suggests that latest fault displacements occurred during the Late Pleistocene to recent. Recorded historical seismic events show that the fault zone is still active. Latest movements of the fault are recorded by: down-to-the-SW offset of calcrete-cemented conglomerate; fluvially modified, asymmetric hanging wall, graben-like structures; at least two left-stepping jogs in the fault trace and structural data from basement rocks in which late-stage crush zones overprint earlier cataclasite. These features provide consistent evidence that the present scarp formed predominantly by normal dip-slip displacement on a NW-striking, steeply SW-dipping master fault with only a minor dextral strike-slip component. Strongly veined cataclastic fault rocks adjacent to the scarp in basement most probably originated at depths of 4–10 km. The conclusion is therefore that recent fault activity has reactivated a pre-existing, much older fault.

Aerial photographic lineaments and similar fault scarps identified NW and SE of the present study area are interpreted as extensions of the same fault structure. Hence the total length of the Hebron Fault is at least 300 km subparallel to the Atlantic margin of southern Africa. Our observations confirm that the Hebron Fault is a neotectonic feature of regional significance that may relate to late Cenozoic and particularly Quaternary neotectonic activity in NE Namibia and NW Botswana.

The Hebron Fault, also referred to by Andreoli *et al.* (1996) as the Kuiseb–Hebron Fault, is a semi-continuous, NW–SE striking structure that can be traced for at least 300 km subparallel to the Atlantic margin of SW Namibia (Fig. 1). The fault lineament is clearly discernible on satellite and aerial photographs and is subparallel to the Great Escarpment, *c.* 15 km to the east, an erosionally modified relict of the Pre-South Atlantic rift shoulder (Stollhofen *et al.* 2000). The latter formed during successive Carboniferous/Permian to Early Cretaceous rifting episodes but has been subsequently modified by post-break-up uplift, most of which occurred between the time of initial break-up (*c.* 130 Ma) and the end of the Eocene (*c.* 34 Ma) (Brown *et al.* 2000; Cockburn *et al.* 2000). The Great Escarpment is a feature that surrounds the entire southern African region and has been related to erosion following substantial post-break-up thermal uplift of southern Africa due to the Plio-Pleistocene superswell (Partridge & Maud 1987; Partridge *et al.* 1995).

The Hebron Fault is expressed in a variety of rocks of different ages, including: Mesoproterozoic granitic and gneissic basement; horizontal to gently dipping Late Proterozoic Nama Group sedimentary strata that unconformably overlie the crystalline basement; Late Miocene to Middle Pliocene Tsondab Sandstone (Pickford & Senut 1999); calcrete-cemented Karpfenkliff Formation conglomerates and capping Kamberg Formation Calcretes of Middle to Late Pliocene age. The fault is marked on the 1:1 000 000 Geological Map of Namibia (Miller & Schalk 1980), and was recognized by Ward (1987) as a young tectonic feature that is still active.

In this paper we document the geomorphological and structural expression of a particularly well exposed 40-km-long section of the Hebron Fault

From: REICHERTER, K., MICHETTI, A. M. & SILVA, P. G. (eds) *Palaeoseismology: Historical and Prehistorical Records of Earthquake Ground Effects for Seismic Hazard Assessment*. The Geological Society, London, Special Publications, **316**, 293–317. DOI: 10.1144/SP316.18 0305-8719/09/$15.00

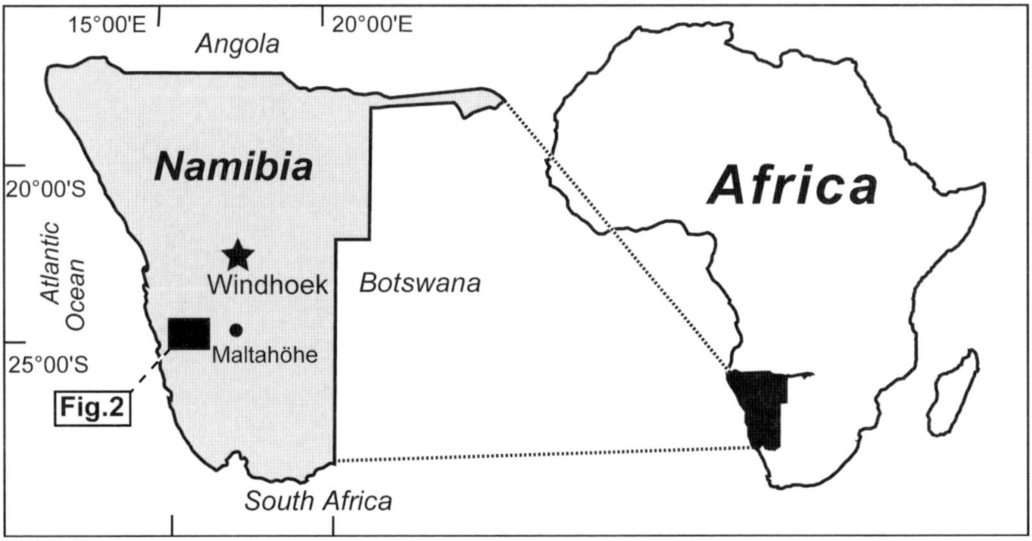

Fig. 1. Map locating the Hebron Fault study area in SW Namibia. Location of satellite image of study area (Fig. 2a) is marked by a black box.

in SW Namibia, west of Maltahöhe and SE of Sesriem (Figs 1, 2). This is the type location of the fault, since the northern end of the fault trace crosses Hebron Farm, from where it is named, on Route C36 about 10 km south of the turnoff to Sesriem and Sossusvlei. Along this section, the Hebron Fault forms a prominent scarp with obvious down-to-the-SW offset (Fig. 3). Andreoli *et al.* (1996) speculated on a throw of up to 65 m based on displacement of Cenozoic deposits of at least this thickness. This may record the cumulative displacement of many earthquakes which were fore-runners of more recent seismic activity registered in historic earthquake events (Jones 1978; Fernandez & Du Plessis 1992; H. Stehmann pers. comm. 1999; Mangongolo & Hutchins 2008).

Viola *et al.* (2005) also identified the Hebron Fault lineament as a significant structural feature of SW Africa, and related its NW-NNW strike and inferred dextral transtensional kinematics to a NNW-directed maximum horizontal compressive stress (S_{Hmax}) during the Cenozoic. In the Orange Basin offshore SW Africa the same stress field has been claimed to be expressed by a 350° striking alignment of mud volcanoes (Ben-Avraham *et al.* 2002) associated with flower structures. Viola *et al.* (2005) proposed that horizontal shearing on both structures is driven by the same NNW-orientated S_{Hmax}, which characterizes the Wegener stress anomaly (WSA) defined by Andreoli *et al.* (1996). The existence of the WSA across a broad zone extending from SW Angola to South Africa is supported by *in situ* stress measurements.

According to Viola *et al.* (2005), the WSA is incompatible with the stress orientation required by plate-scale tectonic constraints, and they speculated that the anomaly could result from ridge-push generated by the South East Indian Ridge. This appears to require special pleading, however, for ridge push from the South East Indian Ridge to be strong enough to create the intraplate WSA while not substantially affecting the NE-directed absolute plate motion of the combined Nubia-Somalia plate, thought to be driven by ridge push from the South Atlantic Ridge. Bird *et al.* (2006), on the other hand, suggested that the WSA might result from within-plate resistance to rotational relative plate motion between the Nubian and (newly forming) Somalian plates. This shows the need for collecting additional structural and stress regime data and detailed studies of major morphotectonic features in Namibia, which occupies a central position within the WSA.

Such geodynamic models provide one source of information for assessing the present-day potential for earthquake activity in SW Africa, but field evidence cited to support the models is weak or inconclusive. For example, the claim by Viola *et al.* (2005) that the supposedly transtensional Hebron Fault is orientated almost parallel to the maximum principal compressive direction (S_{Hmax} or σ_1) is inconsistent with basic fault mechanical principles. This becomes important when considering the present-day earthquake risk in SW Africa, and the significance of recent faulting as an indicator for regional geodynamics. It is particularly important

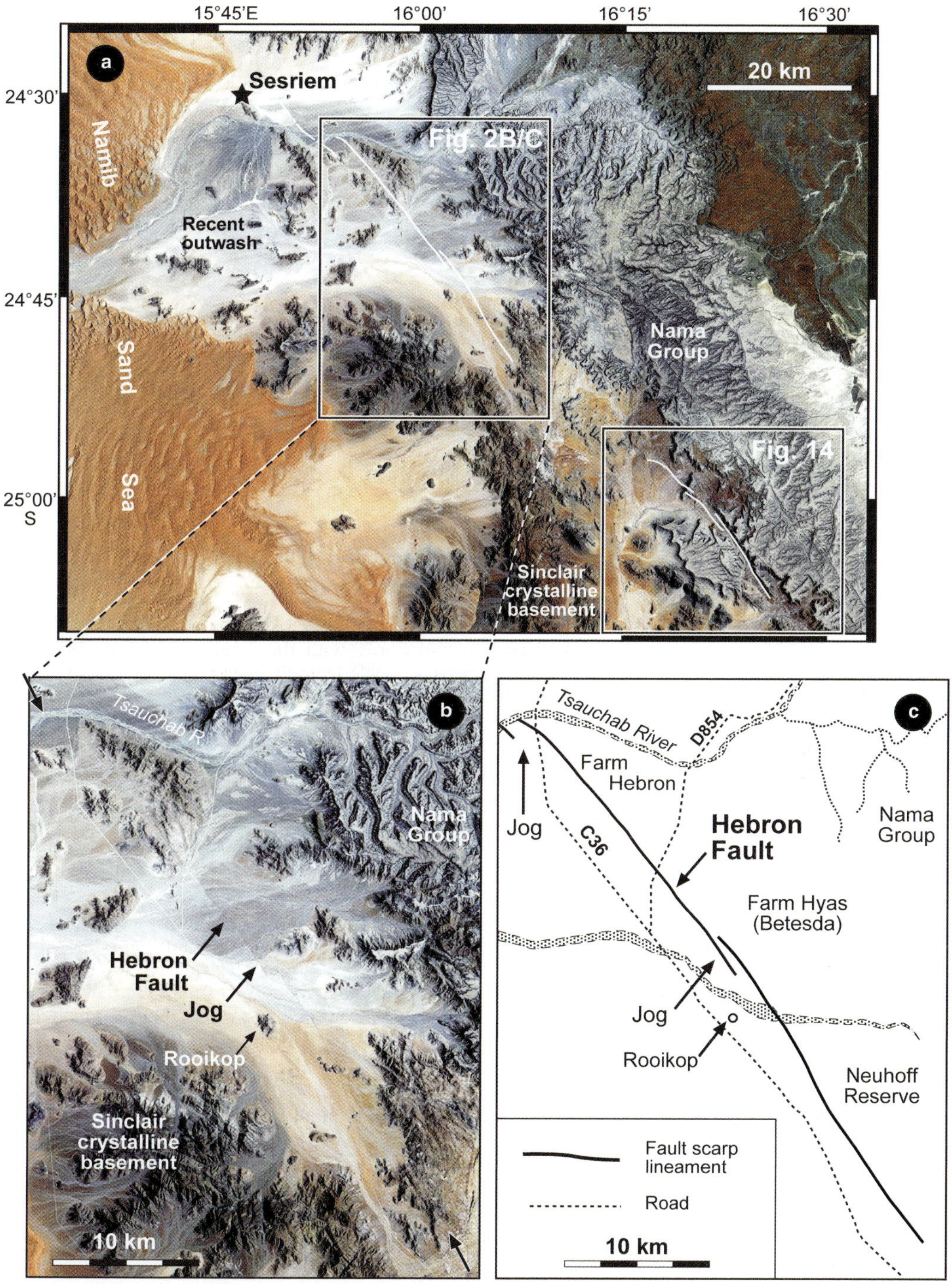

Fig. 2. (**a**) True colour LandSat image of the area SE of Sesriem illustrating the location and trace of the Hebron Fault. Different lithologies and sediment types can be clearly differentiated by changes in colour and texture. (**b**) Enlargement of (a) showing details of the northwestern part of the fault scarp described in this paper. Tips of the fault trace are marked by opposing arrows at top left and bottom right. (**c**) Line drawing of the area in (b), identifying the Hebron Fault segments, roads and major ephemeral rivers.

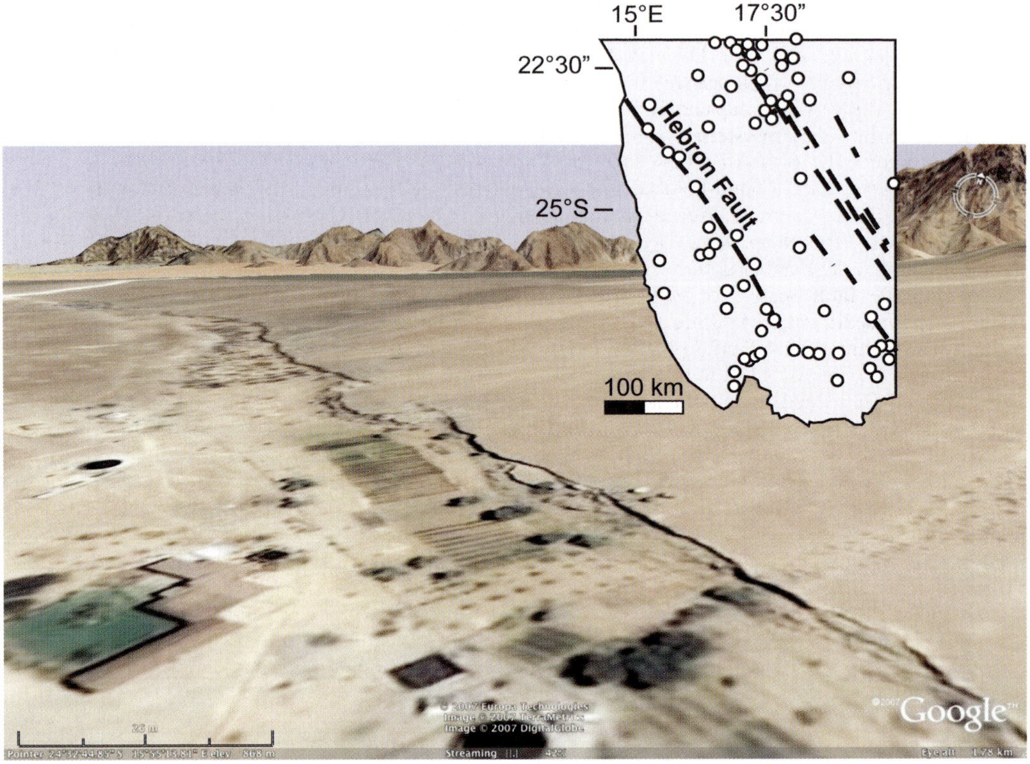

Fig. 3. Oblique model terrain view looking NNW along the northern *c*. 5 km of the Hebron Fault scarp (type section). Apparent viewer elevation is *c*. 970 m above ground, mountains on the northern horizon comprise Sinclair Group crystalline basement, and white strip across the upper left of the image is route C36 (cf. Figs 2c and 4). Dwellings and cultivated land in middle ground are Hebron Farm, and bluish-grey terrain across the far middle ground is the main channel of the westward-flowing ephemeral Tsauchab River. Image from Google Earth^{TM} mapping service kindly provided courtesy of Google.com. Inset shows seismic epicentres in southern Namibia in relation to the post-break-up tectonic framework (compiled from Andreoli *et al.* 1996; Mangongolo & Hutchins 2008).

to show that model predictions are consistent with actual fault kinematic data.

Our aim in the present paper is to constrain as closely as possible, using field and remote sensing data, the character and recent kinematic history of the Hebron Fault. This permits us to make a more rigorous assessment of the stress–strain predictions of various geodynamic models of the region. Together with a survey of known brittle structures, this should improve predictions of seismic activity.

In our evaluation of the Hebron Fault we discuss evidence for: (1) its likely age, displacement and displacement history; (2) its relationship to an older, pre-existing fault and previous episodes of extensional tectonics; (3) the potential for future seismic activity; and (4) its role in the context of neotectonic activity in SW Africa in general.

Although the Hebron Fault is less well known than neotectonic fault lines and earthquakes in the western Cape, e.g. associated with the Worcester Fault at Ceres and Tulbagh, we suggest that the Hebron Fault should be considered one of the important neotectonic structures in SW Africa. Earthquakes related to western Cape faults caused considerable damage in September 1969 (Green & Bloch 1971), and emphasized the need to reassess neotectonic activity in SW Africa, a region that has been commonly viewed as essentially atectonic following South Atlantic opening.

Methods

A combination of field observations, together with aerial photograph and Landsat satellite image analysis was used to map the trace of the fault (Figs 2, 3, 4). Direct observations of the morphology and measurements of structural features associated with the Hebron Fault scarp were made along most of its 40 km strike length from latitude

24°32.04′S, longitude 15°54.12′E (Hebron Farm) to approximate latitude 24°47.40′S, longitude 16°07.85′E (Neuhoff Reserve). Over most of this distance, the trace of the fault scarp is indicated on the 1:50 000 topographic maps 2416CA Donker Gange, 2416CC Hammerstein, and 2415DB Sesriem Canyon. The position of the topographic scarp was checked with a hand-held GPS receiver at ±10 m minimum accuracy. Aerial photographs and Landsat satellite images were used to identify key geomorphological and structural features associated with the fault scarp, for example stream response to synsedimentary faulting and the nature of fault scarp terminations. Aerial photographs and satellite images were also used to trace other faults and suspected fault lineaments, which relate to the Hebron Fault, for up to 50 km north and 100 km south of the present study area (Fig. 2).

Detailed scarp morphologies, measurements of the total scarp heights and maximum scarp slope angles were quantified at ten localities by measuring topographic profiles across the scarp using theodolite surveying equipment. Locations of surveys perpendicular to the trace of the Hebron Fault were selected (Fig. 4) to illustrate both typical scarp morphologies, and to highlight special fault-related features at a number of localities. The minimum length of each survey profile was constrained by the distance from the scarp at which a representative datum could reliably be gauged.

Structural data from slickensided and lineated fault planes and other fractures were recorded from locations where the scarp crosses outcropping basement spurs. Samples of fault rocks exposed in the footwall of basement exposures of the scarp were taken for thin-sectioning and microstructural evaluation of synkinematic petrophysical conditions.

Geomorphic expression of the fault scarp

Figure 3 provides an oblique, 3D model view to the NNW along the northern *c.* 5 km of the fault scarp where it traverses Hebron Farm, showing clearly the linear trace of the fault, the down-to-the-SW sense of displacement of the land surface, and slumping and erosional modification of the scarp. It also shows a preference for natural vegetation and crop cultivation on the SW side of the fault, possibly indicating that the water table is closer to the surface there. This inference is supported by the preponderance of boreholes also located on the SW side of the fault (e.g. Fig. 6d).

Geomorphologically and structurally the 40 km long trace of the Hebron fault is divided by a left-stepping jog into NW–SE trending sections each <20 km long (Figs 2, 4). About 10 km south of the turnoff to Sesriem and Sossusvlei on route C36, a pronounced ramp in the road at 24°32.30′ S, 15°54.80′E near Hebron Farm (Fig. 3) marks the point where the road crosses the northernmost part of the Hebron Fault scarp mapped in this study. The scarp continues west of the road on a strike of 130° ± 10° for at least another 1.5 km, with the height gradually decreasing from *c.* 4 m at the road to <1 m at the Tsauchab River (Figs 2, 4).

East of the road, the scarp trends SE (145–150°) across an almost flat calcrete plain, which rises gradually towards an inselberg of granitic basement rocks. The transition from calcrete-cemented conglomerate scarp to basement scarp occurs at 7.5 ± 0.5 km SE of the road (Fig. 4). The basement scarp, locally with a thin ≤1 m cover of weakly to moderately calcrete-cemented talus, continues SE for a further *c.* 3.5 km before merging again into the calcrete plain south of the inselberg (*c.* 24°37.25′S, 15°59.45′E). Along much of this section, the scarp is moderately to strongly degraded. SE of this latter location, the Hebron Fault scarp is almost exclusively expressed in calcrete-cemented conglomerates, except where it crosses a low spur of basement rocks projecting through the calcrete (Figs 2, 4).

At its southern end (24°47.40′S, 16°07.85′E), the fault scarp runs into a major exposure of basement. Two distinctive NW–SE trending photo lineaments, one of which is colinear with the calcrete-cemented conglomerate scarp, continue south from the scarp termination through the basement rocks (Figs 2, 4); however, no field evidence was found for a convincing fault scarp coinciding with either of these lineaments. No continuation of a scarp was found in the calcrete surface directly south of this basement exposure.

The Hebron Fault scarp analysed in detail

The Hebron Fault is defined by a scarp that varies in height from <1 m to almost 10 m, developed principally in otherwise relatively featureless, calcrete-cemented conglomerates and an overlying thinly laminated hardpan layer. These units are stratigraphic equivalents of the Karpfenkliff Conglomerate Formation and capping Kamberg Calcrete Formation (Fig. 5) defined by Ward (1987), where they form the high-lying terraces of the nearby Sesriem Canyon. Expression of the fault scarp in basement rocks (Hyas Farm and Neuhoff Reserve; Figs 4, 6e) is discussed separately in a later section.

Along the northernmost *c.* 3 km of the scarp, the Karpfenkliff Conglomerate contains dominantly rounded, locally imbricated cobbles of fluvially transported, dark grey Kuibis (Nama Group) limestone. Sedimentologically the deposit is virtually

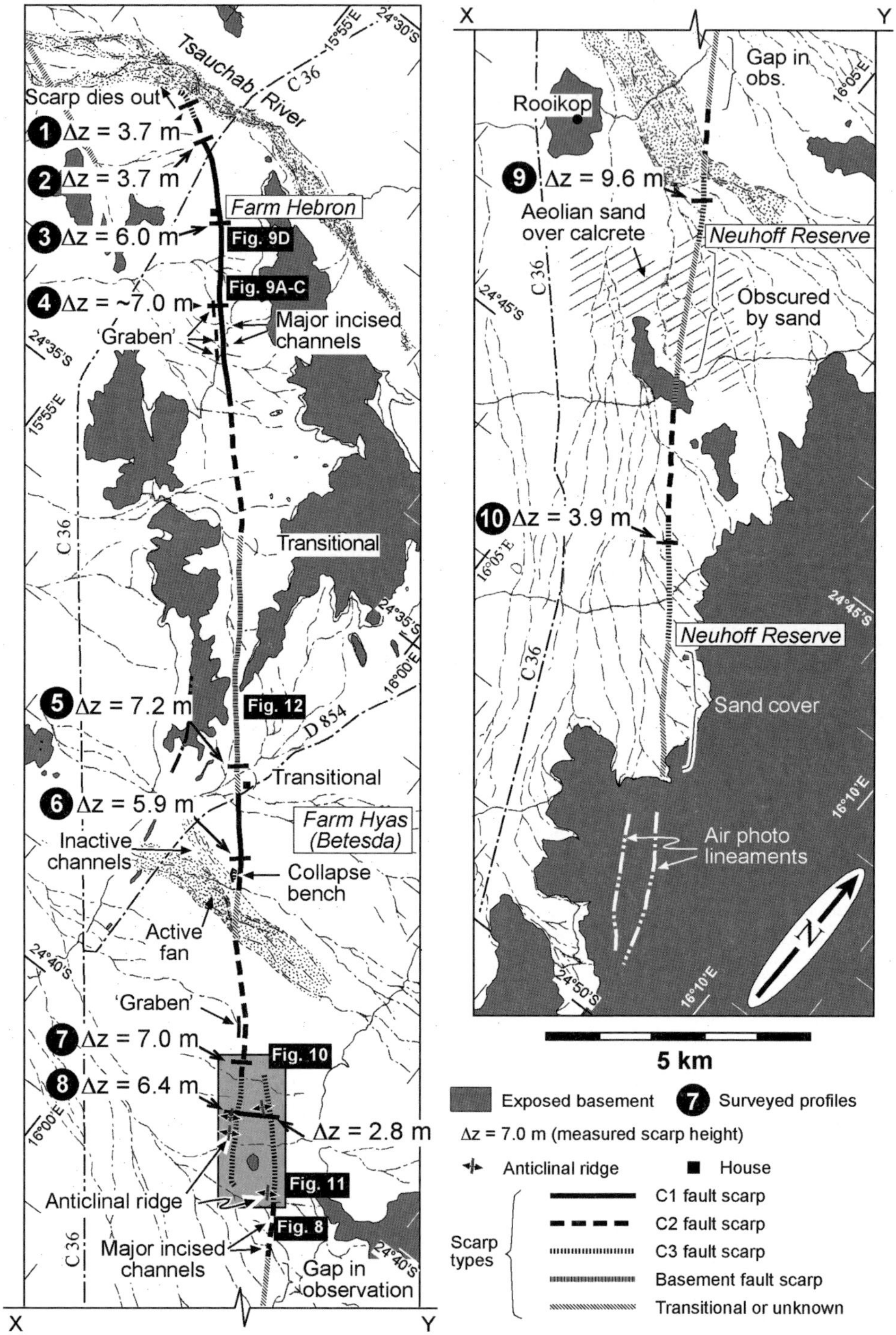

Fig. 4. Structural map, constructed from aerial photographs and field observations, of the Hebron Fault trace covering the area between Hebron Farm (north) and Neuhoff Reserve (south). Fluvial drainages and other geomorphologic features are highlighted, along with locations of surveyed profiles (shown in Fig. 7), different scarp types, scarp heights, and secondary fault-related features. Areas shaded grey mark exposures of basement rocks.

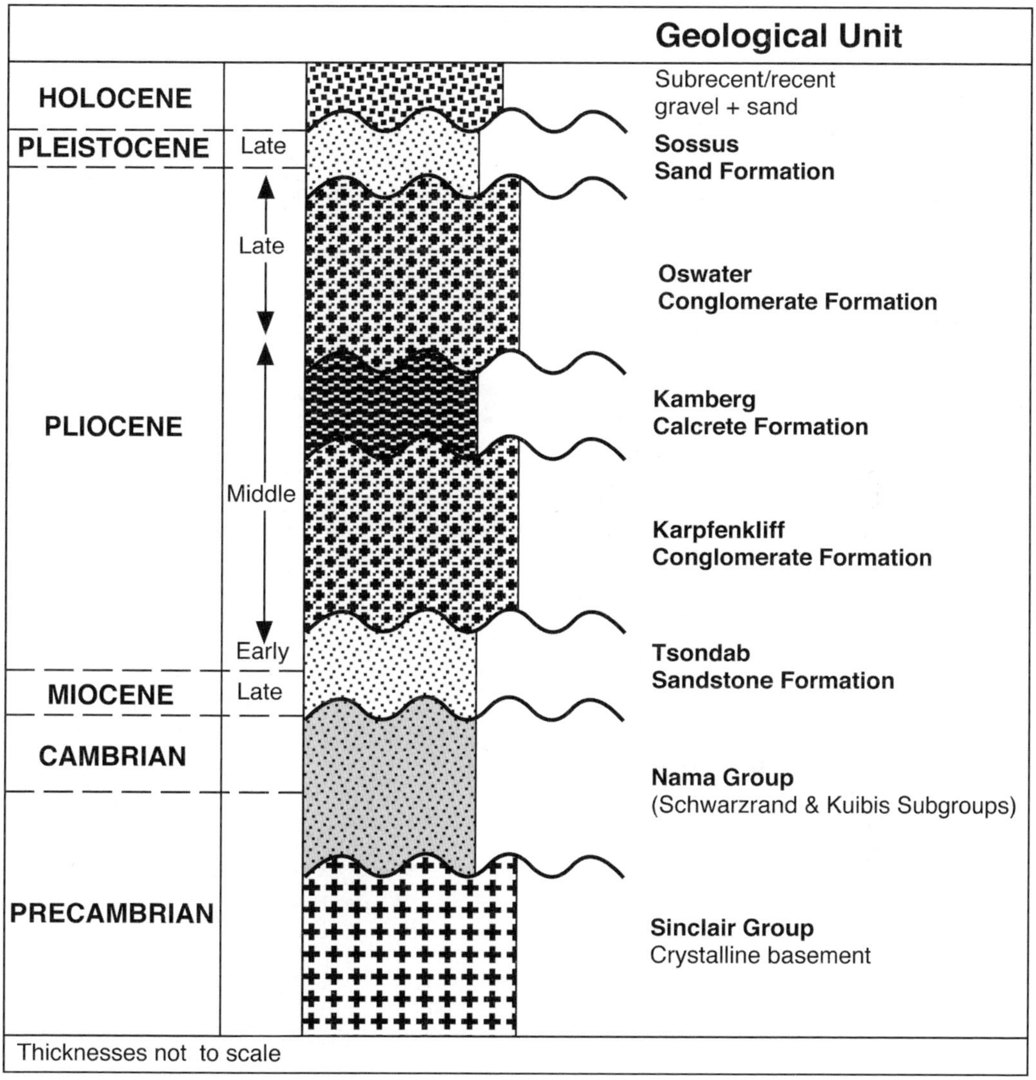

Fig. 5. Stratigraphic framework of the Hebron Fault area in SW Namibia, based on Ward (1987) but considering new age constraints provided by biostratigraphy (Tsondab Sandstone Formation: Pickford & Senut 1999), palaeoclimate studies (Karpfenkliff Conglomerate Formation: Mulder & Ellis 2000; Dupont *et al.* 2005), cosmogenic isotope measurements (Oswater Conglomerate Formation: Van der Wateren & Dunai 2001) and sequence-stratigraphic correlation combined with biostratigraphy (Oswater Conglomerate Formation/coastal +30 m package: Ward & Corbett 1990; Pickford & Senut 1999).

indistinguishable from modern sediments in the bed of the Tsauchab River (Fig. 2). Along the remaining 35 km of the scarp, except where it is formed in basement, the conglomerate comprises dominantly granite clasts with minor dolerite, amphibolite, diorite, metaquartzite, vein quartz and porphyritic volcanic clasts, with only an accessory limestone clast component. Locations close to the mountains include scattered boulders up to 1 m diameter. The compositional change from limestone- to granite-dominated clasts coincides with an increase in clast angularity, and both changes reflect increasing proximity to basement inselbergs and mountains from which the granitic components are derived.

The calcretized, moderately to poorly sorted Karpfenkliff Conglomerate is mostly massive to crudely horizontally stratified (Fig. 6b) with sporadically developed clast imbrication indicating

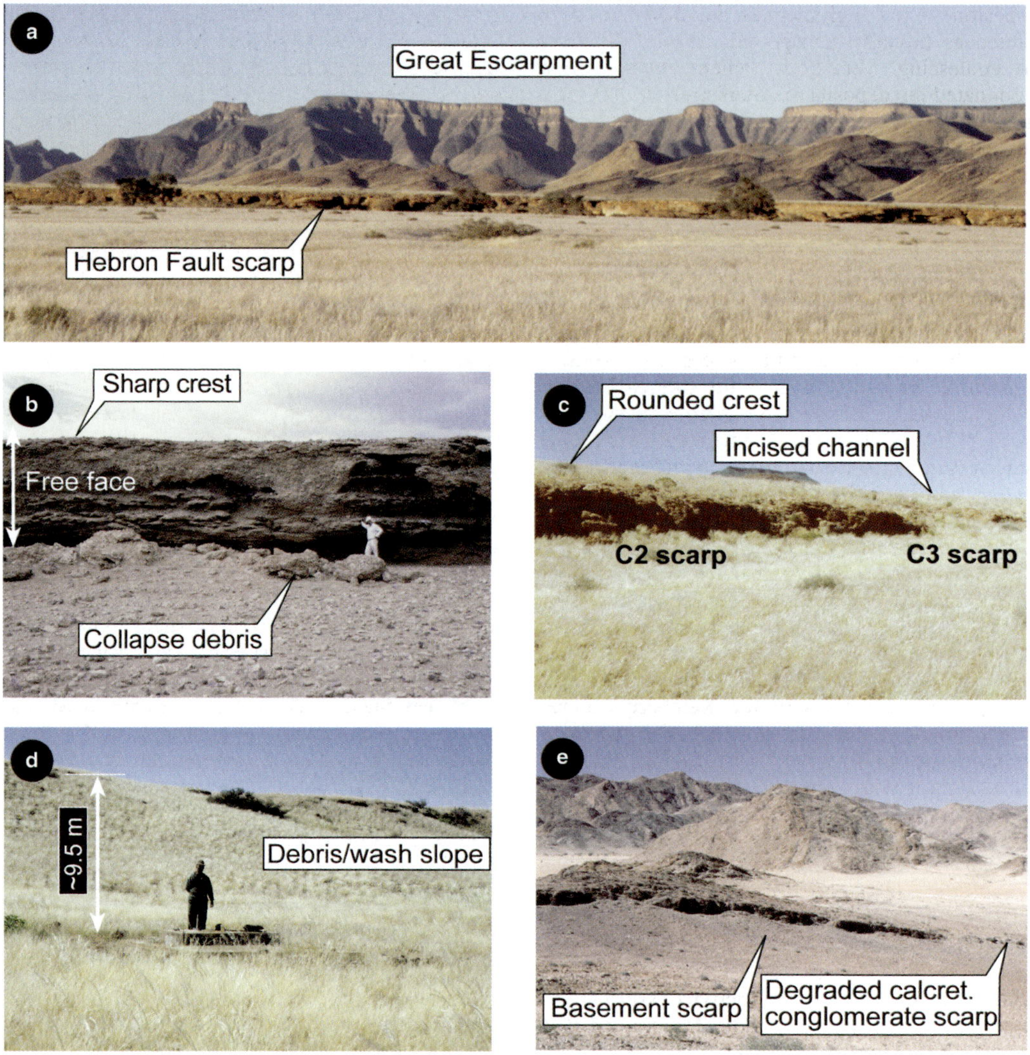

Fig. 6. Field photographs illustrating the variable degree of Hebron Fault scarp degradation along strike. (**a**) Eastward view, Hyas Farm, with linear vertical fault scarp displacing the subhorizontal Kamberg Calcrete hardpan layer and underlying Karpfenkliff Conglomerates. Average fault scarp height is 5 m. The Great Escarpment rises in the background and exposes granitic and gneissic basement, capped by flat-lying Nama Group limestone. (**b**) Example of 'juvenile' C1 fault scarp in calcrete-cemented Karpfenkliff Formation conglomerates with a prominent free face and sharp crest on Hebron Farm. Collapse debris started to accumulate at the base of the scarp. (**c**). 'Intermediate' C2 scarp transitional into C3 scarp, Hyas Farm with remnant free face and rounded crest and well developed, vegetated debris and wash slope. (**d**) 'Mature' C3 scarp, the highest measured, on Neuhoff Reserve. Note advanced rounding of crest and absence of free face which is overtaken by completely vegetated debris and wash slope development accompanied by a decrease in slope angle. Borehole in the foreground yields thermal water. (**e**) Fault scarp crossing a spur of Sinclair Group basement rocks, Neuhoff Reserve.

westward-directed sediment transport. With few exceptions, all conglomerates are clast-supported. The lack of muddy deposits in the Karpfenkliff Formation together with a restricted spread in palaeocurrent directions compares to modern alluvial fan–gravelly braided stream environments. The areal extent of limestone-rich conglomerate coincides closely with the courses of modern, medium to large ephemeral river systems, in particular the Tsauchab River (Fig. 2). The limestone-rich conglomerates are therefore interpreted as fluvially transported material deposited as coalescing, very low-gradient, braided river dominated fan deposits (cf. Stanistreet & McCarthy 1993) whereas the granite-rich conglomerate forms low-angle piedmont (talus) slope deposits at the foot of the mountains (Fig. 2).

The scarp exposes four successions of pedogenic calcrete, each developing an upwardly increasing carbonate cementation within the conglomerate. Each basal part is made up of separate, non-coalescing calcrete nodules <1 to 4 cm in diameter. Above this is developed a honeycomb calcrete in which the nodules are in the process of coalescing and clasts frequently float in the calcrete. A thin, wavy laminated hardpan layer then caps the pedogenic profile. The degree of cementation varies considerably and locally may have a significant bearing on the preservation of the fault scarp. The lower two pedogenic calcretes are less well developed and thus form the least weathering-resistant lower half of the scarp profile. A very thin calcrete layer follows above and the uppermost pedogenic calcrete then is the thickest and most weather-resistant hardpan layer, which regionally demarcates the upper boundary of the Karpfenkliff Conglomerate (Ward 1987). Such duricrusts of the Kamberg Calcrete Formation are widely developed in the uppermost several metres of the Karpfenkliff deposits, forming resistant cappings, and their location in the section closely mark palaeoland surfaces (Yaalon & Ward 1982).

Age constraints

Establishing an age for the youngest strata offset by the Hebron Fault is essential for establishing the latest probable time of fault activity. Ward (1987) originally designated a Middle Miocene Age to the host-rock Karpfenkliff Conglomerate, correlating it with the Arries Drift Gravel Formation (Ward & Corbett 1990) that contains a rich early Middle Miocene vertebrate fossil fauna (Corvinus & Hendey 1978). Pickford & Senut (1999), however, postulate a Late Pliocene to Early Pleistocene age for the Karpfenkliff Conglomerate, based on the occurrences of fossil ostrich egg shells of various *Namornis* and *Diamantornis* struthious species as well as eggs of Late Miocene to Middle Pliocene *Struthio daberasensis* in the Tsondab Sandstone below the Karpfenkliff Conglomerate (Fig. 5).

In addition Pickford & Senut (1999) argue that the Karpfenkliff Formation conglomerates indicate the presence of a more vigorous fluvial regime with wetter climate conditions than those during the Miocene and most of the Pliocene. Such a marked increase in humidity is also indicated by a grass leaf phytolith study of Karpfenkliff samples (Mulder & Ellis 2000) and by the existence of Kamberg Calcrete on top of and in the uppermost several metres of the Karpfenkliff Conglomerate, since these form only under semiarid conditions (Yaalon & Ward 1982). This relatively humid episode coincides with a northward advance of the polar fronts, themselves related to an increase in winter rainfall in SW Africa and a Late Pliocene (*c.* 3.1 to 2.2 Ma) episode of high sedimentation rates recorded by ODP sites offshore Namibia (Wefer *et al.* 1998; Dupont *et al.* 2005).

The Karpfenkliff and Kamberg Formations certainly pre-date a major period of river incision and subsequent deposition of the Oswater Conglomerate (Ward 1987). Cosmogenic isotope measurements indicate that this late Neogene episode of accelerated denudation in the central Namib and incision of deep canyons started around 2.81 ± 0.11 Ma ago (Van der Wateren & Dunai 2001). The Oswater Conglomerate is thought to correlate with the coastal $+30$ m raised marine terrace package (Ward & Corbett 1990), as defined in SW Namibia, that hosts *Donax rogersi* mollusc shells and a distinct warm water marine fauna (Pickford & Senut 1999). The latter is thus considered older than 2.2 Ma when perennial river discharge disappeared and decreasing sea-surface temperatures at the coast started during the later part of the Pliocene (Dupont *et al.* 2005). If all these age constraints are integrated into a stratigraphic section, Karpfenkliff and Kamberg Formations should be placed in the Middle to earliest Late Pliocene whereas the Oswater Conglomerate Formation is essentially Late Pliocene (Fig. 5). Because the Hebron Fault displaces the above-mentioned strata, we can state that its latest phase of activity is younger than Late Pliocene.

Tsondabvlei, *c.* 75 km NW of Hebron Farm, is an unusually linear, NW-trending pan at the termination of the Tsondab River in the Namib Sand Sea. Similar lineaments cross-cutting the inactive Kalahari dunes of the eastern Namibian Eiseb area are well documented as recent fault displacements (Andritzky 1996; Wanke 2005). Miller & Schalk (1980) show a lineament traced NW from the Hebron Fault at Hebron Farm towards Tsondabvlei and beyond. On the Meob Bay 1:250.000 geology sheet 2414 (Schreiber 2000) the Karpfenkliff Formation is truncated by the lineament. This Tsondab lineament is the NW extension of the Hebron Fault into the presently inactive dunes (Sossus Sand Formation, Fig. 5), whose maximum age is constrained by finds of Middle Stone Age artefacts below (Vogelsang 1998). Thus, the latest activity of the Hebron Fault post-dates the Middle Stone Age considered at *c.* 40–200 ka (Late Pleistocene) in southern Africa (Mitchell 2002).

Scarp profiles

Wallace (1977) has shown that surface rupture along a fault produces a fault scarp comprising a free face and a sharp crest with the free face initially reflecting a close approximation of the fault plane itself. Depending on the degree of consolidation of the displaced sediments, debris immediately begins to spall off the free face and accumulates at the base of the scarp as a debris slope at the angle of repose of the material, 34–37° (Yeats *et al.* 1997). Below the debris slope, a wedge of alluvium, the wash slope, overlaps the debris slope and the lower original surface that was offset by faulting. Over time the fault crest becomes rounded, the debris slope overtakes the free face and both are overtaken by the wash slope, all accompanied by a decrease in the maximum slope angle. The rate of scarp degradation is likely to be affected by how strongly indurated or cemented the scarp materials are, with weaker materials degrading more rapidly. The extent of scarp degradation is nevertheless a useful indicator of relative scarp age in the range from a few hundred years to about 20 000 years (Machette 1989).

Multi-event fault scarps will have the maximum scarp-slope angle of the most recent event and a scarp height that is the cumulative result of more than one event (Machette 1982). Bucknam & Anderson (1979) recognized that the relationship between scarp height and maximum scarp slope angle can be approximated by a logarithmic curve.

Related to the Hebron Fault, three principal scarp types, C1, C2 and C3, are distinguished in calcrete-cemented conglomerate (Fig. 7), based on the cross-sectional shape, the maximum scarp slope angle and the degree of degradation of the scarp. Qualitative visual scarp characterization was confirmed by quantitative topographic surveying of each scarp type. Locations of a number of the surveyed scarp profiles were chosen to illustrate type examples of each of the three scarp morphologies (Fig. 7).

C1, 'juvenile' scarp (Fig. 7, profiles 3, 4, 6): fault trace sectors 2–4.5 km long are characterized by a subvertical, SW-facing scarp typically 4.5–6 m high and dipping 60–80°. Further diagnostic features are a pronounced free face and a sharp scarp crest (Figs 6a,b). Some boulder debris has accumulated at the foot of the scarp, most of which originated from gravitational collapse of the uplifted footwall, but there is no substantial development of a debris wedge. The base of the vertical scarp is commonly undercut by up to 1 m in the lower third of the total scarp height (Fig. 6b). Within the range of scarp profiles observed, C1 scarps represent one end-member with the lowest maturity.

C2. 'intermediate' scarp (Fig. 7, profiles 2, 5, 7): fault trace sectors 2.5–4.5 km long alternate with C1 and C3 scarps and combine elements of C1 and C3. A remnant SW-facing, subvertical free face with a rounded crest is preserved over the upper *c.* 1.5–2 m of the total scarp height, while the lower 3–4 m of the scarp consists of a moderately to highly sloping debris and wash slope, extending from the base of the subvertical face to the down-faulted calcrete surface (Fig. 6c). Slope angles are considerably lower compared to C1 scarps, typically ranging from 20 to 50°. The vertical face is typically undercut to a depth of several tens of centimetres.

C3, 'mature' scarp (Fig. 7, profiles 9, 10): the third scarp type consists entirely of a moderately to highly vegetated, uniformly sloping ramp between the calcrete surfaces offset by the fault (Fig. 6d). In most cases the fault crest is well rounded and the free face has been entirely overtaken by a debris slope and in particular a downslope overlapping wash slope. Slope angles are low and consistently in the range of 15 ± 2°. C3 scarps thus represent the other end-member variety of scarp types. Small, vertical scarp faces only rarely remain and most probably indicate multiple-event fault scarps. C3 fault scarps typically extend ≥2.5 km, but these dimensions may be imprecise because parts of the southern <8 km of the fault trace are covered by recent aeolian sand.

The change along-strike from one scarp type to another is typically abrupt, occurring over several tens of metres at most. In several instances, the transition from C2 to C3 scarp was observed across the width of a single incised channel (Fig. 6c).

All sections of the calcrete-cemented conglomerate fault scarp can be readily assigned to one of these three types and the basis for doing so is primarily morphological. As discussed below, however, these morphological differences may also reflect significant along-strike differences in displacement history or kinematics.

Interpretation of scarp morphology

Notwithstanding the degree of cementation, steep scarps of any origin tend to degrade to lower slope angles approaching a natural angle of repose (Yeats *et al.* 1997). This is particularly the case for earthquake fault scarps where aftershocks or later earthquakes contribute significantly to scarp degradation through triggering of gravitational collapses. The gradient of the final slope depends upon material characteristics, such as cementation, clast size, shape and angularity, and on moisture content of the sediment. The survival time of a single earthquake scarp formed in unconsolidated sediment is estimated to be 10^4–10^5 years (Mayer 1986).

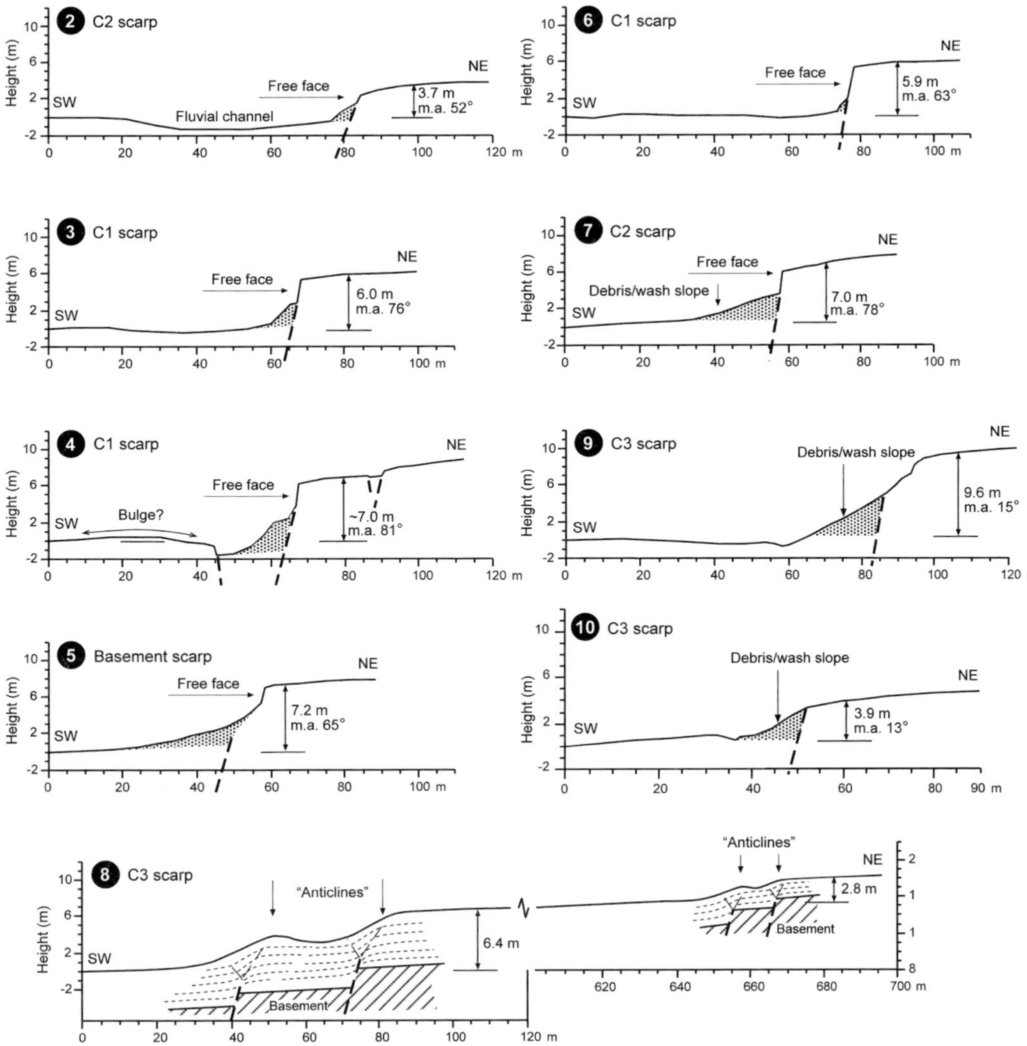

Fig. 7. Topographic profiles 2 to 10 surveyed perpendicular to the Hebron Fault scarp. Vertical exaggeration is 2.5 times. Abbreviation 'm.a.' indicates maximum scarp slope angle. See Figure 4 for profile locations and measured scarp heights. Profile 1 is similar to profile 2, and is omitted from this figure. Inferred subsurface positions of faults are shown with heavy dashed lines. Shaded areas are inferred cross-sections of debris and wash slopes. Note variable scarp slopes and fault scarp degradation. Profile 8 traverses both strands of the fault trace across the left-stepping jog. Subsurface construction in profile 8 shows possible fault-bend folding development. Sections 3, 4, 5, 6 and 7 involve prominent free faces which contrasts with sections 9 and 10 where no free face remains, the crests are markedly rounded and the scarps are mainly debris slope and wash slopes. Stepped scarp slope profile 4 probably involves crests of rejuvenated scarps.

Although carbonate cementation helps to stabilize clastic sediment at higher slope angles than unconsolidated sediment, the vertical to undercut, C1 scarps are inferred to be inherently unstable. This inference is supported by the observation that, at most places, a poorly organized heap of blocky debris has accumulated at the foot of the scarp. This scarp type is, therefore, interpreted to be relatively immature, or juvenile, and subject to sporadic collapse.

In contrast, type C3 scarps have a slope angle close to or less than the empirically expected natural angle of repose for a poorly sorted sediment composed of angular, predominantly cobble-sized clasts. This, and the continuity of the scarp slope connecting the upper and lower calcrete surfaces, suggests that these scarps have degraded to a relatively stable and mature state (Yeats *et al.* 1997).

Rounded fault scarps can also result from fault-bend folding above a buried fault tip (e.g. Price & Cosgrove 1991, p. 249). This possibility is discussed in a later section describing features of the central jog section of the fault trace, but we think it is unlikely to be the primary origin of the majority of C3 fault scarps we have observed.

C2 scarps, which are characterized by elements of both end-member types are, therefore, inferred to represent a state of intermediate scarp maturity.

Juvenile, C1 scarps are present exclusively along the NW section of the Hebron fault trace, and C3 scarps occur almost exclusively along the SE section of the fault trace. Rare exceptions to this are: (1) at the northern tip of the fault, west of route C36, where the <3 m high scarp has degraded to a uniform slope (or potentially lies above a buried fault tip dying out to the NW); and (2) at the left-stepping jog south of route D854, where overlapping sections of the fault trace are also characterized by gentle slope angles (Fig. 7, profile 8). Intermediate, C2, scarps occur on both major sections of the fault trace.

Scarps in calcrete-cemented conglomerate pass seamlessly along-strike into exposed basement scarps, without strike-perpendicular offsets of the fault trace. Therefore, it is inferred that the calcrete-cemented conglomerate scarp everywhere passes directly down-dip into a continuous basement fault. The fact that scarps in both calcrete-cemented conglomerate and basement rocks are laterally continuous also confirms that, despite erosional modification, the free face can be viewed as a close approximation of the fault plane in subsurface.

Modification of the scarp by erosion and incision

Decimetre-scale fluvially incised gullies, and also major ephemeral rivers, cut across the calcrete fault scarp at several locations (e.g. Fig. 4). Most of the incision occurs during high rainfall events, and the orientation of the incised gullies, subperpendicular to the trend of the fault scarp, is consistent with the overall, gentle east-to-west gradient of the Kamberg hardpan layer surface. Only along the southern 6 km or so of the scarp are the surface gradient and drainage pattern subparallel to the azimuth of the scarp.

The fact that incised gullies cut down to the level of the downthrown surface is significant because it suggests that at present the average rate of incision is equal to or greater than the average rate of vertical fault displacement. If average fault displacement exceeded average rate of downcutting, a knick point would be expected at, or a short distance headward of, the SW-facing scarp.

Gullies incised into C1 scarps have a narrow V-shape and a steep gradient. In contrast, gullies incised across C3 scarps are wider, with a less pronounced V-shape, and extend farther in a headward direction from the scarp than gullies cut across C1 scarps. This difference suggests that gullies incised through C3 scarps are more evolved, and this in turn supports the earlier inference that C3 scarps are themselves more mature than C1 scarps. No systematic right or left sense of displacement of river channels at the fault scarp was observed.

The fault scarp is almost totally degraded where major ephemeral river courses cross the scarp close to profiles 6 and 9 (Figs 4, 7). At profile 6 the active channel of the river turns NW to follow the base of the scarp for a few hundreds of metres before resuming its westerly course. The river channel deflection is another example of drainage localized along the base of the scarp (see above), although at this location there is no remaining evidence, apart from the scarp itself, for additional proactive tectonic control on drainage diversion.

The way the modern Tsauchab River drainage attempts to accommodate fault-induced topographic lows is well illustrated 2.4 km NW of Hebron Farm, where the scarp starts dying out towards the NW (Fig. 4). There, subsidiary channels of the otherwise westward-directed river drainage are deflected towards the SE, depositing southeast-ward prograding sand wedges and debris fans, at least 1.2 m thick, on the downthrown side of the fault. Along those parts of the C1 fault which were not reached by the prograding debris fans, small elongate ephemeral slidelakes were observed to exist at least several days after heavy rainfalls.

Structural features in the calcrete-cemented conglomerates

No actual fault plane was located anywhere along the calcretized conglomerate scarp, irrespective of scarp type or inferred maturity. Furthermore, there is a distinct lack of evidence for deformation in the calcretized conglomerates. For example, few cracked or microfaulted cobbles were found, and there are few secondary fractures penetrating the calcrete as a whole. However, observations of several structural features, together with evidence from the basement scarps, provide convincing arguments for the kinematics of the Hebron Fault.

Displaced strata

Key to this analysis is that the Kamberg Calcrete hardpan layer on both sides of the fault scarp is broadly time-equivalent. On Hyas Farm just south of a left-stepping jog, a gully incised perpendicularly across the calcrete scarp (Fig. 8, located on Fig. 4) exposes crudely stratified, limestone clast-bearing, carbonate-cemented conglomerate layers overlying weathered basement. Here the C3 type scarp is not more than 1.5 m high with only a remnant free face and rounded crest remaining but a well developed debris-wash slope. Other incised gullies also expose sections through stratified calcrete, but only at this locality is the incision across the scarp deep enough to expose the same stratigraphy on either side of the scarp for comparison.

On the NE, upstream, side of the scarp, three gravel layers in the Karpfenkliff Conglomerate are identified by variations in colour, clast size, presence in the lower layer of imbricated cobbles,

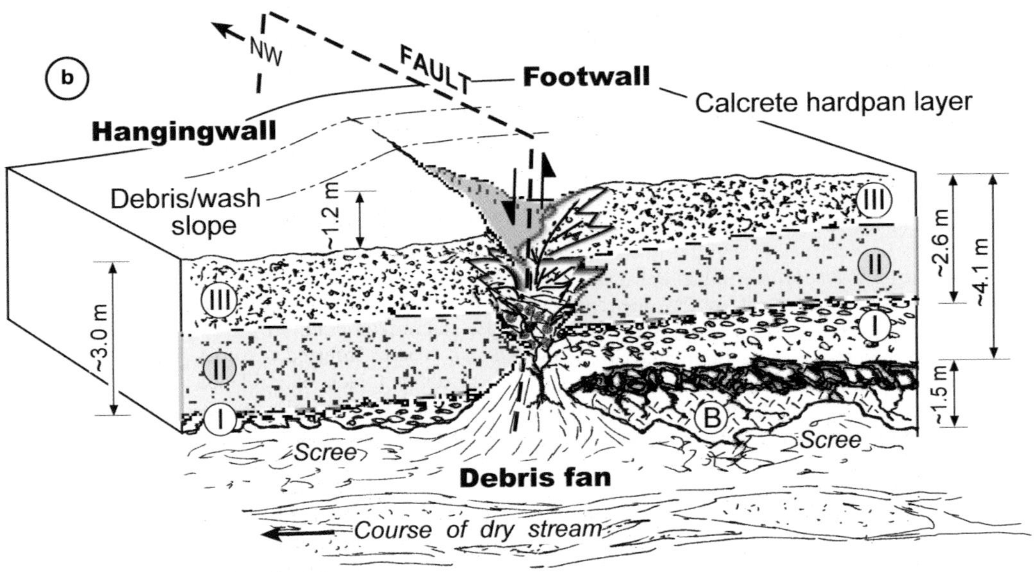

Fig. 8. (**a**) Photograph of fault-offset calcrete-cemented Karpfenkliff Conglomerate layers exposed in a gully incised through the Hebron fault scarp. See Figure 4 for location. (**b**) Interpretative line drawing of the same locality illustrating the internal stratigraphy of the Karpfenkliff Conglomerate with graded layers I, II and III; B marks underlying basement rock. Note 15% thickening of conglomerate layers II and III on hanging-wall block, suggesting multiple synsedimentary faulting.

and distribution grading. The lower layer rests unconformably on *in situ*, weathered, calcrete-cemented, monomict granite breccia that grades downward into fractured and weathered granitic basement rock. Downstream, on the SW side of the scarp, basement is not exposed because it lies below the level of the stream bed. However, the same three conglomerate layers were identified, 1.0–1.5 m lower than on the upstream side of the scarp. Carbonated cemented gravel layers II and III (Fig. 8b) are also about 15% thicker on the down-faulted side of the scarp.

The offset of the calcrete layers coincides with the projected surface trace of the fault scarp and the magnitude of the offset of the calcrete layers is similar to the height of the scarp. The actual position of the fault itself is covered by a debris fan occupying a minor gully that cuts back along the line of the fault (Fig. 8).

Graben-like structures

At several localities along the NW section of the fault trace a shallow, elongate ditch is developed at the foot of the main fault scarp (Fig. 4, Fig. 7 profile 4, Fig. 9a–c). In each case, the ditch trends parallel to, and is confined between, opposing scarps. Maximum height of the main scarp forming the east side of the depression is 7.0 m, while the secondary scarp that forms the west side of the depression is not more than 1.5 m high. The minimum continuous length of the longest of these ditches is 350 m. Immediately SW of the depression, the calcrete is upwardly convex, bulging nearly 1 m above its expected level compared to the average gradient of the surface.

This distinctive feature raises a number of questions: Why are only these selected sections of the calcrete scarp eroded in this way? Why does water here flow parallel to the scarp, when the overall surface gradient at these localities is oblique to the scarp? What is the significance of the slight upward bulge of the footwall calcrete surface? To account for these, we suggest that the scarp-parallel ditches originated as structural grabens localized between opposing, asymmetric, normal antithetic faults (e.g. Fig. 7, profile 4). The slight upward flexure of the downthrown calcrete surface may reflect a gentle footwall roll-over. Thus, scarp-parallel fluvial channels most probably exploit a pre-existing fault-related structure. In late 2000, after unseasonable rain, water was observed running freely in the depression. Fluvial modification of the scarp is evident elsewhere (see above), and this suggests that the depression may, at least partly, be deepened by fluvial erosion along the base of the scarp.

Figure 9d shows the displacement profile across a section of C1 scarp close to surveyed profile 3 (Figs 4 and 7) at Hebron Farm. Displacements were inferred from correlation of a Kamberg Calcrete hardpan cap. The original surface is displaced down-to-the-SW in a series of at least two fault steps, and a small graben-like structure is inferred where the hardpan surface drops below the mean level of the undisturbed hardpan surface that continues west of the scarp.

Left-stepping jogs

The best defined left-stepping jog, south of Hyas Farm, is *c.* 600 m wide and nearly 2 km long (Figs 2a, 4, 10). Over this distance, the overlapping sections of the fault trace are characterized by low to moderate angle, C3-type scarps (Fig. 7, profile 8). Immediately outside the area of the overlap, both sections of the fault trace revert to C2 profiles. Hence, there is a spatial, and possibly causal, relationship between the type of fault scarp and its location with respect to the left-stepping jog. Tips of the overlapping sections of the fault trace curve slightly inward toward the opposing section of the fault.

The total scarp height at the immediate NW end of the jog is *c.* 7.0 m. The height of the scarp then decreases gradually to the SE along this section of the fault toward its termination. Similarly, the total scarp height at the immediate SE end of the jog is *c.* 6.5 m, and the height of this section of the scarp also decreases gradually in the direction of its NW tip. At any perpendicular profile across the jog, the sum of the heights of the parallel scarps is approximately equal to, or slightly greater than, the height of a single scarp outside the jog. Therefore, vertical fault displacement appears to have been partitioned between the two sections of the fault in the area of the jog.

'Anticlines'

At four localities within the left-stepping jog south of Hyas Farm, the calcrete surface on the downthrown side of the C3 scarp rises noticeably upward into an elongate structure having the form of a doubly plunging 'periclinal anticline' (Figs 4, 11). The crest of each 'anticline' trends parallel to, and is 20–25 m west of, the primary fault scarp.

Surveyed profile 8 in Figure 7 shows that the bottom of the shallow depression between the crest of the elongate bulge and the fault scarp is significantly higher than the calcrete surface SW of the bulge. In this case, therefore, the geometry of the structure is inconsistent with the depression originating as a 'graben' between opposing, asymmetric normal fault scarps as suggested for other sections of the Hebron Fault scarp (cf. Fig. 9). We suggest

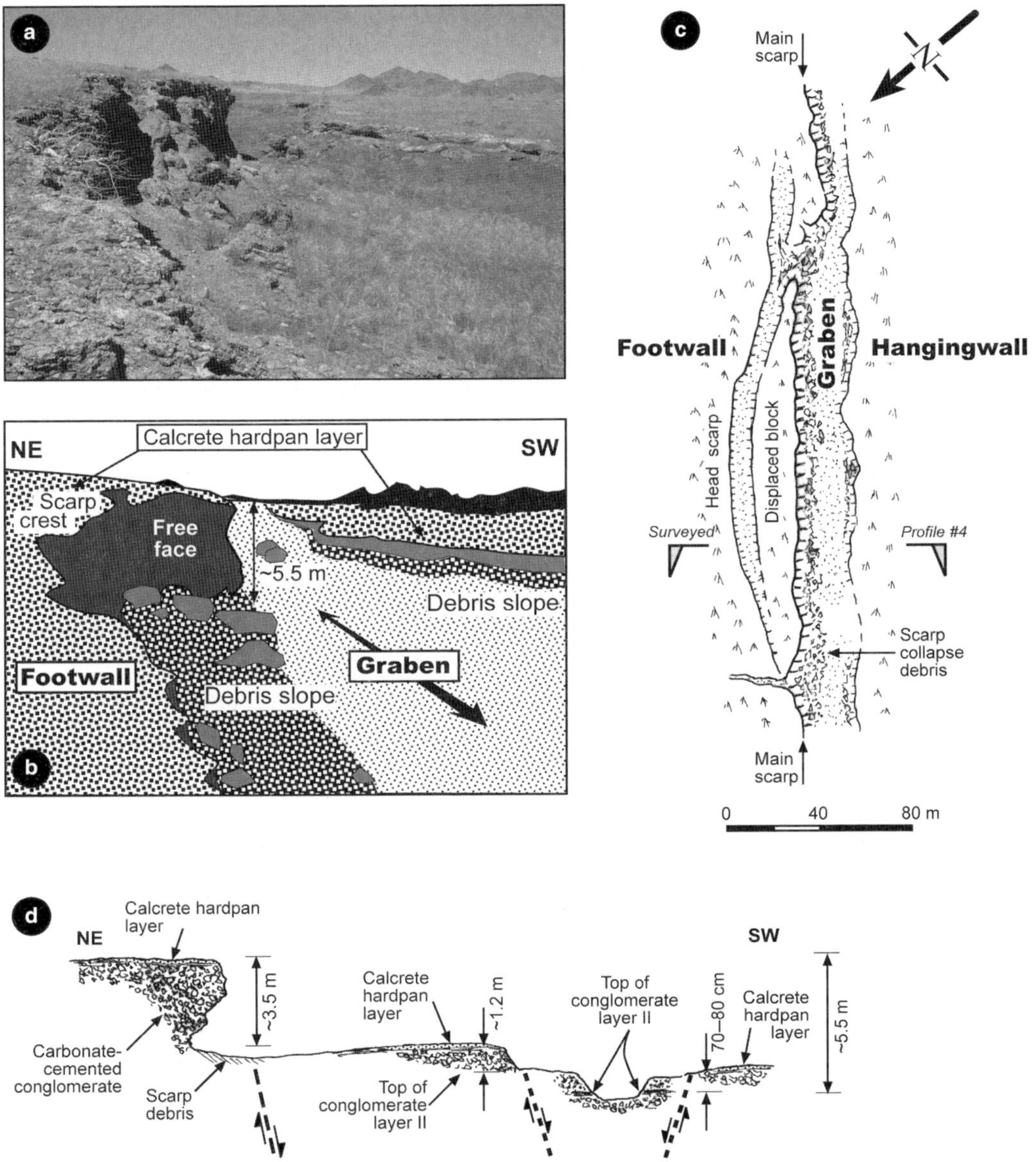

Fig. 9. (**a**) Photograph looking SE along the trough of one of the inferred grabens on Hebron Farm. See Figure 4 for location. (**b**) Interpretative sketch of the oblique view in (a). (**c**) Annotated field map of the graben-like structure shown in (a) and (b). Location of profile 4 (Figs 4, 7) is indicated. (**d**) Field sketch showing stepped C1 scarp, Hebron Farm. This section has been surveyed close to profile 3 (cf. Fig. 4). The offset of the Kamberg calcrete hardpan layer and the subsurface level of the calcrete-cemented conglomerate layer II surface in a small pit suggest a minor graben-like structure. View is to the SE.

instead that the 'anticline' may have originated as a fault-bend fold (monocline) when moderately or poorly consolidated conglomerates were sheared during normal fault displacement of the underlying crystalline basement (i.e. the fault tip fails to breach the ground surface; Fig. 7, profile 8) (Price & Cosgrove 1991). Unfortunately no natural cross-sectional exposures through these features were found, so their internal structure remains unconfirmed. The depression between 'anticline' crest and scarp crest may have been deepened by fluvial erosion.

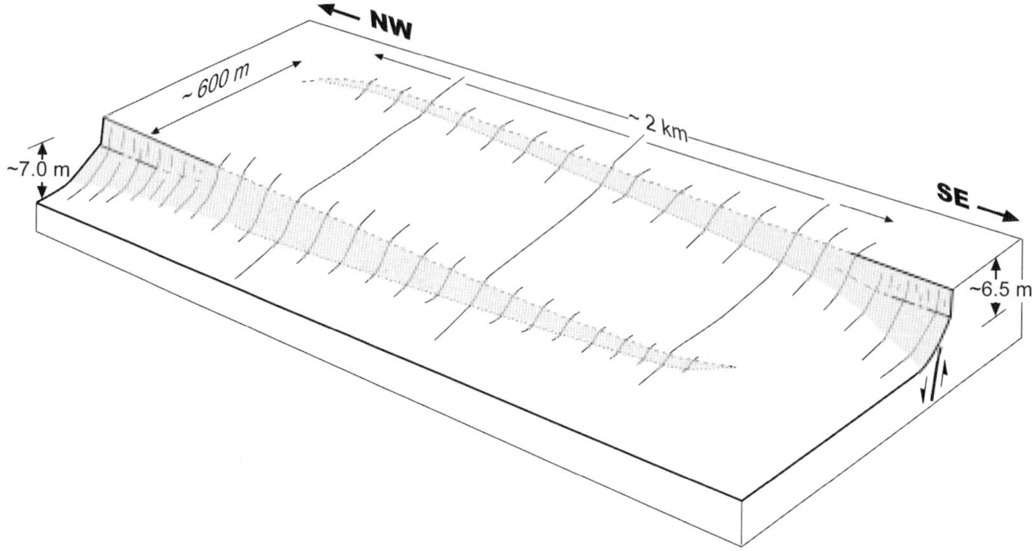

Fig. 10. 3D schematic model of the left-stepping jog on Hyas Farm. See Figure 4 for location. The sketch is significantly vertically exaggerated.

A possible origin of low C3 scarps as fault-related monoclinal flexures provides an alternative explanation for C3 scarps elsewhere along the Hebron Fault trace. It is our view, however, that post-displacement modification by erosion remains the most likely explanation for the origin of C3 scarps greater than about 1–2 m in height, and we think it highly unlikely that the nearly 10 m high C3 scarp along the southeastern section of the fault trace formed solely, or even primarily, by fault-bend folding.

The fault scarp in basement

Scarp profile

Where the fault scarp is formed in exposures of basement rocks, a sharp, planar scarp profile

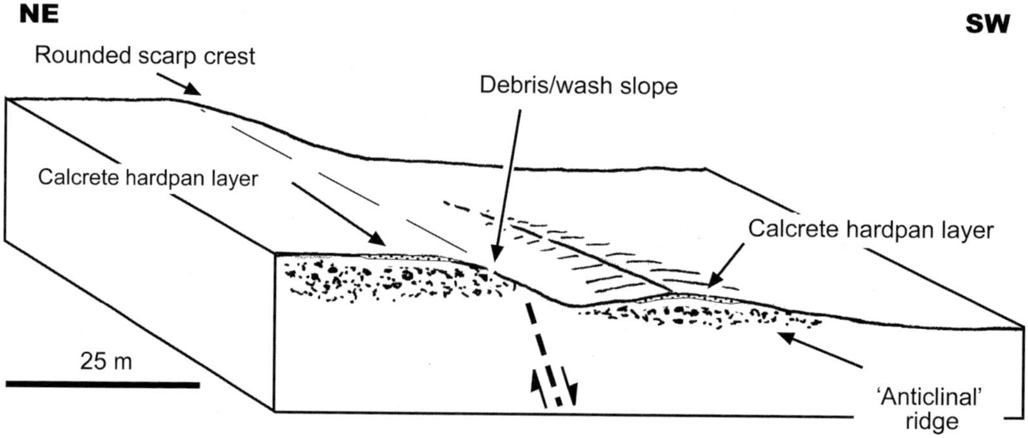

Fig. 11. Block diagram based upon field sketches showing the relationship of a 'periclinal anticline' adjacent to the C3 fault scarp in the area of the left-stepping jog. See Figure 4 for location. Note that the sketch is vertically exaggerated. Potential subsurface positions of faults are shown with thick dashed lines. A model interpreting the structure as a pair of fault-bend folds is shown in Figure 7, profile 8.

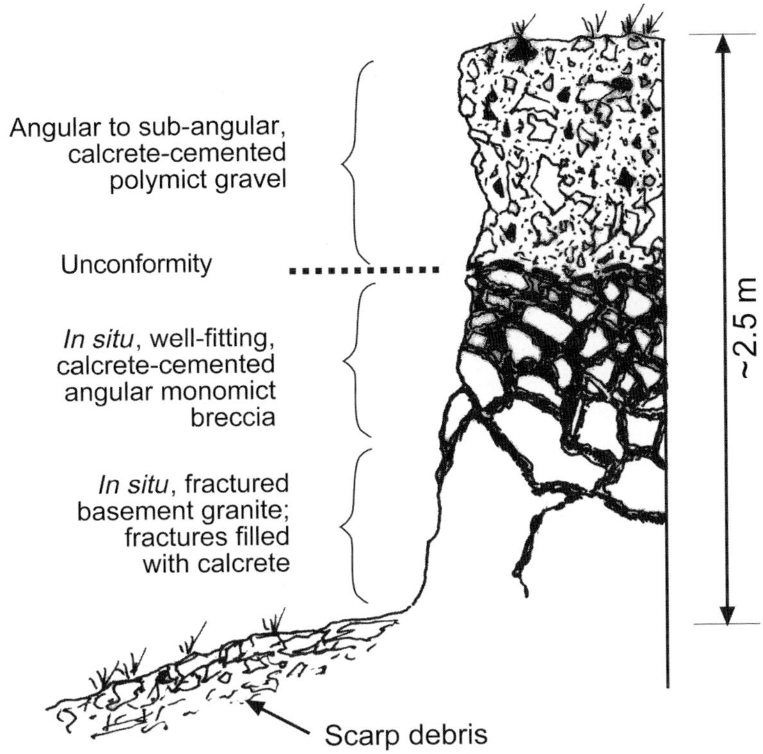

Fig. 12. Stratigraphic profile of a basement scarp exposure, Hyas Farm. See Figure 4 for location. At this locality, *in situ* weathered late Proterozoic basement rocks exposed in the scarp are overlain by a 1.5 m thickness of calcrete-cemented, polymict gravel. The calcrete gravel cap is thinner or absent at other localities.

commonly results. The shape of the scarp is similar to C2 carbonate-cemented conglomerate scarps, whereby the upper half to one third of the total scarp height comprises a subvertical scarp face (Fig. 12). The lower part of the scarp is a debris wedge sloping 15–20°SW from the base of the upper part of the scarp. Unlike calcrete-cemented conglomerate scarps, basement scarps displace sloping or uneven terrain, so the location of the toe of the debris wedge was difficult to pinpoint in some places. This means that the total scarp height is less certain at these localities. Basement rocks in the scarp typically comprise from bottom to top (Fig. 12): (1) slightly weathered, intact or fractured, hydrothermally veined granite, which grades up into (2) carbonate-cemented, monomict, *in situ* weathered granite breccia. The latter is locally overlain by (3) a cover of angular to subangular, polymict calcrete-cemented conglomerate typically <1 m thick.

Fault rocks

Granitic basement up to 5 $\pm$ 1 m away from the scarp face is strongly fractured. Fracture density increases with proximity to the scarp, so that at the scarp face itself, silicic veining may constitute >70 vol% of the rock. Fractures are mostly filled with quartz, although thin, pink veins of remobilized K-feldspar and grey veins of ultracataclasite are increasingly common closer to the scarp front. Approaching the scarp front, cataclasis of the granite between the quartz veins increases progressively.

Feldspar has deformed at the grain scale by brittle fracture mechanisms, such as microfaulting and veining. The majority of quartz clasts are also deformed by brittle fracturing, but internally the polycrystalline quartz shows evidence of undulose extinction, sutured grain boundaries, and limited development of elongate grain-shape fabrics. These latter microstructures, also locally present in vein quartz, are indicative of intracrystalline dislocation creep and incipient, or actual, dislocation glide and climb (e.g. White 1973; Knipe 1989).

Thus, the majority of grain-scale deformation mechanisms indicate that basement fault rocks have been exhumed from temperature–pressure conditions characteristic of the range for cataclastic deformation. A subset of microstructures suggests that conditions, at least locally, verged on those conducive to crystal plastic deformation.

Ordinarily, crystal plastic deformation of quartz is activated at temperatures ≥250°C (Lloyd & Freeman 1994; Stöckhert *et al.* 1999). However, crystal plastic behaviour in quartz may also be activated at temperatures <250°C if quartz is hydrolytically weakened, and especially if the grain size has already been reduced by cataclasis (Mitra 1984; Simpson 1986; Guermani & Pennacchioni 1998).

The intense silicic veining in these basement rocks from the Hebron Fault indicates the presence of abundant aqueous fluids during deformation, so that hydrolytic weakening, and syntectonic temperatures ≤250°C, are plausible. It seems unlikely, however, that these fault rocks originated anywhere under near-surface T-P conditions. Hence, the evidence for intense cataclasis and hydrothermal activity is interpreted to indicate that these rocks were deformed at depths greater than *c.* 4 km and shallower than *c.* 10 km, assuming a typical continental geothermal gradient of 20–25°C/km.

Variously orientated, centimetre-scale crush zones locally cross-cut the granitic cataclasite at the scarp front. A few crush zones up to 10 cm thick, located 10–20 m back from the scarp, were also recognized in basement rocks exposed in a deeply incised channel close to the Betesda homestead (Hyas Farm). Rock in the crush zones is pebble- to cobble-sized, weakly cemented fault breccia. Larger clasts in the crush zones comprise finely comminuted granitic cataclasite. The finer fault breccia matrix is inferred to be composed of the same parent material. The cross-cutting relationships clearly show that these crush zones post-date the phase of cataclasis and veining that characterizes the bulk of the basement fault scarp rocks. The rather friable nature of the crush zone breccia suggests that deformation occurred at a shallow depth under low confining pressure.

Hence, indurated cataclastic basement rocks proximal to the fault are interpreted to have been exhumed from shallow to almost mid-crustal depths, while the cross-cutting brecciated crush zones may be related to more recent tectonism responsible for the present-day Hebron Fault scarp.

Structural fabric analysis

The network of basement fissures and cracks is important for water provision in the Hebron area, where the water boreholes are selectively drilled on the downthrown side of the structure. Figure 6d shows one such borehole which yields water at elevated temperatures.

Poles to joints measured from the fault scarp exposure in basement rocks on Hyas Farm fall into three main groups (Fig. 13a). (I) About 30% of the data define a tight cluster of poles corresponding to a set of joints consistently dipping steeply SW. The mean strike of these joints is identical to the overall trend of the fault scarp. (II) A small set of poles corresponds to joints that dip moderately to steeply NE. These NE-dipping joints are interpreted as antithetic extension fractures related to a steeply SW-dipping normal fault. (III) The remainder of the data form a loose cluster corresponding to fractures with NNE to ENE strikes and moderate to steep SE dips. Figure 13b shows these fracture sets in block diagram.

Mesoscopic faults, inferred faults, and crush zones in basement rocks at Hyas Farm mostly strike NW and dip moderately to steeply SW (Fig. 13c). The mean strike is parallel to the overall trend of the macroscopic fault scarp, and the mean orientation of these fractures may reflect the orientation of a subsurface master fault. Some of these SW-dipping fractures have a normal component of displacement. The sense of displacement of other, similarly orientated, fractures was not determined in the field, but slip lineations on these plunge WNW, consistent with normal oblique slip with a minor to moderate dextral component. Only a few fractures preserve slip lineations with apparent sinistral or reverse sinistral displacement.

No outcrop evidence was found for major through-going faults corresponding to the aerial photographic lineaments in basement rocks south of the calcrete fault scarp on Neuhoff Reserve (Fig. 4). A small number of slickensided and/or lineated faults were found, however. The fractures mostly strike NE and dip moderately SE (parallel to the set of SE-dipping joints recorded at Hyas Farm), and the predominant sense of displacement is reverse (Fig. 13d).

Kinematic interpretation

The majority of Hebron Fault kinematic data, particularly those from basement rocks, are broadly consistent with normal displacement on a NW-striking, steeply SW-dipping master fault (Fig. 13). We infer that these sets of NW-striking joints and faults originated, together with the breccia zones, during a late-stage phase of deformation associated with creation of the modern surface scarp. These interpretations are consistent with down-to-the-SW displacement of conglomerate strata (Fig. 8) and the development of graben-like structures (Fig. 9). The smaller number of mainly SE-dipping reverse and reverse oblique-slip faults, and possibly also the set of SE-dipping joints, may belong to an earlier phase of deformation, potentially that

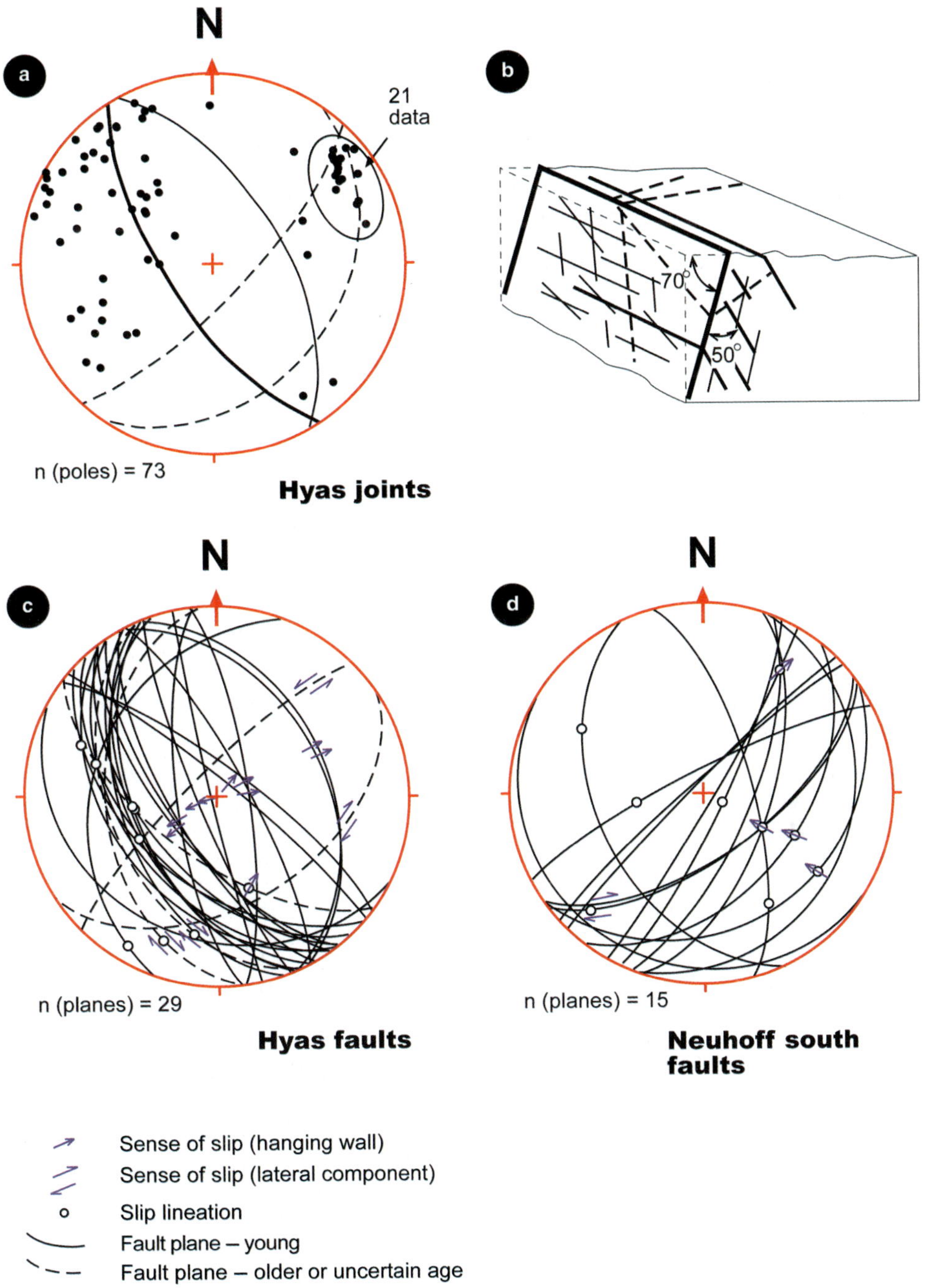

Fig. 13. Equal area, lower hemisphere plots of structural data measured from basement exposures of the Hebron Fault. (**a**) Poles to scarp face joints, Hyas Farm. Heavy solid great circle 147/70°SW; thin solid great circle 150/60°NE. Thin dashed great circles are mean planes for NW-plunging poles. (**b**) Block diagram showing the relative angular relationships of the mean joint plane orientations identified in (a). (**c**) Mesoscopic faults within 10 m of the scarp face, Hyas Farm. Solid great circles are assumed young faults; dashed great circles are inferred older fault planes. Open circles are slip lineations (n = 9). Arrows show movement of the hanging wall and opposing half arrows show the sense of lateral offset, where this was determined. (**d**) Mesoscopic faults or inferred faults in basement rocks due south of the calcrete fault scarp, Neuhoff Reserve. Symbols as for (c), except that the age of the fractures is unknown.

responsible for intense cataclasis and silicic veining of the host rock.

We have found evidence for only a minor component of dextral strike-slip displacement on this section of the Hebron Fault indicated by steeply plunging, oblique-slip lineations. We have no field evidence to suggest systematic lateral offset of incised gullies or debris fans at the fault scarp. In some cases there is a minor diversion of the stream course as it crosses the scarp, but neighbouring gullies are as likely to be diverted in the opposite sense as in the same sense.

In contrast, Viola *et al.* (2005) claimed to have observed lateral offsets of alluvial fans and used this plus the side-stepping geometry of the fault trace to suggest transtensional kinematics with a large component of strike-slip. Our observations do not support this interpretation. Furthermore, it seems unwarranted to use the sense of side-stepping of a fault trace as primary evidence for kinematics.

SE continuation of the Hebron Fault

Approximately 20 km south of Neuhoff Reserve (Fig. 2c), the Kamberg Formation hardpan layer on Asbaakies Farm (*c.* 24°55.0′S, 16°20.0′E) is offset in a manner similar to that of the Hebron Fault scarp segments described above. The lineament on Asbaakies Farm (Fig. 14) trends SE and crosses a ridge of basement rocks (Sinclair Group and older rocks) that projects west from the escarpment. These rocks are overlain unconformably by horizontally bedded sedimentary rocks of the Nama Group. The fault lineament makes a small right-stepping jog across the basement ridge, and then continues SE across modern fan surfaces north and south of the Zarishoogte Pass road (Fig. 14).

The fault scarp is not as well developed as in the Hebron Farm area farther north, but relief across the scarp is still on the order of 1 to 5 m. The unconformity between crystalline basement and Nama Group rocks is offset down to the west by at least 20 m on a normal fault. The difference between displacement of the basement unconformity and the height of the calcrete scarp suggests at least 15 m of throw on this fault segment prior to formation of the Kamberg Calcrete. This earlier displacement would explain discrete thickening of the Karpfenkliff Conglomerate across the fault (Fig. 8) suggesting at least a Middle Pliocene displacement history.

South of Zarishoogte Pass road the fault makes a left-stepping jog and continues at least 15 km further SE along a prominent linear valley (Fig. 14b). A number of other subparallel lineaments, defined by unusually straight drainages, are evident in Nama Group limestone east and west of the main valley lineament. Thus, there is good evidence in this area for a SE-striking structural grain, and this fault lineament is interpreted to be a discontinuous extension of the Hebron Fault in this area (cf. Fig. 2a). The down-to-the-SW sense of displacement of the basal Nama Group unconformity is the same as that for the Hebron Fault between Hebron Farm and Neuhoff Reserve.

Discussion

Summary of Hebron Fault features

The Hebron Fault is evidently a young, extensional tectonic structure. (1) It features a prominent scarp up to 9.6 m high that cuts across and displaces Late Pliocene calcrete-cemented alluvial and talus slope deposits. (2) Sections of the scarp appear to be relatively fresh and 'juvenile', in the sense that they appear not to have degraded much, while other sections have degraded to a near-natural angle of repose. (3) Localized hanging wall grabens are developed adjacent to the fault scarp. (4) The scarp in the calcrete-cemented conglomerates is collinear and continuous with an unmistakable fault scarp developed in crystalline basement rocks. (5) The scarp in the basement rocks preserves unequivocal evidence for late-stage tectonic overprinting of older cataclastic and submylonitic textures along the fault. (6) Structural data measured from basement scarp exposures support the existence of a NW-striking, steeply SW-dipping master fault, rejuvenation of which produced the present scarp. (7) A lineament exists as the fault is traced northwards, where it displaces aeolian dunes, attributed a Late Pleistocene/Holocene? age.

In our view, these features are consistent with interpretation of the Hebron Fault as a dominantly normal fault. There is no geomorphic evidence for a major component of lateral displacement. The height of an earthquake scarp is related to the number and magnitude of earthquakes producing a surface rupture, and scarps greater than a few metres in height are likely to represent the cumulative offset of several earthquakes (Yeats *et al.* 1997). Therefore the Hebron Fault scarp documented in this study is most probably the result of multiple fault displacements.

Reconciliation with previous work

Viola *et al.* (2005) assume dextral transtensional Hebron Fault motion based on the right-stepping en echelon fault geometry and right-lateral displacement of alluvial fans. They also state that there is no seismic or geomorphic evidence for Holocene fault activity.

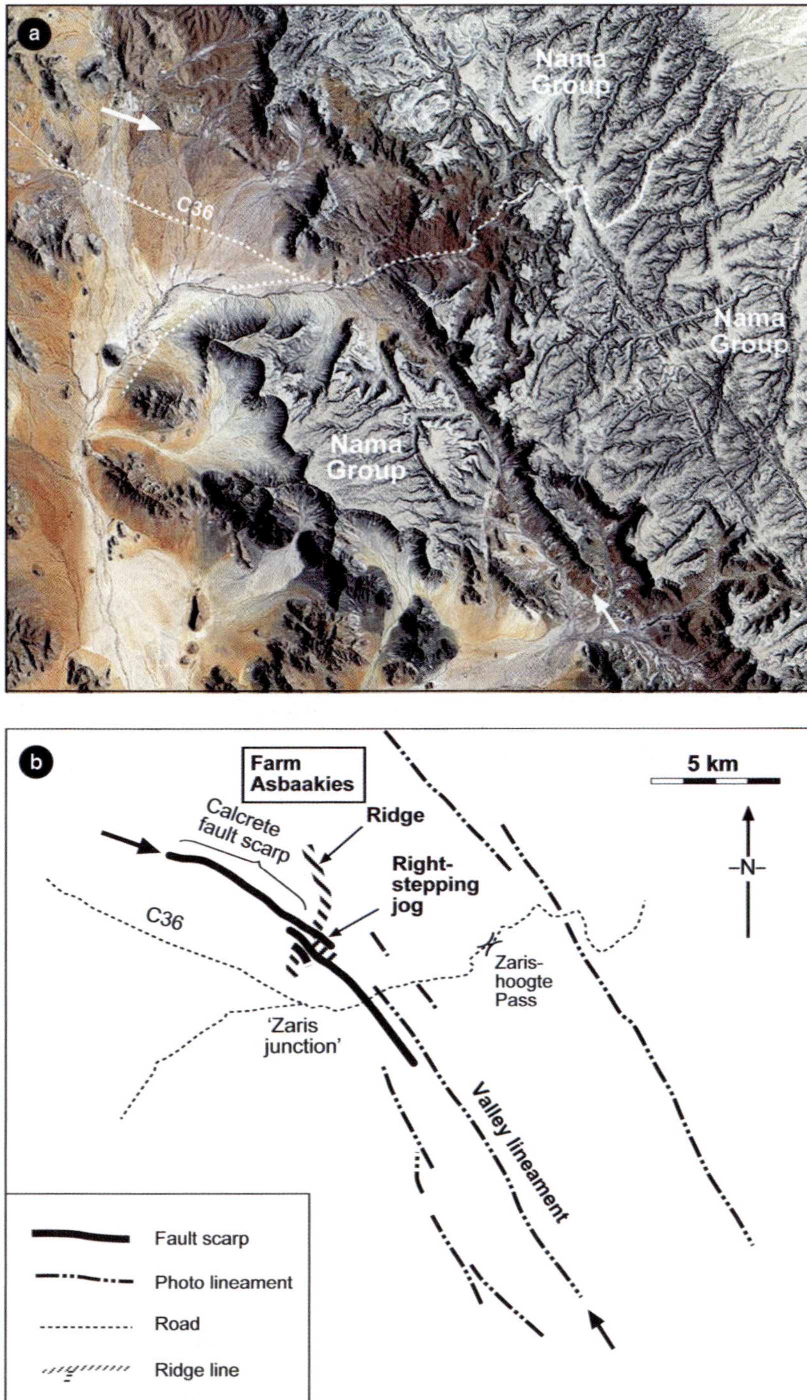

Fig. 14. (**a**) True colour LandSat image showing the SE extension of the Hebron Fault trace in the area near Zarishoogte Pass. See Figure 2a for location. Tips of the fault trace are marked by opposing arrows at top left and bottom right. (**b**) Line drawing of the area in (a), identifying the Hebron Fault lineament, other prominent lineaments in Nama Group rocks, and roads.

These conclusions are at odds with our field-based structural and stratigraphic data and appear, in fact, to also be at odds with the regional stress model that Viola *et al.* (2005) themselves propose. Evidence presented here shows that the Hebron Fault is extensional with predominantly normal dip-slip kinematics. Small lateral components on normal faults are dextral. Diversions of water courses, however, have no systematic sense of polarity. The Hebron Fault has both left- and right-stepping jogs and the sense of step-over alone is not a reliable criterion for determining kinematics.

We concede that net displacement may be oblique, in the sense that the steeply plunging net slip vector probably trends more westerly than the dip of the fault. However, the orientation of regional σ_1 suggested by Viola *et al.* (2005), bisecting the angle between NW-striking (e.g. Hebron Fault) and NNW-striking fracture sets, would induce a component of fault-normal shortening across the Hebron Fault, not extension.

Furthermore, several lines of evidence lead us to disagree with the conclusion of Viola *et al.* (2005) that the Hebron Fault is exclusively early Pleistocene. Instead, the field evidence points to a more prolonged episodic history including that of the Pliocene and evident at the present day. That fault activity still continues is proved by small epicentres (Fig. 3) along the fault line recorded by Fernandez & Guzman (1979) and Fernandez & Du Plessis (1992) and by short rumblings that momentarily disturb animals and rattle crockery in cupboards of nearby farms (H. Stehmann pers. comm, 1999). This present activity of the Hebron Fault explains the enhanced water yields on the fault downthrow, together with localized provision of thermal waters. Elsewhere in Namibia, small-magnitude earthquakes have been recorded next to the escarpment (Korn & Martin 1951), along the southwestward continuation of the Waterberg fault in the Omaruru area (Range 1914; Sieberg 1914; Fairhead 1977; Klein 1980), along the Windhoek and Eiseb Graben structures (Andritzky 1996; Wanke 2005) and in the southern Namibian Namaqualand area (Fernandez & Guzman 1979; Andreoli *et al.* 1996).

Relationship to regional stress field

Attempting to integrate onshore and offshore Namaqualand neotectonic structural features with the Hebron Fault line, Viola *et al.* (2005) propose a NNW-directed principal horizontal stress (σ_1) for SW Africa based on their interpretation of an offshore line of fault-related mud volcanoes as a shear direction complementary to the Hebron Fault. This regional σ_1 was correlated with the maximum horizontal compressive stress orientation of the Wegener Stress Anomaly (WSA; Andreoli *et al.* 1996).

As discussed above, however, this interpretation seems inconsistent with the field data, and may be overly model-dependent. An alternative explanation for dextral, normal displacement on the Hebron Fault appears to be required. We interpret the Hebron Fault as due to coastal gravitational collapse as a direct consequence of the instability associated with the development of the African Superswell (Partridge *et al.* 1995). Partridge & Maud (1987) record evidence of up to 900 m uplift above this thermal anomaly during the Pleistocene, leading to gravitational instability, particularly at the continental margin. We find no conclusive evidence that faulting is controlled by a subhorizontal, NNW-directed maximum principal stress.

This reinterpretation is supported by the shell finite element models of Bird *et al.* (2006) who found that southern Africa is not in a state of horizontal compression, although it is surrounded by spreading ridges. They propose that it is generally in a state of horizontal extension because its high elevations lead to density movements exceeding those of spreading ridges. Concerning the WSA anomaly Bird *et al.* (2006) also inferred from modelling, that the NW–SE band of NW–SE directed greatest compressive horizontal principal stress (the WSA) is real but may be caused elsewhere more by SW–NE horizontal tension (σ_{2H}) than it is by NW–SE compression (σ_{1H}).

Conclusions

The fact that the Hebron Fault scarp is collinear with a basement scarp that preserves evidence for previous tectonic activity is significant, because it suggests that the young Hebron Fault has reactivated an older structure. Fault reactivation is a recurring theme in studies of young tectonic activity affecting older terrains (e.g. Daly *et al.* 1989), where it is inferred that, if pre-existing faults are suitably orientated with respect to the contemporary stress field, they tend to fail in preference to nucleation of new structures (Sibson 1991).

The southwestern margin of Africa was the locus of large-scale extensional tectonics during a sequential Carboniferous/Permian to Lower Cretaceous rifting history that led eventually to the formation of the South Atlantic Ocean (Erlank *et al.* 1984; Light *et al.* 1992; Gladczenko *et al.* 1998; Stollhofen 1999). Thus, in this area, we find the root zones of generally NW-trending normal faults related to pronounced basement anisotropies and Late Palaeozoic/Early Mesozoic rifting (Stanistreet & Charlesworth 2001). The Hebron Fault is subparallel to the Atlantic coastline of SW Africa, and to the erosionally modified Great Escarpment.

However, it follows the general trend of those earlier rift-related faults (e.g. Light *et al.* 1992).

Multiple phases can be distinguished during which fault movements were taking place along the structural weakness, used ultimately by the Hebron Fault. (I) The youngest phase affects sedimantation of and displaces the Middle Pliocene Karpfenkliff and Kamberg Formations and younger (Late Pleistocene–Holocene) dune accumulations of the Namib Sand Sea. (II) The middle phase displaces the basal Nama unconformity by about 20 m. (III) The oldest phase displaces basement rocks by a considerable amount at depths of 4–10 km. The oldest displacement relates to juxtaposition of pre- and mid-Sinclair rock against Late Sinclair rocks of the Aubures Formation with a throw of almost 2800 m (Miller 1969). The middle 20 m displacement we relate to pre-South Atlantic rifting episodes recorded elsewhere by Karoo and Etendeka strata. The latest and younger faulting represents the most recent reactivation of a longstanding zone of structural weakness.

Even this latest phase in fault history appears to have been episodic, whereby fault movements do not affect the entire fault trace at one time. The C1, C2 and C3 morphologies indicate differential fault movement in both time and space. The C1 morphology suggests latest segments of fault displacements, whereas C3 would suggest earlier such segments, subsequently degraded by erosion. C3 scarps occur almost exclusively along the southeastern section of the fault trace that we have mapped in detail, and a possible implication of this observation is that fault activity propagated progressively from SE to NW.

We further conclude that the Hebron Fault has been historically a locus for earthquake activity, warranting a seismic monitoring programme aimed at establishing (1) whether locally felt groundshaking indeed originates on the Hebron Fault, and (2) the level, magnitude, frequency and foci of background seismicity.

This study was undertaken under the auspicies of the Geoscience Graduiertenkolleg at the University of Würzburg, Germany. Funding by the German Science Foundation (DFG) is gratefully acknowledged. Special thanks are due to the Namibian Geological Survey for help with logistics, to Namibian Nature Conservation for permission to carry out research in the Sesriem-Sossusvlei area and in particular to the Swart, Rust and Stehmann families at Hebron, Betesda and Neuhoff farms and to the crews of Witwater and Sosses Desert camps for their hospitality and allowing access to Hebron Fault exposures. Carmen Krapf processed the satellite images for us; the paper benefited greatly from discussions with Carmen Krapf, Jürgen Kempf and Nick Lancaster. Wilfried Jooß assisted in the field with surveys of scarp profiles. We thank Roy Miller, Klaus Reicherter, Eulàlia Masana and Tim Little for their thorough reviews of the manuscript, and Karl-Heinz Hoffmann and Thomas Becker for their useful discussions. We also thank Google.com for their kind permission to use the image in Figure 3.

References

ANDREOLI, M. A. G., DOUCOURÉ, M., VAN BEVER DONKER, M., BRANDT, D. & ANDERSEN, N. J. B. 1996. Neotectonics of southern Africa – a review. *Africa Geoscience Review*, **3**, 1–16.

ANDRITZKY, G. 1996. *Fracture zones, mineral deposits and magmatic features in Namibia*. German-Namibian Cooperation Mineral Prospecting Promotion Project, Investigation Series **10**, 1–28.

BEN-AVRAHAM, Z., SMITH, G., RESHEF, M. & JUNGSLAGER, E. 2002. Gas hydrate and mud volcanoes on the southwest African continental margin off South Africa. *Geology*, **30**, 927–930.

BIRD, P., BEN-AVRAHAM, Z., SCHUBERT, G., ANDREOLI, M. & VIOLA, G. 2006. Patterns of stress and strain in southern Africa. *Journal of Geophysical Research*, **111**, B08402.

BROWN, R. W., GALLAGHER, K., GLEADOW, A. J. W. & SUMMERFIELD, M. A. 2000. Morphotectonic evolution of the South Atlantic margins of Africa and South America. *In*: SUMMERFIELD, M. A. (ed.) *Geomorphology and Global Tectonics*. Wiley, Chichester, 255–281.

BUCKNAM, R. C. & ANDERSON, R. E. 1979. Estimation of fault-scarp ages from a scarp-height-slope-angle relationship. *Geology*, **7**, 11–14.

COCKBURN, H. A. P., BROWN, R. W., SUMMERFIELD, M. A. & SEIDL, M. A. 2000. Quantifying passive margin denudation and landscape development using a combined fission-track thermochronology and cosmogenic isotope analysis approach. *Earth and Planetary Science Letters*, **179**, 429–435.

CORVINIUS, G. & HENDEY, Q. B. 1978. A new Miocene vertebrate locality as Arrisdrift in South West Africa (Namibia). *Neues Jahrbuch für Geologie und Paläontologie, Monatshefte*, **4**, 193–205.

DALY, M. C., CHOROWICZ, J. & FAIRHEAD, J. D. 1989. Rift basin evolution in Africa: the influence of reactivated steep basement shear zones. *In*: COOPER, M. A. & WILLIAMS, G. D. (eds) *Inversion Tectonics*. Geological Society, London, Special Publications, **44**, 309–334.

DUPONT, L. M., DONNER, B., VIDAL, L., PÉREZ, E. M. & WEFER, G. 2005. Linking desert evolution and coastal upwelling: Pliocene climate change in Namibia. *Geology*, **33**, 461–464.

ERLANK, A. J., MARSH, J. S., DUNCAN, A. R., MILLER, R. MCG., HAWKESWORTH, C. J., BETTON, P. J. & REX, D. C. 1984. Geochemistry and petrogenesis of the Etendeka volcanic rocks from SWA/Namibia. *In*: ERLANK, A. J. (ed.) *Petrogenesis of the Volcanic Rocks of the Karoo Province*. Geological Society of South Africa, Special Publications, **13**, 195–245.

FAIRHEAD, J. D. 1977. The seismicity of southern Africa and incipient rifting. *20th Annual Report, Research Institute of Africal Geology*. University of Leeds, 42–46.

FERNANDEZ, L. M. & DU PLESSIS, C. 1992. *Seismic hazard maps for Southern Africa.* Government Printer, Pretoria.

FERNANDEZ, L. M. & GUZMAN, J. A. 1979. *Earthquake Hazard in Southern Africa.* Geological Survey of South Africa, Seismological Series, **10**, 1–38.

GLADCZENKO, T. P., SKOGSEID, J. & ELDHOM, O. 1998. Namibia volcanic margin. *Marine Geophysical Researches*, **20**, 313–341.

GREEN, R. W. E. & BLOCH, S. 1971. The Ceres, South Africa, earthquake of September 29, 1969. *Bulletin of the Seismological Society of America*, **61**, 851–859.

GUERMANI, A. & PENNACCHIONI, G. 1998. Brittle precursors to plastic deformation in a granite: An example from the Mont Blanc massif (Helvetic, western Alps). *Journal of Structural Geology*, **20**, 135–148.

JONES, K. D. 1978. *A study of seismicity in the Damara Province.* MSc thesis, University of the Witwatersrand, Johannesburg.

KLEIN, J. A. 1980. *Pleistocene to recent faulting in the area west of Omaruru (SWA/Namibia).* Geological Survey of Namibia, Regional Geology Series, Open File report **RG 4**, 1–26.

KNIPE, R. J. 1989. Deformation mechanisms – recognition from natural tectonites. *Journal of Structural Geology*, **11**, 127–146.

KORN, H. & MARTIN, H. 1951. The seismicity of South-West Africa. *Transcations of the Geological Society of South Africa*, **54**, 85–88.

LIGHT, M. P. R., MASLANYJ, M. P. & BANKS, N. L. 1992. New geophysical evidence for extensional tectonics on the divergent margin offshore Namibia. *In*: STOREY, B. C., ALABASTER, T. & PANKHURST, R. J. (eds) *Magmatism and the Causes of Continental Break-up.* Geological Society, London, Special Publications, **68**, 257–270.

LLOYD, G. E. & FREEMAN, B. 1994. Dynamic recrystallization of quartz under greenschist facies conditions. *Journal of Structural Geology*, **16**, 867–881.

MACHETTE, M. N. 1982. Quaternary and Pliocene faults in the La Jencia and southern part of the Albuquerque-Belen basins, New Mexico: Evidence of fault history from fault-scarp morphology and Quaternary geology. *Field Conference Guidebook.* New Mexico Geological Society, **33**, 161–170.

MACHETTE, M. N. 1989. Slope-morphometric dating. *In*: FORMAN, S. L. (ed.) *Dating Methods applied to Quaternary Geologic Studies in the Western United States.* Utah Geological and Mineral Survey, **89-7**, 30–42.

MANGONGOLO, A. & HUTCHINS, D. G. 2008. Seismicity of Namibia from 1910 to 2006. *In*: MILLER, R. (ed.) *Geology of Namibia* Geological Survey of Namibia, Windhoek, Vol. 3. 27-1–27-7.

MAYER, L. 1986. Tectonic geomorphology of escarpments and mountain fronts. *In*: Geophysics Study Commitee (eds) *Active Tectonics.* National Academy Press, Washington D.C., 125–135.

MILLER, R. MCG. 1969. The Aubures Formation of the Bethanie District, South West Africa. *Bulletin of Precambrian Research*, Unit **2**.

MILLER, R. MCG. & SCHALK, K. E. L. 1980. *South West Africa/Namibia geological map 1:1,000,000.* Geological Survey of Namibia.

MITCHELL, P. 2002. The Archaeology of Southern Africa. Cambridge Africa Collection, Cambridge University Press, Cambridge.

MITRA, G. 1984. Brittle to ductile transition due to large strains along the White Rock thrust, Wind River mountains, Wyoming. *Journal of Structural Geology*, **6**, 51–61.

MULDER, C. & ELLIS, R. P. 2000. Ecological significance of South-West African grass leaf phytoliths: A climatic response of vegetation biomes to modern aridification trends. *In*: JACOBS, S. W. L. & EVERETT, J. (eds) *Grasses: Systematics and Evolution.* CSIRO, Melbourne, 251–260.

PARTRIDGE, T. C. & MAUD, R. R. 1987. Geomorphic evolution of southern Africa since the Mesozoic. *South African Journal of Geology*, **90**, 179–208.

PARTRIDGE, T. C., BOND, G. C., HARTNADY, J. H. & DE MENOCAL, P. B. 1995. Climatic effects of Late Neogene tectonism and volcanism. *In*: VRBA, E., DENTON, G., PARTRIDGE, T. C. & BURCKLE, L. (eds) *Paleoclimate and Evolution with Emphasis on Human Origins.* Yale University Press, New Haven, 331–355.

PICKFORD, M. & SENUT, B. 1999. Geology and palaeobiology of the Namib Desert Southwestern Africa. Geological Survey of South West Africa/Namibia, Memoirs, **18**, 1–155.

PRICE, N. J. & COSGROVE, J. W. 1991. *Analysis of Geological Structures.* Cambridge University Press, Cambridge.

RANGE, P. 1914. Erdbeben in Deutsch-Südwestafrika. *Mitteilungen aus den deutschen Schutzgebieten*, **27**, 206–208.

SCHREIBER, U. M. 2000. *Geological map of Namibia, 1:250,000 Geological Series, Sheet 2414 Meob Bay.* Geological Survey of Namibia.

SIBSON, R. H. 1991. Loading of faults to failure. *Bulletin of the Seismological Society of America*, **81**, 2493–2497.

SIEBERG, A. 1914. Die Erdbeben in Deutsch-Südwestafrika während der Zeit von Juni bis September 1913. *Mitteilungen der Zentralbureaus der Internationalen Seismologischen Association*, **1914**, 64–65.

SIMPSON, C. 1986. Fabric development in brittle-to-ductile shear zones. *Pageoph*, **124**, 269–288.

STANISTREET, I. G. & CHARLESWORTH, E. G. 2001. Damaran basement-cored fold nappes incorporating pre-collisional basins, Kaoko belt, Namibia and controls on Mesozoic supercontinental break-up. *South African Journal of Geology*, **104**, 1–12.

STANISTREET, I. G. & MCCARTHY, T. S. 1993. The Okavango Fan and the classification of subaerial fan systems. *Sedimentary Geology*, **85**, 115–133.

STÖCKHERT, B., BRIX, M. R., KLEINSCHROT, R., HURFORD, A. J. & WIRTH, R. 1999. Thermochronometry and microstructures of quartz – a comparison with experimental flow laws and predictions on the temperature of the brittle-plastic transition. *Journal of Structural Geology*, **21**, 351–369.

STOLLHOFEN, H. 1999. Karoo Synrift-Sedimentation und ihre tektonische Kontrolle am entstehenden Kontinentalrand Namibias. *Zeitschrift der Deutschen Geologischen Gesellschaft*, **149**, 519–632.

STOLLHOFEN, H., STANISTREET, I. G., ROHN-DAVIES, R., HOLZFÖRSTER, F. & WANKE, A. 2000. The

Gai-As lake system, northern Namibia and Brazil. *In*: GIERLOWSKI-KORDESCH, E. & KELTS, K. R. (eds) *Lake Basins through Space and Time*. Studies in Geology, American Association of Petroleum Geologists, **46**, 87–108.

VAN DER WATEREN, F. M. & DUNAI, T. J. 2001. Late Neogene passive margin denudation history – cosmogenic isotope measurements from the Central Namib desert. *Global and Planetary Change*, **30**, 271–307.

VIOLA, G., ANDREOLI, M., BEN-AVRAHAM, Z., STENGEL, I. & RESHEF, M. 2005. Offshore mud volcanoes and onland faulting in southwestern Africa: neotectonic implications and constraints on the regional stress field. *Earth and Planetary Science Letters*, **231**, 147–160.

VOGELSANG, R. 1998. *Middle-Stone-Age-Fundstellen in Südwest-Namibia*. Heinrich-Barth-Institut, Cologne.

WALLACE, R. E. 1977. Profiles and ages of young fault scarps, north-central Nevada. *Bulletin of the Geological Society of America*, **88**, 1267–1281.

WANKE, H. 2005. The Namibian Eiseb Graben as an extension of the East African Rift: evidence from Landsat TM5 imagery. *South African Journal of Geology*, **108**, 541–546.

WARD, J. D. 1987. The Cenozoic succession in the Kuiseb Valley, Central Namib Desert. Geological Survey of South West Africa/Namibia, Memoirs, **9**, 1–45.

WARD, J. D. & CORBETT, I. 1990. Towards an age for the Namib. *In*: SEELY, M. (ed.) *Namib Ecology: 25 Years of Namib. Research*. Transvaal Museum Monograph, **7**, 17–26.

WEFER, G., BERGER, W. H., RICHTER, C. *ET AL*. 1998. *Proceedings of the Ocean Drilling Program: Initial Reports*, Vol. 175. Ocean Drilling Program, College Station, TX.

WHITE, S. 1973. The dislocation structures responsible for the optical effects in some naturally deformed quartzites. *Journal of Material Sciences*, **8**, 490–499.

YAALON, D. H. & WARD, J. D. 1982. Observations on calcrete and recent calcic horizons in relation to landforms in the central Namib. Desert. *In*: *Proceedings of VIth SASQUA Conference*. Palaeoecology of Africa, **15**, 183–186.

YEATS, R. S., SIEH, K. & ALLEN, C. R. 1997. *The Geology of Earthquakes*. Oxford University Press, New York.

Index

Figures are shown in *italic* font, tables in **bold**